Animal Physiology
Adaptations in Function

Animal Physiology
Adaptations in Function

F. Reed Hainsworth
SYRACUSE UNIVERSITY

ADDISON-WESLEY PUBLISHING COMPANY
Reading, Massachusetts * Menlo Park, California
London * Amsterdam * Don Mills, Ontario * Sydney

This book is in the Addison-Wesley Series in
THE LIFE SCIENCES

Library of Congress Cataloging in Publication Data

Hainsworth, F   R
    Animal physiology.

    Bibliography: p.
    Includes index.
    1. Physiology. 2. Adaptation (Physiology)
I. Title. [DNLM: 1. Adaptation, Physiological.
2. Animals—Physiology. 3. Physiology,
Comparative. 4. Environment. QP82 H153a]
QP31.2.H34   591.1   80-13692
ISBN 0-201-03401-8

ISBN 0-201-03401-8
ABCDEFGHIJ-HA-8987654321

To Diane, Charlotte, and Emily

Preface

This is a book about *how* animals function; it includes the physical and chemical principles governing gas exchange, feeding, digestion, temperature regulation, locomotion, water and solute balance, reproduction, information processing, and movement. The book is intended for introductory physiology courses that stress understanding of basic principles in a comparative biological and environmental context. Although the book is written at an introductory level, an attempt has been made to stress quantitative relationships and to examine alternative hypotheses that can form a basis for more advanced treatments.

This is also a book about *why* animals function as they do and of alternative solutions to the problems of survival. Physical and chemical principles are developed, and they are then used to examine a diversity of animals and environments. My interests are in environmental (ecological) and comparative studies to understand the similarities and differences in the functions of animals.

The book is organized around an emerging, major, unifying theme for comparative physiology. The theme is an economic one. Adaptations in animal functions are considered in terms of the *benefits* (advantages) and *costs* (disadvantages) for animals to function in different ways. This theme is developed in the context (point of view) of resource use, that is, the environmental *supplies* of important resources and the *demands* of animals for them.

Natural selection has resulted in adaptation in the ways animals function. Adaptations are attributes that favor survival and reproductive success of individuals. The supplies of certain resources and the demands of animals for them have influenced the evolution of functional adaptations in animals. Certain resources are lost by animals and must be gained from their environments. Obtaining resources provides benefits, but the mechanisms for obtaining them involve costs. The basic premise for an economic interpretation of animal physiology is that survival and reproductive success depend on the degree of *net benefit* (benefits minus costs). Those individuals that obtain the greatest net benefits (within certain constraints) will be more likely to survive and to produce offspring with similar traits.

Net benefits involve what is usually called "efficiency" in the use of resources. In physics, efficiency is measured as a ratio of useful output from a system to the total input into a system, such as the efficiency of conversion of heat to electricity. In statistics, efficiency is measured as the extent of variation. In economics (and biology), it is the *difference* between benefits (intakes) and costs (expenditures) that influences whether an animal has sufficient resources for survival and reproduction or is effective in the ways it deals with resources. We can say that an animal will be "efficient" when supplies of required resources are sufficient to meet demands for them, and it will be "inefficient" if it fails to match resource supplies with demands such that it fails to survive or reproduce.

It is my hope that an economic theme for comparative physiology will have considerable value to students of biology. Perhaps more than any other natural science, biology seems filled with many facts that students must learn. An unfortunate approach to this overwhelming amount of material is to memorize facts without necessarily paying attention to their meaning. However, by stressing what *should* be important for function based on simple economic principles of relative costs and benefits for resource use, students can begin to grasp the theory needed to organize facts into a useful framework.

An economic interpretation for comparative physiology is not the only possible framework in which to view animal functions. The most common alternative interpretations for differences between animals are what I call "historical" hypotheses. These are considered throughout the book as alternatives, and they are usually rejected on the basis of current evidence.

Historical hypotheses have their foundations in taxonomic or phylogenetic relationships among animals. Taxonomy demonstrates obvious and important differences between animals, but it does not provide a consistent rationale with which to explain differences in function. Most functional historical hypotheses are based on the premise that different living animals represent different "degrees" of evolutionary development ("primitive" versus "advanced" species). These hypotheses fail to provide a logical framework to explain adaptations, and they usually neglect important environmental factors by stressing "species" differences. For these reasons I have deliberately avoided a taxonomic organization for this book. Instead, basic physical, chemical, and biological principles are stressed, and economic interpretations of function are contrasted with more traditional historical explanations.

Although I have deliberately not followed a taxonomic organization, I have included discussion of a diversity of animal functions. The functions of individual cells, invertebrates, fishes, amphibians, reptiles, birds, mammals, and even some plants are considered in order to illustrate adaptations and the variety of alternative solutions to problems of survival posed by resource supplies and demands.

The book is divided into different parts, based on different major resources. These resources include oxygen (respiration and circulation), energy and nutrients (metabolism, temperature regulation, locomotion, and feeding), and water (osmoregulation). Moreover, reproductive demands require additional resource supplies, as do demands for avoiding predators (and obtaining resources) by means of rapid integration and effector mechanisms.

The book is introduced with a chapter that discusses the general features of resource *exchange*, *regulation*, and *control*. In subsequent chapters the characteristics of animals and their environments that influence the net exchange (losses and gains) for particular resources are considered. The importance of achieving regulation, i.e., a degree of constancy in the net content of resources, is discussed for each resource, followed by discussion in each chapter of the control mechanisms by which regulation is achieved.

Each chapter of the book is concluded with a summary. In this way the major ideas in each chapter are presented in three contexts: first, in major discussions within the text; second, in brief descriptions along with figures; and third, in the summary. This organization was selected to provide a maximum opportunity for students to assess ideas and judge their importance.

Each part of the book has a brief untitled introduction and is concluded with annotated references. Each introduction is intended as a nontechnical orientation to the problems to be considered in that part. The references provide the major source material for the ideas discussed in a way that will permit a reader to exercise some judgment over which references should be consulted.

I owe a debt of thanks to several individuals who have assisted me. Drs. John F. Anderson, University of Florida, William Dawson, University of Michigan-Ann Arbor, and Samuel J. Velez, Dartmouth College, read and commented on the entire text. Drs. Albert Bennett, University of California-Irvine, Henry Horn, Princeton University, Donald Jackson, Brown University, Edward Stricker, University of Pittsburgh, and John Vernberg, University of South Carolina, read and commented on separate parts of the book. Their comments have considerably improved my efforts. However, they share no responsibility for remaining errors or omissions. I would very much appreciate hearing from those who use this book so that future improvements can be made.

The quality of both content and appearance of the final product depends to a large degree on the publisher. I thank the staff of Addison-Wesley for their continued encouragement, assistance, and support on decisions influencing the quality of production.

Syracuse, New York   *F.R.H.*
January 1981

Contents

Detailed

Contents

Part I

REPRODUCTION **443** *Part* V

**INFORMATION PROCESSING FOR
REGULATION** **489** *Part* VI

Introduction

Resource Exchange, Regulation, and Control

There are certain basic characteristics of living animals that we will have occasion to discuss in some detail in each of the coming chapters. These include the processes of exchange, regulation, and control. Some attention to the general features of these processes will help in understanding their application in the specific situations considered throughout the book.

EXCHANGE

All living organisms carry out almost continuous exchanges of a large number of substances between themselves and their environments. Those substances exchanged are resources, which are both lost and gained. The study of physiology is largely a study of these exchanges.

There are three questions about exchange that we will consider for each major resource: (1) What is exchanged? (2) What is the *net* exchange under various conditions, that is, the gain together with the loss? (3) What physical, chemical, and biological factors are involved in determining and controlling net exchange?

Demands of animals for resources arise from use or losses of resources. Supplies of resources ultimately come from external environments. The extent to which gains of resources are sufficient to meet demands of animals for them will determine whether an animal survives and reproduces. Thus the net exchange of resources is influenced both by animals, from use of resources (loss), and by environments, from effects on supply of resources (gain).

Oxygen is used in cells of many animals and must be supplied from the external environment at rates that match the rates of use. This use is accompanied by production of carbon dioxide and requirements for its loss (disposal) at rates that match rates of production. The exchange of respiratory gases occurs through the respiratory and circulatory systems (Chapters 1-4).

Energy and nutrients must be found in amounts that match demands for their use. In animals this is accomplished by the capture, digestion, and assimilation of chemicals in food. In addition, heat is a product of chemical energy use, and animals exchange heat as well as chemical energy. Temperature depends on the net exchange of heat between animals and their environments (Chapters 5-11).

The functions of cells take place in water, and water is exchanged between animals and their environments. The losses of water must be balanced by gains from ingestion and metabolism, and excess gains must be balanced by losses from evaporation and/or excretion (Chapters 12-14).

It is important in biology that animals survive to reproduce. In terms of exchange, it can be thought of as a relationship for a population between the number of individuals that are lost (from deaths) and the number gained (from births). Reproduction also involves increased demands within individuals for the exchange of physical resources. Certain resources must be gained at higher rates than their rates of loss during the time when other animals are being produced (Chapter 15).

Finally, survival partly involves the avoidance of becoming a resource for the survival and reproductive demands of other animals. The interactions between predators and prey determine which animals become *sources* of resource supplies and which survive to *use* those sources to replace losses of their own resources and to reproduce other individuals (Chapters 15-18).

REGULATION

For survival and reproductive success, there must be a specific balance in the net exchange of certain resources so that the *contents* of the resources within an animal remain relatively constant over a period of time. The specific values of resource contents vary among animals and environments, and the precision with which resource contents remain constant also changes, but the variation in content of regulated resources within animals is generally less than the variation in those resources in external environments.

Regulation is manifested as *relative constancy* in the net exchange of resources. If the loss of a resource increases, net content will remain constant if there are mechanisms to counteract the increased loss with a higher gain. Conversely, if gains increase, net content will remain constant if there are mechanisms to counteract the increase with a higher loss.

It is important to note that the time periods over which the net content of a resource remains relatively constant vary with different resources and among animals. For example, some seasonal hibernators become "obese" from very high rates of net energy gain that is stored as fat in late summer and fall. Over this time period it may appear that energy is not precisely regulated, since net energy content is increasing instead of remaining constant. However, the use of this stored (internal) high net gain of energy occurs over winter, when the rate of gain of energy from external sources is very low or nonexistent. Over a long seasonal time period energy is regulated in these animals; i.e., there is a balance between intakes and losses.

In general, there is a relationship between the availability of different resources to animals and the time periods over which the resources are regulated. For example, oxygen is usually more available than water in most environments, and water is usually more available than food (on a global scale). Regulation of oxygen is usually very short-term; most animals cannot survive more than minutes without it. The regulation of water is usually over an intermediate period of time; survival without it may be over periods of hours or days. Regulation of energy (food) is usually over a longer term; most animals can survive longest without it. Thus regulation of the resource least available and most difficult to obtain occurs over longer periods.

"Efficiency" and the Economics of Net Resource Exchange

We will be concerned with these characteristics of regulation in each section of the book. There are many interesting exceptions to the general pattern outlined above that provide interesting insights into the process of regulation. The discussion of each major resource will start with a consideration of why supplies and demands must be balanced (regulated) for survival and of the animal and environmental factors that influence net exchange of the resources. We will then examine the *economics* of regulation to assess the consequences to animals of balancing the exchange of resources.

We will attempt to explain the adaptations of animals in terms of economic principles of "efficiency." Adaptations are characteristics of individuals that favor their survival and reproductive success. Economic efficiency depends on the difference between benefits (gains) and costs (losses), i.e., the net difference in exchange of resources (see Preface). To be adaptive, a characteristic should be economically efficient with regard to resource use. Within each chapter the *costs* of resource gains needed to balance losses will be related to the *benefits* to animals of achieving a degree of regulation.

The importance of an economic basis for understanding animal adaptations can be illustrated with a simple analogy to what is considered most important for "survival" in a business. Profits are ultimately most important, and profits are the difference between income and expenditures. We will examine regulation in economic terms to try to assess what constitutes "efficient" resource use by animals. Except when noted otherwise, efficiency will mean a function of differences between intakes (supplies) and losses (demands) of resources.

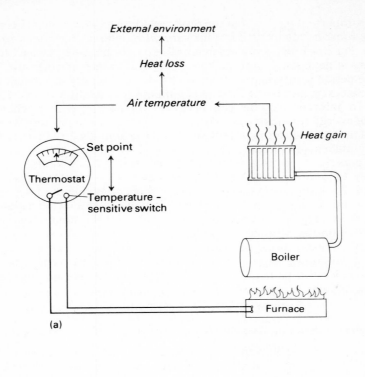

(a)

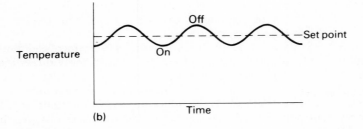

(b)

Figure I.1

(a)   Schematic illustration of the components involved in thermostatic control of room air temperature. Heat loss from the room decreases air temperature, which is detected at a thermostat. When the decrease is below a reference ("set point") temperature, a switch is closed, activating heat production and ultimately increasing room air temperature.

(b)   As a consequence of thermostatic control of heat production, room air temperature will be regulated or will remain relatively constant over time. The precision of regulation depends on the extent to which room air temperature varies above and below the constant set point as the source of heat gain is turned on and off.

CONTROLS FOR REGULATION

The word "control" is often used to mean "regulation," but there are very good reasons to distinguish between them. Controls result in regulation of resources, and regulation is not possible without controls. However, what is controlled is seldom what is regulated.

Controls involve the mechanisms by which regulation is achieved. The most commonly asked question about controls is, "*How* is regulation accomplished, i.e., managed?" The most commonly asked question about regulation is, "*Why* is a degree of constancy in net resource content important for survival?"

A frequently used example illustrating control and regulation is the regulation of the air temperature of a room by thermostatic control. It is shown schematically in Fig. I.1 for a case where the environment surrounding the room is colder than the room. If air temperature at the thermostat decreases because of heat loss from the room, an electrical switch is closed, turning on a furnace. This event ultimately results in heat being added to the air, raising room air temperature and, in turn, eventually resulting in the opening of the electrical circuit in the thermostat, which will shut off the furnace. Note that what is regulated is room air temperature; what is controlled is the extent of fuel combustion in the furnace. They are related, but they are not the same value.

We will consider many distinctions between regulations and controls. For example, regulation of oxygen involves controls that respond to hydrogen ion concentrations in air-breathing animals and pH is only indirectly related to demands for oxygen through carbon dioxide production and the interaction of carbon dioxide with water to produce hydrogen ions. Regulation of energy involves an intricate set of controls that respond to a variety of characteristics of food related to energy value but which are not themselves direct measures of energy. These characteristics include chemical composition (assessed from taste and smell), osmotic concentrations, mechanical distension of digestive organs, and pH. Many of the fascinating intricacies of animal physiology involve interpretations of how control systems result in regulation of a resource, where what is being regulated and what is being controlled are different but related.

Controls through Negative Feedback

All of the regulatory controls we will consider have some common features, illustrated in Fig. I.2. Some change or rate of change in an environment is detected. The environmental change that is detected is called a *stimulus*, and a detector is called a *receptor*.

A receptor generates information within a control system about the environmental change that influences net exchange of a resource. In the example illustrated in Fig. I.1, the thermostat contains temperature detectors that receive information related to air temperature in the room. Often the information detected at receptors concerns changes in environments from certain "preset" values rather than from absolute values. The thermostats in rooms and in animals, for example, respond to a decrease or increase in temperature from a reference or "set point" value that may be adjustable (Fig. I.1).

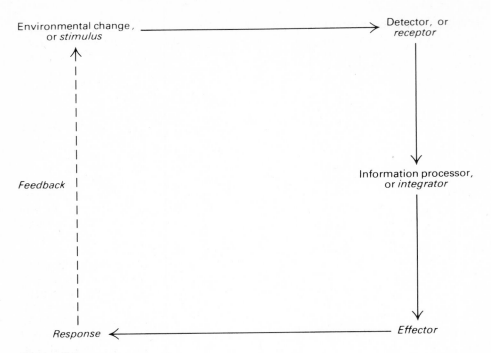

Figure I.2
The basic components of a negative-feedback control system. A stimulus or environmental change related to the exchange of a resource is detected, the information is processed, and an effector is activated to produce a response. In regulatory controls the response is opposite in direction to the originally detected change (dashed arrow).

The information from a receptor is processed within a control system. Processing can involve complicated comparisons with other control systems, or it may simply involve the closing of a switch with respect to a reference value as in Fig. I.1. Information processing is called *integration* (Fig. I.2).

As a consequence of integration, some effector is activated. In Fig. I.1 it is the furnace; in animals, muscles or glands are activated. The effectors produce a *response*, i.e., some action directed toward the environment.

In regulatory controls, responses act in directions opposite to originally detected environmental changes. In Fig. I.1 the response is increased heat production, which acts to reverse a decrease in room temperature. In respiration, increased ventilation of respiratory tissues (from increased rate and force of contraction of muscles associated with gills or lungs) is a response that counteracts a decrease in supply of oxygen as compared with tissue demands for oxygen (see Chapters 1 and 2).

The connection between the response and the environmental change is called a *feedback*. Since regulatory-control feedbacks are opposite to original environmental changes, they are called *negative* feedbacks (usually depicted with a dashed arrow as in Fig. I.2.).

There are some functions where nonregulatory *positive* feedback is important. In this case the response *adds* to the input stimulus instead of acting against it. Obviously, there must be some limit or "shutoff point" for a positive-feedback system; otherwise, the responses of a system would be diverted more and more to increasing the amount of the stimulus. Two common examples of limited positive-feedback controls in physiology are the generation of action potentials in neurons (see Chapter 16) and sexual behavior (see Chapter 15). In these cases positive feedback serves to concentrate resources synchronously up to a shutoff limit.

Several different types of regulatory controls can be distinguished, based on the manner in which a response is produced when a stimulus is detected. A response can be produced that depends on these conditions: (1) whether a particular stimulus is present regardless of its magnitude (*on-off control*); (2) whether a stimulus has changed by a certain amount from a set point (*proportional control*); (3) whether a stimulus has changed over time by a certain amount (*integral control*); or (4) whether a stimulus shows an instantaneous rate of change by a certain amount (*differential control*).

Proportional controls are the most common types in physiology. When we discuss mechanisms for adjusting losses or gains compared with the extent (degree) of a response needed for regulation, we will usually be dealing with proportional controls, where the magnitude of a response depends on the extent (proportion) of a loss or gain. Integral and differential controls per se are much less common in physiology. Some types of proportional controls may show features of differential or integral control, but the proportional characteristics of responses to meet demands for resources are usually the most important in understanding the mechanisms for regulation. On-off controls usually occur when rapid, maximal responses to stimuli are important for survival, and we will consider some examples when neural integration is discussed (Chapter 17).

Stimulus reception and information integration in animals often involve the function of cells associated with nervous systems (Chapters 16-18). Animals have evolved sensory mechanisms to detect environmental stimuli that influence the exchange of resources important for survival. In some cases *rapid* detection and information processing are important, such as the detection of predators or decreases in oxygen supplies. In these cases nerve interactions result in rapid detection, integration, and response to environmental changes. In other cases *longer* intervals are important for processing information related to changes in environmental resources, such as timing reproduction during growth to the most favorable set of seasonally available resources. Long-term integration and response usually involves hormones.

In some cases regulatory controls occur without either nerve or hormonal mediation. For example, hemoglobin molecules change when their local temperature or chemical environment changes so that release of oxygen from hemoglobin produces a proportionally increased supply of oxygen when tissue demands for oxygen increase (Chapter 3). Moreover, control of some nutrient supplies compared with demands does not involve specific detectors that respond to the exchange of each nutrient (Chapter 11). Nevertheless, *all* control mechanisms that result in resource regulation can be described according to the very general components summarized in Fig. I.2. The nature of the elements involved in detection, integration, and response varies from one regulatory system to another, between different animals, and between environments, but the functional relationships for the components are similar for each case we will consider.

PRECISION AND INTERACTIONS IN REGULATION

One very important feature of resource use that will quickly become apparent is that effective regulation for survival changes within and among animals and environments. If regulation suggests constancy, how do we deal with changes in what should remain constant?

Constancy in regulated systems is never perfect but is a matter of degree; this is why the expression "relatively constant" is used so often in describing regulation. The degree to which a regulated system remains precisely constant depends on the costs to maintain precision of regulation versus the benefits derived from a greater degree of precision. Economic interpretations based on costs and benefits are thus very important in explaining patterns of regulation and variation in the degreee of precision of regulated systems among animals and environments.

The precision of regulation for one resource also depends on supplies and demands for other resources; i.e., there are important *interactions* between systems of regulation. For example, the precision of temperature regulation (the degree of constancy of temperature over time) for animals in hot, terrestrial environments depends on the availability of water supplies, since water evaporation is a mechanism for controlling an increase in body temperature (see Chapter 14).

This type of complexity is the challenge and fascination of the study of animal physiology. The complexity becomes understandable when the economic factors of resource use are taken into account. With sufficient information about factors influencing both demands for and supplies of resources, it is possible to understand a great deal about the diversity of animal functions. A major theme in the coming chapters will be to understand the similarities and differences in the functions of animals in terms of the economic factors of resource supplies and demands that influence survival.

ANNOTATED REFERENCES

Brobeck, J.R. (1965). Exchange, control, and regulation. In W.S. Yamamoto and J.R. Brobeck (eds.), *Physiological Controls and Regulations*. Saunders, Philadelphia. (A clear distinction between the concepts of regulation and control, with several examples.)

Cannon, W.B. (1929). Organization for physiological homeostasis. *Physiol. Rev.* 9:399-431. (Cannon coined the term "homeostasis" to describe the constancy typical of regulated systems.)

Grodins, F.S. (1963). *Control Theory and Biological Systems*. Columbia University Press, New York. (Detailed discussion of various feedback controls in living organisms.)

Langley, L.L. (1965). *Homeostasis*. Reinhold, New York. (Introductory discussion of the principles of regulation and control, with brief examples for various resources.)

McFarland, D.J. (1977). Decision making in animals. *Nature* 269:15-21. (General discussion of economic "decisions" from an evolutionary point of view—the consequences of costs and benefits of different alternatives for survival.)

Riggs, D.S. (1970). *Control Theory and Physiological Feedback Systems*. Williams & Wilkins, Baltimore. (Detailed mathematical treatment of different types of physiological controls involving feedback.)

Schoenheimer, R. (1942). *The Dynamic State of Body Constituents*. Harvard University Press, Cambridge. (Early description of the extent of resource exchange in animals, using radioactive tracers to quantify exchanges.)

Supply and Demand through Respiration and Circulation

Part I

We take breathing for granted. Except under special circumstances we are seldom aware of the automatic events involved in obtaining oxygen and removing carbon dioxide. Yet among the resources we and other animals depend on for survival, oxygen is most important in an acute or short-term way. If we fail to breathe for more than a few minutes, brain damage or death occurs. We can survive for relatively long periods without food or water but not without a supply of oxygen.

Oxygen fuels the "fire of life" for many animals. Its lack will as quickly extinguish life as a candle flame in a sealed jar is extinguished when all the oxygen is consumed. This acute dependency on the resource is reflected in the function of a great diversity of animals that obtain oxygen from different sources and release carbon dioxide in different environments. Cells must be supplied with oxygen whether they occur in invertebrates, fishes, reptiles, amphibians, birds, or mammals and whether the animals breathe water or air.

In these first four chapters we will explore the problems that different animals face in obtaining sufficient continuous supplies of oxygen to meet their different demands for survival. In each case, whether we consider gills, lungs, the heart, vascular systems, or gas transport in blood, the evolution of respiratory and circulatory functions reflect a close relationship between environmental supply and the demands of animals for oxygen.

This close relationship will become apparent when we consider a diversity of respiratory and cardiovascular adaptations in other animals as compared with ourselves. The ability of a fish to efficiently breathe water or a diving whale to remain submerged for hours without breathing reflect these animals' functional evolution in relation to particular environmental conditions. Understanding these adaptations helps us to appreciate both the unity and diversity of respiratory functions: unity in the common dependence of many animals on short-term supplies of oxygen, and diversity in the extent to which different natural "experiments" have resulted in variations in the patterns with which oxidative metabolism is maintained among animals.

We will encounter many problems of providing sufficient oxygen supplies to meet different demands for survival. We will consider major changes in respiration and circulation that are required as the size of animals increases making the problems of oxygen delivery to and carbon dioxide removal from cells more complex. We will consider the major effects that environmental supplies of oxygen in water versus air have on respiratory function. We will examine differences in lung and blood function that reflect different demands for supplies of oxygen among different animals. Moreover, we will examine the basis for why some animals can survive for longer periods without oxygen than can others.

For each of these problems of survival, understanding adaptations in respiration and circulation depends on understanding the relationships between oxygen supplies and demands. The characteristics of the environment influence availability of supplies, while the characteristics of animals dictate demands.

Aquatic Respiration

I

There is a hierarchy (a difference in degree) in availability of resources in the external environments of animals. For an animal that utilizes the general reaction

$$O_2 + Fuel \longrightarrow CO_2 + H_2O + \text{``Waste''} + Energy,$$

oxygen and "fuel" (food) ultimately must be obtained entirely from the external environment. Some water must also be obtained from the external environment if production from internal oxidation is not sufficient to meet demands (it often is not; see Part IV).

   Of these three resources, oxygen is generally most available in the external environment on a global scale, followed by water (which is more variable in distribution), with food usually being the least available and most patchily distributed resource. We are all aware that air contains oxygen, as does water, and there are certainly relatively plentiful supplies of both these resources on earth, but food is usually harder to obtain because of its patchy distribution and the effort needed to acquire it. Thus it is easy to visualize why it can be important for animals to be economically efficient in obtaining food if they are to survive and reproduce. Predators consume prey, and there is obvious competition among animals for food resources. There is also "competition" among animals for oxygen but in an indirect way that is not as obvious.

Even though oxygen can be plentiful in the external environment of an animal, the efficiency of obtaining oxygen depends not only on its external supply but also on matching or equalizing *supplies* with *demands*. The demands for oxygen can increase or decrease within and among animals under a variety of circumstances; consequently, more or less oxygen must be obtained from the external environment if supplies are to equal demands. Moreover, animals must expend energy and food nutrients to obtain oxygen. Lungs, gills, hearts, blood, arteries, veins, and capillaries are all structures that have evolved in part to transfer oxygen from the external environment to the cells where it is used. The development and maintenance of these respiratory and circulatory systems require energy and nutrients. Since food is a difficult resource to obtain, an animal that expends a *minimum* amount of it to obtain and distribute a sufficient quantity of oxygen to meet its demands will be more effective and more likely to survive. As a consequence, more of a limited food resource will be available for other activities such as movements, temperature regulation, excretion, reproduction, and obtaining and processing food. Figure 1.1 shows some of the maintenance and reproductive functions of animals that depend on energy and nutrients. Minimizing expenditures for any one, that is, the evolution of effective function, will help ensure sufficient supplies for others.

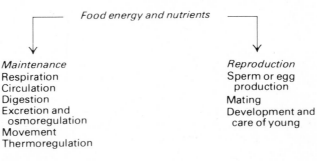

Food energy and nutrients

Maintenance
Respiration
Circulation
Digestion
Excretion and
  osmoregulation
Movement
Thermoregulation

Reproduction
Sperm or egg
  production
Mating
Development and
  care of young

Figure 1.1
Some of the maintenance and reproductive functions of animals that depend on energy and nutrients from food. Minimizing requirements for any one can contribute to sufficient supplies for others.

In addition to the interaction between oxygen and food energy and nutrients, there is an interaction between obtaining oxygen and the exchange of water. We will find that the physical characteristics influencing uptake of oxygen from the external environment influence the loss and uptake of water and some other important chemicals. Thus respiration involves efficient function not only with respect to demands for oxygen but also with respect to other important resources.

Living organisms function and behave in an integrated way; that is, the effects of the function of one physiological system (e.g., respiration) are reflected in how other physiological systems function, such as those listed in Fig. 1.1. A large part of this book will be concerned with the interactions between supply-demand problems for a variety of resources (oxygen, energy, nutrients, water, solutes, reproductive demands, etc.), and you do not need to be concerned at this point with understanding these relatively complicated interactions. We will develop the rationale for understanding them through the principles governing resource supply-demand problems.

In this chapter we will identify the physical principles governing gas exchange and how they influence mechanisms for efficiently obtaining oxygen and removing carbon dioxide, particularly in aquatic environments. The principles of efficiency that we will deal with involve delivery of sufficient quantities of oxygen to meet demands while simultaneously *minimizing losses* or *changes* in other important resources. We will first establish the principles governing *regulation* of oxygen supply so that it equals demands; we will then discuss some negative-feedback *control* systems that provide the mechanisms for regulation (see Introduction). As a first step it is important to understand some basic physical principles governing movement of oxygen and carbon dioxide.

PHYSICAL PRINCIPLES OF GAS EXCHANGE

Respiration involves the exchange of gases between animals and their environments. Exchange involves movement, and *diffusion* provides an important part of the movement of gases over small distances. A chemical will move by diffusion from a region of higher concentration to a region of lower concentration from random thermal collisions between molecules. The rate of diffusion of a chemical from one location to another depends on the magnitude of the difference in concentration of the chemical at the two locations, the area or space through which diffusion can occur, and the distance between the two locations. Rate of diffusion also depends on the size of molecules; smaller molecules diffuse more rapidly, in inverse proportion to the square root of molecular weight. For any molecule, the rate of diffusion can be described by the following equation that summarizes these factors:

$$\frac{\Delta Q}{\Delta t} = \frac{-DA(c_1 - c_2)}{x},$$

(1.1)

where

ΔQ = change in quantity of a diffusing chemical,

Δt = change in time,

D = diffusion coefficient (a characteristic of the size of molecules and the environment in which they are diffusing),

A = area for diffusion,

c_1 = concentration of chemical at location 1,

c_2 = concentration of chemical at location 2, and

x = distance between location 1 and location 2.

The diffusion coefficient (D) depends on the size of the molecules and the characteristics of the medium in which they are moving that offer resistance to movement. For example, diffusion in liquids is slower than in gases, since a moving molecule is more likely to collide with another molecule, which can slow its net progress. The cells of animals are bathed in fluid, and the oxygen and carbon dioxide that move by diffusion must do so through water. This condition places a limit on the rate of supply of respiratory gases by diffusion in water as compared with their diffusion in air.

Oxygen is used in cells, and carbon dioxide is produced as a consequence of oxidation. Thus a *gradient*, a difference in concentration for the two gases ($c_1 - c_2$), is established. Less oxygen will be available in cells where it is used, and carbon dioxide will be higher in concentration in those cells. Thus diffusion will result in both movement of oxygen into cells and movement of carbon dioxide out of cells. The quantity of gases that move in this way depends on actual differences in concentrations over the distances the molecules diffuse; thus to understand gas movements by diffusion, we have to be able to measure concentrations in different locations.

Measurement of Gas Concentrations

The concentration of a chemical is the amount, measured in moles (mol), of the substance in a given space or volume measured in liters (l). Gases can be compressed or confined to a volume under pressure; thus gas concentrations in a gas phase are measured by the pressure they exert in a volume at a given temperature. This relationship is usually expressed with a simple equation for an "ideal" gas:

$$PV = nRT,$$

or

$$P = \frac{nRT}{V},$$

where

P = pressure of gas,

V = volume,

n = moles of gas,

R = universal gas constant, and

T = temperature (in absolute, or Kelvin, degrees).

When temperature is constant, n/V is directly related to pressure. The higher the pressure, the more gas molecules in a volume will collide to produce the measured pressure.

Pressure can be measured by the height of a liquid column supported by a gas. For example, the pressure of a gas will force a fluid up a tube, and the distance the fluid is displaced can be used to measure that gas pressure. By convention, the fluid used to measure gas pressures is mercury (Hg); since it is relatively dense, gas pressures are measured as millimeters of mercury (mmHg). One millimeter of mercury pressure is also called one *torr* of pressure.

The oxygen used by animals comes from plants that release oxygen during metabolism. It is added to the atmosphere where it mixes with other gases, including the carbon dioxide produced by animals and plants. At sea level at 0°C (standard temperature) the atmosphere exerts a pressure equivalent to a column of mercury 760 mm high; thus one atmosphere of pressure = 760 mmHg. Occasionally you will find pressures expressed as atmospheres or fractions of an atmosphere, and these units can be interconverted. Appendix I provides a discussion of a variety of different units used in physiological measurements, including the Système International (SI), the International System of Units. All systems are related; one can easily be converted to another simply by multiplying or dividing by the appropriate constants given in Appendix I, Table AI.2.

Partial Pressures

The atmosphere is a mixture of gases; we are particularly interested in oxygen and carbon dioxide. Table 1.1 gives the composition of dry air. Gases such as nitrogen and argon are usually combined as a single value and called "nitrogen" since their physiological effects are similar. To independently determine the pressure of oxygen in air, we need to know what proportion of the total pressure of the mixture is exerted just by oxygen; it involves multiplying the pressure of the total mixture by the fraction composed of oxygen (at a constant temperature and total pressure, usually corrected to 0°C and 760 mmHg—*standard temperature and pressure*). Thus for dry air at sea level at standard temperature, the pressure from oxygen is

$$760 \text{ mmHg} \times 0.2095 = 159.2 \text{ mmHg},$$

or

$$1 \text{ atm} \times 0.2095 = 0.2095 \text{ atm}.$$

This value is called the *partial pressure* of oxygen, or the P_{O_2}, which denotes the part of the total pressure contributed by oxygen. The partial pressures for other gases in dry air at standard temperature and pressure are shown in the upper left part of Fig. 1.2. Each partial pressure is the fraction contributed by each component to the total.

Table 1.1 The composition of dry air (from Otis, 1964*)

MOLECULE OF GAS	PERCENT COMPOSITION	
Oxygen	20.95	
Carbon dioxide	0.03	
Nitrogen	78.09	} 79.02
Argon	0.93	

*Credit references for tables and figures are near the end of the book before the index, as are the figure copyright acknowledgments.

	Dry air	Pressure *higher alt.*
$O_2 = 20.95$	$P_{O_2} = 159.3$	$P_{O_2} = 62.8$
$CO_2 = 0.03$	$P_{CO_2} = 0.23$	$P_{CO_2} = 0.09$
$N_2 = 79.02$	$P_{N_2} = 600.5$	$P_{N_2} = 237.1$
Total 100	760 mmHg	300 mmHg

Air with 3% water vapor

$O_2 = 20.32$	$P_{O_2} = 154.4$	$P_{O_2} = 60.96$
$CO_2 = 0.03$	$P_{CO_2} = 0.22$	$P_{CO_2} = 0.09$
$N_2 = 76.64$	$P_{N_2} = 582.5$	$P_{N_2} = 229.9$
$H_2O = 3.0$	$P_{H_2O} = 22.8$	$P_{H_2O} = 9.0$
Total 100	760 mmHg	300 mmHg

Composition, %

Figure 1.2
Illustration of the major effects of total pressure (rows) and gas composition (columns) on the partial pressures of gas components in air.

In addition to the effects of temperature, the partial pressure of a component in a gas mixture can change in two ways. If the total composition stays the same but the total pressure changes, partial pressures will change, since the same fractional amount is multiplied by a different total quantity. This relationship is shown by comparing the upper parts of Fig. 1.2. An example is an increase in altitude where atmospheric pressure decreases; a pressure of 300 mmHg occurs at about 7000 meters (about 23,000 feet) above sea level. It represents a significant change in the environmental supply of oxygen for animals that live at high altitudes (see Chapter 2).

The second way partial pressures in air change is if another gas is added to the mixture or a different total amount of gas is enclosed in the same volume. For example, atmospheric air is seldom completely dry but instead contains variable amounts of water vapor. The example in the lower part of Fig. 1.2 is for 3% water vapor in air. P_{H_2O} is 3% of total pressure, or 22.8 mmHg at 760 mmHg total pressure. Thus P_{O_2} is 20.95% of the remaining mixture, or 20.95% of $(760 - 22.8) = 154.4$ mmHg. In this situation, n has changed relative to V in the equation relating pressure to amount of gas in a volume.

Gases in Water versus Air

When a liquid and a gas are in contact, the gas enters the liquid until an equilibrium is reached, when as many gas molecules leave the liquid as enter it. The total quantity of a gas in a volume of liquid at this equilibrium—the concentration of the gas in the liquid—depends on *two* factors: (1) the *pressure* (or partial pressure) of the gas in the gas phase, and (2) the *solubility* of the gas in the liquid. The only liquid in which we are interested is water, and the gases in which we are interested are in air.

Table 1.2   Solubility coefficients (α = milliliters gas/liter water at 760 mmHg gas-phase pressure) for the common gases in air (from *CRC Handbook of Chemistry and Physics*)

TEMPERATURE (°C)	OXYGEN	CARBON DIOXIDE	NITROGEN
0	48.9	1,713	23.5
5	42.9	1,424	20.9
10	38.0	1,194	18.6
15	34.1	1,019	16.8
20	31.0	878	15.4
25	28.3	759	14.3
30	26.1	665	13.4
35	24.4	592	12.6
40	23.1	530	11.8

Table 1.2 presents information on the "solubility coefficients" (= α values) of the gases in air for pure water. These values were measured for pure gases at a pressure of 760 mmHg. The units on α are milliliters of gas/liter of water at 760 mmHg pressure. Thus at 10 °C, 38.0 milliliters of oxygen will dissolve in one liter of water when the gas phase is pure oxygen at a pressure of 760 mmHg (Table 1.2). The solubility coefficients can be used to determine the quantity of a gas in a liquid from the following relationship (Henry's law):

$$\text{Concentration of gas in water} = \alpha_{g,°C}\left(\frac{P_g}{760}\right), \text{ in ml gas per l } H_2O,$$

where

$\alpha_{g,°C}$ = solubility coefficient (Table 1.2) for a particular gas (g) at a particular temperature (°C), and

P_g = partial pressure of the gas (g) in the gas phase.

For example, dry air at an atmospheric pressure of 760 mmHg has a P_{O_2} = 159.2 mmHg (Fig. 1.2). At 5 °C, $\alpha_{O_2,5°C}$ = 42.9 ml O_2/l H_2O, so

$$\frac{\text{ml } O_2/\text{l } H_2O}{\text{at 5 °C}} = (42.9)\left(\frac{159.2}{760}\right)$$

$$= 8.99 \text{ ml } O_2/\text{l } H_2O.$$

Two features of the solubilities of gases in water are important for interpreting environmental effects on supply. First, note from Table 1.2 that carbon dioxide is 25 to 35 times more soluble in water than is oxygen (depending on temperature). Second, note from Table 1.2 that, for any gas, the concentration in water is higher at lower water temperatures. For example, a change in an aquatic environment from a temperature of 0 °C to 30 °C (with the same P_{O_2}) involves a reduction in available oxygen by (48.9 – 26.1) = 22.8 ml O_2/l water (Table 1.2), or (22.8/48.9) = 47 %.

Another environmental variable that influences oxygen availability is the effect of salts in water on the solubility of gases. The salts in seawater reduce the solubility coefficients for oxygen by about 20%; thus at any temperature the solubility coefficients for oxygen will be only 80% of the values listed for pure water in Table 1.2 (e.g., $\alpha_{O_2,5°C} = (0.8 \times 42.9) = 34.3$ ml O_2/l seawater). The effects of salts on gas solubilities can be illustrated when you salt your beer, a common way to "raise a head" by forcing more carbon dioxide out of solution. Since the solubility coefficients for carbon dioxide are so much larger than for oxygen (Table 1.2), salts are believed to have little impact on aquatic respiration with respect to release of CO_2, but their impact on oxygen availability can be appreciable, particularly when water temperatures are high.

DIFFUSION LIMITATIONS IN RESPIRATION

Cells are surrounded by a membrane, and oxygen must diffuse through the membrane and reach the point in the cell farthest from the membrane where oxidation is taking place, whereas carbon dioxide must diffuse through the membrane in the opposite direction. Therefore, the membranes of cells must be *permeable* to oxygen and carbon dioxide, i.e., not offer appreciable resistance to their diffusion movements. This characteristic is important because permeability to gases is associated with permeability to water. Water will move into and out of cells across membranes when there are concentration differences in total solutes between the inside and the outside of cells (see Chapter 12). In some environments water will be gained or lost as a result of having membranes permeable to gases. (See Chapters 12, 13, and 14 for an extensive discussion of these problems in different environments and their interactions with the requirements of respiration.)

What is the impact of a change in the dimensions of a cell on diffusion? A number of cells are spherical in shape; we can examine consequences for them. For these cells, membrane surface area is minimum compared with the volume of the cell (Fig. 1.3). For example, when a cell is spherical in shape, the area for diffusion (A in Eq. 1.1) is at a minimum relative to the volume of the cell in which oxidation is taking place. For a sphere, the maximum distance from the surface is the radius (r in Fig. 1.3), representing the maximum distance over which diffusion (x in Eq. 1.1) must occur to supply oxidative demands.

Figure 1.3 illustrates the interrelationships between r (x in Eq. 1.1), surface area for diffusion exchange (A), and volume (V) for a sphere. When r increases, A increases as the square of r, and V increases as the cube of r. Thus an increase in radius by two times results in an increase in surface area by *four* times but an increase in volume by *eight* times (Fig. 1.3). The volume of the cell contains all the metabolic machinery that must be supplied with oxygen and from which carbon dioxide must be removed. Since the volume of a sphere increases more than the surface area does when the radius increases, the disproportionate relationship between A and V sets an upper limit on the size of spherical cells because of limitations on rate of supply and removal for respiration.

The maximum distance for diffusion of oxygen and carbon dioxide (r) can also set a limit on cell size due to physical limits on rate of supply of oxygen and

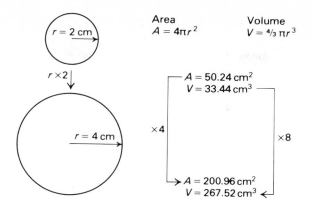

Figure 1.3
Illustration of the interactions between the radius of a
sphere and its surface area (A) and volume (V) when
the size of a sphere changes (after McCauley, 1971).

rate of removal of carbon dioxide. Oxygen diffuses rapidly only over short
distances. For example, in water it will diffuse over 0.1 mm in about 1 second,
but it will take 100 times longer to diffuse over a distance 10 times as great. It
would take oxygen about 100 seconds to diffuse 1 mm and almost 3 hours to
diffuse 1 cm. The continuous demands for oxygen at the farthest point from the
membrane of a cell dictate that oxygen be supplied relatively rapidly. This rapid
supply is achieved in large cells from the folding of membrane surfaces (to
decrease r) and/or by internal mixing, that is, the circulation of cellular protoplasm. Cells that are larger than about 0.5 mm in radius usually exhibit protoplasmic streaming, which transports resources to and from the cell membrane. This type of movement mechanism may be required if distances for
oxygen and carbon dioxide diffusion are relatively large. However, the movement of protoplasm requires energy (see Chapter 18); the increase in the size of
a cell will thus dictate a greater expense for supplying the demands of respiration.

   Multicellular animals larger than about 1 mm must also have mechanisms
to deal with the distance problems for the rate of the diffusion supply of oxygen
and removal of carbon dioxide. It has been accomplished through the evolution
of two classes of mechanisms for *bulk flow* of fluids functioning together with
diffusion over small distances. The general pattern is illustrated for oxygen in
Fig. 1.4; the situation is reversed for carbon dioxide. Some structures, such as
muscles associated with lungs or gills, produce bulk movement of a respiratory
medium over relatively *large* distances to a respiratory exchange surface,
where diffusion occurs over much *smaller* distances.

   Bulk flow of a respiratory medium over an exchange surface is called *ventilation*. At the exchange surface, oxygen diffuses into blood and carbon dioxide
diffuses out. The blood is circulated between respiratory organs and the cells in
tissues utilizing oxygen and producing carbon dioxide. Like respiration, cir-

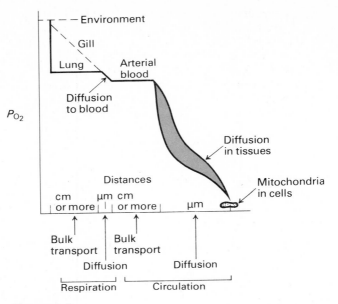

Figure 1.4
Schematic illustration of the major bulk transport and
diffusion-exchange characteristics that result in delivery of
oxygen from the external environment to the cells utilizing
oxygen. Diffusion exchange occurs only over small
distances, while bulk transport occurs over larger distances
(modified from Wood and Lenfant, 1977).

culation involves bulk transport of fluid over distances of centimeters or more
(Fig. 1.4; see also Chapter 4). Diffusion of oxygen into and carbon dioxide out of
cells occurs in capillaries where distances for diffusion are small.

There are four major mechanisms for achieving bulk transport and diffusion for respiratory exchange between an animal and its environment: (1) use
of "digestive" systems and external body surfaces; (2) use of gills; (3) use of
lungs; and (4) use of tubes (trachea). The first two classes of respiratory adaptations are most prevalent among animals that obtain oxygen from and release
carbon dioxide to water. The last two classes of adaptations are most prevalent
among air-breathing animals. Within each major environmental division of
water breathers *versus* air breathers, there is a relationship between demands
for oxygen and the complexity and mode of function of respiratory ventilation
systems.

AQUATIC RESPIRATION

"Digestive" and Surface Respiratory Ventilation

Multicellular animals with only one, two, or three cell layers obtain oxygen as a
consequence of their feeding and/or obtain food as a consequence of respiratory

ventilation. These animals include a large number of invertebrate aquatic animals, such as sponges (Porifera); hydroids, sea anemones, and corals (Coelenterata); and the flatworms (Platyhelminthes), including planarians and flukes.

Figure 1.5 illustrates the design of the combined digestive and respiratory systems for some typical aquatic animals with three or fewer cell layers. In simple or complex sponges, cells with cilia (thread-like projections) produce a water current through the animal in one direction. In addition, local water currents around the animals favor movement of water through them in one direction.

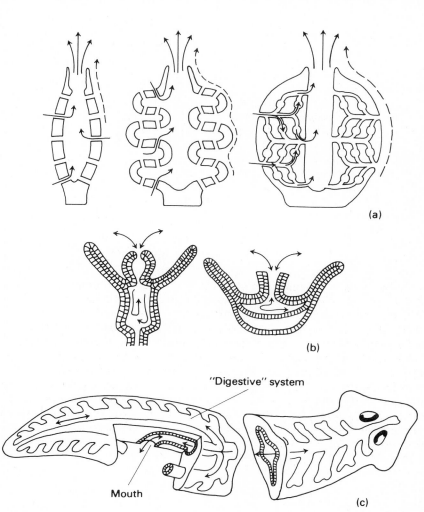

Figure 1.5
Schematic illustration of the pattern of bulk water ventilation of the digestive systems for sponges (a), coelenterates (b), and a planarian (c) (from Buchsbaum, 1948).

This movement establishes a bulk flow of water, allowing the oxygen that diffuses from water to the cells to be rapidly and continuously renewed and to remain at a relatively high concentration just outside the cell membranes. Similarly, in coelenterates and flatworms fluid is continuously moved through the "digestive" system. Particles of food are trapped, and the respiratory medium is continuously renewed. Diffusion distances are small because of the limited number of cells and their organization in the animals. The functions of feeding and respiration are combined since they are both continuous processes in these animals.

In addition to bulk flow through the internal "digestive" system, cells on the outer surfaces of these animals obtain oxygen from and release carbon dioxide to their environments. Some bulk flow of the environment with respect to the external surface occurs from water currents established by ciliary beating, from water currents around the animals, or from their movements through their environments.

The link between obtaining food and obtaining oxygen from an aquatic environment is an effective system for both feeding and respiration, but there is little information about the degree to which bulk flow of water may be independently related to feeding versus respiration. If food is the least available resource in the external environment, bulk flow of water may be primarily for the purpose of feeding, with only secondary importance for respiration. This possibility is suggested because many filter-feeding animals show a daily rhythm of bulk-flow activity associated with periods of food availability. Requirements for oxygen may be relatively continuous but may be satisfied with less bulk flow than is required for obtaining food. It would be interesting to obtain more information on relationships between these processes, such as the minimum bulk flow for providing sufficient diffusion of oxygen and carbon dioxide to meet respiratory demands in different animals.

The extent to which supply of oxygen can be met by relatively simple bulk flow of water through a "digestive" system depends in part on the size or the number of cells that require relatively continuous delivery of oxygen and removal of carbon dioxide. An increase in body size leads to diffusion supply-demand problems even with "digestive" bulk flow. In addition, an increase in body size requires the processing of more food (see Chapter 10); therefore, digestive function will have to be *dissociated* from respiratory function. The upper limit appears to occur for invertebrates (such as some tube-dwelling brachiopods) that are larger than flatworms but that process large quantities of water for filter feeding and respiratory ventilation. Moreover, when large size is coupled with activity, further demands are placed on the supply of oxygen. Many filter-feeding invertebrates are relatively inactive exept for their bulk movements of water. Larger, more active animals have evolved *separate* structures that serve a more exclusively respiratory function in the largest and most active species.

Gill Respiration

The gill is the most common aquatic respiratory structure in larger animals. The epithelium of a gill is thin so that water and blood will flow in close proximity.

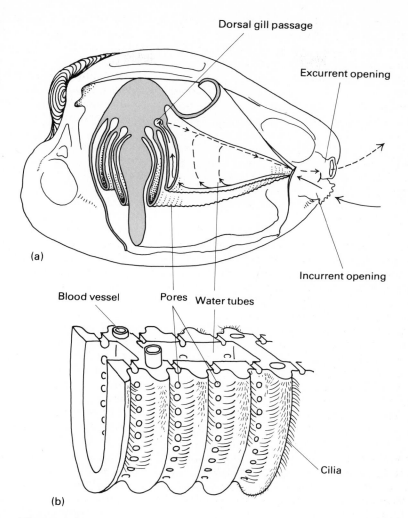

Dorsal gill passage

Excurrent opening

(a)

Incurrent opening

Blood vessel   Pores Water tubes

Cilia

(b)

Figure 1.6
Structure and pattern of water ventilation for the gill of a clam.
Diagram (a) illustrates the arrangement of the gill with respect to
water flow in the animal. Diagram (b) illustrates both the folding of
the gill surfaces and the surface cilia that filter food (from
Buchsbaum, 1948).

Figure 1.6 illustrates the gross (a) and fine (b) structure for the gill of a clam.
Water is pumped into the gill area and passes through many small holes in the
gill from bottom to top and then out of the animal. The surface of the gill is
divided into many small ridges, which increase the effective area for diffusion
exchange of gases (A in Eq. 1.1); also, distances between blood and water are
small. For some aquatic animals the gill also serves a food-gathering function.

For example, the ridges on the gill of the clam have small, hairlike projections that trap food particles (Fig. 1.6). However, in most larger aquatic animals the primary function for the gill is respiratory gas exchange.

Large body size together with localization of respiration in the area of the gills requires relatively large, localized bulk flow of water across the gills. Several characteristics of water as a respiratory medium influence the efficiency of this gill ventilation. Table 1.3 lists four properties of water and air that influence ventilation. Water is 1000 times more dense than air; water is 100 times more viscous, that is, more resistant to flow; at 20°C water contains, per ml, about 1/30 the oxygen present in air; and finally, oxygen will diffuse about 10,000 times slower in water than in air (Table 1.3).

Table 1.3 Some characteristics of air and water that influence gas exchange

CHARACTERISTICS	AIR	WATER
Relative density of medium	0.001	1.0
Relative viscosity of medium	0.01	1.0
Oxygen concentration in environment	209.5 ml/l at sea level	Fresh = 6.5 ml/l (20°C) Salt = 5.2 ml/l
Diffusion coefficient for oxygen	0.196 cm^2/s	0.0000183 cm^2/s

All of these factors make water a less effective source of oxygen supply than air; the mechanisms of respiration in relatively large aquatic animals must be efficient to extract a maximum amount of oxygen from the environment relative to the effort required for respiratory ventilation. For example, 1 ml of water at 20°C contains about 0.0065 ml of oxygen and weighs 1000 times more than 1 ml of air containing 0.21 ml of oxygen (Table 1.3). Thus to obtain access to 0.21 ml of oxygen an aquatic animal must move a mass of water weighing over 30,000 times the mass required for air. This effort requires an increased expenditure of energy for each ml of oxygen in the environment, and an effective respiratory system should involve extracting a maximum amount of oxygen for the effort expended.

The pattern of bulk-flow ventilation and its importance for effective gill respiration has been most completely studied in teleost fishes. The gills are located on several gill arches in the opercular cavity (Fig. 1.7). Bulk flow of water across the gills is almost continuous due to the action of two "pumps" (Fig. 1.8; compare this with the flow diagram in Fig. 1.7), one a suction pump and the other a pressure pump.

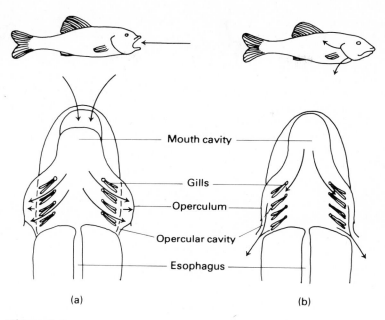

Figure 1.7
Schematic illustration of the pattern of water flow during respiration
in a teleost fish. Water movement when the opercular covers expand
during the phase when the mouth is open is shown in (a). Water
movement when the mouth pump forces water over the gills is
shown in (b) (after McCauley, 1971).

Teleost fishes alternately open and close their mouths to ventilate the gills.
When the mouth opens, the outer part of the opercular cavity expands, creating
a negative or "suction" pressure on the outer surface of the gills that draws
water across the gills (Fig. 1.7a; Fig. 1.8a1, 1.8b1). The mouth then closes. For
a brief period the mouth and the opercular gill covers are both closed (Fig.
1.8a2), but there is still a small pressure gradient for water flow over the gills
(Fig. 1.8b2). The outer opercular covers then open while the oral cavity is compressed (Fig. 1.8a3; Fig. 1.7b), and positive oral pressure is maintained to produce flow over the gills for a relatively long period (Fig. 1.8b3). Finally, water
flow is reversed for a very brief period when the mouth and opercular covers
open (Fig. 1.8a4, 1.8b4) prior to the initiation of another cycle.

For an inactive fish the entire cycle for gill ventilation takes about 0.7 second. During this time water flow occurs across the gills from inside to outside
for about 0.6 second, and flow occurs in the opposite direction for only 0.1 second (Fig. 1.8a4), so for about 86% of the time water flows in *one direction*. This
flow is extremely important because it provides the basis for effective gas exchange between water and blood.

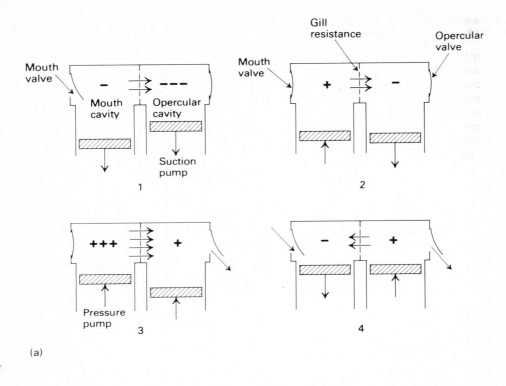

(a)

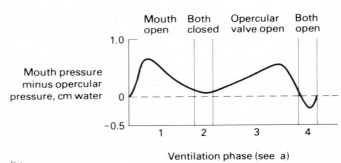

(b)

Figure 1.8
Diagram of the action of the mouth pump and opercular pump during the four phases
of ventilation for a teleost fish (a). The + and − symbols indicate pressures in the
mouth and each opercular cavity. Graph (b) indicates the magnitude of the total
pressure gradient between the mouth and each opercular cavity (across the gills) for
each phase of the respiratory cycle (after Hughes and Sheldon, 1958; Hughes, 1964).

Countercurrent Gas Exchange: Demonstrating Respiratory Efficiency with an Experiment

Figure 1.9 illustrates the relationship between the structure of the gills and water ventilation during the major portion of a respiratory cycle, when water flow is in one direction. Each gill arch contains many gill filaments projecting from it, and each gill filament contains a series of ridges projecting perpendicularly from the filaments. These ridges are called the *secondary gill* ✕ *lamellae* and are where gas exchange takes place. The secondary lamellae are thin and collectively provide a large surface area for diffusive gas exchange (*A* in Eq. 1.1).

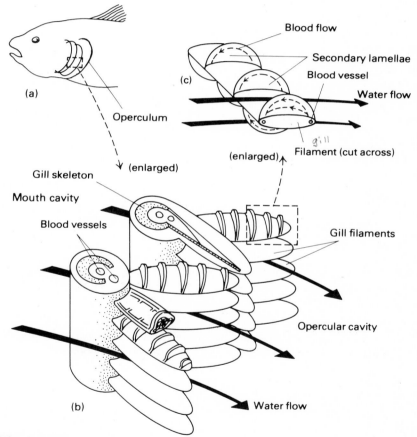

Figure 1.9

Schematic diagram of the anatomy of the gills in relation to water flow for the major portion of the respiratory cycle, when water flow is countercurrent to blood flow at the exchange surfaces of the secondary lamellae.

Arterial blood flows from the gill arches along the outer edges of the gill filaments and enters small capillary vessels in the secondary lamellae; the flow of blood where exchange takes place is *countercurrent*, or opposite in direction, to the flow of water across the gills (Fig. 1.9). Thus when exchange first takes place for blood in the secondary lamellae, oxygen-poor blood meets water from which some oxygen has been removed by diffusion earlier in the passage over the gill. As the blood moves across the secondary lamellae it gains oxygen (or becomes more saturated with oxygen), but it will, by about the same amount, always contain less oxygen than water. This fact is illustrated graphically in Fig. 1.10(a).

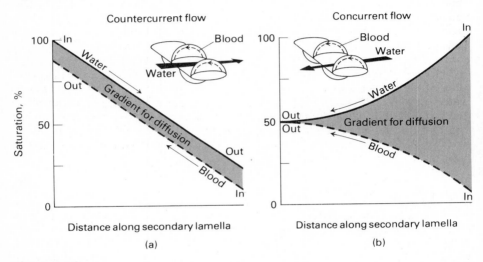

Figure 1.10
Illustration of the extent of the diffusion gradient for oxygen movement from water to blood (shaded areas) when water flow is countercurrent to blood flow (a), and when water flow is concurrent with blood flow (b) (modified from Hughes, 1964).

The important feature of the countercurrent exchange between blood and water is the relatively *constant* gradient, or difference in concentration, of oxygen between water and blood along the entire length of the secondary lamellae. Diffusion to blood will occur when the concentration of oxygen in blood is less than in water ($c_1 - c_2$ in Eq. 1.1). With countercurrent flow, uptake of oxygen through diffusion should occur at all positions in the gill exchanger, even when the blood has gained some oxygen. Continual uptake has been hypothesized to be effective for respiration in water, since it will maximize the amount of oxygen extracted from water compared with the energy expended for bulk flow of water and blood.

This hypothesis has been tested with an experiment. The alternative to countercurrent flow is concurrent flow (flow in the same direction). The flow of blood cannot be reversed, but the flow of water can easily be reversed experimentally by forcing water in the opposite direction through the gills at pressures greater than those produced by the fish for ventilation.

Concurrent flow produces the result shown in Fig. 1.10(b). When blood first enters the secondary lamellae, a very large gradient for diffusion exists between blood and water, but it occurs only for a short distance. As blood gains oxygen and water loses oxygen the concentration difference between them (c_1 − c_2) will decrease, so that, by the time both blood and water exit the secondary lamellae, less *total* oxygen will have entered the blood than with countercurrent flow (Fig. 1.10a). For the example shown in Fig. 1.10, countercurrent exchange results in blood achieving about 80% saturation compared with only about 50% saturation of blood for concurrent exchange. The difference in oxygen entering the blood, about 1.6 times that of concurrent exchange, represents the efficiency of having flow of blood and water opposite in direction during the major portion of a respiratory cycle, enabling a constant difference in oxygen concentration between blood and water to be maintained.

Measurements have been made of the extent to which water loses oxygen when it flows over gills in various fishes. From 48% to 80% of the oxygen in water diffuses into blood in triggerfish, carp, trout, and eels breathing water equilibrated with atmospheric air. The variation in the amount extracted depends on the ventilation volume relative to the total surface area for gas exchange in the gills, and considerably less would be extracted if water ventilation was not countercurrent to blood flow.

Skin Ventilation in Vertebrates

The pattern of gas exchange between the external body surface of an animal and an aquatic environment, described above for animals with three or fewer cell layers, is also present in some larger vertebrates. Ventilation of the external surfaces involves modifications to adjust the magnitude of the required exchange to the increased demands dictated by larger size and activity. Several species of amphibians are primarily aquatic. Some, such as *Necturus*, have external gills as adults; others, such as *Cryptobranchus, Amphiuma,* and *Euproctus*, are aquatic and have no gills. Their mouth regions may be highly vascularized and lungs may be present, but most gas exchange occurs at the surface of the skin. This has been demonstrated for *Euproctus* by reducing cutaneous respiration by covering the skin with Vaseline. Individuals treated in this way quickly suffocate.

The skin of aquatic amphibians is highly vascularized; either the movements of the animals in their environments and/or local water currents produce bulk flow for skin ventilation. In general, carbon dioxide readily diffuses through the skin; most interest has concerned the uptake of oxygen. We will find (Chapter 2) that most amphibians exchange some gases with water through the skin even if they primarily respire air. The importance of the skin for aquatic respiration in vertebrates can be illustrated by an amphibian with a very high degree of aquatic cutaneous gas exchange: the Lake Titicaca frog (*Telmatobius culcus*). In the next chapter we will study gas exchange for amphibians in which the primary route of supply is from air.

Figure 1.11
A Lake Titicaca frog (*Telmatobius culcus*). The photograph illustrates the extent of skin folding for increasing skin respiration in water. Photo courtesy of Victor Hutchison, University of Oklahoma.

Lake Titicaca is 3812 meters above sea level, between Peru and Bolivia in the Andes Mountains. The temperature of the water below the surface is relatively constant at about 10 °C, and it is equilibrated with ambient oxygen at about 100 mmHg P_{O_2} (5 ml O_2/l water) because of continuous mixing of water from currents and strong wind and wave activity. Lake Titicaca frogs occur at all depths (down to 281 meters below the surface) and have been taken with fishing nets dragged along the lake bottom.

Telmatobius has lungs, but they are only one-third the size of the lungs of terrestrial frogs of the same size. The mouth cavity is richly vascularized, but the skin shows extensive modifications for respiratory gas exchange (Fig. 1.11). The skin is loose and abundantly folded, increasing the surface area for exchange (*A* in Eq. 1.1). There are large numbers of capillaries in the skin, with fine divisions passing close to the surface. The blood contains a large number of very small red blood cells, and the respiratory properties of the blood show several adaptations for obtaining oxygen from an environment with a low P_{O_2} (see Chapter 3 for details). In addition, the rate of use of oxygen by *Telmatobius* is low compared with other amphibians of the same size.

When water contains 5 ml O_2/l the frogs do not surface to breathe, and skin respiration is sufficient to provide for oxygen demands. If the availability of oxygen is experimentally reduced by about 20-40% (to 3-4 ml O_2/l), they will surface and take air into the lungs. When surfacing occurs, the rate of oxygen use increases by 50-60%. We will find (Chapter 2) that this type of change is typical for animals that can breathe air or water. Access to the richer supply available from air is associated with a higher rate of use of oxygen, although the supply from water is sufficient if demands for oxygen are less.

If the frogs are prevented from surfacing when water oxygen availability is reduced, they increase ventilation of the skin. They periodically "bob" by kicking their hind legs and pushing off the bottom. As they sink back through the water to the bottom, the large skin folds ripple, providing increased bulk flow for skin ventilation. An increase in respiratory ventilation also occurs in aquatic animals that use gills when the supply of oxygen decreases compared with demands.

CHANGING GILL VENTILATION WHEN SUPPLIES OR DEMANDS CHANGE

Some species of large aquatic animals are quite active and have relatively high demands for oxygen, while others of similar size are more sedentary in habit with lower demands for oxygen. Measurements of gill surface areas (A in Eq. 1.1) of different species indicate that more active species have increased surface areas for gas exchange (Fig. 1.12). In addition, within any animal that breathes water, demands for oxygen will change with activity, and the supply of oxygen will change relative to demands if activity results in a change in environmental temperature or salt concentration. For *regulation* to occur, the bulk flow of water over gills must change; supply is adjusted to equal demands in a manner similar to the increase in skin ventilation seen in the Lake Titicaca frog when environmental supply was decreased relative to demands.

Changing the ventilation supply of oxygen involves changing the number of milliliters of water that pass over the gills every minute. This is called the *minute volume* of gill ventilation. Minute volume (ml/min) is the product of two quantities: the number of breaths taken per minute, the *respiratory rate*, and the volume per breath, the *tidal volume*.

$$\begin{aligned} \text{Minute volume} \quad &= \text{Respiratory rate} \times \text{Tidal volume} \\ \text{(ml/min)} \quad &= \text{(breaths/min)} \times \text{(ml/breath)} \end{aligned}$$

The "breaths" on the right-hand side of the equation cancel through multiplication.

For some fishes that have been studied, minute volume changes involve both changes in the amount of water breathed per minute and the rate of breathing. Figure 1.13 shows how each component of the gill minute volume contributes to total ventilation for the tench. In this fish, changes in respiratory demands are met by approximately equal proportional changes in respiratory rate and tidal volume.

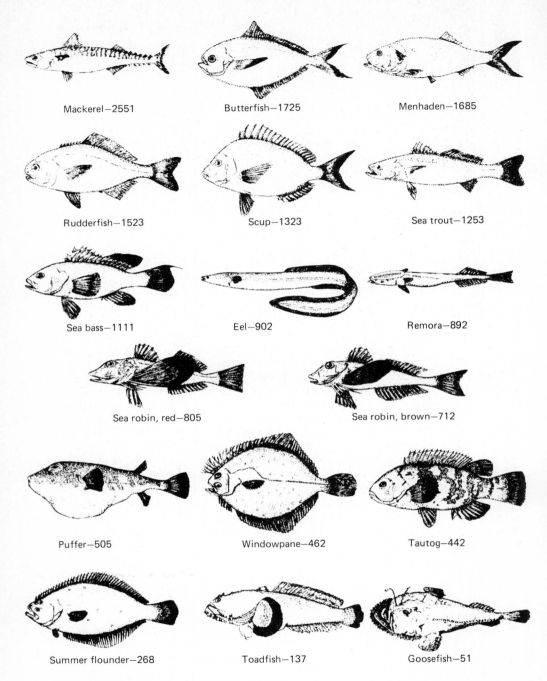

Figure 1.12
The gill surface area per gram body weight (in arbitrary units) for gas exchange is greatest in fishes with high demands for oxygen and lowest for fishes with relatively sedentary habits (from Schmidt-Nielsen, 1979; based on Grey, 1954).

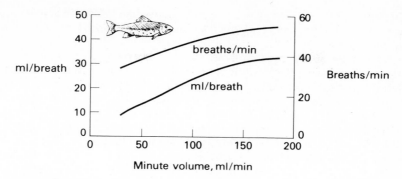

Figure 1.13
In the tench, a teleost fish, minute volume for gill ventilation
changes as a consequence of changes in both respiratory rate and
tidal volume (after Hughes, 1964).

Some fishes achieve proportionally increased gill ventilation when they
are active simply by keeping their mouths open as they swim forward. In
essence, they take one huge breath at a rate dependent on swimming speed.
This technique has been called "ramjet ventilation" in contrast to the more
periodic ventilation involving a rhythmic rate of breathing. Use of the energy
expended for forward movement to ventilate the gills will reduce the energy
cost for separate bulk transport of water over the gills through mouth and opercular pumping, but this mechanism involves a cost; a fish will be less streamlined
(and, will therefore have a higher drag; see Chapter 9) if it swims with its mouth
open. However, a number of aquatic animals that swim almost continuously,
such as mackerel and some sharks, utilize ramjet ventilation and will suffocate
if they cannot swim forward. For these species with high demands for oxygen,
the cost from increased drag is less than the cost involved for rapid pump ventilation of the gills.

Not all aquatic animals regulate respiration by proportionally changing
minute volume if oxygen supplies change. Those that do not change minute
volume generally have relatively low demands for oxygen and/or live in environments where supplies are always sufficient to meet demands without
minute volume changes. For example, species of lobsters that have evolved to
live in the cool waters of northern oceans do not change the rate of gill ventilation if the supply of oxygen in their environment is experimentally reduced.
They have never been exposed to such a situation, so there has been no selection for a mechanism to change rate of supply if the environment changes.

Control of Aquatic Ventilation

For species that do change minute volume when supply-demand situations
change, there must be a negative-feedback control system to adjust supplies
and/or demands to new values. It will involve a stimulus, a receptor to detect
changes in the stimulus, and effectors, i.e., mechanisms to produce a response

(see Introduction). For respiration, the response is a proportionally changed minute volume so that more or less water flows over the gills as supplies and/or demands for oxygen vary.

The details of the mechanisms for the operation of control systems in aquatic animals are poorly understood, and most information comes from responses of fishes to experimental changes in the P_{O_2} or P_{CO_2} in the external environment. Most fishes respond to a decrease in P_{O_2} by increasing minute volume. The location of the receptors for this change has not yet been pinpointed. There is some evidence that the receptors may be located in the area of the gills. We will find that receptors for detecting changes in blood P_{O_2} in air-breathing vertebrates are located in the aorta and carotid arteries, with nerves that pass to an area of the brain called the medulla oblongata. We will also find (Chapter 2) that many air-breathing vertebrates respond to a change in blood pH produced from changes in blood P_{CO_2} to control respiratory ventilation. A change in P_{CO_2} does not appear to be as important for respiratory control in aquatic animals. Carbon dioxide is much more soluble in water than is oxygen (Table 1.2), so it is difficult for a change in P_{CO_2} to serve as an index of a changed demand for oxygen. The possible relationships between components of a general negative-feedback control system for gill ventilation are shown in Fig. 1.14 for a situation controlling minute volume with increased demands for oxygen relative to supplies. Note that some components of these relationships are guesses based on limited information.

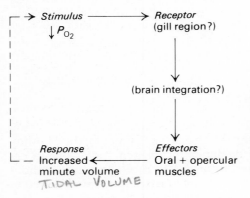

Figure 1.14
Schematic diagram of some possible relationships in a negative-feedback control of ventilation in an aquatic vertebrate. The situation represents a response to decreased supply of oxygen relative to demands.

RESPIRATION AND ACID-BASE REGULATION

Just because an aquatic animal may not show a ventilation response to changes in P_{CO_2} does not mean that the exchange of carbon dioxide is not important for efficient function in these animals. Net carbon dioxide content plays a crucial role in the regulation of the *acidity* of the blood and cellular fluids of all animals. This is a consequence of a reaction of carbon dioxide with water to form a weak acid, carbonic acid, according to the following chemical reaction.

$$CO_2 + H_2O \rightleftharpoons H_2CO_3 \rightleftharpoons H^+ + HCO_3^- \tag{1.2}$$

The carbonic acid readily dissociates into hydrogen ions (H^+) and bicarbonate ions (HCO_3^-).

By convention, pH is defined as "the inverse of the logarithm of hydrogen ion concentration," or $1/Log\ [H^+]$. It is important to note that pH is a logarithmic function, that is, a function based on powers of 10. When the pH of a solution changes from 5 to 4, the hydrogen ion concentration increases from 1×10^{-5} to 1×10^{-4} moles of hydrogen. Because pH is an inverse function, a *decrease* in pH involves an *increase* in hydrogen ion concentration.

Why is pH important? The function of animals is influenced by the ionic environment, the electrical charge, which affects certain chemicals in cells. It appears to be particularly important for the activities of *enzymes* that catalyze (promote) important cellular reactions. Enzymes are complicated structures of amino acids linked together in particular sequences. The amino acid units represent the *primary* structure of an enzyme, and their linear sequence of linkage provides a *secondary* structure. Many enzymes are not effective in speeding reactions if they do not have a particular secondary structure because of the three-dimensional (or *tertiary*) structure of an enzyme in solution. The chain of amino acids folds on itself in a particular way that permits effective interaction with a substrate to speed a reaction. The three-dimensional folding depends on relatively weak bonds (hydrogen bonds and van der Waals forces) between the units in the secondary structure, and a change in enzyme ionic composition or electrical charge forms or disrupts these weak bonds and changes the effectiveness of enzymes.

The importance of pH for enzyme function can be studied by measuring the rate of a reaction catalyzed by an enzyme when the concentration of hydrogen ions is varied (Fig. 1.15). Normally, there is a narrow range of pH where a particular reaction rate is obtained. Since it is important that the reactions catalyzed by enzymes occur at particular values (not necessarily the maximum), mechanisms have evolved to maintain specific pH values or optimal tertiary structures for enzyme function.

For most animals the optimal pH for enzyme function is slightly alkaline with respect to the neutral pH of water. A neutral pH exists when the number of hydrogen ions and hydroxyl (OH^-) ions are equal. Thus by definition, pure water always has a neutral pH. The total possible range of pH at 25 °C is from 1 to 14, and the product of hydrogen ion concentration and the concentration of hydroxyl ions is

$$[H^+]\,[OH^-] = 10^{-14}.$$

A neutral (pure water) solution will have no difference in the number of hydrogen ions and hydroxyl ions, or when

$$[H^+] = [OH^-] = 10^{-7} \text{ at } 25\,°C,$$

so at 25 °C water has a neutral pH of 7.0.

Most enzymes have optimal tertiary folding for catalyzing reactions when there is a net negative charge on the enzyme. This charge occurs when the pH is greater (more alkaline) than 7.0 at 25 °C, and the optimal pH at 25 °C for many enzymes is about 7.8.

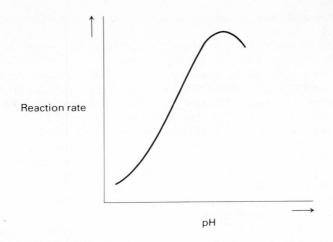

Figure 1.15
For most biological enzyme-substrate interactions at a
particular temperature, reaction rate increases to a maximum as pH increases. As a consequence, a particular
reaction rate will occur only at a specific value of pH.

For many water-breathing animals it is important to note that pH changes
with temperature. When temperature increases, more hydrogen ions and hydroxyl ions dissociate from pure water; thus the neutral pH of water *decreases*
as temperature *increases*. The change in the neutral pH of water when temperature changes is shown in Fig. 1.16. Note that water has a neutral pH of 7.0 only
at 25 °C.

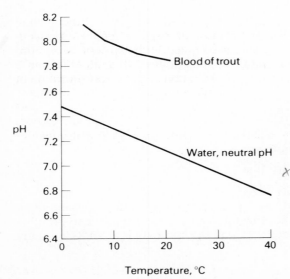

Figure 1.16
Because the dissociation of water
depends on temperature, the
neutral pH increases when water
temperature decreases. Rainbow
trout that change in body
temperature also change blood pH
to maintain the same degree of
net negative charge for optimal
enzyme function (after Randall
and Cameron, 1973).

Figure 1.16 also shows measured values for the pH of the blood of rainbow trout kept at different temperatures. Most water-breathing animals are "ecto-therms" and have body temperatures that are similar to the temperatures of their environments (see Chapter 6). Note that the pH of the blood also increases when temperature decreases so that the blood is maintained at the same degree of alkalinity with respect to the changing neutral pH of water. This condition provides effective enzyme function by maintaining an appropriate, relatively constant net ionic enzyme charge as temperature changes. Similar regulation of the relative alkalinity of blood has been observed in at least eight other species of fishes including carp, salmon, lungfish, and electric fish.

Control of pH in Aquatic Animals

How is this regulation of acid-base balance achieved? There are several ways that depend on the reaction of carbon dioxide with water to form hydrogen ions and bicarbonate ions (see Eq. 1.2). The law of mass action states that a revers-ible chemical reaction can be driven in either direction by changing the concen-tration of the reactants or products. Thus pH will increase (fewer hydrogen ions) by decreasing the concentration of the reactant carbon dioxide, resulting in less carbonic acid formation and less dissociation to hydrogen and bicarbonate ions. Alternatively, an increase in the product bicarbonate ions will force the reaction toward carbonic acid and carbon dioxide, reducing hydrogen ion concentration. The opposite sequences (increased CO_2; decreased HCO_3^-) will result in a de-crease in pH (more hydrogen ions).

Water-breathing animals appear to regulate acid-base balance primarily by changing the concentration of bicarbonate ions in the blood. Recall that car-bon dioxide readily leaves the exchange surface of an aquatic animal because of its high solubility in water. Measurements of the partial pressure of carbon diox-ide in the blood of rainbow trout at different temperatures indicate that P_{CO_2} stays constant at about 2 mmHg. However, bicarbonate concentration in-creases when temperature decreases.

Part of the supply of bicarbonate ions comes from the external environ-ment of fishes (and part comes from internal changes in blood protein buffers; see Chapter 3). Cells in the gills expend energy to actively transport ("pump") some bicarbonate ions into the blood. We will examine this type of mechanism for regulation in some detail in Chapter 12. We will also return to the impor-tance of carbon dioxide and pH in blood when we discuss the oxygen and carbon dioxide transportation properties of blood (Chapter 3). For present pur-poses it is sufficient to note that use of bicarbonate ions results in precise regula-tion of acid-base balance in aquatic animals. In the next chapter we will find that air-breathing animals solve the problem of regulation of acid-base balance in quite a different way.

SUMMARY

Efficiency in respiration involves obtaining sufficient quantities of oxygen (and releasing the carbon dioxide produced) to meet demands for energy use while simultaneously minimizing losses of other important resources (energy, water, solutes, nutrients). Effective movements of oxygen and carbon dioxide occur over small distances by diffusion from locations of high concentration to locations of low concentration. Gas concentrations in air are measured from the partial pressures of its components at a given temperature, i.e., the fractional composition times the total pressure. Gas partial pressures change if either total pressure or air composition change. The concentration of a gas in water depends on its partial pressure in the gas phase and the solubility of the gas in water. Solubility (milliliters of gas/liter of water) depends on water temperature; carbon dioxide is 25-35 times more soluble in water than is oxygen. Solubility also depends on salts dissolved in water; oxygen is 20% less soluble in sea water than in pure water.

Diffusion movement depends on the magnitude of concentration differences between two locations, the area for diffusion, the distance of diffusion, and the "diffusion coefficient," which incorporates characteristics of the molecules (such as size) and environmental influences on the random movements of diffusion. Cells are surrounded by water, and diffusion coefficients are much less for water than for air. Cells are also surrounded by membranes that must be permeable to oxygen and carbon dioxide, dictating some permeability to water. Demands for respiration thus influence movements of water as a resource. Demands for oxygen uptake and release of carbon dioxide are dictated by the volume of cells containing the metabolic machinery using oxygen. A spherical shape minimizes surface area for exchange relative to volume, and the radius of a sphere represents the maximum distance of diffusion from the cell surface.

When the volume of a sphere increases, the surface area increases less than the volume, which means that there is an upper limit to spherical cell volume based on diffusion exchange across cell membranes. The maximum distance for diffusion also sets a limit on cell size because the rate of diffusion depends on distance. For cells larger than a fraction of a millimeter, there must be either some mechanism for circulation inside the cell (e.g., protoplasmic streaming) or increased surface area and decreased radius from folding of the cell membrane because the rate of delivery of oxygen by diffusion could take minutes or hours instead of seconds. Size and distance limitations also influence movements of gases for multicellular animals; therefore, respiration involves a *combination* of bulk transport of a respiratory medium (air, water, and/or blood) over relatively large distances together with diffusion over small distances (at respiratory surfaces and between blood and cells).

Bulk flow of a respiratory medium over an exchange surface is called ventilation. Four major mechanisms of ventilation have evolved in multicellular animals. Animals with only three or fewer cell layers ventilate as a consequence of bulk movement of water used for feeding. Continuous water movement renews oxygen concentration in the "digestive" system and near external surfaces, where diffusion distances are small because of limited numbers of cells in few layers. Animals with larger numbers of cells have evolved gills to exchange gases in water, or lungs or tubes (trachea) to exchange gases in air.

Localization of gas exchange at a gill involves effective extraction of oxygen from water. Water contains a lower supply of oxygen than does air, and more energy is required to move it in bulk. Gill ventilation in fishes involves mouth and opercular pumps, which produce water flow in the same direction for 86% of a respiratory cycle. This flow results in efficient oxygen extraction, since blood flow in the secondary lamellae of the gills is *countercurrent* to water flow. A constant gradient in oxygen concentration is established along the entire length of the exchange surface. If water flow is experimentally reversed, the gradient for oxygen concentration is initially larger but decreases rapidly along the exchanger, resulting in less *total* oxygen being extracted from a volume of water.

Some relatively large aquatic vertebrates use skin for respiration. Some aquatic amphibians have no gills but instead have highly vascularized skin surfaces. The Lake Titicaca frog illustrates this mode of respiration. Although it has a small lung, it seldom surfaces for air. When it is submerged it has a low rate of oxygen use, and sufficient gas exchange occurs at a highly folded and vascularized skin surface. If the oxygen supply is experimentally reduced, the frogs increase ventilation of the skin by moving in the water.

Other aquatic animals that normally experience an increased demand and/or reduced supply of oxygen also proportionally increase respiratory ventilation. Ventilation is measured as minute volume (ml/min), i.e., the product of respiratory rate and tidal volume. Some fishes change minute volume proportionally by changing both respiratory rate and tidal volume. Others increase ventilation by keeping their mouths open as they swim, but it increases their drag and the energy required for swimming. A few species do not increase ventilation when supplies of oxygen are reduced, but these species have evolved in environments where supplies are not normally limited relative to demands for oxygen.

A proportional change in ventilation requires a control mechanism that will detect the change in supply and/or demand and produce an appropriate response. Little is known of details of negative-feedback controls for respiration in aquatic animals, although decreases in P_{O_2} may be detected in the gills and may ultimately result in changes in tidal volume or respiratory rate or both. Variation in P_{CO_2} normally has little effect on gill ventilation, although carbon dioxide is an important component in the regulation of acid-base balance.

Carbon dioxide reacts with water to produce carbonic acid, which dissociates to form hydrogen ions (H^+) and bicarbonate ions (HCO_3^-). The net electrical charge on enzymes in cells has an important effect on the efficiency of cellular reactions because it influences optimal enzyme tertiary structures to produce particular rates for important reactions. The optimal pH (concentration of hydrogen ions) is such that there is a slight negative charge on an enzyme. Moreover, the optimal pH changes with temperature. The blood pH is controlled in fishes through changes in the concentration of negatively charged bicarbonate ions, which can be pumped in through the gills. Thus as temperature changes, blood pH in fishes changes in a way that maintains the same degree of net negative charge on enzymes.

Air Respiration

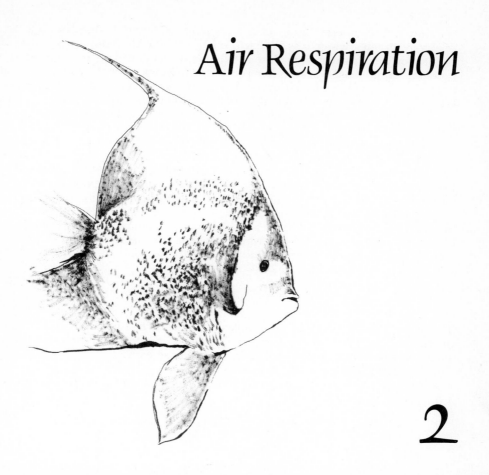

2

There are several advantages to air as a supply of oxygen compared with water, including its lower density, lower viscosity, higher oxygen availability, and the higher rate of diffusion of oxygen in air (see Table 1.3). One consequence is that a sufficient oxygen supply can be provided as demands increase without large increases in losses of energy. For example, trout use about 18% of their total energy expenditure to obtain oxygen, and measurements for some other fishes give values as high as 40% to 50% of total energy expenditure used for respiration during activity. Air-breathing vertebrates use less than 5% of their total energy expenditure to obtain oxygen, with a maximum usually less than 10%. Thus air-breathing animals can have a larger proportion of total energy resources available for other important processes, in part because of characteristics associated with obtaining oxygen supplies from air. In addition, advantages associated with increased ability to use oxygen, such as for higher levels of activity, can be supported more effectively in air-breathing animals (see Part II).

Although air respiration is more energy effective, a disadvantage is a relatively greater water loss. We will examine how the consequences of air respiration affect water and solute regulation in detail in Part IV. In general, a problem of increased water loss occurs when water evaporates from an organism to the air in its environment. The characteristics of gas exchange dictate that the respiratory exchange surfaces of animals be kept moist for diffusion of gases across membranes, and these surfaces represent a large total surface area.

Another problem associated with air respiration concerns the partial pressure of carbon dioxide. The high solubility of carbon dioxide in water results in its relatively rapid removal from the blood of aquatic animals; it remains at relatively low concentrations. However, carbon dioxide and oxygen are equally "soluble" in air; thus P_{CO_2} in air-breathing animals is much higher.

This situation can be illustrated with a relatively simple calculation. If one quantity (mole) of carbon dioxide is produced for each quantity (mole) of oxygen consumed, the P_{CO_2} in expired water depends on the ratio between oxygen solubility and carbon dioxide solubility and the amount of oxygen extracted from water. For pure water at 20°C,

$$P_{CO_2} \text{ expired in water} = \frac{0.04}{1.15} \left(\frac{\text{moles } CO_2}{\text{moles } O_2} \right) (P_{O_2} \text{ in } - P_{O_2} \text{ out}).$$

The ratio of moles CO_2 produced/moles O_2 consumed is called the *respiratory quotient* (R.Q.). The R.Q. can vary (see Chapter 5), but for simplicity we will consider the case where it equals 1.0. O_2 solubility is expressed as 0.04 milliliter O_2/liter water per mmHg P_{O_2}; 1.15 is CO_2 solubility per mmHg P_{CO_2}. If inspired water has a P_{O_2} = 150 mmHg and 50 mmHg of oxygen are removed for an R.Q. = 1.0,

$$P_{CO_2} \text{ expired in water} = \frac{0.04}{1.15} (1.0) (150 - 100) = 1.74 \text{ mmHg}.$$

In air, carbon dioxide production has a much more pronounced effect on expired P_{CO_2}, since the "solubility coefficients" of oxygen and carbon dioxide are equal for air. Consequently, their ratio will be 1.0, and

$$P_{CO_2} \text{ expired in air} = (1.0) \left(\frac{\text{moles } CO_2}{\text{moles } O_2} \right) (P_{O_2} \text{ in } - P_{O_2} \text{ out}).$$

For equal moles of CO_2 and O_2, removal of 50 mmHg O_2 gives

$$P_{CO_2} \text{ expired in air} = (1.0) (1.0) (150 - 100) = 50 \text{ mmHg}.$$

Figure 2.1 presents a comparison of the P_{CO_2} in expired water and air when different quantities of oxygen are removed from the inspired medium. The medium and blood are in equilibrium at the exchange surface, so the quantities of CO_2 in the blood will be much higher in air-breathing animals than in water-breathing animals. We will find that important differences in the control of ventilation and the regulation of acid-base balance are involved because of higher values of P_{CO_2} in air-breathing animals.

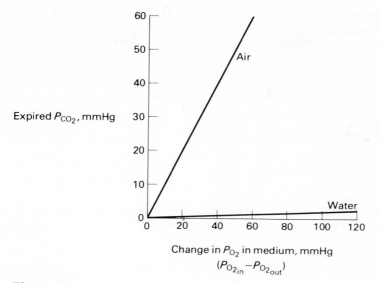

Figure 2.1
The solubility of carbon dioxide in water results in much lower
P_{CO_2} in expired water compared with expired air for a given
change in P_{O_2} in the respiratory medium. The example shown is
for equal amounts of CO_2 produced relative to O_2 consumed.
Since the respiratory medium and blood are in equilibrium at the
exchange surface, these differences have pronounced effects on
blood P_{CO_2} (modified from Rahn, 1966).

Two major types of structures have evolved for ventilation and extraction
of oxygen from air: lungs and trachea. There are two types of lungs: *tidal* lungs,
where air flows alternately in and out of blind sacs in the respiratory system;
and *flow-through* lungs, where air flows continuously through the respiratory
exchange system. Function in both lung types reflect differences in demands
for oxygen compared with supplies.

We will examine the function of tidal lungs in amphibians, reptiles, and
mammals, and flow-through lungs in birds. Tracheal, ("tubular") respiration
occurs in insects and represents an adaptation for minimizing both water loss
and energy expenditure, while simultaneously providing sufficient oxygen supply and carbon dioxide removal. However, before examining respiration in
animals that breathe *only* air, it is instructive to examine some supply-demand
relationships for respiration in those animals that can breathe *either* water *or*
air.

AIR-AND-WATER BREATHERS

A number of aquatic vertebrates can breathe either air or water. At least 18
species of fishes have respiratory structures for extracting oxygen from air, including the garpike (*Lepisosteus*), the bowfin (*Amia*), some catfish

(*Plecostomus; Clarias*), the mudskipper (*Periophthalmus*), the electric eel (*Electrophorus*), and three genera of lungfishes (*Lepidosiren, Neoceratodus,* and *Protopterus*). In addition, most amphibians can extract oxygen from air or water to varying degrees, and some aquatic reptiles, such as the Nile turtle (*Trionyx triunguis*), also extract oxygen from either water or air.

The type of respiratory exchange surface and the amount of oxygen removed from air versus water varies considerably from species to species. Among fishes air respiration occurs in the mouth (electric eel), stomach (catfish), skin (mudskipper), swimbladder (bowfin; garpike), or lung (lungfish). Among amphibians most air respiration occurs in a tidal lung (see below), but some exchange takes place across moist, vascularized skin. The proportion of gas exchange that occurs from breathing air varies from as little as 10% of total exchange to 80% to 90% of total exchange, depending on the species and the environmental circumstances.

What accounts for this diversity in reliance on two different sources for oxygen supply? In general, there is a good relationship between the distribution of different species in aquatic versus terrestrial habitats and the extent to which they obtain oxygen from water versus air. Thus those species that have evolved in habitats permitting access to both water and air tend to rely on one or the other as a primary source of supply.

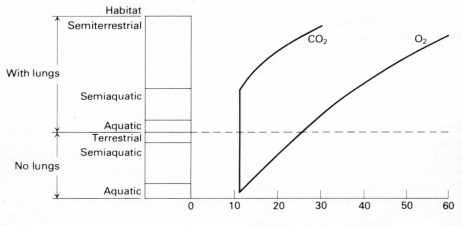

Figure 2.2
The proportion of total gas exchange in different species of salamanders from different habitats that occurs with air (versus water) increases for more terrestrial species (modified from Lenfant *et al.*, 1970).

Figure 2.2 shows this relationship for different species of salamanders. Those species without lungs that have a more terrestrial distribution obtain up to 28% of the required oxygen from air through the skin and mouth (versus up

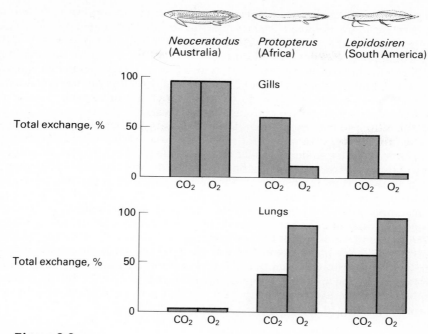

Figure 2.3
The proportion of total gas exchange for oxygen and carbon dioxide via gills (upper) and lungs (lower) for the three species of lungfish (modified from Lenfant *et al.*, 1970).

to 60% from air in those terrestrial species with lungs), with the remainder obtained from water via the skin. There is less of a relationship involving loss of carbon dioxide and habitat because of the high solubility of CO_2 in water, and the most terrestrial species with lungs lose a maximum of only 30% of carbon dioxide to air, with the remainder lost to water through the skin (Fig. 2.2).

Even among the three genera of lungfishes there is variation in the proportion of total oxygen and carbon dioxide exchange that occurs from gills (with water) or lungs (with air) when the fish have access to both water and air as a respiratory medium (Fig. 2.3). This variation is a consequence of the evolution of these species in environments with different degrees of change in the availability of oxygen from different sources. The Australian lungfish (*Neoceratodus*) is not exposed to environmental changes that would require extraction of oxygen from air for long periods, whereas the African (*Protopterus*) and South American (*Lepidosiren*) species have evolved respiratory mechanisms to survive relatively long periods of desiccation that do occur regularly.

If air is potentially a more effective source of oxygen supply than water, why don't all those species that can breathe air rely more heavily on it for obtaining oxygen? Other costs associated with getting oxygen from the two different sources are involved besides just the respiratory efficiency of gas exchange. For example, a fish that has access to the oxygen in water most of the time would have to expend extra time and energy to swim to the surface to get air. This will make air less effective than water as an oxygen source despite the lower availability of oxygen in water. However, if the supply of oxygen in water decreases relative to total demands, the costs for obtaining oxygen from air will be lower, relative to the total demands for survival.

According to this hypothesis, the overall importance of air and water for gas exchange depends mainly on the extent of availability of oxygen from the two sources *relative to* demands for oxygen. This has been studied in some air and water breathers by changing both supplies of and demands for oxygen by changing water temperatures. An increase in body temperature for aquatic ectotherms increases demands for oxygen (see Chapter 6), and an increase in water temperature reduces oxygen solubility. When the bowfin (*Amia calva*) is exposed to water of different temperatures it will take less of its higher total required oxygen from water at higher temperatures (Fig. 2.4). At lower temperatures the relatively higher availability of oxygen in water is sufficient for lower demands. Similarly, the quantity of oxygen obtained from lungs (air) versus skin (water) for the spotted salamander (*Ambystoma maculatum*) depends on demands (that increase with temperature) relative to supplies from either respiratory route (Fig. 2.5).

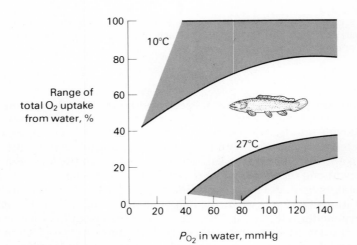

P_{O_2} in water, mmHg

Figure 2.4
When temperatures are low (10 °C) and demands for oxygen are thereby less, most oxygen is obtained by the garpike (*Amia*) from water, particularly when availability of oxygen (P_{O_2}) from water is high. At higher temperatures demands increase and more oxygen is obtained from air at a given availability of oxygen from water (modified from Lenfant *et al.*, 1970).

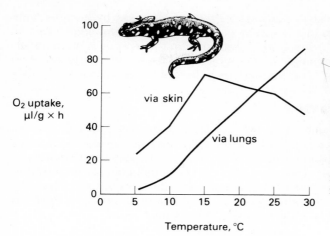

Figure 2.5
Oxygen uptake via skin (primarily aquatic) and lungs (air) for the salamander *Ambystoma maculatum* when supplies and demands are changed because of changes in temperature. Increased temperature is associated with increased oxygen up-take from air (modified from Whitford and Hutchison, 1963).

The demands of some animals for oxygen also change if the oxygen supply source changes. For example, Fig. 2.6 shows the amount of oxygen obtained by African lungfish via air or via water when both supply sources are available and also when *only* water is available. Note the drastic reduction in total rate of oxy- ✗

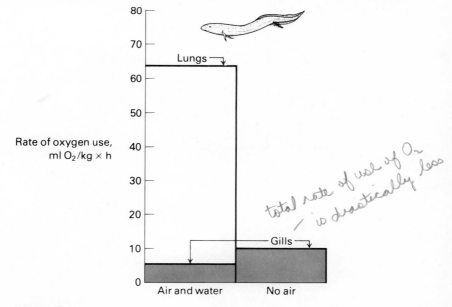

total rate of use of O₂ is drastically less

Figure 2.6
The amount of total oxygen use supplied from both lungs (air) and gills (water) or from just gills in an African lungfish. Despite increased supply from gills, with no air available the total rate of use of oxygen is drastically less (modified from McMahon, 1970).

gen use when no air is available despite the increased uptake by the gills. This rate change is a common characteristic of animals that obtain oxygen from a primary source (either air or water) when access to that source is interrupted. We will return to an examination of this subject and its importance for survival when we discuss "diving" and adaptations to lack of oxygen (Chapter 4).

TIDAL LUNG RESPIRATION

Lung ventilation occurs when differences in pressure between the external environment and the lungs result in air movement in or out. Pressure differences for inspiration are established either from forcing air into the lungs using positive pressure or from sucking air into the lungs using negative pressure. The former mechanism occurs in amphibians, and the latter occurs in reptiles, birds, and mammals. In amphibians, reptiles, and mammals lung ventilation is tidal, while in birds ventilation is relatively continuous.

Postive-pressure Ventilation

Figure 2.7 shows the sequence of air passage during a respiratory tidal cycle for a frog. With the nose open and the glottis (pharynx) closed, air under suction pressure passes into the mouth (buccal) cavity when the throat expands outward and downward (Fig. 2.7a). The glottis then opens and the elastic lungs ex-

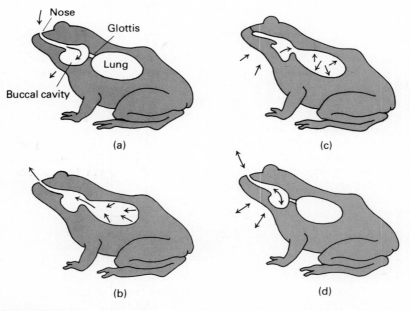

(a)

(b)

(c)

(d)

Figure 2.7
Pattern of positive-pressure ventilation in tidal respiration of a frog (after Gans et al., 1969).

pel air from the previous breath above the fresh air in the buccal cavity and out the nose (Fig. 2.7b). The nose then closes and the throat region constricts, forcing air from the buccal cavity into the lungs under positive pumping pressure (Fig. 2.7c). The glottis then closes and the throat region expands and contracts, rapidly flushing the buccal cavity while gas exchange takes place in the lungs (Fig. 2.7d).

This mechanism for pumping air into the lungs requires a relatively large mouth cavity. The buccal cavity of frogs serves other functions besides respiration that make it suitable for positive-pressure ventilation. Frogs consume their prey whole, and the buccal cavity is utilized for communication by means of sound production. Mammals, reptiles, and birds have evolved negative-pressure ventilation mechanisms that do not depend on a mouth pump but instead require different mechanisms to expand the lungs during inspiration. The mechanism for tidal ventilation is relatively well understood for mammals, and there is considerable information about mammals on changes in ventilation caused by changes in supplies of or demands for oxygen.

Negative-pressure Ventilation

In mammals the lungs are contained within a separate cavity (the thoracic cavity), which is surrounded by a rib cage and separated from the abdomen by the diaphragm muscle. The pressure within the thoracic cavity is slightly less than atmospheric pressure, which aids in keeping the lungs inflated. Puncture of the thoracic cavity (pneumothorax) will deflate the lungs and cause drastically increased energy expenditures for ventilation.

Pressure within the lungs decreases during inspiration when the diaphragm contracts downward and the ribs move up and out. Both movements expand the thoracic cavity, drawing air into the elastic lungs. During expiration, most of the air in the lungs is forced out due to the elastic recoil of the lungs; negative pressure is reduced from relaxation of the inspiratory muscles. Additional air can be forced out if contraction of a second set of rib muscles occurs, forcing the ribs inward, or if contraction of abdominal muscles pushes the diaphragm upward, reducing thoracic cavity volume.

Figure 2.8 shows the gross and fine structure of the respiratory system of mammals. Air passes from the external environment through a series of tubes of decreasing diameter and length (from trachea to terminal bronchioles) to a series of blind sacs where gas exchange takes place (the *alveoli*, Fig. 2.8). The alveoli cumulatively provide a large surface area for gas exchange, and the distance between air in the alveolar spaces and blood flowing through the lung capillaries is only a fraction of a micrometer (Fig. 2.8); thus diffusion suffices for alveolar-capillary gas exchange.

Changing Lung Ventilation when Supplies or Demands Change

Minute volume changes as a consequence of changes in either rate of respiration or tidal volume. However, not all of the air involved in ventilation of a tidal lung can exchange gases with blood in the alveoli. For each breath, some air remains in the tubes of the trachea and bronchioles. Because no gas exchange can

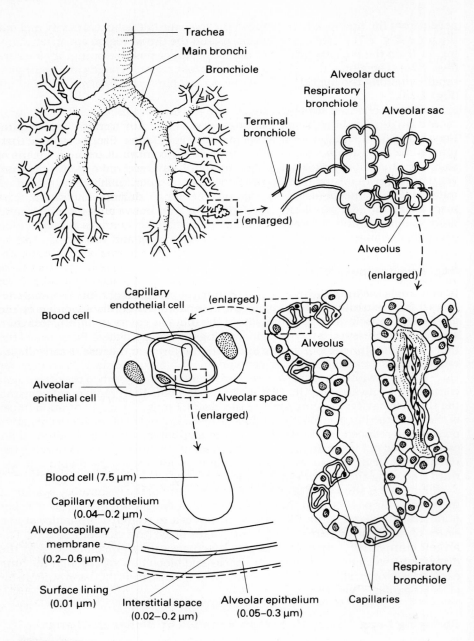

Figure 2.8
Gross and fine morphology of the respiratory system of mammals.

From *Animal Physiology* by Roger Eckert and David Randall, W.H. Freeman and Company.
Copyright © 1978.

take place here, this volume of air is called the "dead-space" volume. *Alveolar minute volume*, therefore, is the product of respiratory rate and total ventilation volume minus dead-space ventilation volume:

Alveolar minute = Respiratory × (Tidal volume – Dead space volume)
   volume   rate

ml/min = breaths/min × ml/breath.

The demands of different species of mammals for oxygen increase with increasing size (see Chapter 5), so the evolution of larger size must be accompanied by mechanisms to increase ventilation supplies of oxygen. For most mammals, resting tidal volume and dead-space volume change in direct proportion to weight. For inactive mammals, tidal volume in milliliters is about 7.7 times weight in kilograms and dead-space volume in milliliters is about 2.8 times weight in kilograms. However, respiratory rate for inactive animals is higher for smaller species.

These relationships are shown for tidal volume and respiratory rate in mammals in Fig. 2.9. Note that both y axes and the x axis of this graph are logarithmic, that is, based on powers of ten. When the Log of y is a function of the Log of x, y changes as some power (exponent) of x (see Appendix II for details). The relationship between tidal volume and weight for mammals includes an exponent on weight in kilograms of 1.0, and the relationship between respiratory rate and weight for mammals includes a *negative* exponent on weight of -0.25 (Fig. 2.9).

Total minute volume is the product of respiratory rate and tidal volume, and when exponential equations are multiplied the exponents are added, so for the equations in Fig. 2.9,

Minute volume = Tidal volume × Respiratory rate

ml/min = $7.7 \ (kg)^{1.0} \times 53.5 \ (kg)^{-0.25}$

     = $(7.7 \times 53.5) \ (kg)^{1.0 \ - \ 0.25}$

     = $419 \ (kg)^{0.75}$.

The demands of resting mammals of different weights for oxygen are also related to weight raised to a power of 0.75 (see Chapter 5):

ml O_2 consumed/min = $11.6 \ (kg)^{0.75}$.

In general, mammals of different weights extract about the same fraction of the oxygen in each ml of inspired air (about 24%), and dead-space volume also has an exponent of 1.0 with respect to weight (ml/breath = $2.8 \ (kg)^{1.0}$). Therefore, changing demands associated with mammals of different weights are met primarily by changes in volumes and rates of respiration, with one factor (tidal volumes) *increasing* as demands increase and the other (respiratory rate) *decreasing* as demands increase.

These comparisons involve measurements of components of ventilation for a large number of different species of mammals while they are resting quietly. We will explore the basis for the nature of the 0.75 exponent for oxygen demands and some of its consequences in Chapter 5 when we discuss size determinants of energy requirements.

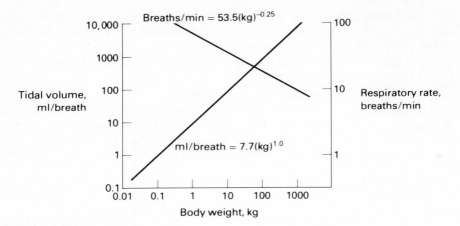

Figure 2.9
Relationships between tidal volume (left axis), respiratory rate (right axis), and body size in mammals (equations from Stahl, 1967).

For any individual that becomes more or less active there are also mechanisms to proportionally increase or decrease ventilation to provide higher or lower rates of oxygen supply. The mechanisms involve changes in both tidal volumes and respiratory rates. In addition, there are differences among species in lung structure and function that reflect different supply-demand problems for obtaining oxygen.

Changing Supplies Proportionally with Demands in Individuals

Figure 2.10 shows how tidal volumes change in humans. Resting tidal volume is usually about 10% of total lung volume. It can be increased by increasing inspired volumes from inspiratory reserve volume and expired volumes from expiratory reserve volume (Fig. 2.10). The maximum possible tidal volume is called "vital capacity." Vital capacity is never equal to total possible lung volume in humans because all the air in the lungs cannot be expired; a residual volume of about 25% of the total volume is left (Fig. 2.10). Proportional changes in minute volume in humans involve changes in both tidal volumes and respiratory rates (Fig. 2.11).

The volumes shown in Fig. 2.10 for humans are subject to considerable changes among different species of mammals as a consequence of different supply-demand problems of obtaining oxygen. For example, an 80-pound sea otter has a total lung capacity that is twice that of an adult human. This difference results in a greater capacity for lung ventilation in this species that obtains food when it dives in shallow water. While it is diving, a primary source of oxygen supply comes from the relatively large tidal volume taken in prior to a dive. Other species of diving animals, such as the porpoise, have flexible rib cages, allowing residual volume to be reduced almost to zero. These deep-diving animals expire air prior to a dive, reducing gas volumes in the lungs that

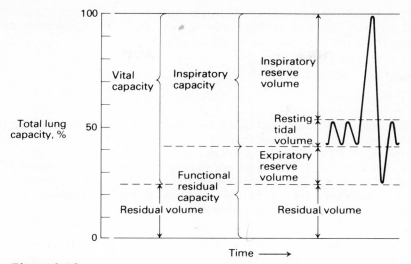

Figure 2.10
Volumes involved in respiration of the tidal lung of humans. Tidal volume
is inspired and expired volume. Inspired volume can be increased from
resting values by increases in the use of inspiratory reserve volume, and
expired volumes can be increased by the use of expiratory reserve volume.

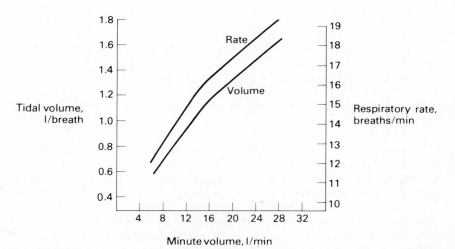

Figure 2.11
For humans, changes in minute volume occur that result from changes in
both tidal volume and respiratory rate (modified from Vander *et al.*, 1975).

could enter the blood under high water pressure and result in the "bends," i.e., decompression sickness, during surfacing. (See Chapter 4 for further discussions of respiratory adaptations in diving animals.)

Control of Ventilation

The mechanisms underlying proportional changes in lung ventilation for mammals are relatively well understood. The most effective stimulus under normal circumstances is a change in P_{CO_2}. When alveolar P_{CO_2} in humans is increased by breathing a mixture of gases high in P_{CO_2}, minute volume increases proportionally (Fig. 2.12) from a combination of increased tidal volume and respiratory rate (Fig. 2.11). Under normal circumstances an increase in P_{CO_2} is associated with an increased demand for oxygen, since CO_2 is a product of oxidative metabolism. Thus a control system sensitive to changes in P_{CO_2} can regulate oxygen supplies relative to demands.

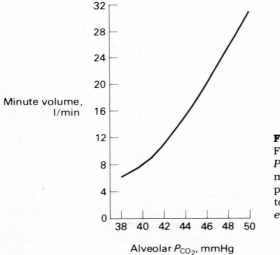

Figure 2.12
For humans, changes in alveolar P_{CO_2} ultimately result in changes in minute volume; as a result, supplies of oxygen are adjusted relative to changed demands (after Vander *et al.*, 1975).

Despite this, the actual effective stimulus for control of ventilation is not P_{CO_2} but pH, which is related to P_{CO_2}. This fact was demonstrated by experiments where the pH of fluid bathing the brain cells of turtles was changed independently of changes in P_{CO_2}. When pH was increased minute volume decreased, and when pH was decreased minute volume increased. The response was proportional to a change in pH. Normally, the production of carbon dioxide influences blood pH through carbonic acid dissociation (see Chapter 1), and it is the chemical product of the dissociation reaction, a change in pH, that serves as the most important stimulus for proportional control of ventilation.

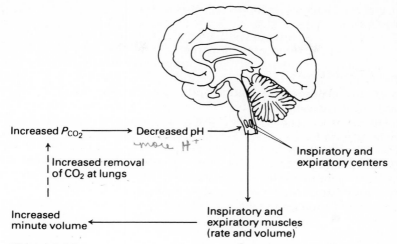

Figure 2.13
A schematic diagram of the negative-feedback control system involved
in control of respiratory ventilation in mammals. The case illustrated is
for an increased demand for oxygen associated with increased P_{CO_2}.

The receptors that detect changes in pH are located in the brain stem just
above the spinal cord (Fig. 2.13). Other nerve cells for the control of respiration
are located in two areas: one that activates muscles involved in inspiration (the
inspiratory area), and the other that activates muscles involved in expiration
(the expiratory area). These two centers alternate activation of separate sets of
muscles involved in ventilation.

Changes in pH of the fluid that bathes receptor cells results in changes in
the frequency and force of contraction of both inspiratory and expiratory
muscles, resulting from effects on inspiratory and expiratory centers. This action changes minute volume, changing the quantity of carbon dioxide removed
from alveolar air, which in turn will have a negative-feedback effect on the pH
of blood, causing it to remain at or return to a relatively constant value. The
relationships of the components of this control system are illustrated in Fig.
2.13, which shows the control of increased ventilation from a decrease in pH.

Distribution of Ventilation and Blood Flow within Tidal Lungs

Air and blood must be in close proximity for effective gas exchange by diffusion.
Diffusion occurs within alveoli, but different degrees of alveolar ventilation can
occur within a tidal lung. Pressures and resistances to air flow within a lung
produce *stratification* of air flow; different amounts of air flow to different parts
of a lung. In humans, as much as four times more inspired air is distributed to
the top part of a lung than to the base. Thus for effective oxygen uptake and carbon dioxide release, blood flow should also be stratified within the lung, with

the greater proportion of blood flow distributed to areas receiving the higher supplies of air.

The relationship between ventilation and blood perfusion of different parts of a tidal lung has been studied in the carpet python (*Morelia spilotes variegata*). This snake has a pair of long "alveolar" lungs connected posteriorly with lung sacs (Fig. 2.14b). As in most reptiles, ventilation is triphasic (in-pause-out) with variable periods between inspiration and expiration depending on the level of activity or demands for oxygen. Gas exchange occurs in the alveolar lungs where surface areas are high because of extensive folds of the respiratory epithelium (Fig. 2.14b).

Following inspiration, most inspired air in the carpet python is distributed primarily to the anterior part of the lungs. However, this is accompanied with a stratification of blood flow in a similar manner. The pattern of blood perfusion is shown in Fig. 2.14(a). As a consequence of stratification of *both* air and blood flow, gas exchange is more effective compared with situations where the two would not be matched.

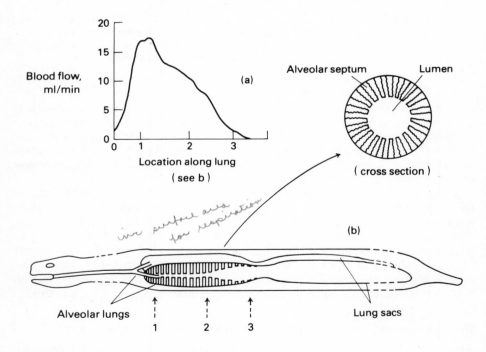

Figure 2.14

(a)   The curve illustrates the distribution of blood flow to different areas of the lung of the carpet python.

(b)   Schematic illustration of the respiratory system of the carpet python, where most ventilation occurs in the anterior portions of the alveolar lungs (areas 1 and 2) (after Donnelly and Woolcock, 1977).

FLOW-THROUGH LUNG VENTILATION

In tidal lungs, inspired air mixes with dead-space air containing relatively low amounts of oxygen and relatively high amounts of carbon dioxide. Moreover, when air reaches an alveolus there is no obvious countercurrent arrangement between air and blood, since the air in an alveolus forms a "uniform pool" from which oxygen is extracted. Studies of respiration in birds suggest that a potentially more effective exchange system has evolved (Fig. 2.15).

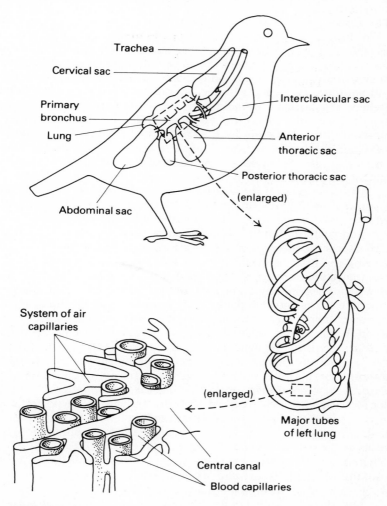

Figure 2.15
Diagram of the gross and fine structures of the respiratory system of birds. Diffusion gas exchange occurs between the systems of air capillaries and blood capillaries (from Waterman, 1971).

Birds have higher demands for oxygen compared with other vertebrates. Their resting rate of use of oxygen is somewhat higher than for most mammals of similar sizes (see Chapter 5), and their demands for oxygen increase dramatically during flight, which is the most energy-demanding form of movement for rate of use of oxygen (see Chapter 9). Although some mammals can sustain similar rates of oxygen use for brief periods, birds experience high rates of oxygen use for prolonged flights. Furthermore, flight may occur at relatively high altitudes where oxygen supplies are lower because of the decrease in total atmospheric pressure.

The anatomy of the avian respiratory system is extremely complex (Fig. 2.15). The lung is a rigid collection of finely divided tubes, or bronchi. The smallest divisions where gas exchange takes place are called "air capillaries" (Fig. 2.15). Larger tubes connect the main lung with a set of air sacs (interclavicular, anterior thoracic, posterior thoracic, cervical, and abdominal in Fig. 2.15). In addition, there is a smaller "secondary" lung posterior to the main lung, where the primary bronchus divides extensively before reaching the posterior thoracic and abdominal air sacs. This secondary lung is called the neopulmonary lung; the main lung is called the paleopulmonary lung.

Diffusion gas exchange takes place between air capillaries and blood capillaries in the lungs; the air sacs serve as reservoirs of air. The passage of air through the respiratory system has been studied by placing small measuring devices in various parts of the system to detect directions of air flow. The flow can be summarized by referring to Fig. 2.16, which shows the system of air sacs divided into two functional sets: anterior (cervical, interclavicular, and anterior thoracic) and posterior (abdominal, and posterior thoracic). The shaded part of Fig. 2.16 shows the passage of one breath through the system.

During the first inspiration the air sacs expand, drawing most of the air through the neopulmonary system to the posterior air sacs while a smaller quantity of air goes directly to the posterior part of the paleopulmonary system. Pressure differences between the primary bronchus and the anterior part of the lungs prevent inspired air from entering the anterior air sacs. During the first expiration the air in the posterior air sacs passes back through the neopulmonary system and into the paleopulmonary system (Fig. 2.16). Pressure differences also prevent this air from passing back up the primary bronchus; the "expiration" doesn't involve air from this breath leaving the body.

During the second inspiration air from the first breath is drawn into the anterior air sacs while new inspired air is drawn back to the posterior air sacs. During the second expiration the air from the first breath finally leaves through the trachea while the air from the second breath is passing through the air capillaries of the paleopulmonary lung (Fig. 2.16).

There are two important features to this pattern of air flow. First, there is little mixing of inspired air with expired air; even though there may be a relatively large anatomical dead space associated with the trachea, primary bronchi, and air sacs, P_{O2} is not greatly reduced at the exchange surfaces. Second, note that air flow in the paleopulmonary lung is always in the *same direction* (posterior ⟶ anterior) (Fig. 2.16). This is a primary prerequisite for utilizing the efficiency of a countercurrent exchange system similar to the gills of fishes. However, blood and air flow in the avian lung is *not* countercurrent.

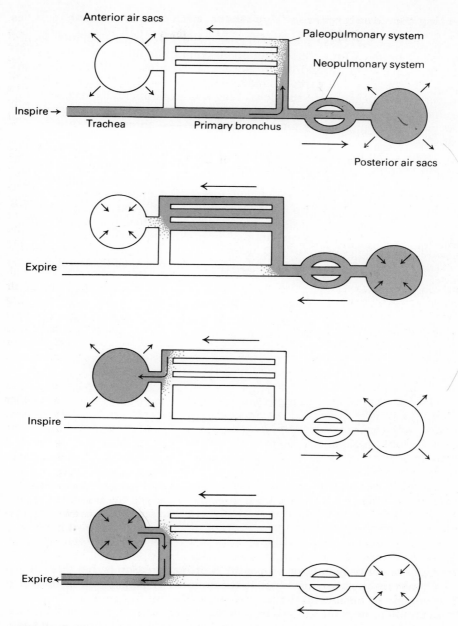

Figure 2.16
Movement of one breath through a schematic bird respiratory system. Shaded
areas represent the location of a single breath during the two respiratory cycles
required for complete passage (modified from Bretz and Schmidt-Nielsen, 1971;
Piiper and Scheid, 1973).

Testing the Countercurrent Hypothesis with an Experiment

As was the case with gas exchange in gills, if countercurrent exchange occurs, reversal of air flow within the paleopulmonary lung should result in less extraction of oxygen from air and less removal of carbon dioxide from pulmonary blood. The oxygen and carbon dioxide partial pressures of expired air were measured for ducks where the flow of air was experimentally reversed. The results showed that it made *no difference* for the efficiency of gas exchange if air flow in the paleopulmonary lung was posterior ⟶ anterior, or anterior ⟶ posterior.

Despite the inescapable conclusion that blood and air do not exchange gases in a countercurrent system, respiration in birds is still more effective than in mammals. Birds extract about 31% of inspired oxygen with a higher P_{O_2} in arterial blood (lower concentration difference), but mammals extract about 24% of inspired oxygen with a lower P_{O_2} in pulmonary blood (higher concentration difference). Moreover, birds such as ducks are able to maintain high extraction values and adequate oxygenation of arterial blood when they are exposed to simulated altitudes of up to 6000 meters, whereas mammals are less effective at extracting oxygen and maintaining oxygenation of blood if the P_{O_2} of inspired air decreases.

What hypothesis could account for a relatively efficient exchange system that is not countercurrent? The model that is currently being tested is called a "crosscurrent" exchange system (Fig. 2.17). Exchange between blood and air is postulated to occur in a series of channels that produce blood flow *across* the air capillaries.

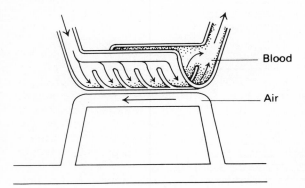

— Blood

— Air

Figure 2.17
Schematic diagram of the "crosscurrent" model of avian gas exchange, where blood is hypothesized to pass in separate channels across air capillaries (after Piiper and Scheid, 1973).

Why was this hypothesis proposed? A primary reason is that this pattern of exchange results in no difference in the quantity of oxygen and carbon dioxide exchanged if air flow in the lung is reversed. This fact must be accounted for. In addition, exchange might be more effective compared with alveolar exchange if there is a relatively constant gradient for exchange along each channel. This pattern would not be as effective as a completely countercurrent system, but it could be more effective than a concurrent or uniform-pool (alveolar) system. Figure 2.18 compares the degrees of effectiveness postulated for the three major gas exchange types in different respiratory systems.

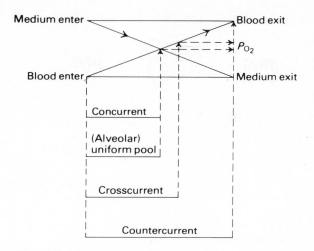

Figure 2.18
A comparison of the relative effectiveness of gas exchange in different respiratory systems. Effectiveness is measured from the degree of uptake of oxygen by blood (right-hand values for P_{O_2}). Alveolar tidal respiration would be equivalent to a "uniform pool" mode of exchange. When blood exits, P_{O_2} is highest for countercurrent exchange and lowest for concurrent exchange (modified from Piiper and Schied, 1973).

ACID-BASE REGULATION AND CONTROL IN AIR-BREATHING VERTEBRATES

Recall for aquatic animals with different temperatures that it is not pH per se but rather the degree of alkalinity that is regulated because the neutral pH changes inversely with temperature. For animals with a relatively constant body temperature (endotherms; see Chapter 7), pH is a direct reflection of degree of alkalinity and should remain relatively constant. For example, for mammals with a constant body temperature of 37 °C we expect pH to remain close to 7.4. However, at a different body temperature we expect pH to be different if acid-base status is regulated (see Fig. 1.16).

Figure 2.19 shows measurements of blood pH from toads, turtles, and frogs that normally undergo changes in body temperature. The pH of blood increases with a decrease in body temperature in a way that maintains a constant alkalinity of blood compared with the neutral pH of water. These air-breathing animals regulate acid-base balance as do aquatic animals (Chapter 1). However, they *control* relative alkalinity in a different way, and this provides some additional information about the mechanisms of ventilation control.

Figure 2.19 also shows measurements of blood P_{CO_2} for the same animals. Changes in blood P_{CO_2} are opposite to changes in blood pH; as pH increases, P_{CO_2} decreases, and vice versa. This type of change is in contrast to the relatively constant P_{CO_2} in the blood of aquatic animals caused by the high solubility of

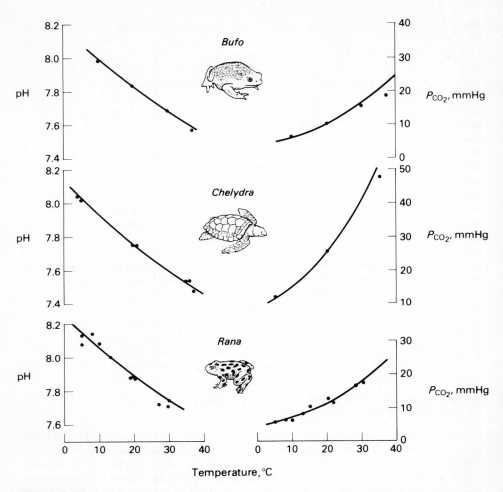

Figure 2.19
For toads (*Bufo*), turtles (*Chelydra*), and frogs (*Rana*), changes in blood pH are associated with changes in blood P_{CO_2} in such a way as to maintain constancy of the degree of alkalinity with respect to the neutral pH of water (modified from Reeves, 1977).

carbon dioxide in water. Recall for aquatic animals that bicarbonate ions play a major role in changes of pH as temperature changes. For air-breathing vertebrates, changes in pH are more closely associated with changes in the net amount of CO_2 that combines with water to form carbonic acid (see Eq. 1.2).

How do air-breathing animals such as those in Fig. 2.19 change blood P_{CO_2} for acid-base regulation when temperature changes? As temperature increases, P_{CO_2} increases. Blood P_{CO_2} is influenced by the rate of CO_2 production from metabolism. More of this carbon dioxide will stay in the blood if lung ventilation temp↑ : CO₂↓ : H⁺↓ : pH↑

decreases *relative to* metabolic rate when body temperature increases. Thus
blood pH varies appropriately with regard to variation in the difference between
CO_2 production and CO_2 loss. The animals *hypo*ventilate compared with CO_2
production as their temperature increases, and they *hyper*ventilate compared
with CO_2 production as their temperature decreases.

Physiologists used to believe that changes in pH in the blood of fish, amphibians, and reptiles indicated a "failure" to achieve regulation because
animals with relatively constant body temperatures (such as mammals and
birds) control pH at constant values. Fish, amphibians, and reptiles were considered to be more "primitive" in their physiological abilities compared with
more "advanced" mammals and birds because they were not taxonomically far
enough "along" in the evolutionary process. However, this point of view was
based on the premise that pH (hydrogen ion concentration) *alone* was what
should be regulated for efficient function. We now know that pH should change
when body temperature changes in order to maintain effective enzyme function. Animals with variable body temperatures are not in the least "primitive"
with respect to acid-base regulation but instead regulate with a high degree of
precision. This point of view also tells us something about what is important in
receptor function for the control of ventilation in all air-breathing animals.

The receptors involved with the control of ventilation in animals with
variable body temperatures should operate to keep protein net negative charge
relatively constant when temperatures change. For example, ventilation should
be controlled compared with CO_2 production so that pH changes by an appropriate amount when body temperature changes. Thus information on the effects of a change in temperature on protein net negative charge, not just information on a change in pH, should be utilized to control ventilation.

Perhaps the simplest mechanism to achieve this control is to have receptors detect and respond to changes in degree of protein alkalinity as temperature changes. Some proteins are particularly sensitive to changes in their three-
dimensional structure when the protein net charge varies. These proteins include the *histidine imidazole* group of amino acids in their primary structure.
The proteins can play an important role in detection of changes in acid-base
status by means of conformational changes; proportional adjustments are made
using ventilation relative to Co_2 production to keep protein net negative charge
constant when temperature varies.

TRACHEAL RESPIRATION

Insects are relatively small, and we will find that animals of small size face acute
problems of water loss, since surface area is large relative to the volume of the
body containing water (see Chapter 13; also see Fig. 1.3 for a summary of
surface-volume relationships). Respiration in insects occurs through a tubular
tracheal system that reduces water lost from gas exchange, and the mechanism
for tracheal ventilation also involves a minimum expenditure of energy to produce bulk flow of air.

The general morphology of the tracheal system is shown in Fig. 2.20. A
series of tubes passes throughout the body from openings in the surface cuticle

called *spiracles*. Each major segment of an insect contains a pair of spiracles, and they are opened or shut by contraction of muscles associated with the lips of the spiracle valves (Fig. 2.20). The tubes divide and branch extensively so that each cell of the body is ultimately in close proximity (within a micrometer) to a thin tracheal tube. This arrangement permits diffusion exchange between the cells and trachea, but the distances along the tubes to the external environment are too long (on the order of millimeters or centimeters) in all but the smallest insects for respiratory exchange to occur entirely by diffusion of oxygen and carbon dioxide. Thus a mechanism is required for bulk flow of air in the tracheal system.

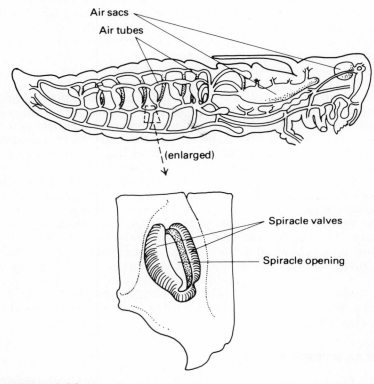

Air sacs

Air tubes

(enlarged)

Spiracle valves

Spiracle opening

Figure 2.20
Diagram of the tracheal system for respiration in a grasshopper and the spiracle valves providing entry to the system (from Buchsbaum, 1948).

A number of insects that are active have flexible expanded regions, or air sacs, in the tracheal system associated with major contractile muscles (Fig. 2.20). Insects that walk or fly have air sacs located in the muscles of the legs or wings. Contraction and relaxation of the muscles creates positive and negative pressures, producing ventilation that changes as demands from activity change. This ventilation provides a supply of oxygen (and removal of carbon dioxide) sufficient for the rate of oxygen use in some large flying insects that is similar to the rates of oxygen use by some birds and mammals. The association of air sacs with muscles used for locomotion provides proportional ventilation without added energy expense in a manner similar to ramjet ventilation during swimming in some fishes.

Discontinuous Respiration in Insect Pupae

Even completely inactive insects achieve considerable bulk flow of air for ventilation without any movement other than opening and closing spiracle valves. Several insect species form a puparium between the larval and adult stages of development (see Chapter 15). The pupae are in an "arrested" state of development, with low requirements for oxygen. Intake of food or water does not occur for an extended period, and a hard cuticle provides an effective barrier to water loss. Nevertheless a pupa is still alive and consumes oxygen and releases carbon dioxide without any obvious ventilation movements. The mechanisms for this have been studied by an ingenious series of experiments to measure gas pressures inside the tracheal system of silkworm pupae (*Hyalophora cecropia*) by placing small capillary tubes in individual spiracles.

Carbon dioxide release in silkworm pupae is "discontinuous"; that is, release occurs at intervals as "bursts" of expired gas. Carbon dioxide release is related to movements of the spiracle valves. The bursts occur when the spiracles open (Fig. 2.21), and this only occurs for brief periods over relatively long intervals in pupae.

Despite the periodic release of carbon dioxide, oxygen uptake by the pupae is continuous. For oxygen to move in and carbon dioxide to move out, pressure differences must be established for the two gases between the inside of the body and the external atmosphere. The total pressure inside the trachea relative to atmospheric pressure is shown in the upper part of Fig. 2.21. The respiratory cycle can be divided into three phases: (a) a period of spiracle constriction, during which the pressure inside the trachea drops below atmospheric pressure, creating a partial vacuum; (b) a period of "fluttering" of the spiracle valves, i.e., very brief opening and closing of the valves at frequent intervals; and (c) the brief period of the "burst" when the valves stay open while carbon dioxide moves out (Fig. 2.21).

The vacuum established during phase (a) of spiracle closure results from oxygen use in cells and the accumulation of carbon dioxide in the fluid surrounding the cells. Since carbon dioxide is highly soluble in water, removal of oxygen from tracheal gas and the formation of carbon dioxide creates a vacuum, since the CO_2 produced enters solution and does not replace the removed oxygen in the gas phase.

Even though the spiracle valves are "closed," air is sucked in from the low pressure established in the trachea. The P_{O_2} of tracheal air falls during the period of spiracle closure (Fig. 2.21), since more oxygen is used than is entering

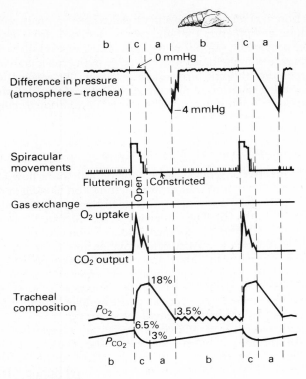

Figure 2.21
Relationships between tracheal pressures, spiracle movements, gas exchange, and tracheal gas compositions during the three phases of a ventilation cycle for an insect pupa (from Levy and Schneiderman, 1966).

as a part of the air. The P_{O_2} eventually reaches a low "critical" value that stimulates the valves to begin the fluttering phase. If pupae are placed in an atmosphere of high P_{O_2}, the P_{O_2} in tracheal gas will not decrease to the critical value and fluttering will not occur.

The fluttering period is the longest period for normal respiratory ventilation. Each flutter of the spiracle valves involves a brief opening and closing of the valves. When they are closed, a vacuum develops that sucks air in. This bulk transport of air into the trachea impedes outward carbon dioxide movement. Some carbon dioxide moves out when the valves are open for brief periods, but less moves out than is being produced. Consequently, the P_{CO_2} in the trachea gradually builds up all during phase (a) and (b) (lower part of Fig. 2.21). When the valve is open, oxygen diffuses in four times faster than carbon dioxide diffuses out, since the pressure difference between inside and outside is about four times greater for oxygen than for carbon dioxide. Each flutter of the valves is stimulated by the P_{O_2} in tracheal gas reaching a lower "critical" value while the valves are closed (on-off control; see Introduction). This low value stimulates a brief opening followed by closure; the P_{O_2} in tracheal gas oscillates close to the lower critical value during the fluttering period (Fig. 2.21).

As carbon dioxide builds up in the fluid it is in equilibrium with the tracheal gas at the gas-fluid interface; thus tracheal P_{CO_2} increases during the flutter period and eventually reaches an upper "critical" value that is associated with a burst (phase c), that is, extended opening of the spiracles. If pupae are placed in an atmosphere of high P_{CO_2}, the spiracles will remain open for prolonged periods in exaggerated bursts.

At the start of a burst the pressure in the trachea is equal to atmospheric pressure (O mmHg difference in Fig. 2.21), and P_{CO_2} inside the trachea is high while P_{O_2} is low (Fig. 2.21). Thus there is a gradient for diffusion of oxygen in and diffusion of carbon dioxide out during the burst. This gradient results in the discontinuous release of CO_2 (in solution in fluid) out of the trachea. Once P_{CO_2} inside has decreased and P_{O_2} inside has increased at the end of a burst (Fig. 2.21), the spiracles close and the sequence is repeated.

Note that the control of tracheal ventilation involves both P_{O_2} and P_{CO_2}. Neither the receptors for the detection of changes in gas concentrations nor the actual effective stimuli have been described, but both gases are involved with the control of ventilation. During most of the cycle, P_{O_2} controls the opening of the spiracles, which occurs at the end of a spiracle closure (phase a) and during the flutter period (phase b). Changes in P_{CO_2} determine the timing of bursts (phase c). If either P_{O_2} or P_{CO_2} changes, the ventilation cycle will change to modify the minute volume of tracheal ventilation.

The only energy expenditure for tracheal ventilation other than from normal cell metabolism involves contraction of spiracle muscles, and the suction pressures developed during spiracle closure (-4 mmHg) are similar to the negative pressures produced in mammals from diaphragm and rib movements. Reduced use of energy resources for ventilation is a consequence of the direct link of the trachea to the cells using oxygen; pressures developed from cellular oxygen use for other purposes are used for ventilation. In addition, the passage of water out of the trachea is limited to the period of bursts, which reduces the rate of water loss compared with more continuous ventilation mechanisms (see also Chapter 13).

ACID-BASE REGULATION AND CONTROL IN INVERTEBRATES

Most information on the pH of body fluids in invertebrates comes from studies of crabs that breathe water and a few species that breathe air using modified gill systems. There is also some indirect information on the control of acid-base balance in terrestrial insects and its role in the control of tracheal ventilation.

The temperatures of many insects change when environmental temperatures change, and some species maintain relatively constant temperatures in certain parts of their bodies under some conditions (see Chapters 6 and 7). Does the pH of the fluid that bathes cells of insects change when their temperatures change? If so, what mechanisms produce the change, and how are these reflected in the pattern of respiratory ventilation?

It would be very interesting to examine these problems for terrestrial insects in detail. If acid-base balance is controlled to provide a relatively constant degree of protein net negative charge, pH should increase if temperature decreases in a manner similar to fishes, amphibians, reptiles, and crabs.

There is some evidence to suggest that ventilation in insects is controlled with respect to acid-base balance, in a manner similar to other animals. Tracheal ventilation changes, and bursts of CO_2 release occur at different "critical" levels of tracheal P_{CO_2} when temperatures change. Measurements of tracheal gas partial pressures for silkworm pupae at three temperatures show that the P_{CO_2} decreases as temperatures decrease (Table 2.1) in a manner similar to other air-breathing animals (see Fig. 2.19). This decrease should increase pH when temperature decreases, since a lower P_{CO_2} would result in fewer hydrogen ions dissociated from carbonic acid.

Table 2.1 Effect of temperature on tracheal P_{CO_2} for *Cecropia* pupae (from Levy and Schneiderman, 1966)

TEMPERATURE, °C	P_{CO_2}, mmHg	
	MINIMUM	MAXIMUM
8.5	20	33
15	23	35
25	73	77

These measurements also suggest that on-off control of bursts is not triggered by P_{CO_2} in tracheal gas per se but by ionic effects of carbon dioxide. It may occur through a histidine imidazole mechanism similar to what has been proposed for control of relative alkalinity by ventilation in vertebrates. Receptors with these amino acids are sensitive to protein tertiary folding from changes in the protein net charge and may be the basis for control of ventilation in many species of animals, regardless of the source of supply of oxygen or the mechanisms used to produce ventilation.

SUMMARY

Although air respiration involves supplying oxygen with relatively low energy expenditure, water losses are increased and blood P_{CO_2} is higher compared with water respiration. Animals that breathe air use internal respiratory exchange mechanisms involving either lungs or trachea, where ventilation occurs with the external environment through a series of tubes.

A number of species of fishes, amphibians, and reptiles exchange gases with either air or water. The extent of total gas exchange that occurs with either respiratory medium depends on the availability of oxygen from either water or air relative to total demands for oxygen. If supplies of oxygen from water are sufficient, most exchange will occur with water. If supplies of oxygen from water decrease and/or demands for oxygen increase (e.g., from increased temperature), species that exchange with either medium increase air respiration.

Tidal lung ventilation occurs in amphibians, reptiles, and mammals. Frogs ventilate lungs by positive pressure from buccal pumping; mammals ventilate primarily by negative pressure from expansion of the thoracic cavity. Air passes through a series of tubes to blind alveolar sacs, where the surface area for gas exchange is large and distances for diffusion exchange are small. However, not all air in a tidal lung can be expired; fresh air mixes with air remaining in the dead-space volume.

Tidal alveolar ventilation varies proportionally if supplies of or demands for oxygen change. This occurs because of variation in either tidal volumes or respiratory rates. For mammals, oxygen demands increase for different species if weight increases. Demands are related to weight raised to an exponent of 0.75. Tidal volume increases with weight raised to an exponent of 1.0, while respiratory rate decreases with weight raised to an exponent of -0.25. Therefore, the product of tidal volume and respiratory rate meets demands associated with different weights. Ventilation also changes proportionally within individuals from changes in tidal volumes and respiratory rates if supplies and/or demands change. There are considerable differences in volumes involved in lung ventilation among species of mammals that are associated with either increased supplies (e.g., larger tidal volumes in shallow divers) or different demands dictated by environmental constraints (e.g., lower residual volumes in deep divers).

Air is not distributed evenly within tidal lungs; usually more air ventilates the anterior parts of lungs. Studies of blood perfusion in the lungs of carpet pythons indicate that stratification of blood flow is matched to stratification of air flow so that exchange will be most effective between alveolar air and blood.

Control of ventilation in mammals (with relatively constant body temperatures) is achieved from receptors in the brain that detect changes in pH resulting from increases or decreases in carbon dioxide production. For example, increased P_{CO2} in humans (associated with increased demand from increased conversion of oxygen to carbon dioxide) results in decreased pH, which in turn results in increased ventilation from increases in both tidal volume and respiratory rate. However, in air-breathing vertebrates with variable body temperatures it is not pH but rather the degree of protein alkalinity that is important in control of ventilation. Ventilation changes relative to metabolic rate and results in maintenance of the same degree of alkalinity of blood when temperatures change (increased pH with decreased temperature). The receptors for this may involve tertiary folding of proteins with amino acids that are sensitive to net ionic charge.

Flow-through lungs provide the potential for decreased mixing of inspired and expired air and more effective gas exchange. Flow-through respiration has been demonstrated in birds, where air flows continuously in one direction through capillary exchange sites between major air reservoirs in the posterior and anterior air sacs. It takes two ventilation cycles for air to pass through the respiratory system, and air flow in one direction suggests the opportunity for countercurrent gas exchange.

An experiment in which air flow in the avian lung was reversed demonstrated that air and blood do *not* exchange gases in a countercurrent system, since reversal had no effect on the gas partial pressures of expired air. Gas exchange in birds is more effective than it is in mammals because more oxygen is extracted (31% versus 24%) and arterial blood P_{O_2} is higher. The hypothesis proposed to account for these observations involves "crosscurrent" flow of blood with respect to air.

Problems of water loss are most severe for very small animals, and insects respire with a tracheal system that minimizes both water loss and energy required for ventilation. Tracheal tubes pass from spiracle valves throughout the body so that each cell is in close proximity with trachea for diffusion exchange. Tracheal ventilation for active insects is accomplished with a minimum of additional energy expenditure because the air sacs are located within major muscles that are involved in movement. Even for inactive insects, the only added energy expense of ventilation involves contraction of muscles associated with the spiracle valves. This phenomenon has been studied in diapausing pupae of silkworms.

Tracheal respiration in silkworm pupae involves three stages. During closure of the spiracle valves, the pressure inside falls below atmospheric because oxygen is used while carbon dioxide goes into solution in fluid. The suction pressure provides movement of air into the trachea, but less oxygen enters than is being used, so P_{O_2} in the trachea falls. At a lower "critical" P_{O_2} the spiracle valves begin to "flutter," or open and close rapidly. Some carbon dioxide diffuses out during a flutter when the valves are open, but more is produced than leaves; thus P_{CO_2} inside gradually increases until an upper "critical" value is reached and the valves then open in a prolonged "burst." During a burst, carbon dioxide diffuses out and oxygen diffuses in. Water loss is limited to the short burst periods, and the normal metabolism of the cells provides the forces for ventilation.

The control of ventilation in insects may involve timing bursts of carbon dioxide release so that different amounts are released at different temperatures. This explains the variation of "critical" P_{CO_2} in tracheal gas at different temperatures, and it will result in regulation of the degree of protein net negative charge as temperatures change.

Blood Gas Transport

3

Once oxygen has diffused into blood in a respiratory structure it must be transported to the tissues where use occurs, and carbon dioxide must be transported from the tissues to the respiratory exchange site. The circulatory system provides the mechanisms for this internal bulk flow of respiratory gases (see Fig. 1.4).

The quantity of a substance transported by blood (g/min) is the product of the rate of movement of blood (ml/min) and the concentration of a substance in the blood (g/ml), or

$$\text{Transport (g/min)} = \text{Blood flow (ml/min)} \times \text{Concentration (g/ml)}.$$

We will be concerned with this important relationship in two ways. In this chapter we will consider some important determinants of oxygen and carbon dioxide concentrations in blood (g/ml or ml gas/ml blood) related to supply-demand problems of respiration. In the next chapter we will consider determinants of blood flow (ml/min) within circulatory systems.

Even though we will treat these two aspects of transport separately, it is important to keep in mind that they are related. Moreover, although we will concentrate on the respiratory transport properties of blood, other important substances are also transported via the circulatory system. These substances include nutrients (enzymes, water, salts), energy substrates, excretory products other than carbon dioxide, hormones, specialized cells (such as white blood cells), and heat. We will encounter these throughout the remainder of the book, particularly in Parts II, III, and IV.

The supplies of oxygen to tissues must be sufficient to meet demands set by rate of use of oxygen, i.e., that necessary for survival.

$$\text{ml } O_2 \text{ delivered/min} = \text{ml } O_2 \text{ required/min}$$

The delivery (transport) of oxygen can be accomplished either from circulatory function, influencing the number of milliliters of blood pumped per minute, or from the amount of oxygen carried in each milliliter of blood.

The fluids circulated by animals are composed of water and varied additional components that influence the *oxygen carrying capacity* of blood. Animals with very low demands for oxygen, such as unicellular species or species with few cell layers (such as sponges or planarians), achieve sufficient transport supplies of oxygen by circulating "blood" that is essentially water. Protoplasmic streaming of cellular water from the inner surface of a cell membrane, for example, transports 0.005 ml O_2/ml fluid (with P_{O_2} = 159 mmHg in the gas phase at 30°C) throughout a cell. However, larger animals with higher transport demands have to circulate relatively large quantities of fluid unless the amount of oxygen transported in every milliliter is increased. Moreover, special pumps and blood vessels are required to circulate blood in larger animals (see Chapter 4). Energy is required to move fluid through a circulatory system, and the amount of gas transported relative to the energy required for transport, the economic *efficiency of transport*, can be maximized by increasing the amount of oxygen carried in each milliliter of blood.

RESPIRATORY PIGMENTS TO INCREASE CARRYING CAPACITY

Animals with high demands for oxygen possess chemicals called "respiratory pigments" that increase oxygen transport by increasing oxygen carrying capacity of the blood. Respiratory pigments are proteins with iron or copper components that have the important property of *reversibly binding oxygen*. They include hemocyanin (containing copper), hemerythrin, chlorocruorin, and hemoglobin (the latter three containing iron). The most thoroughly studied of these is hemoglobin. Each molecule of hemoglobin (with 4 heme units per molecule) will reversibly bind up to four molecules of oxygen,

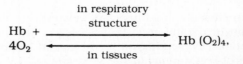

in such a way that oxygen is bound to hemoglobin in a respiratory structure and released in tissues.

Theoretically, each gram of hemoglobin in vertebrates will carry up to 1.38 ml O_2, so by having just a small amount of hemoglobin in blood an animal will realize a large increase in oxygen transport compared with water. This increase will reduce the total energy cost of transport by requiring fewer milliliters of blood to be circulated per minute to supply demands.

The concentration of hemoglobin in blood varies among species depending on demands set by rates of oxygen use, illustrated for four major groups of vertebrates in Table 3.1. In general, fishes have lower demands for oxygen than do amphibians and reptiles, which, in turn, have lower demands than do mammals (see Chapter 5). With an average hemoglobin concentration of 0.038 g/ml and with each gram of hemoglobin carrying up to 1.38 ml O_2, fish blood will carry a maximum of 0.053 ml O_2/ml blood compared with about 0.185 ml O_2/ml blood for mammals (Table 3.1). Both amounts are higher than the amount that would be carried in physical solution without hemoglobin. Note from Table 3.1 that blood flow is also related to demands for oxygen; thus the product of blood flow and carrying capacity (hemoglobin concentration in Table 3.1) provides different *total* transport abilities depending on demands.

Table 3.1 Summary of the general relationships between factors influencing oxygen transport for different major groups of vertebrates; the groups are listed in order of increasing demands for oxygen in tissues (modified from Lenfant et al., 1970)

ANIMALS	AVERAGE SYSTEMIC BLOOD FLOW (ml/kg × min)	×	AVERAGE HEMOGLOBIN CONCENTRATION (g/ml)	—	AVERAGE O_2 TRANSPORT ABILITY (g Hb/kg × min)
Fishes (7)*	17		0.038		0.646
Amphibians (4)	25		0.070		1.75
Reptiles (5)	45		0.072		3.24
Mammals (7)	75		0.132		9.90

*Numbers refer to number of separate measurements.

This supply-demand interpretation for differences between species suggests that blood flow and hemoglobin concentrations interact to provide sufficient oxygen supplies. This point can be illustrated further by examining how the amount of blood flow per milliliter of oxygen use changes when hemoglobin concentrations change in different species (Fig. 3.1). Note that blood flow increases relative to use as hemoglobin concentrations decrease. The extreme case of the Antarctic icefish in Fig. 3.1 "proves" the rule. This fish lives in cold waters where oxygen availability is relatively high and where the use of oxygen is relatively low. Its blood has no hemoglobin, and blood flow is much higher relative to demands than in fish with hemoglobin. Without a respiratory pigment, change in blood flow is the only way to provide for demands (see also Chapter 4). For the icefish, demands may be sufficiently low that the cost for higher "blood" flow is less than the cost for manufacturing and maintaining hemoglobin.

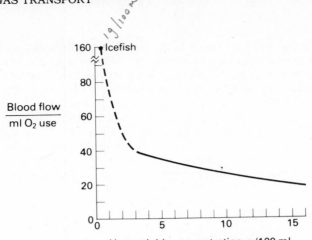

Figure 3.1
Change in blood flow required to maintain one milliliter
of oxygen use as hemoglobin concentrations change.
Data for the line based on measurements from a number
of fishes, amphibians, reptiles, and mammals. Note the
very high blood flow required for the icefish, which has
no hemoglobin (modified from Lenfant *et al.*, 1970).

Why are hemoglobin concentrations not higher than those actually
observed in different species if a reduction in energy required for transport can
occur? This is the opposite problem to that posed by the icefish. It is due to the
transport functions of blood other than the supply of oxygen. Nutrients must be
transported to and waste products removed from tissues, and other substances
must be transported via blood in which special "carrier" molecules do not
enhance transport. Thus some minimum flow of blood through tissues must be
maintained, and hemoglobin concentrations result in maximizing efficiency of
oxygen transport for those rates of flow.

There is a great diversity of hemoglobins, and differences in their function also reflect differences in supply-demand problems for oxygen transport. Table 3.2
lists hemoglobins from different animals according to their molecular weights. Note
also that most hemoglobins of very high molecular weight are found in solution in
blood plasma, while most hemoglobins of low molecular weight are packaged in
red blood cells. Why do we see these differences? Most hypotheses approach this
problem by attempting to explain the evolution of red blood cells.

RED BLOOD CELLS: WHY AND WHY NOT?

One hypothesis to explain why some hemoglobins are in cells proposes that if
they were not, the blood would be more viscous (more resistant to flow), requiring more energy to produce flow through tissues. However, this is not the case.
Figure 3.2 shows measurements of the change in relative viscosity of blood with
changes in hemoglobin concentration for hemoglobins of different molecular
weights. Note that the low-molecular-weight hemoglobins, which are found in
cells (Table 3.2), have little influence on blood viscosity when they are in solu-

Table 3.2 For species with hemoglobins, there can be differences in the types of hemoglobins, as reflected in differences in molecular weights, as well as differences in whether the hemoglobins are in solution or packaged in red blood cells (modified from Schmidt-Nielsen, 1979)

HEMOGLOBINS IN PLASMA	MOLECULAR WEIGHT
Oligochaetes (*Lumbricus*)	2,946,000
Polychaetes (*Arenicola*)	3,000,000
Mollusks (*Planorbis*)	1,539,000
Insects (*Chironomus*)	31,400
HEMOGLOBINS IN CELLS	
Cyclostomes (*Lamprey*)	19,100
Polychaetes (*Notomastus*)	36,000
Echinoderms (*Thyone*)	23,600
Mollusks (*Arca*)	33,600
Insects (*Gastrophilus*)	34,000
Birds	68,000
Mammals	68,000

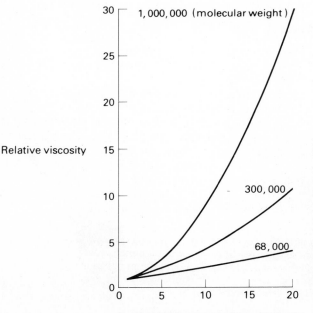

Figure 3.2
The change in viscosity, i.e., resistance to flow, of a solution of hemoglobin depends on the concentration of hemoglobin, but high-molecular-weight hemoglobins in solution have much more effect on viscosity than do low-molecular-weight hemoglobins (from Snyder, 1977).

tion. Experiments in which red blood cells were broken to release hemoglobin into solution have confirmed that it has little effect on the viscosity of blood. However, note that high-molecular-weight hemoglobins, which *are* found in solution, *would* have a pronounced effect on blood viscosity if they were present at higher concentrations (Fig. 3.2).

Blood viscosity is important. The more viscous a solution is, the more energy is required to produce flow (see Chapter 4). Thus we can predict that animals with higher-molecular-weight hemoglobins in solution should only have them present at low concentrations where viscosity is low. For example, a hemoglobin with a molecular weight of 1,000,000 has a viscosity effect at 5 grams/100 ml that is equivalent to a hemoglobin with a molecular weight of 68,000 at a concentration of 15 grams/100 ml (Fig. 3.2). As a consequence, the oxygen-*transport* properties of blood with different concentrations of hemoglobin of different molecular weights will vary. This variation is shown in Fig. 3.3. High-molecular-weight hemoglobins transport oxygen at maximum amounts when hemoglobin concentration is less than that of low-molecular-weight hemoglobins. Maximum oxygen-transport ability is inversely related to molecular weight.

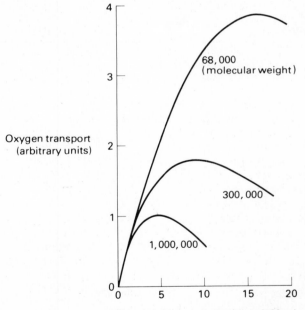

Figure 3.3
Hemoglobins of low molecular weight (e.g., 68,000) will transport more oxygen per unit than will high-molecular-weight hemoglobins (e.g., 1,000,000), and the maximum transport ability occurs at a higher concentration for low-molecular-weight hemoglobin (from Snyder, 1977).

Note in Fig. 3.3 that high-molecular-weight hemoglobins always transport less oxygen than low-molecular-weight hemoglobins, particularly at concentrations equal to or greater than the maximum for oxygen transport. Animals with high-molecular-weight hemoglobins (earthworms, some polychaetes, and some mollusks) have lower demands for oxygen compared with species with low-molecular-weight hemoglobins. The increase in oxygen carrying capacity for these animals (as compared with transport via blood with no hemoglobin) is sufficient to meet their demands. In addition, as a consequence of having high-molecular-weight hemoglobins, these species do not require the complicated machinery needed to package hemoglobin in cells; they are spared this cost of oxygen transport.

Why are low-molecular-weight hemoglobins in cells? Several hypotheses have been proposed. First, since these hemoglobins are relatively small, they would be filtered from blood in the kidneys if they were not confined within membranes (see Chapters 12 and 13). Filtration loss would increase the cost of maintaining the most effective hemoglobin concentrations in blood since filtered molecules would have to be resorbed (with an energy expense), or additional molecules would have to be produced to replace those lost. Second, if the hemoglobin was released from cells it would increase the osmotic concentration of blood (see Chapters 4 and 12), drawing water into the blood from cells, decreasing the efficiency of cellular and cardiovascular functions, and increasing the cost as compared with packaging hemoglobin in cells. In addition, with hemoglobin packaged in cells the local environment of the hemoglobin (involving pH and enzyme concentrations) can be adjusted for effective gas transport (see below).

In summary, the differences among species in relation to hemoglobins and their packaging in cells become understandable when supplies are interpreted relative to demands. Species with low demands increase oxygen transport sufficiently with high-molecular-weight hemoglobins at low concentrations without incurring the added expense of cellular packaging. Species with higher demands for oxygen achieve higher supply rates with lower-molecular-weight hemoglobins. It incurs a cost of cellular packaging to maintain effective concentrations, but the cost is less than costs for alternative mechanisms, such as a high rate of manufacture to replace filtered molecules. We can now turn to an examination of the property of reversible binding of oxygen with hemoglobin in the context of providing sufficient oxygen when either demands for or supplies of oxygen change.

OXYGEN ASSOCIATION AND DISSOCIATION WITH HEMOGLOBIN

Each hemoglobin molecule (with 4 heme units per molecule) will reversibly bind up to four molecules of oxygen. This reaction is concentration-dependent, so the higher the P_{O_2} (at a constant temperature), the more oxygen will *associate* with hemoglobin. At low P_{O_2}, oxygen *dissociates* from hemoglobin. The association of oxygen with hemoglobin occurs in a respiratory structure where P_{O_2} is high, whereas dissociation from hemoglobin occurs in tissues where P_{O_2} is lower as a consequence of the metabolic use of oxygen.

The pick-up and delivery aspects of oxygen transport are described by an oxygen "association-dissociation" curve. An example, the blood of humans under resting conditions, is shown in Fig. 3.4. The hemoglobin in blood represents a population of molecules, some of which will contain four or fewer oxygen molecules, and the probability that *all* hemoglobin molecules will be "saturated" with four oxygen molecules increases when P_{O_2} increases. This characteristic generates the typical "S" shape for a hemoglobin-oxygen dissociation curve (Fig. 3.4). It is common to express this relationship in terms of the "percent saturation" of hemoglobin with oxygen, but recall that different species may have different total concentrations of hemoglobin. Thus the same percent saturation can mean differences in absolute transport quantities (Table 3.1).

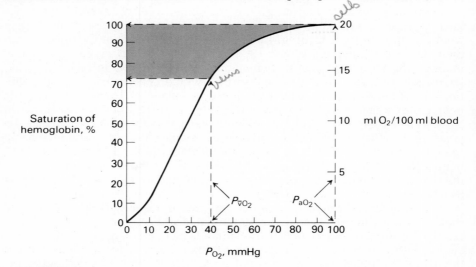

Figure 3.4
An oxygen dissociation curve for a resting human. As a result of the "S" shape and position of the curve, blood will be nearly 100% saturated in the lungs (P_{aO_2}), and some oxygen will dissociate (shaded area) at blood P_{O_2} values typical of tissues ($P_{\bar{v}O_2}$).

By the time oxygen has diffused from an alveolus in a human into the fluid of blood, P_{O_2} is reduced from 150 mmHg in inspired air (at sea level) to about 100 mmHg. The reduction occurs from changes in gas composition as air passes to an alveolus. Fresh air is mixed with dead-space air (with higher P_{CO_2}) and is humidified (P_{H_2O}), and both processes reduce P_{O_2}. Thus blood in the gas exchange site (P_{aO_2} in Fig. 3.4) has hemoglobin exposed to a P_{O_2} of about 100 mmHg. For human blood this is sufficient to almost completely saturate hemoglobin with oxygen (right-hand dashed arrow in Fig. 3.4).

When blood passes through tissue capillaries, oxygen diffuses from the fluid of blood to the cells where it is used. This action reduces P_{O_2} in blood, so by the time it reaches the veins the average P_{O_2} for an inactive human is about 40 mmHg ($P_{\bar{v}O2}$ in Fig. 3.4). Human hemoglobin exposed to this P_{O2} will dissociate (deliver) about 25% of its oxygen (shaded area in Fig. 3.4), that is, about 5 ml of oxygen for every 100 ml of blood passing through the capillaries, where diffusion to cells takes place.

There is tremendous diversity in this general pattern of hemoglobin function within and among animals, and all of the variation can be interpreted with respect to changes in *environmental oxygen supplies* compared with *tissue demands* for oxygen. Examples of differences in hemoglobin function provide some of the best evidence for adaptations in respiratory function.

Changing Oxygen Delivery when Supplies and/or Demands Change

The *position* of the oxygen dissociation curve will change in ways that maintain, or regulate, the supply of oxygen by means of blood transport. Two types of changes occur that depend on whether (1) demands in tissues increase relative to supplies, or (2) environmental supplies of oxygen decrease relative to tissue demands. These changes are illustrated in Fig. 3.5 and are referred to as "shifts" in the oxygen dissociation curve to either the *right* or the *left*.

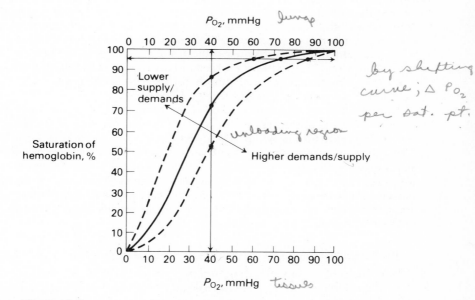

Figure 3.5
General summary of the effects of shifts in an oxygen dissociation curve to the right or left. The points indicate levels of saturation of blood either in lungs (upper horizontal arrow at 95% saturation) or in tissues (vertical arrow at P_{O2} = 40 mmHg). A shift to the right delivers more oxygen at P_{O2} = 40 mmHg; a shift to the left produces 95% saturation at a lower P_{O2} in a respiratory structure.

When an oxygen dissociation curve shifts to the right, more oxygen dissociates from hemoglobin at a given P_{O_2}; it can be seen in Fig. 3.5 by comparing the percent saturation for a normal and right-shifted curve at a P_{O_2} of 40 mmHg. For the normal curve, about 25% of oxygen is dissociated; for the right-shifted curve, about 50% of oxygen is dissociated. There will be a slightly lower total saturation for a right-shifted curve at high P_{O_2} (in respiratory structures), but it is usually minor (the exception is the Root effect; see below) because the S-shaped curves are rather flat at higher values for P_{O_2} (Fig. 3.5). A right-shifted curve has a more pronounced effect in the middle part of the curve (the "unloading region") where the slope is steeper, and this effect occurs at values of P_{O_2} that are typical for blood passing through tissue capillaries. In general, then, a shift to the right increases the supply of oxygen to tissues (if demands increase) without greatly reducing uptake of oxygen at respiratory exchange sites.

When an oxygen dissociation curve shifts to the left, hemoglobin becomes saturated at a respiratory exchange site at a lower P_{O_2} (Fig. 3.5). For the example in Fig. 3.5, the normal curve would give 95% saturation at $P_{O_2} = 72$ mmHg, while the left-shifted curve would give 95% saturation at $P_{O_2} = 60$ mmHg. Note, however, that a left-shifted curve results in less oxygen dissociation at a given lower P_{O_2} than is shown by the normal curve, setting a restriction on the extent to which a left shift can occur. For example, in Fig. 3.5 the left-shifted curve gives a dissociation of only about 10% at $P_{O_2} = 40$ mmHg. Unloading of oxygen can be maintained if P_{O_2} in tissue capillaries decreases or if a left-shifted curve has a steeper slope at P_{O_2} values for blood in tissues. Within these limits, a shift to the left will maintain tissue demands for oxygen if environmental supplies decrease.

Under what circumstances do these shifts occur? There are three general categories of curve shifts that occur over different time scales: (1) short-term shifts within individuals when tissue demands change relative to supplies; (2) long-term (evolutionary) shifts associated with the evolution of species in environments with different supplies and/or associated with different demands among species; and (3) intermediate-term (acclimation) shifts within individuals as a consequence of exposure to environments with different supplies of oxygen relative to demands. Table 3.3 summarizes these types of shifts (discussed below) and some factors producing them.

Short-term Shifts

Within individual organisms, demands for oxygen in tissues increase rapidly with increased activity. Three factors associated with transient increased use of oxygen will shift an oxygen dissociation curve to the right; increased blood temperature, increased blood P_{CO_2}, and increased blood acidity (Table 3.3). Each of these will change the three-dimensional structure of hemoglobin proteins and change the affinity of oxygen for hemoglobin.

Oxygen is used to oxidize substrates within cells to yield energy (see Chapter 5). The energy is ultimately converted to heat; an increase in use of oxygen results in an increase in the temperature of blood passing through tissues. This increase shifts the oxygen dissociation curve to the right; the par-

TABLE 3.3 Summary of the different types of shifts in hemoglobin-oxygen dissociation curves, according to factors influencing supply of or demands for oxygen

---

1. SHORT-TERM SHIFTS

    A. *Right shifts* (higher demand/supply)

        1) Increased blood temperature $\downarrow O_2$
        2) Increased blood P_{CO2}
        3) Decreased blood pH

    B. *Left shifts* (lower supply/demand)

        1) Decreased blood temperature $\uparrow O_2$
        2) Decreased blood P_{CO2}
        3) Increased blood pH

2. LONGER-TERM (EVOLUTIONARY) SHIFTS

    A. *Right shifts* (higher demand/supply)

        1) Body-size-related demands between species (higher tissue demands in smaller species)
        2) Activity-related demands between species (higher tissue demands in more active species)

    B. *Left shifts* (lower supply/demand)

        1) High-altitude species
        2) Hypoxic aquatic versus air-breathing species
        3) Fetal versus adult forms

3. INTERMEDIATE-TERM SHIFTS WITHIN INDIVIDUALS

    Acclimation changes in hemoglobin-association characteristics

---

tial pressure for 50% saturation (or the P_{50}, a useful point of reference) shifts by about 1 mmHg for every change in temperature by 1.0°C. Thus a short-term increase in demand resulting from activity will result in an increased supply of oxygen to the tissues, and when activity and temperature decrease, the dissociation curve will shift back to "normal" values.

Increased use of oxygen will also result in a transient increase in blood P_{CO2}. This increase influences oxygen dissociation in two ways. First, carbon dioxide will preferentially bind to hemoglobin at relatively high P_{CO2}, which will result in some release of oxygen. Second, recall that carbon dioxide combines with water to form carbonic acid, which dissociates to hydrogen ions and bicarbonate ions (see Chapter 1). The change in pH from changes in P_{CO2} also influences oxygen dissociation by influencing the three-dimensional structure of hemoglobin (in a manner similar to effects of pH on enzymes). The effect of differences in P_{CO2} on oxygen dissociation is shown in Fig. 3.6. Most of the shift is due to changes in pH; about 17% of the right shift is due to CO_2 binding to hemoglobin.

The shift of an oxygen dissociation curve with a change in pH is called the *Bohr effect*. The Bohr effect is different for different species, depending on oxygen supply-demand problems. For example, small mammals have a greater

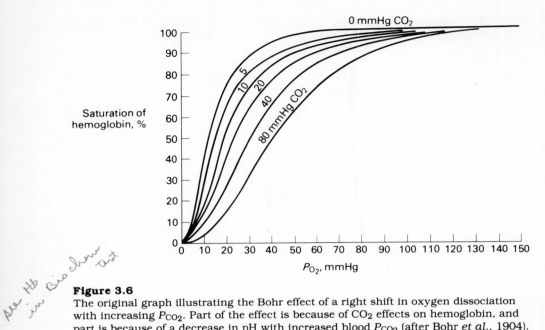

Figure 3.6
The original graph illustrating the Bohr effect of a right shift in oxygen dissociation
with increasing P_{CO_2}. Part of the effect is because of CO_2 effects on hemoglobin, and
part is because of a decrease in pH with increased blood P_{CO_2} (after Bohr *et al.*, 1904).

shift in oxygen dissociation to the right for the same change in pH than do larger
mammals. We will find that the demands for oxygen are higher in the tissues of
small mammals than in tissues of larger mammals (Chapter 5; also see below),
and a more pronounced Bohr effect helps to ensure sufficient oxygen delivery at
high rates during activity, when carbon dioxide production increases.

The most pronounced Bohr effect occurs in the blood of some fishes. For
some species, the dissociation curves not only shift to the right, the hemoglobin-
oxygen carrying capacity also *decreases*; hemoglobin cannot achieve full
saturation when pH decreases. This pronounced effect of pH on the blood of
fishes is called the *Root effect*, and the effect on oxygen carrying capacity is illustrated in Fig. 3.7.

The Root effect serves a function quite separate from oxygen supply for
respiration. It is important for supply of oxygen to the swimbladders of fishes,
which are used to achieve neutral buoyancy and reduce the energy cost for
movement (see Chapter 9). Acid production by special cells in the swimbladder
results in large quantities of oxygen dissociating from hemoglobin because of
the Root effect at the swimbladder. In addition, there are two types of hemoglobin in the blood of some fishes: acid-sensitive and acid-insensitive. The former
types are thought to be important for oxygen supply to the swimbladder,
whereas the latter types maintain oxygen supplies to other tissues.

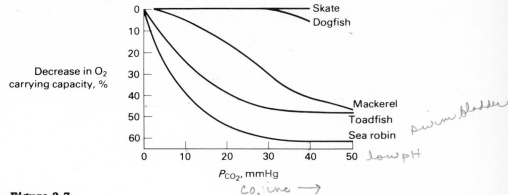

swim bladder

low pH

CO_2 *inc* →

Figure 3.7
Illustration of the impact of the Root effect, an extreme right shift in oxygen dissociation, on the maximum oxygen carrying capacity of blood in several species of fishes. Those fishes with the largest decreases in oxygen carrying capacity have swimbladders (after Wood and Lenfant, 1977).

Figure 3.8 summarizes the factors influencing short-term supply-demand changes from right shifts in oxygen dissociation curves. The example is for a situation involving increased tissue demands. This relationship has all the characteristics of a negative-feedback proportional control system. The "receptors" involve interactions of hydrogen ions, carbon dioxide molecules, and temperature effects on hemoglobin-oxygen affinity. The response is reflected in a dissociation-curve shift that results in increased supplies of oxygen to meet increased demands.

It is interesting and important to note that although the control relationship shown in Fig. 3.8 results in regulation of oxygen supplies to meet demands, the transport functions of hemoglobin play no role in the regulation of respiratory ventilation. It has been demonstrated many times by tragic deaths from cellular oxygen lack caused by carbon monoxide poisoning. Carbon monoxide has a much higher binding affinity for hemoglobin than oxygen; even relatively small amounts in the lungs will saturate hemoglobin and prevent effective transport of oxygen to tissues. As saturation occurs, there is little effect on respiratory ventilation. This situation is what makes carbon monoxide poisoning so dangerous; there are no signs that it is occurring (no indications of reduced supply) until a victim passes out from a lack of sufficient oxygen supply to the brain. It vividly illustrates the dangers humans face as a consequence of environmental changes for which there has been no evolutionary selection for response mechanisms. Except for the advent of modern machinery, there has been no reason for respiratory regulation mechanisms to respond to hemoglobin transport by means of changes in ventilation.

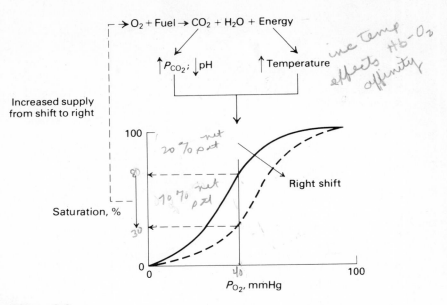

Figure 3.8
Summary of the short-term right shifts in oxygen dissociation
from changes in pH, P_{CO_2}, and temperature; these shifts produce
a negative-feedback (dashed line) type of proportional control for
supply of oxygen to meet increased demands.

Longer-term (Evolutionary) Shifts

Species that have evolved characteristics dictating higher oxygen supplies to
tissues have oxygen dissociation curves shifted to the right, while species that
have evolved in environments with relatively low supplies of oxygen have
dissociation curves shifted to the left (see below). A right or left shift can evolve
in one or both of two ways: a different type of hemoglobin can evolve with different oxygen dissociation characteristics; or the same hemoglobin can be
utilized, but its dissociation characteristics can be affected by a change in the
chemical environment of the hemoglobin.

Chemicals called *organic phosphates* influence the position of an oxygen
dissociation curve. Organic phosphates are present in red blood cells. The most
common one in mammals is 2,3-diphosphoglycerate (DPG), in birds and turtles
inositol hexosephosphate (IHP) is common, and in amphibians and some other
reptiles adenosine triphosphate (ATP) is most common. Some of each is found
in most species. The effect of DPG on the oxygen dissociation curve of human
blood is shown in Fig. 3.9. "Pure" hemoglobin has a dissociation curve shifted
to the *far* left. As the concentration of DPG is increased, the dissociation curve
shifts more to the right. Adjustments in the concentration of organic phosphates provides a mechanism for many differences observed between species in
supply-demand relationships involving oxygen transport.

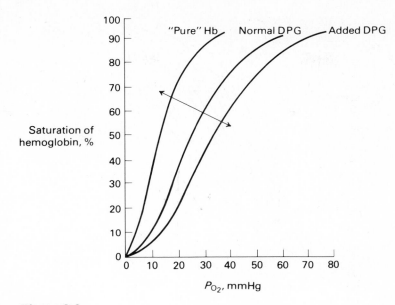

Figure 3.9
Oxygen dissociation curves for human hemoglobin. "Pure" hemoglobin has a left-shifted curve with respect to hemoglobin having
normal concentrations of DPG. If DPG is increased to a concentration greater than normal, the curve shifts further to the right (modified from Laver *et al.*, 1977).

Right shifts (higher demand/supply) between species   Although smaller
mammals require less *total* oxygen to support the metabolism of fewer cells,
their tissues consume oxygen at a higher *rate*; that is, each gram of tissue of a
small mammal uses more oxygen per minute than a gram of tissue of larger
mammals. We will examine the basis for this in Chapter 5. Figure 3.10 shows
the oxygen dissociation curves for the blood of several mammals of different
sizes; note that the curves for smaller species are shifted to the right. This shift
provides a greater degree of oxygen delivery to their tissues to match their
higher demands. The mechanism involves differences in the concentrations of
DPG between species, and small rodents have relatively high DPG concentrations in their red blood cells.

Smaller species of mammals also have more pronounced right shifts from
the Bohr effect, partly mediated by an enzyme called *carbonic anhydrase*. This
enzyme catalyzes the formation of carbonic acid from carbon dioxide and
water. Without the enzyme, blood in tissues would not change in pH within a
short enough period for the dissociation curve to shift to the right. With the enzyme, increases in P_{CO_2} in capillary blood result in very rapid pH changes, since
the enzyme accelerates carbonic acid formation when P_{CO_2} increases. It is contained within red blood cells, where it has a direct effect on oxygen dissociation;
the red blood cells of a mouse contain about three times more carbonic anhydrase than is in the red blood cells of an elephant.

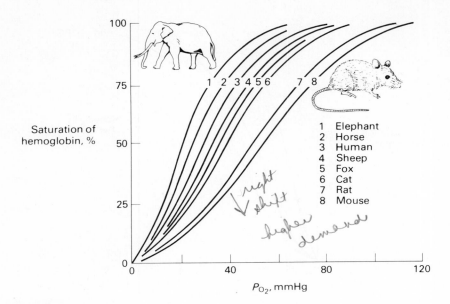

Figure 3.10
Oxygen dissociation curves for blood from mammals of different sizes. The right shift for species of smaller size provides increased oxygen delivery to tissues in which demands for oxygen are higher (from Schmidt-Nielsen, 1970).

Demands for oxygen in tissues also differ among species, depending on whether a species is relatively active or relatively sedentary in its habits. These habits provide an interpretation for some differences in oxygen dissociation curves for blood from different species of fishes (Fig. 3.11). The mackerel is normally highly active with high demands for oxygen delivery to tissues; the toadfish is more sedentary. The toadfish is also exposed to water with a relatively low P_{O_2}, so the left position of its dissociation curve reflects lower environmental supply of oxygen relative to demands. Experiments with carp indicate that exposure to environments with low P_{O_2} produce left shifts resulting from decreases in concentrations of organic phosphates.

Left shifts (lower supply/demand) between species   There can be considerable variation in the availability of oxygen in different environments that influences the degree of oxygen saturation of blood in respiratory structures. This variation is associated with changes in altitude of air-breathing animals and changes in availability of oxygen from water versus air for animals that may breathe either medium (see Chapter 2). Variation in environmental oxygen availability can occur during the life of individuals that move between different environments, or it can represent a permanent change for species that have evolved in habitats with lower oxygen concentrations. The former situation in-

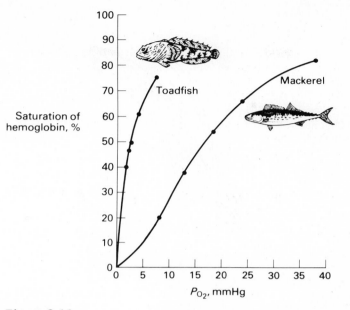

Figure 3.11
Oxygen dissociation curves for an active fish (mackerel) and a
relatively sedentary fish from a hypoxic aquatic environment
(toadfish). The right shift for the blood of mackerel increases
supplies of oxygen to tissues. Note also that, for both species,
blood is saturated at a lower P_{O_2} as compared with air-
breathing species (modified from Hall and McCutcheon, 1938).

volves intermediate-term (acclimation) shifts in dissociation curves (see below).
The latter involves natural selection for a left-shifted dissociation curve that is
more stable.

Figure 3.12 shows oxygen dissociation curves for the llama and vicuña,
two relatives of the camel that have evolved at high altitudes on the altiplano of
the Andes in South America. These mammals typically live at elevations of
3000 meters or more, where P_{O_2} in inspired air is reduced by about one-third
compared with sea level. The left shift in the oxygen dissociation curves permits
saturation of blood in the lungs at a lower P_{O_2} than is found in mammals that
have evolved in environments where supplies of oxygen are higher (shaded area
in Fig. 3.12). Even llamas and vicuñas born in zoos at lower elevations have left-
shifted dissociation curves, so the difference appears to be genetically determined. Relatively low concentrations of DPG are responsible for the left shift
with respect to other species of mammals.

Water may have a lower P_{O_2} in some cases (hypoxic aquatic environments),
and this is reflected in the position of the oxygen dissociation curves for some
aquatic species. Moreover, in species that undergo a transition from breathing
water that is hypoxic to breathing air, the position of the dissociation curves

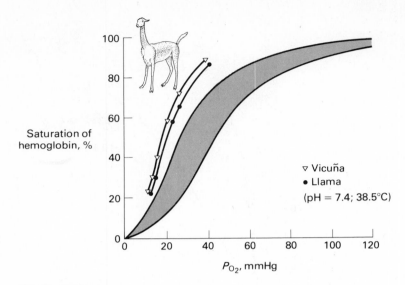

Figure 3.12
Oxygen dissociation curves for the llama and vicuña are shifted to
the left compared with mammals from lower elevations, which
means that blood is saturated at a lower P_{O_2} in the lungs (modified
from Chiodi, 1971).

reflect the changes in environmental supplies of oxygen, which are illustrated
for the blood of tadpoles and adult frogs in Fig. 3.13. The dissociation curves for
tadpole blood are left-shifted with respect to the curves for adult blood. Note also
that the dissociation curves for tadpole blood are essentially insensitive to
changes in pH, while there is a pronounced Bohr shift in the blood of adults.
This difference may be a consequence of the lack of pH change in tadpoles as influenced by blood P_{CO_2} because of the solubility of CO_2 in water.

The differences between tadpoles and adults is partly due to the *type* of
hemoglobin produced during development. During metamorphosis, an "adult"
hemoglobin is produced that has a different dissociation curve and is more sensitive to blood pH changes. A similar change in the type of hemoglobin produced
is seen in mammals during development. Prior to birth, the fetus has
hemoglobin with a dissociation curve shifted to the left with respect to the
hemoglobin of the mother. The fetus lives in a relatively hypoxic (lower P_{O_2}) environment as compared with the mother. The fetus must be hypoxic as compared with the mother because oxygen will diffuse to fetal blood across the
placenta only if the P_{O_2} in fetal blood is less than the P_{O_2} in maternal blood (c_1 −
c_2 in Eq. 1.1). The left shift of fetal hemoglobin ensures sufficient oxygen supplies for the hypoxic environment.

Adult hemoglobin is produced when the fetus undergoes "metamorphosis" from a hypoxic aquatic environment to air breathing. The position of
the dissociation curves of both fetal and adult hemoglobins (Fig. 3.14) depends

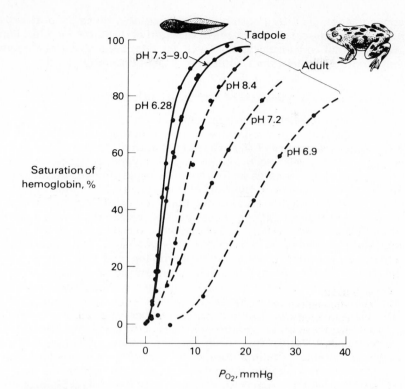

Figure 3.13
Oxygen dissociation curves for tadpole blood are to the left of those
for adult frog blood, and the dissociation curves for tadpole blood are
less sensitive to pH changes (from Riggs, 1951).

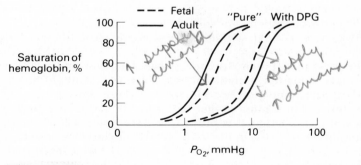

Figure 3.14
Although "pure" fetal hemoglobin has a dissociation curve
to the right of adult hemoglobin, with normal concentrations of DPG the fetal dissociation curve is to the left of the
adult curve, resulting in saturation of fetal hemoglobin at a
lower P_{O_2} (modified from Tyuma and Shimiza, 1970).

on the concentration of DPG. If both human hemoglobin types are purified, the position of the fetal dissociation curve is to the *right* of the curve for adult hemoglobin. With normal concentrations of DPG, the fetal hemoglobin dissociation curve shifts to the right less than adult hemoglobin does (Fig. 3.14).

Intermediate-term (Acclimation) Shifts within Individuals

Differences between species in concentrations of organic phosphates and differences in hemoglobin types represent genetic differences, which have been selected for as a consequence of the evolution of animals in environments posing different problems relating to the supply of oxygen. Short-term shifts in dissociation curves are mechanisms for relatively rapid changes in rates of oxygen supply to tissues. However, the position of oxygen dissociation curves can also change over intermediate periods of hours or days in animals that experience intermediate-term changes in supplies of oxygen relative to demands.

Although changes in activity-associated blood temperature result in short-term increased supplies of oxygen to meet demands, many aquatic animals have temperatures similar to their environments for relatively long periods of hours or days (see Chapter 6). An increase in temperature results in an increased demand for oxygen in tissues, since rates of oxygen use increase when tissue temperatures increase. However, many animals with changeable temperatures modify the rates of cellular reactions when temperatures change over long periods ("acclimation" to a temperature change; see Chapter 6). After a period of time, rates at higher temperatures decrease. The function of hemoglobin transport changes in a manner that matches these changing demands for oxygen.

Figure 3.15 shows oxygen dissociation curves for bullheads that have been acclimated to water temperatures of either 9 °C or 24 °C for a period of weeks. As expected, the dissociation curve for the higher-temperature individuals is to the right of the curve for the lower-temperature individuals. However, if some of the individuals in the 9 °C group are moved to 24 °C, their dissociation curve initially shifts to the right more than it does for the individuals acclimated to 24 °C for a longer period (Fig. 3.15). This indicates that supply of oxygen through hemoglobin transport is "fine-tuned" to changes in demands for oxygen that occur within individuals over either short or long time periods.

When humans live at high altitude for several days or weeks, the oxygen dissociation curve shifts to the right and not to the left, as might be expected from the reduced environmental P_{O_2}. The right shift is a consequence of the increased importance of unloading oxygen in tissues compared with the ability to load hemoglobin in the lungs. It is difficult to have a dissociation curve with both a high affinity for oxygen in the lungs and the ability to unload oxygen in tissues at low P_{O_2}. The slope of such a curve would be extremely steep. In humans, the demands of tissues for oxygen during acclimation are best met by the right shift, involving less saturation in the lungs. Because both increased saturation in the lungs and increased unloading in tissues cannot be achieved, the tissues take priority.

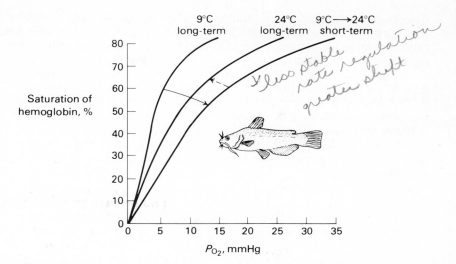

Figure 3.15
Acclimation from long-term exposure of individual bullheads to 24 °C results in lower demands for oxygen compared with short-term exposure to that temperature. The position of blood oxygen dissociation curves reflects these changes in demands relative to supplies (modified from Grigg, 1969).

Intracellular Supplies of Oxygen

Most hemoglobin is found in blood, and the adaptations we have discussed can all be interpreted as providing a sufficient gradient for the diffusion supply of oxygen from blood to tissues. However, under special circumstances even the supply of oxygen via blood through "normal" mechanisms is not sufficient. This situation occurs during extreme activity, when muscular contraction produces very large demands for energy use. In addition, some animals are active in environments where oxygen is not available. The best examples are animals that "dive"; that is, they breathe one medium (air or water) and periodically enter the other (see Chapter 4).

For animals that either dive or must be active (or both), there are two mechanisms to maintain supplies of energy: (1) utilization of special cellular oxygen storage supplies, and/or (2) switching from use of oxygen to other ways of utilizing energy substrates without oxygen. The latter mechanism is called *anaerobic metabolism*, and we will discuss this strategy in detail in the next chapter when we consider the set of adaptations evolved by diving animals to survive for certain time periods without any environmental oxygen supplies.

Even though there are ways to obtain energy without using oxygen, we will find that many animals maintain oxidative metabolism under most circumstances, since it is more energy-efficient than anaerobic metabolism; that is, more energy is obtained from a gram of "fuel" (substrate) within cells by use of oxidative metabolism (see Chapter 4). This energy difference provides an explanation for the presence of special respiratory pigments within cells that have high demands for oxygen. The pigment most studied to date has been *myoglobin*.

Figure 3.16 shows oxygen dissociation curves for myoglobin and hemoglobin. Note that the curve for myoglobin is normally shifted far to the left of the curve for blood hemoglobins, meaning that myoglobin remains essentially completely saturated with oxygen unless the P_{O_2} in the cells where myoglobin is found decreases to very low values (generally less than $P_{O_2} = 20$ mmHg; see Fig. 3.16). Thus the oxygen associated with myoglobin serves as an "emergency" supply for situations when an increase in oxygen demand results in a very low intracellular P_{O_2}.

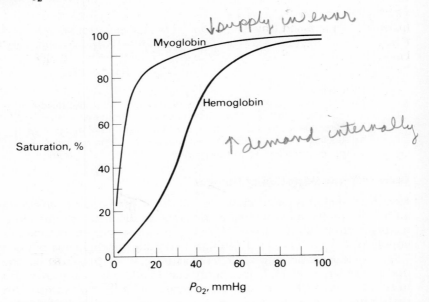

Figure 3.16
The dissociation curve for myoglobin results in delivery of oxygen from this intracellular supply only when intracellular P_{O_2} is reduced to very low values.

Myoglobin is found in certain types of muscles called "red" muscles, because this respiratory pigment combined with oxygen produces a color seen as red by our eyes. There are several kinds of muscles (see Chapter 18); they are described as "high-oxidative, fast," "high-oxidative, slow," and "low-oxidative, fast." Myoglobin is found primarily in the two "high-oxidative" types, where its most important function is to maintain an oxygen supply for the energy needed for contraction particulary when demands exceed the usual supplies provided by oxygen diffusion from blood.

CARBON DIOXIDE TRANSPORT

The transport of oxygen to tissues must be accompanied by simultaneous
transport of carbon dioxide from tissues to the respiratory exchange site. The
amount of carbon dioxide transported by blood depends on three processes that
depend on the partial pressure of carbon dioxide: (1) carbon dioxide dissolved in
solution in blood plasma water; (2) formation of bicarbonate in aqueous solution; and (3) association of carbon dioxide with hemoglobin. Each of these processes can be described by a functional relationship between the quantity of
carbon dioxide in these various components of the blood and P_{CO_2}.

   Figure 3.17 shows the relationship between P_{CO_2} and ml CO_2/100 ml blood
as *dissolved* CO_2 in the water of blood at a temperature of 37 °C. The linear relationship is determined by the solubility coefficient for carbon dioxide at 37 °C,
which depends on P_{CO_2} (see Chapter 1) for blood water with dissolved salts
(plasma). When blood leaves the tissue capillaries in a resting human, it has a
P_{CO_2} = 46 mmHg (line marked "V" in Fig. 3.17), and diffusion of CO_2 out of
blood in the lungs reduces this to about 40 mmHg in arterial blood (line marked
"A" in Fig. 3.17).

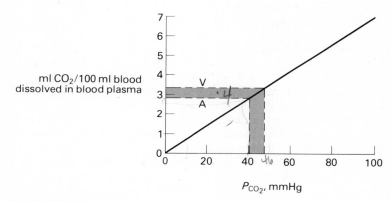

Figure 3.17
A "dissociation" curve for the transport of carbon dioxide via
changes in quantities of CO_2 dissolved in blood plasma water.
Venous blood ("V") coming to lungs gives up a small quantity of
CO_2 (shaded area) when P_{CO_2} decreases in lungs ("A") because of
diffusion of carbon dioxide out of blood.

   The difference between the P_{CO_2} of venous and arterial blood results in
release of a relatively small quantity of CO_2 from what is dissolved in solution in
blood plasma. For the example in Fig. 3.17, only about 0.4 ml CO_2/100 ml blood
"dissociates" from aqueous solution. For a resting human, about ten times this
amount of CO_2 must be released in the lungs during carbon dioxide transport to
match the production of CO_2 in tissues.

The majority of carbon dioxide transport in humans and other animals occurs as bicarbonate. Recall that CO_2 and water react according to the reaction

$$CO_2 + H_2O \rightleftharpoons H_2CO_3 \rightleftharpoons H^+ + HCO_3^-,$$

which we have discussed in terms of the importance of hydrogen ion concentrations (pH; see Chapters 1 and 2). The amount of carbon dioxide carried as HCO_3^- in blood depends on the fate of the hydrogen ions that are also produced. According to the law of mass action, if hydrogen ions are removed from the right side of the above reactions, HCO_3^- formation is favored. This principle forms the basis for the quantitatively most important component of carbon dioxide transport.

If hydrogen ions are removed as fast as they are produced, pH will not change, and any chemical that "accepts" hydrogen ions as they are formed is called a *buffer*. The proteins in blood act as buffers by combining with hydrogen ions. For example, hemoglobin ("Hb") will accept hydrogen ions:

$$Hb^- + H^+ \rightleftharpoons HbH,$$

and this reaction is particularly favored for hemoglobin that has released oxygen (deoxyhemoglobin). Hemoglobin provides about half the buffering ability of blood in humans, with other blood proteins providing the remainder.

The dissociation of H_2CO_3 to H^+ and HCO_3^- occurs relatively slowly except under the influence of the enzyme carbonic anhydrase. This enzyme is found at the highest concentration within red blood cells, where hemoglobin functions as a buffer. The HCO_3^- diffuses from the red blood cells to the blood plasma because of differences in concentration; thus most transport of CO_2 as HCO_3^- occurs in blood plasma. The diffusion of HCO_3^- out of red blood cells is accompanied by the inward movement of Cl^-, which is called the "chloride shift." It maintains electrical neutrality within the red blood cell and is an integral part of the mechanism of carbon dioxide transport.

Figure 3.18 shows a "dissociation" curve for CO_2 transport via HCO_3^-. At very low P_{CO_2} virtually all the hydrogen ions produced from dissociation of H_2CO_3 (produced in smaller amounts at low P_{CO_2}) are buffered, and as P_{CO_2} increases the curve for carbon dioxide (as HCO_3^- in blood) gradually flattens as the blood protein buffers become saturated with hydrogen ions. However, note that the magnitude of CO_2 as HCO_3^- is considerable compared with the amount of CO_2 in solution (Fig. 3.17). For a resting human, the difference between arterial ("A") and venous ("V") P_{CO_2} results in "dissociation" of about 3 ml CO_2/100 ml blood from bicarbonate for the example in Fig. 3.18.

The third mechanism for carbon dioxide transport is provided by the association of CO_2 with hemoglobin. Carbon dioxide combines with a different part of the hemoglobin molecule (the terminal amino group) than does oxygen, and the amount of CO_2 that combines with hemoglobin depends on whether the hemoglobin is carrying oxygen. Blood with no oxygen on hemoglobin has a higher CO_2 carrying capacity than does blood saturated with oxygen (referred to as the *Haldane effect*; see Fig. 3.19). However, the magnitude of CO_2 transport via hemoglobin is relatively small. For a resting human, about 0.4 ml CO_2/100 ml blood are transported this way, an amount similar to CO_2 transport via solution in blood plasma water.

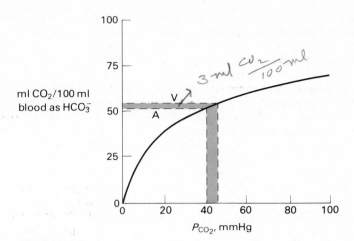

Figure 3.18
A "dissociation" curve for the transport of carbon dioxide
via changes in amounts of bicarbonate that occur when
P_{CO_2} changes between venous ("V") and arterial ("A")
blood. The shaded area indicates the amount transported
by means of bicarbonate.

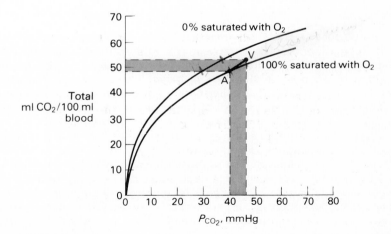

Figure 3.19
A "dissociation" curve for the *total* transport of carbon dioxide.
The line between "V" and "A" represents the quantity released
from blood as a consequence of changes in P_{CO_2} between venous
and arterial blood for a resting human.

Both the buffering effect of hemoglobin and its ability to combine with CO_2 depend on whether hemoglobin is carrying oxygen. Blood that is 100% saturated with oxygen will buffer fewer hydrogen ions as well as accept less CO_2 in association with hemoglobin. For this reason, the *total* CO_2 transport in blood is depicted as in Fig. 3.19. The point marked "A" indicates total CO_2 content (ml CO_2/100 ml blood) of arterial blood leaving the lungs and essentially 100% saturated with oxygen at P_{CO_2} = 40 mmHg. Blood entering the lungs at P_{CO_2} = 46 mmHg will not be 0% saturated (upper curve in Fig. 3.19) but will contain less oxygen at some intermediate point ("V" in Fig. 3.19). The difference between "A" and "V" accounts for the total transport of CO_2, with the majority (about 90% in resting humans) transported by means of bicarbonate and the remainder transported about equally in solution and by association with deoxyhemoglobin.

Changes in Carbon Dioxide Transport with Changes in "Supplies" and/or "Demands"

An increased production of CO_2 in tissues results in increased transport and release of CO_2 in several ways. Increased CO_2 production increases venous P_{CO_2} above 46 mmHg. In addition, venous blood will be less saturated with oxygen as a consequence of increased use of oxygen by tissues. Both these effects shift the "V" point in Fig. 3.19 further up toward the upper curve, resulting in more venous transport of CO_2 to an exchange site. At the exchange site increased ventilation will lower P_{CO_2} of arterial blood and shift the "A" point in Fig. 3.19 further down on the lower curve. The net result is an increased difference in total CO_2 content between arterial and venous blood to match demands for increased transport.

Carbon Dioxide and Acid-base Regulation: A Summary

"Demands" for CO_2 transport are not only related to tissue production but also to the regulation of pH. There can be important differences involved in transport properties of blood, depending on the use of CO_2 "supplies" for regulation of the relative alkalinity of blood. The transport of carbon dioxide depends on P_{CO_2}, carbonic acid formation and dissociation, blood protein buffers, and temperature effects. All of these factors can now be summarized as they influence both P_{CO_2} and hydrogen ion concentrations.

Figure 3.20 shows five important effects in the interaction of carbon dioxide with water in animals. First, the amount of CO_2 available to combine with water is determined *both* by rate of ventilation and by rate of tissue CO_2 production. The difference between them determines the net amount of CO_2 in blood. Thus one mechanism to adjust pH as temperature changes is to hypoventilate in relation to production as temperature increases and to hyperventilate in relation to production as temperature decreases in a manner such that pH changes by 0.017 units/°C (see Fig. 2.19).

The second factor involves the equilibrium for dissociation of water into ions (Fig. 3.20). It is temperature-dependent; dissociation increases (pH de-

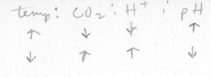

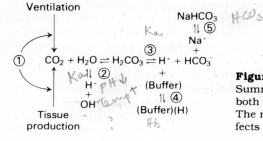

Figure 3.20
Summary of the related factors influencing both CO_2 transport and regulation of pH. The numbers refer to temperature-related effects discussed in the text.

creases) as temperature increases, producing a linear relationship between the neutral pH of water and temperature, also with a slope of 0.017 pH units/°C (see Fig. 1.16).

The third factor involves the dissociation equilibrium for carbonic acid. This is also temperature-dependent (with a slope of 0.005 pH units/°C), and it is strongly influenced by carbonic anhydrase, which favors dissociation of carbonic acid.

The fourth factor involves the equilibrium for association of hydrogen ions with buffers. This reaction is also temperature-dependent in a way that buffers accept hydrogen ions at higher values of pH as temperature decreases (with a slope of 0.017 pH units/°C for protein buffers).

The fifth factor involves mechanisms influencing the amount of bicarbonate. It can involve temperature-dependent equilibria with salts or mechanisms such as energy expenditures to "pump" bicarbonate into or out of blood (see Chapter 1).

All of the factors shown in Fig. 3.20 are related because they are interconnected with various chemical equilibria. Thus it is possible to "trace" the effects of a change in one factor on others by utilizing the law of mass action. In animals, ventilation in relation to CO_2 production is a major mechanism for controlling enzyme (protein) net negative charge. The observed increase in pH with decreasing temperature is what is expected if the degree of alkalinity remains constant ($\triangle$ pH/°C = 0.017, or the same as the change in neutral pH of water). For many nonprotein buffers the change in pH with decreasing temperature is less than the above (usually $\triangle$ pH/°C = 0.005); thus nonprotein buffer effects alone are insufficient to maintain protein net negative charge when temperature changes.

SUMMARY

The transport of oxygen to and carbon dioxide from tissues is the product of blood flow (ml/min) and the amount of the two gases in blood (ml gas/ml blood). For species with low demands for oxygen, oxygen concentrations in blood water may be sufficient for supplies, but the energy efficiency of oxygen transport (ml O_2 transported per unit of energy expended in circulation) is increased by increasing the oxygen carrying capacity of blood through the use of respiratory pigments (hemocyanin, hemerythrin, chlorocruorin, and hemoglobin). A hemoglobin molecule will reversibly bind up to four molecules of oxygen, and each gram of hemoglobin can carry up to 1.38 ml O_2. Therefore a small amount of the pigment greatly increases oxygen carrying capacity as compared with the capacity of water by itself.

Even though hemoglobin increases oxygen transport, not all species have the same quantities of hemoglobin in blood, and not all hemoglobins are identical. These differences can be interpreted from differences in supply-demand problems in oxygen transport. For example, different major groups of animals (fishes, amphibians, reptiles, and mammals) have different demands for oxygen, which are reflected in differences in hemoglobin concentrations and blood flow. Blood flow within an animal can be determined by demands for transport of other substances (nutrients, excretory products, heat), and hemoglobin concentrations are related to maximizing oxygen transport within the constraints of meeting these other demands. Different types of hemoglobin may also reflect differences in costs for providing optimal oxygen transport. Large-molecular-weight hemoglobins provide sufficient transport at low concentrations (where viscosity effects are minimal) without the expense of cellular packaging of hemoglobin. Low-molecular-weight hemoglobins provide greater transport capacities for species with higher demands. These hemoglobins must be packaged in cells; if they were not, they would be filtered from blood and/or produce problems from increased osmotic pressure. Cellular packaging involves a cost, but it is less than the alternatives for providing hemoglobin concentrations sufficient to meet oxygen-transport demands.

The association and dissociation of oxygen with hemoglobin is described by an oxygen dissociation curve. The curve for most hemoglobins is sigmoid (S-shaped) as a consequence of changes in the probability of all hemoglobin molecules having the maximum of four molecules of oxygen (100% saturation) as the partial pressure of oxygen increases. Blood is usually close to 100% saturated at partial pressures of oxygen typically found in respiratory structures, and some dissociation occurs at lower P_{O_2}, which is typical for blood flowing through tissue capillaries. However, the P_{O_2} in respiratory structures and the demands for oxygen in tissues can change, and blood dissociation curves will shift either to the left or the right as a consequence of factors related to changed supplies or demands. A shift to the right results in more oxygen dissociation in tissues at a given P_{O_2}; a shift to the left results in enhanced ability to achieve saturation at lower P_{O_2} in respiratory structures. These curve shifts occur over three time scales: (1) short-term shifts, (2) long-term (evolutionary) shifts, and (3) intermediate-term (acclimation) shifts.

Demands for oxygen in tissues increase rapidly within individuals when they become active. Increased use of oxygen increases blood temperature, blood P_{CO_2}, and blood acidity (lower blood pH), and each of these factors produces some right shift in a dissociation curve over short time periods. The right shift with decreased pH is called the Bohr effect. Increased carbon dioxide production will decrease pH, and the speed of this process is increased because of the presence of the enzyme carbonic anhydrase in red blood cells. Moreover, the enzyme is present in higher concentrations in smaller species of mammals that have higher tissue demands for oxygen delivery than do larger mammals. The most pronounced right shift from decreased pH occurs in the blood of some fishes (the Root effect) for reasons related to the gas supply to a swimbladder, used for achieving neutral buoyancy. Although short-term right shifts in oxygen dissociation provide a type of negative-feedback proportional control of oxygen supply to tissues, the mechanisms controlling oxygen supplies through respiratory ventilation operate independently of information on hemoglobin-oxygen transport properties of blood. For example, decreased hemoglobin-oxygen transport during carbon monoxide poisoning does not result in effective proportional compensating effects in respiratory ventilation.

A more permanent right or left shift in oxygen dissociation curves results from a change in organic phosphate concentrations maintained in red blood cells and/or from the evolution of a different type of hemoglobin. Organic phosphates (DPG, IHP, ATP) cause a right shift as concentrations increase. Smaller species of mammals with higher tissue demands for oxygen have higher concentrations of DPG in red blood cells. In addition, species of fishes with relatively high levels of activity have oxygen dissociation curves shifted to the right as compared with more sedentary species from more hypoxic environments. The left shift in the dissociation curves for llamas and vicuñas, which have evolved at high altitudes, results in hemoglobin saturation in lungs at lower P_{O_2}, and the shift appears to be because of lower concentrations of DPG. Species that have evolved in aquatic environments having a lower P_{O_2} also have dissociation curves shifted to the left, and species that change from hypoxic aquatic to more terrestrial environments during development have evolved "fetal" and "adult" types of hemoglobin that reflect changes in environmental P_{O_2}. In mammals, oxygen dissociation curves of the fetus are to the left of the dissociation curves for the hemoglobin of the mother. This difference is a consequence of both a difference in hemoglobin type and the effects of DPG.

An individual that has evolved a particular type of hemoglobin and DPG concentration can still experience some change in oxygen-transport function over relatively long periods as a consequence of acclimation to changes in supplies relative to demands. In humans, exposure to low environmental P_{O_2} results in increased red blood cell DPG concentrations because of increased demands in tissues relative to transport supplies. In fishes that may acclimate (change) demands for oxygen because of changes in temperature, oxygen dissociation characteristics are "fine tuned" to the changes required in oxygen transport.

Even though hemoglobin functions exhibit considerable adaptive modifications, some individuals still experience demands for oxygen that exceed the capability of blood transport to supply it. Extreme activity or a change in environment resulting in lack of oxygen ("diving") are examples. Muscles with relatively high demands for oxygen (high-oxidative types) have intracellular respiratory pigments (e.g., myoglobin) that dissociate only at low intracellular P_{O_2}. These pigments provide an emergency supply of oxygen for situations when blood supplies are insufficient relative to tissue demands.

Carbon dioxide transport from tissues to a respiratory exchange site occurs via three avenues, depending on P_{CO_2} of blood. A small amount is transported as dissolved CO_2 in the plasma water of blood, accounting for about 5% of total transport requirements in resting humans. A relatively large quantity of CO_2 is in the form of bicarbonate ions as a consequence of the buffering, that is, hydrogen ion acceptance, of blood proteins, including hemoglobin. About 90% of total CO_2 transport in resting humans occurs from association and dissociation of CO_2 with bicarbonate. The remaining 5% of carbon dioxide transport occurs by means of association with hemoglobin. Both buffering of hydrogen ions and CO_2 association with hemoglobin depend on whether blood is oxygenated; the *total* transport dissociation curve for CO_2 represents differences between oxygenated (arterial) and partially deoxygenated (venous) blood.

Changes in venous and arterial P_{CO_2} and the extent of blood oxygenation will change the amount of CO_2 delivered when demands change. Moreover, the amount of CO_2 as bicarbonate can change because of temperature effects on blood protein buffers. Decreased or increased temperatures produce a variety of effects on the interaction between CO_2 and water. Included are effects on CO_2 production compared with ventilation loss, temperature effects on the dissociation of water and carbonic acid, temperature effects on blood protein buffers, and temperature effects on bicarbonate concentrations. The result of all of these related factors is the adjustment of CO_2 transport to demands that are associated with regulation of enzyme net negative charge when temperatures change.

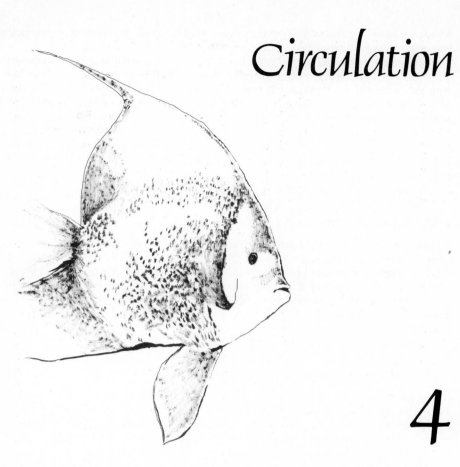

Circulation

4

Bulk fluid flow provides an important component of internal transport. Recall that total transport (amount/min) is the product of substance concentration (amount/ml) and fluid flow rate (ml/min). We have seen in the last chapter how gas amounts in blood change when supplies or demands change. Changes in blood flow rate also contribute to matching supplies with demands, and blood flow represents a quantitatively important component of total transport. For example, in some animals demands for oxygen increase 10 to 20 times during activity. Changes in the oxygen dissociation characteristics of blood can increase transport about 3 to 4 times, which means that increases in blood flow rate must make up the difference if supplies to tissues are to be sufficient to meet demands.

Energy is required to move fluids, and the circulatory systems of animals should show characteristics of efficiency for bulk fluid flow throughout the body. The principle of efficiency in fluid circulation involves minimizing energy expenditures for transport compared with demands dictated by the requirements of tissues. The principles of efficiency for circulation are analogous to those we have discussed for respiratory ventilation; our emphasis has now shifted to problems of internal fluid flow.

There are several ways to view circulatory supply-demand relationships. First, we will examine some general *patterns of circulation* in a variety of animals to see how the structure and function of circulatory systems can be interpreted relative to demands of different animals with regard to circulatory supplies. Second, we will examine some characteristics of the *pumping mechanisms* used in different animals to produce the pressures required for fluid flow. Third, we will examine the characteristics of *fluid flow through tubes* of a circulatory system. Finally, we will examine a set of circulatory and respiratory adaptations evolved by animals to survive periods of lack of oxygen by means of a discussion of the *diving reflex*.

PATTERNS OF CIRCULATION

In general, there is a relationship between differences in demands of animals and differences in circulatory systems that function to provide sufficient fluid flow to meet demands. This relationship can be illustrated at several levels of organization: (1) external versus internal circulation; (2) open versus closed circulation; and (3) differential pulmonary (respiratory) and systemic (other tissues) circulations. Each of these levels involves a change in general circulatory design that results in increased flow (ml/min) to meet increased demands for circulatory transport.

External versus Internal Circulation

The characteristics of diffusion exchange dictate that distances be small between fluids carrying oxygen and the cells where it is used. For small unicellular organisms, diffusion movement of oxygen and nutrients from an external fluid into the organism is sufficient to meet demands, but large cells and multicellular animals must expend energy to produce bulk fluid flow (see Chapter 1). Whether this is viewed as ventilation or circulation is a matter of definition, particularly since the bulk flow of fluid through animals such as coelenterates (see Fig. 1.5) serves a variety of transport functions, including feeding.

Internal circulation of fluids is apparent even in certain unicellular organisms. Streaming movements within the cytoplasm (called "cyclosis") in relatively large cells transports respiratory gases and nutrients throughout the cell. In the flatworms (Fig. 1.5) there is no organized tubular circulatory system, but fluid between cell layers is circulated by movements of the animals. In this case contraction of muscle-like cells produces sufficient bulk fluid flow to meet demands for internal transport, so there is little added expense for circulation per se.

Open versus Closed Circulation

The body cavity of an animal (the coelom) contains a fluid that can be circulated among internal organs. Relatively slow and undirected flow will occur as a result of the movements of an animal or from movements of cilia that line the cavity. However, an increased demand for transport because of increased size and/or activity requires flow to be increased and *directed* to the tissues with the highest demands. The direction of flow is controlled by restricting flow within tubes; flow rate within an animal is increased by pumping fluid through tubes with a heart or hearts.

Among annelid worms (bristleworms, earthworms, and leeches), an organized tubular vascular system is lacking in very small species in which movements of coelomic fluid suffices for internal transport. In larger species the circulatory system consists of a series of vessels and wider spaces (blood sinuses) through which blood is pumped with a series of tubular hearts (Fig. 4.1a). The blood vessels in the earthworm branch to supply digestive organs,

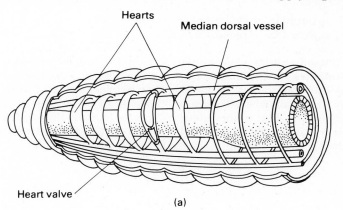

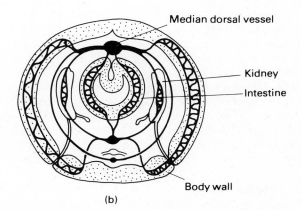

Figure 4.1
The circulatory system of an earthworm is enclosed and has five anterior hearts pumping blood to tissues including the skin, where gas exchange takes place (from Buchsbaum, 1948).

excretory organs, and the skin (where gas exchange takes place), but there is no obvious separation of circulation of blood carrying oxygen to and from the skin (pulmonary circuit) and blood distributed to other tissues (systemic circuit) (Figure 4.1b). The five pairs of hearts of an earthworm are located in the anterior segments, and the hearts and larger blood vessels contain valves that direct the flow from the hearts.

Depending on demands, the circulatory systems in mollusks are either open or closed in a tubular system. Closed circulation produces greater flow rates because pressure produced from contraction of a heart is transmitted throughout the system. The cephalopods (squids and octopods) are the largest and most active mollusks (highest demands), and their circulatory systems are closed.

Part of the circulatory system of an octopus is shown in Fig. 4.2. Contraction of the ventricle forces blood through an aorta to the tissues. Part of the blood enters the separate "gill hearts" (Fig. 4.2), contractile tissue located just prior to the gill circulation. Contraction of the gill hearts forces blood through the small capillaries of the gills and back to the ventricle. Note that this pattern of circulation involves some separation of flow between systemic and pulmonary tissues by the use of two sets of hearts. This pattern is also seen in some vertebrates (see below). In other more sedentary species of mollusks (such as clams), the heart pumps blood to tissues through open blood sinuses, and it is eventually channeled through the gills and back to the heart.

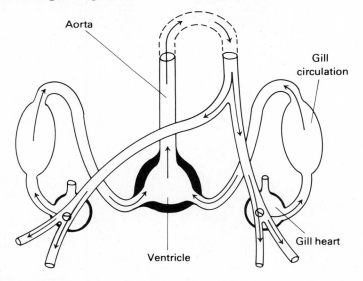

Figure 4.2
The circulatory system of the large, relatively active cephalopod, *Octopus dofleini*, is also enclosed and has two types of heart pumps. The gill hearts pump blood through the gills, and the ventricle pumps blood to the gill hearts and systemic (nonrespiratory) tissues. Arrows indicate directions of blood flow (after Johansen and Martin, 1962).

There is no separate circulatory system in very small arthropods (such as mites) or other invertebrates of very small size. In larger arthropods (with higher demands) sufficient circulation occurs partly as a consequence of movements that occur for other functions. The crustacean fairy shrimp, for example, moves its legs almost continuously for feeding purposes, and fluid is pumped internally from pressure changes in flexible blood sinuses associated with muscles. Movement is also important in pumping blood through the gills of crustaceans such as the lobster (Fig. 4.3a). The dorsal heart pumps blood through arterial channels to tissues, with a relatively large vessel supplying the legs. The arterial blood enters blood sinuses, but blood returning from the legs is channeled into a blood space (Fig. 4.3b) so that movement of the legs forces fluid through the gills back to the heart. This movement provides a mechanism to increase supplies of blood when demands from activity increase.

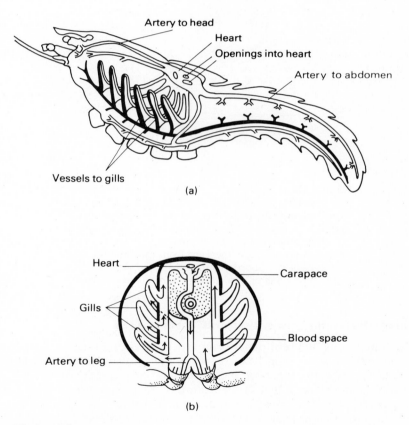

Artery to head

Heart

Openings into heart

Artery to abdomen

Vessels to gills

(a)

Heart

Carapace

Gills

Blood space

Artery to leg

(b)

Figure 4.3
The circulatory system of the crustacean lobster is partially open, with arterial blood entering blood sinuses in the legs. However, movements of the legs forces blood back to the heart through the gills (from Buchsbaum, 1948).

Some of the most active invertebrates include the insects, yet their circulation is open rather than closed (Fig. 4.4). If their demands for oxygen are high, why is the circulatory system not enclosed to provide higher rates of blood flow? The answer is that demands by tissues for oxygen are met via the separate tubular tracheal system (see Chapter 2). The circulatory system serves to transport substances other than respiratory gases (nutrients, waste products, hormones, heat), and transport of these is sufficient at low fluid flow rates. However, increased blood flow is necessary if demands for distribution of these substances increase. During activity, for example, nutrient supplies to tissues must increase, and increased transport of heat from contracting muscles is important (see Chapter 7). In these situations, changes in contraction of the hearts produce increased flow, and movements of the legs or wings also increase blood flow.

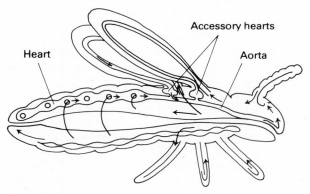

Figure 4.4
The circulatory system of insects is partially open, and oxygen demands are met by the tracheal respiratory system. Transport of metabolites, hormones, and/or heat occurs at relatively low flow rates, and flow rates can increase when demands increase from movements of appendages and/or increases in heart rates. Arrows indicate directions of flow (modified from Wigglesworth, 1972).

Separate Pulmonary and Systemic Circulations

Within and among different major groups of animals, differences in circulatory patterns indicate separation between respiratory and systemic circulations as demands for oxygen transport increase. The circulatory pattern for active species of cephalopods (Fig. 4.2), for example, shows some separation between blood pumped through the gills and blood pumped to other tissues.

Separate circuits provide two important features that influence the efficiency of oxygen transport. First, a separate circulation to a respiratory exchange site maximizes the concentration of oxygen in blood pumped to tissues. Second, separation permits pumping blood in each circuit under separate pressures. Blood pumped to systemic tissues can be forced under higher pressures through a more extensive system of channels than blood pumped through respiratory structures. These general changes in circulatory patterns are illustrated by differences in circulation in the major groups of vertebrates (Fig. 4.5).

The demands of fishes for oxygen are only about 1 percent of the demands of similar sized birds and mammals, and different demands for oxygen are reflected in rates of blood flow to nonrespiratory tissues (Fig. 4.6). In general, normal low rates of flow to systemic tissues in fishes is sufficient to meet demands for oxygen transport, while relatively high demands in mammals require much higher rates of flow.

Major differences in the separation of pulmonary and systemic circulations relative to these different demands are illustrated for the extremes of fishes (a) and mammals (b) in Fig. 4.7. In fishes, blood is pumped from the heart through the gills and then through other tissues before returning to the heart. In mammals, blood is pumped from the right heart through the lungs and back to the left heart for pumping to other tissues under higher pressures and through systemic circuits with lower resistance. Both these factors produce rates of flow in mammals that are higher than in fishes (Fig. 4.5; Fig. 4.7; and see below).

The "intermediate" patterns of circulation exhibited by some amphibians and reptiles (Fig. 4.5) have been the subject of considerable controversy. The controversy has centered on whether blood returning to a heart from a respiratory circuit is *mixed* with blood of lower oxygen content before distribution to both pulmonary and systemic circuits. If mixing occurs, it would indicate less effective oxygen transport (lower oxygen concentrations/ml blood pumped). The anatomy of the hearts for most urodele amphibians indicates that there is a single ventricle and two auricles (Fig. 4.5). On this evidence it has been argued that mixing would occur in the ventricle, and circulation in amphibians and some reptiles (with "incompletely" divided ventricles) used to be described as more "primitive" than circulation in more "advanced" vertebrates.

This hypothesis has been disputed. Results from a variety of experiments where dyes were injected into blood entering the heart from respiratory circuits indicate that most of the blood passes to the systemic circulation. Conversely, most blood returning from the systemic circulation passes to the pulmonary circuit.

Figure 4.5 (next two pages)
The major features of circulatory patterns in vertebrates. "V" refers to ventricle; "At" refers to atrium (auricle). Light areas represent arterial channels, and dark areas represent venous circuits. In fishes, blood is pumped through the gills to systemic tissues (see also Fig. 4.7). In all other vertebrates there is a degree of anatomical separation between pulmonary and systemic circuits. Although amphibians and some reptiles do not show complete anatomical separation within the ventricle, pulmonary and systemic flow still can be separated (modified from Gordon *et al.*, 1972).

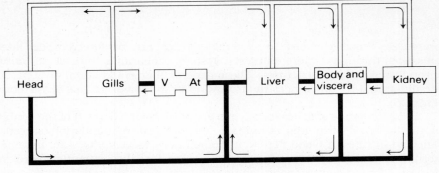

Fish

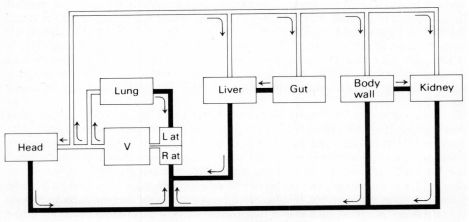

Urodele amphibian

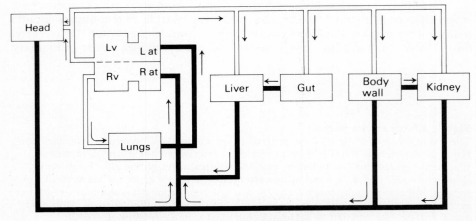

Reptile (noncrocodilian)

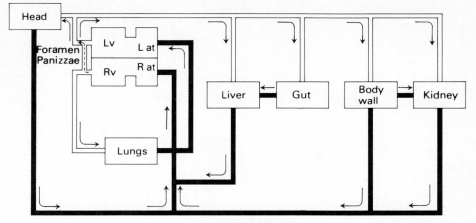

Reptile (crocodilian)

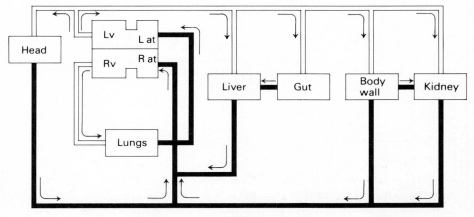

Bird

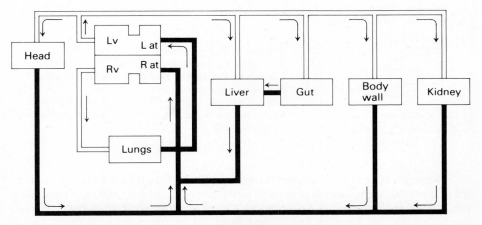

Mammal

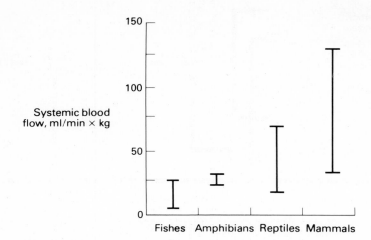

Figure 4.6
The rate of systemic blood flow (indicated by the range of variation among several species) usually increases from fishes to amphibians to reptiles to mammals (modified from Lenfant *et al.*, 1970).

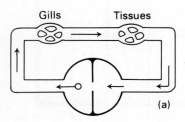

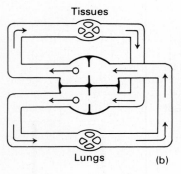

Figure 4.7
Separation of pulmonary and systemic circuits in mammals (b), in contrast to fishes (a), permits higher pressures and lower systemic resistance for higher systemic rates of blood flow (see Fig. 4.6) and a higher degree of oxygen transport to tissues in mammals.

Figure 4.8 shows the anatomy of a frog heart. When blood enters the ventricle from the right auricle it goes to a different part of the ventricle than does blood from the left auricle. In addition, there are local pockets of tissue within the ventricle that serve to keep blood from the two auricles separated. When

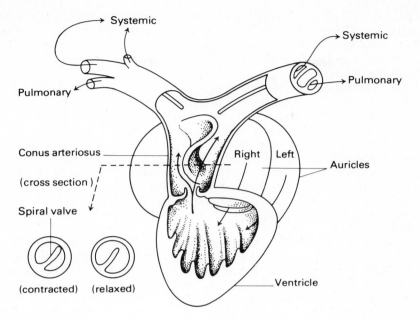

Figure 4.8
Anatomy of the heart of a frog. Blood from the auricles enters the ventricle in different locations and is kept separate in local pockets in the ventricle. The spiral valve within the conus arteriosus serves to produce some separation of pulmonary and systemic blood (modified from Gordon *et al.*, 1972).

ventricular contraction occurs, the blood passes into the first part of the aorta, called the *conus arteriosus*. Within the conus is a fold of tissue called the *spiral valve*. When the conus contracts, the spiral valve produces two channels for blood flow (Fig. 4.8), allowing most of the oxygen-poor blood from the right auricle to move into the pulmonary arteries and most of the oxygen-rich blood from the left auricle to move into the systemic circulation.

A similar interpretation of blood flow has been provided from experiments defining central circulation patterns in reptiles. In noncrocodilian reptiles the ventricular chambers are in anatomical continuity, which led to early hypotheses that oxygen-rich and oxygen-poor blood would mix. However, blood from systemic circuits enters the ventricle and is ejected early during contraction. The resistance to flow is less in the pulmonary circuit during initial ventricular contraction, so most oxygen-poor blood passes to the lungs. Later during ventricular contraction a muscular ridge moves to the wall of the ventricle and serves to separate blood returning from the systemic and pulmonary circuits.

As an alternative to the morphologically based hypothesis that some "lower" vertebrates have more "primitive" patterns of circulation, studies like these for amphibians and reptiles, which were based on actual patterns of central blood flow, support the hypothesis that circulation represents adaptations to differences in demands for oxygen transport relative to supplies. An amphibian or reptile may experience increases or decreases in requirements for circulatory transport, which are reflected in functional characteristics of central circulatory patterns.

One observation that supports this view is the extent to which circulation changes in reptiles as a consequence of the *shunting* of blood flow through different circuits. A circulatory shunt is a movement of blood flow from one channel to another as a consequence of a change in resistance to flow through different channels. If the resistance to flow through one of a series of blood vessels increases, flow through that channel will decrease, and if resistance decreases flow will increase (see below). If demands for oxygen transport to different tissues change, the blood flow to those tissues can change via shunt mechanisms. This type of supply-demand shift in flow has been described in a number of amphibians and reptiles. The crocodile is a particularly well-studied example.

In crocodiles the heart is completely divided into two halves, but there is a connection between the arteries leaving the left and right ventricles (Fig. 4.5) that is called the *foramen Panizzae*. On anatomical evidence alone, this connection is hard to interpret. Does blood from the heart enter both channels? If so, it would suggest mixing of pulmonary and systemic blood. Functional studies indicate that this is not the case; rather, blood is shunted, depending on environmental changes that influence rate of supply of oxygen as compared with demands.

When a crocodile is breathing air, resistance to flow through the pulmonary circuit is low and blood flow from the two ventricles is separated. When it dives (and cannot breathe), resistance to flow through the pulmonary circuit increases and blood from the pulmonary ventricle is shunted to the systemic ventricular arteries. In this way, blood flow to the lungs is reduced when available supplies of oxygen in the lungs are reduced.

In comparing birds and mammals with other animals it is important to note major differences in the demands of animals for oxygen. Birds and mammals are endothermic (see Chapter 7) with high rates of oxygen use, but many other vertebrates are ectothermic (see Chapter 6), with much lower demands for oxygen supplies. These differences are reflected in a number of circulatory functions including systemic blood flow (Fig. 4.6) and total blood oxygen transport ability (see Table 3.1) as well as patterns of central circulation (Fig. 4.5). However, these differences do *not* reflect differences in degree of effectiveness among these animals (e.g., either a more "primitive" or more "advanced" condition) because the characteristics within any group of animals will be effective if supplies are sufficient for demands.

Taxonomic versus Supply-demand Hypotheses

The interpretations of circulatory function based on actual patterns of blood flow illustrate an important point. We will encounter a number of hypotheses in subsequent chapters based on taxonomy, dealing with temperature regulation, hibernation, and nitrogen excretion. We have encountered this viewpoint earlier in the discussion of acid-base regulation in animals with changeable body temperatures (Chapter 2); they used to be thought of as "primitive" (taxonomically less "advanced") because pH was not maintained at a constant value. A taxonomic point of view also forms the basis for the "primitive" versus "advanced" hypotheses for circulatory patterns discussed above for amphibians and reptiles. In essence, the function of an animal is presumed to reflect its degree of "progress" in an evolutionary scheme. Living representatives of less "advanced" species are not expected to be as efficient in function as more "advanced" species.

This is dangerous reasoning because it can keep us from seeking alternative explanations. Moreover, this type of hypothesis is seldom internally consistent in logic. Taxonomically, the crocodile is less "advanced" than birds and mammals, yet its heart shows features of being as advanced in functional design as the hearts of birds and mammals. Furthermore, how is the circulatory pattern in fishes interpreted in a taxonomic framework? They are certainly more "primitive" taxonomically, yet all the blood reaching the systemic circulation is oxygenated, with no anatomical basis for "mixing".

We will find that although taxonomy provides a convenient and important way to classify animals into different groups, it does not as a consequence provide a consistent explanation for function based on some degree of evolutionary development. In studies of circulatory function an alternative approach based on circulatory mechanisms to achieve transport of supplies compared with demands provides ideas testable with experiments for any species. This approach also focuses more attention on the interactions between animals and their environments.

HEART FUNCTIONS

The rate of blood flow (ml/min) to tissues is determined by both the heart, which provides the force to move fluid, and the vessels into which blood is pumped. The heart and blood vessels operate in the circulatory system as an integral unit, and the function of one is influenced by the other. However, for simplicity it is convenient to examine characteristics of cardiac and vascular components separately.

There are a variety of heart types, but all are based on the movement of fluid resulting from muscular contraction. Muscles can contract in peristaltic waves along a tube, or a separate set of muscles may alternately constrict a tube, forcing fluid through it. However, the highest rates of flow compared with the energy expended for contraction occur in heart pumps that are chambered (blind tubes) in which constriction of the chambers expels fluid.

Function of the Mammalian Heart

Most information on chambered hearts comes from studies of mammals, which provide a reference point for viewing heart functions in other animals (see below). Figure 4.9 shows the anatomy of the human heart during a contraction

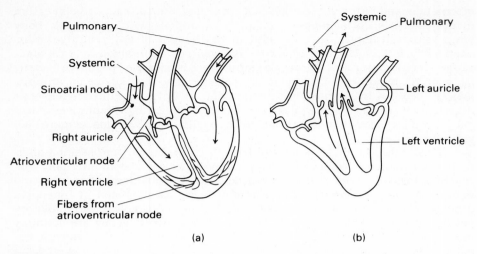

Pulmonary
Systemic
Sinoatrial node
Right auricle
Atrioventricular node
Right ventricle
Fibers from
atrioventricular node

Systemic
Pulmonary
Left auricle
Left ventricle

(a)

(b)

Figure 4.9
Section through the human heart during relaxation (a) and contraction (b). The movement of blood is shown by arrows; the location of the atrioventricular and sinoatrial
nodes are indicated for the relaxed heart (modified from Vander *et al.*, 1975).

cycle. During relaxation (a), blood from the systemic circulation enters the right
atrium (the main chamber of the auricle) from the vena cavae. This blood
passes through the tricuspid atrioventricular valve to the right ventricle. Oxygenated blood from the pulmonary veins enters the left atrium and passes to the
left ventricle through the mitral atrioventricular valve. Contraction of the ventricles (b) forces blood from the right ventricle into the pulmonary arteries and
blood from the left ventricle into the aorta to the systemic circulation, while the
valves prevent flow of blood in other directions.

The contraction of the muscle fibers of the heart must be coordinated for
sufficient pressures to be produced to pump blood, particularly with regard to
ventricular contraction. The muscle cells of the mammalian heart are in close
proximity (see Chapter 18), and a special nervous conducting system within the
heart results in synchrony of cellular contraction. Certain cells of the heart are
"autorhythmic"; that is, they produce activity periodically without any apparent external influence. These are called "pacemaker" cells, since they influence the contraction of other muscle cells of the heart. In humans the
pacemaker cells are located in the right atrial wall near the entrance of the vena
cavae (Fig. 4.9). This collection of cells is the *sinoatrial node*.

Contraction of the atria is initiated from the activity of cells in the
sinoatrial node and spreads across both atria. At the base of the right atrium is
the *atrioventricular node*. Cells in this node are connected with fibers passing
from the atrium throughout each ventricle (Fig. 4.9). This network provides a
rapid conduction pathway for electrical potentials, resulting in simultaneous
contraction of both ventricles (see Chapter 18 for a detailed discussion of the
mechanisms for muscular contraction).

Contraction of the heart results in changes in electrical potentials that can be monitored by placing electrodes on the surface of the skin. Figure 4.10 shows an *electrocardiogram* (ECG) superimposed on a measurement of the force (tension) developed during ventricular contraction. The "P" wave of the ECG corresponds to atrial depolarization and contraction. This is followed in the resting human by a period of about 0.1 second during which activity spreads down the ventricles through the atrioventricular system. The "QRS" complex corresponds to ventricular depolarization or contraction. The "T" wave corresponds to ventricular repolarization or relaxation. The electrical events accompanying atrial relaxation are masked by the QRS complex.

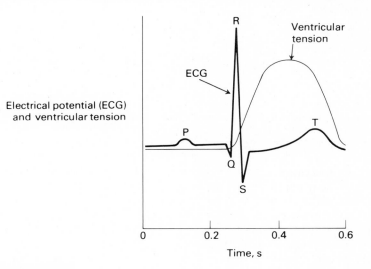

Figure 4.10
The electrocardiogram (ECG) together with measurements of changes in ventricular tension during contraction of the human heart (from Vander *et al.*, 1975).

Mechanical Events of a Contraction Cycle

The period of ventricular contraction is called *systole*, while the period of ventricular relaxation is called *diastole*. Figure 4.11 summarizes the changes in pressure and volume occurring within the left ventricle of a human during a normal (resting) contraction cycle; the events in the right ventricle are similar, although the pressures developed are smaller.

The sequence in Fig. 4.11 starts midway through diastole as blood is entering the ventricle from the left atrium at low pressure (atrial pumping is not necessary for ventricular filling in resting humans). Toward the end of diastole the atria contract (P wave), producing a slight additional increase in ventricular volume and pressure. At the start of systole, ventricular pressure increases

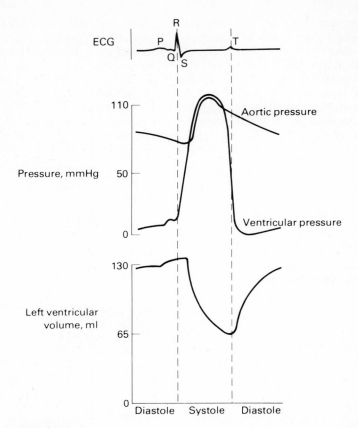

Figure 4.11
Pressure and volume changes in a contraction cycle of the
left side of the human heart. The ECG trace indicates when
electrical events occur. Atrial contraction (P → Q) results
in a slight increase in ventricular volume and pressure.
Ventricular contraction (QRS) increases ventricular
pressure until valves to the aorta open and aortic pressure
begins to increase (from Vander et al., 1975).

dramatically. After a short time, pressure in the ventricle is sufficient to force
open the valves to the aorta, and blood volume decreases in the ventricle while
pressure in the aorta increases. In early diastole, ventricular relaxation produces lower ventricular pressure as the cavity expands, which pulls blood in
from the atrium at an initially rapid rate (Fig. 4.11).

Heart Dimensions, Pressures, and the Law of Laplace

Pressure produces flow of blood from the heart, and the tension, or force/cm,
that must be exerted to produce sufficient pressure for flow is related to the

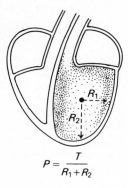

$$P = \frac{T}{R_1 + R_2}$$

Figure 4.12
The ventricle of a heart can be characterized in dimension by two principal radii of curvature (R_1 and R_2). The law of Laplace states that pressure is determined by wall tension divided by the sum of the two radii.

geometry of the heart chambers. This relationship was first stated in another context by the Marquis de Laplace and is illustrated in Fig. 4.12. The dimensions of a ventricle can be characterized by the two principal *radii of curvature,* R_1 and R_2. The pressure (force/cm^2) equals the tension (force/cm) developed in the ventricle wall from muscle contraction, divided by the sum of the two principal radii of curvature (cm).

The work requirements of the heart (energy and oxygen demands) depend on the *tension* developed during contraction to produce sufficient pressure for flow. As R_1 and R_2 decrease during contraction, P will increase for a given tension developed during contraction. The dimensions of the heart (R_1 and R_2) are set by demands for particular pressures; that is, we would not expect to see hearts of inappropriate dimensions with respect to requirements for producing flow pressures.

The importance of dimensions can be illustrated by examining the consequences of changing the dimensions of the heart. For example, if the radii of curvature of a ventricle increase by a factor of 2 (chamber size is doubled), tension must increase by a factor of 4 to produce constancy in pressure. Thus for any given species, heart dimensions should be set by the pressures that are required to produce sufficient flow. This determinant of geometry will minimize energy expenditures required for producing flow; any significant departure from this geometry (e.g., as occurs in heart failure) places excessive demands on the heart to produce required flow.

Changing Cardiac Output in Mammals when Supplies or Demands Change

Note in the example shown in Fig. 4.11 for resting humans that about 70 ml of blood are pumped from the human ventricle during systole in a resting individual. The total quantity of blood pumped by each side of the heart is the same, and the rate of supply (ml/min) from *either* side of the heart is called *cardiac output*. At equilibrium, cardiac output is equal to venous return to the heart through the closed circulatory system. The units of cardiac output are ml/min (flow), and it is analogous to respiratory ventilation, since it is the product of the volume pumped per beat (the *stroke volume*) and the rate at which the heart beats.

| Cardiac output | = | Stroke volume × Heart rate |
| (ml/min) | = | (ml/beat) × (beats/min) |

Cardiac output will change as a consequence of changes in rate of contraction, stroke volume, or both.

The changes observed between species of different weights when demands for oxygen transport change are similar to those that occur in respiratory minute volume (see pp. 55-56 in Chapter 2). For example, recall that total demands in mammals for oxygen increase with increasing species weight (see also Chapter 5). For ventilation, *volumes* are related to weight raised to a power of 1.0, whereas *rates* are related to weight raised to a power of -0.25. Their product (minute volume) is related to weight raised to a power of 0.75, which also describes the relationship between total demands for oxygen and body weight. In mammals, stroke volume is also related to $(weight)^{1.0}$, while heart rate is related to $(weight)^{-0.25}$; their product, or cardiac output, also matches demands that change with weight. Stroke volume, i.e., the size of the ventricle, increases among species as demands increase, while heart rate decreases as demands associated with weight increase.

These relationships between heart dimensions and heart rate match changes in demands for oxygen. Furthermore, there is evidence that heart rate varies among species of different weights in a way that minimizes cardiac work. The velocity of blood leaving the heart (c, cm/s) is the same in species of different weights, and the wavelength of blood pulsations (λ, cm/beat) equals velocity divided by frequency (f, beats/s), or $\lambda = c/f$ (where c is constant among species). Ventricular work at a given frequency is influenced by the ratio of arterial length to pulse wavelength; these values should be matched as arterial dimensions change so that cardiac work is minimized. Because arterial length increases with weight, minimizing cardiac work would involve increasing λ as weight increases. With c constant, an increase in λ would require a decrease in f or a decrease in heart rate as weight increased.

These comparisons include changes observed in cardiac output among mammals of different weights while they are resting (inactive). Demands for blood transport also change within an individual in several circumstances. Increased activity results in increased blood flow to tissues that are using oxygen at higher rates. Increased heat production in tissues requires an increase in blood flow for temperature regulation (see Chapter 7). Assimilation of food from the digestive tract requires a change in cardiac output. For regulation of supplies relative to demands in these situations, cardiac output must be controlled. Controls involve both stroke volumes and heart rates.

Control of Stroke Volume

The volume pumped from a ventricle is the difference between the volume in the ventricle at the end of diastole and the volume at the end of systole (see Fig. 4.11), or

$$\text{Stroke volume} = \left(\begin{matrix}\text{End-diastolic}\\\text{volume}\end{matrix}\right) - \left(\begin{matrix}\text{End-systolic}\\\text{volume}\end{matrix}\right)$$

Increases in stroke volume occur either from increases in end-diastolic volume (increased filling of the ventricle with blood) or from increases in the force of contraction during systole, reducing end-systolic volume.

The importance of end-diastolic volume was first described by the British physiologist Starling and has become known as "Starling's Law of the Heart." As a consequence of increased filling, the muscle fibers of the ventricle are stretched, and it is a characteristic of muscles that the *strength* of a contraction depends proportionally on the degree of muscle stretch (see Chapter 18 for a discussion of this mechanism). Thus increased stretch of ventricular muscles from increased filling during diastole will produce a more forceful contraction and an increased stroke volume. This relationship is illustrated for the heart of a frog in Fig. 4.13. A maximum force of contraction occurs when stretch reaches a certain value; above it the strength of contraction decreases somewhat.

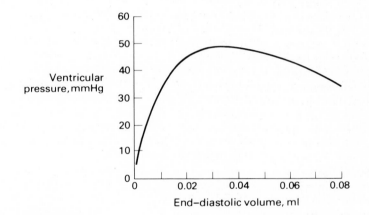

Figure 4.13
Starling's Law of the Heart illustrated for the heart of a frog.
As the volume of the ventricle increases at the end of diastole,
ventricular pressure increases to a maximum (after Doi, 1920).

This control of change in stroke volume is called *intrinsic* control, since it depends on the inherent properties of muscle. It provides a very simple way to control cardiac output proportionally when demands change. For example, increased muscular activity will result in more blood being forced back to the heart from veins in muscles, resulting in increased stretching of ventricular muscles and increased right ventricular stroke volume. This blood passes through the lungs and results in increased filling of the left ventricle, increasing stroke volume in the systemic circulation and providing the muscles with an increased rate of oxygen delivery from increased cardiac output.

The second mechanism to increase stroke volume involves an increase in the "contractility" of heart muscle, or the force of ventricular contraction, which reduces end-systolic volume. In mammals this increase is usually more important for the control of cardiac output than increases in end-diastolic volume. Chemicals released from certain nerves (norepinephrine) or from the adrenal medulla (epinephrine and norepinephrine) influence ventricular muscles to produce a more forceful contraction during activity. These chemicals and others are also involved in the control of cardiac output in mammals by means of effects on the rate of beating of the heart.

Control of Heart Rate

In humans, resting stroke volume is about 70 ml/beat, and increased contraction and venous return can increase this to a maximum of about 140 ml/beat. Any additional increases in cardiac output must occur from increases in the rate of contraction of the heart.

In mammals heart rate is under continuous control of two sets of nerves and one hormone. The nerves influencing heart rate are part of the *autonomic* nervous system governing involuntary functions (see Chapter 17 for a detailed discussion). There are two parts to the autonomic nervous system: the *parasympathetic* and the *sympathetic*. Activity in the parasympathetic nerve fibers to the heart (contained in the vagus nerves) produce a *decrease* in heart rate from release of a chemical called acetylcholine at the sinoatrial node. Activity in the sympathetic nerves produces an *increase* in heart rate (and heart contractility) from release of norepinephrine. These two sets of nerves continuously influence heart rate in mammals because cutting either type changes heart rate even in the resting condition. Whether heart rate increases or decreases from the influence of these nerves depends on the degree to which each set is active at any given time.

The demonstration that parasympathetic nerves release a chemical that slows heart rate resulted in the award of the Nobel Prize to Otto Loewi for an experiment reported in 1921. In his experiment two frog hearts were prepared as shown in Fig. 4.14. Both hearts contracted rhythmically, but when the "donor" heart was stimulated through the vagus nerve it stopped contracting. The only connection between the "donor" and "recipient" hearts was by means of fluid bathing each heart. The "recipient" heart slowed shortly after the "donor" heart stopped contracting. It demonstrated that something released into the fluid from the vagus nerve was responsible for slowing (or stopping) the rate of beating of the hearts. The chemical was later identified as acetylcholine. This experiment not only demonstrated an important aspect of control of heart rate but also provided definitive evidence that nerves influence tissues through the action of chemicals (see Chapter 17).

Another important chemical influences heart rate in mammals. Under some circumstances the adrenal medulla, situated near the kidneys, secretes the hormone epinephrine into the blood. Epinephrine is similar in its effects to the more locally secreted norepinephrine; that is, it has a "sympathetic" effect of increasing heart rate. We will discuss the basis for sympathetic hormones (and

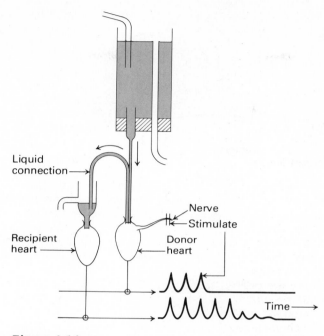

Figure 4.14
Schematic diagram of the Nobel-Prize-winning experiment of
Otto Loewi. The donor heart was stimulated through the
parasympathetic vagus nerve to decrease the rate of beating,
and the chemical released by the nerve (acetylcholine) traveled in solution to the recipient heart, causing it to decrease
its rate of beating shortly afterward (based on Loewi, 1921).

lack of parasympathetic hormones) in Chapter 17. In general, epinephrine is
secreted into blood under conditions of "stress" and serves to maintain increased
cardiac output when there are predictable, continued increased demands for
circulatory transport.

Where are the receptors in this control system for heart rate in mammals,
and how do they interact with the parasympathetic and sympathetic nerves influencing heart rate? The carotid arteries of mammals supply blood to the head.
They branch in the neck, and where they branch the walls of the arteries are
slightly enlarged and contain a large number of nerve fibers. This area is known
as the *carotid sinus*.

The nerves are very sensitive to stretch and provide information on the
magnitude of blood pressure near the heart. Blood pressure is a major determinant of blood flow (see below). Increased blood pressure (increased stretch)
results in a proportional increase in activity of the nerves from the carotid sinus.
These nerves connect to an area of the brain just above the spinal cord called
the medulla oblongata (Fig. 4.15), where they influence the activity of nerves in

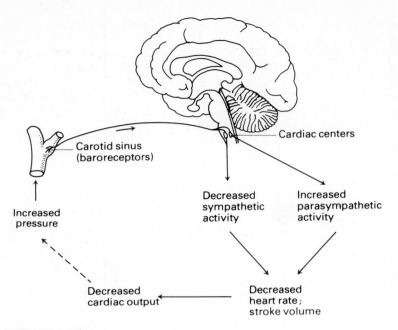

Figure 4.15
Schematic illustration of the negative-feedback control system for
heart rate that will cause a reduction of a detected increase in blood
pressure. Increased pressure is detected at the carotid sinus
baroreceptors, and nerves of the cardiac centers in the medulla
stimulate decreased sympathetic activity and increased parasympathetic activity in nerves of the heart, which decreases heart rate
and cardiac output. The system also operates to raise blood pressure
so that output can be maintained at levels sufficient for different
demands.

the *cardiac centers*. If there is an increase in blood pressure, the parasympathetic nerves to the heart increase in activity while the sympathetic nerves
decrease in activity (Fig. 4.15). Both activity changes slow heart rate. If pressure
at the carotid sinus decreases, the opposite sequence of events occurs; the
negative-feedback control systems act to keep blood pressure within limits, yet
sufficient for supplying demands from appropriate cardiac output. In addition
to pressure receptors (baroreceptors) in the carotid sinus, there is a set of receptors in the bend of the aortic arch that act similarly.

The cardiac centers of mammals are also influenced by other stimuli
(some through the carotid sinus) that result in a change in cardiac output by
means of heart rate changes. Blood P_{O_2}, P_{CO_2} and pH all influence heart rate;
therefore, increased demands in tissues for oxygen result in some increase in
heart rate. Note that these variables are also involved with changing supplies of
oxygen by affecting hemoglobin (see Chapter 3). Changes in these variables also

influence rate of blood flow by affecting fluid passage from the heart through the tubes of the circulatory system (see below).

Figure 4.16 summarizes the contributions of heart rate and stroke volume toward supplying the demands of increased oxygen consumption resulting from exercise in humans. There is an initial increase in stroke volume and an increase near the upper limit of rate of oxygen use. Most of the contribution to increased cardiac output within an individual comes from a fairly linear increase in heart rate as demand increases.

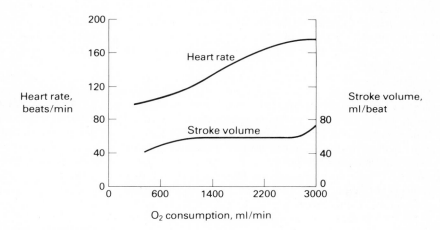

Figure 4.16
Contributions of heart rate and stroke volume that supply the increased demands for oxygen transport during exercise in humans. Note the greater contribution of changes in heart rate (modified from Rushmer, 1960).

Heart Function in Other Animals

Although the principles involved in cardiac output changes are similar in other animals, there are striking differences in the contribution of heart rate and stroke volume to cardiac output and in the nature of the control systems influencing cardiac output. These differences are apparent from studies of heart function in different species of fishes.

The heart of the hagfish is *aneural*—it has no parasympathetic or sympathetic nerve innervations. Thus there is no extrinsic control system for changing cardiac output by changes in heart rate or heart contractility. However, intrinsic control occurs from changes in end-diastolic volume (Starling's Law of the Heart); when activity increases, cardiac output will increase. This control is sufficient to increase cardiac output during activity in hagfish. The hearts of a number of invertebrates are also classified according to the extent to which control is neural or intrinsic. Pupal stages of insects (having low demands for circulatory transport) are characterized by a heartbeat with little or no neural control, but the adults of a number of insect species have neurally controlled hearts.

Other fishes have hearts innervated by a branch of the vagus nerve (parasympathetic), but there appear to be no sympathetic fibers innervating fish hearts. In essence, heart rate is determined only by the degree of inhibition (called "vagal tone"), not inhibition combined with acceleration.

In some species of fishes a decrease in parasympathetic inhibition leads to a considerable increase in heart rate and cardiac output. However, in most fishes cardiac output is determined primarily by changes in stroke volume, shown in Fig. 4.17 for actively swimming trout. Although heart rate increases somewhat, most of the increased cardiac output is due to a dramatic increase in stroke volume. Thus as in hagfish, the characteristics of the heart in fishes are sufficient to provide increased circulatory transport without more elaborate control mechanisms.

Recall that ventilation in fishes involves countercurrent flow of water and blood in the gills (Chapter 1). The opercular and buccal pumps produce ventilation flow in the same direction during most of a ventilation cycle (Fig. 1.9). The contraction of the heart in fishes is synchronized with this ventilation so that a maximum rate of blood flow through the gills will occur when water flow in one direction is maximum. This interaction between ventilation and circulation ensures that diffusion exchange of gases in the gills will be at a maximum, and it is analogous to the matched stratification of blood and air flow in the lungs of air breathers (see Chapter 2).

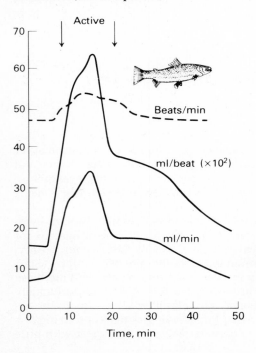

Figure 4.17
Changes in stroke volume (ml/beat) are much more pronounced than changes in heart rate (beats/min) during activity in the rainbow trout (from Randall, 1968).

PRESSURES, FLOW, AND RESISTANCE IN THE TUBULAR VASCULAR SYSTEM

The heart produces the force to produce fluid flow through the tubes of the vascular system. The charcteristics of the tubes influence the rate of flow in various parts of the system and can be described by some simple equations. The equations depend on the *type of flow* within the tubes and the *direction* of the pressure gradient established during flow.

Poiseuille's Equation

Poiseuille developed the equation to describe fluid flow along the length of *rigid* tubes at *low* flow rates:

$$\text{Flow rate (ml/min)} \ = \ \frac{(P_1 - P_2)}{r}, \tag{4.1}$$

where

P_1 = pressure at start of tube,

P_2 = pressure at end of tube, and

r = resistance to flow.

This equation is the hydraulic equivalent of Ohm's law. The determinants of r have been precisely described for low flow rates in rigid tubes:

$$r \ = \ \frac{8\ell\eta}{\pi R^4},$$

where

ℓ = length of a tube,

η = viscosity of fluid (internal resistance to flow), and

R = radius of a tube.

Thus we can write Eq. (4.1) as

$$\text{Flow rate} \ = \ \frac{(P_1 - P_2)\pi R^4}{8\ell\eta}.$$

Within a rigid tube, fluid flow rate is directly proportional to the pressure gradient along the length of the tube, directly proportional to the fourth power of the tube radius, and inversely proportional to the length of the tube and fluid viscosity.

It is common to divide the vascular system into tubes of different diameters and lengths. The relatively large *aorta* from the heart branches into several *arteries* of smaller radius and length that, in turn, branch into several thousand *arterioles* of smaller radius, that branch again into many millions of *capillaries* with the smallest radii and lengths. In the venous system, blood from capillaries passes to *venules* and then to *veins* of increasing radii and lengths.

The fourth power of the radius has an important impact on flow within the changing tube dimensions of the vascular system. A tube that decreases in radius by half exhibits a decrease in flow rate by a factor of 16 (2^4), given a constant length, viscosity, and pressure difference. The most pronounced effect of this factor occurs in the arterioles and venules, which show the greatest change in radius of any of the components of the vascular system.

Figure 4.18 shows the change in flow velocity throughout the vascular system for a dog. Velocity decreases most on the arterial side in arterioles and increases most on the venous side in venules. Velocity is lowest through the capillaries, where diffusion exchange takes place.

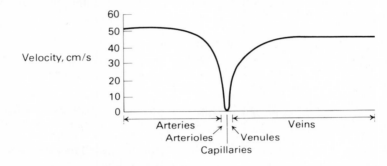

Figure 4.18
Changes in velocity in various parts of the peripheral vascular system (modified from Vander et al., 1975).

Figure 4.19 shows changes in blood pressure that occur in various parts of the vascular system for a standing human at rest. Note that the most pronounced drop in pressure occurs within arterioles, which have a larger decrease in radius relative to their length. Note also that blood pressure in veins may be less than zero. In this case muscles surrounding veins serve as auxiliary "pumps," adding pressure to force blood back to the heart.

Veins contain one-way valves that permit blood flow only in the direction of the heart. You can demonstrate this for yourself. If you make a fist so that the veins on the back of your hand stand out, and if you place your thumb and forefinger on a long segment of vein and press blood toward the heart with your thumb and then remove your thumb, the vein you have compressed will remain compressed; i.e., no blood will flow to the empty region from veins closer to the heart. This experiment was first reported by William Harvey in the seventeenth century. The mechanism for restricting direction of flow is particularly important in veins, where the pressure gradient for flow is low.

Reynolds Equation at High Flow Rate

If the pressure gradient for flow is increased, flow rate will increase linearly until fluid flow becomes *turbulent*. When flow is turbulent, small eddies develop

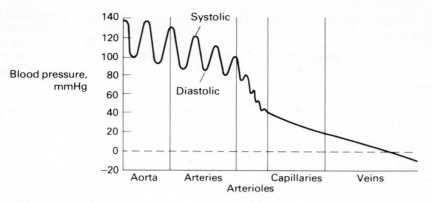

Figure 4.19
Changes in blood pressure that occur in a standing human. Note that the greatest drop in pressure is along the length of the arterioles, and note that pressure within veins can decrease below zero.

within a tube, requiring greater pressures to produce increases in flow rate. At lower flow rates the movement of fluid is *laminar*, i.e., parallel streams within a tube. Poiseuille's equation applies only to laminar flow.

Osborne Reynolds described turbulent flow in tubes. The velocity at which flow *changes* from laminar to turbulent is described by

$$V_t = \frac{K\eta}{\varrho R},$$

where

V_t = transition velocity from laminar to turbulent flow,

η = fluid viscosity,

ϱ = fluid density,

R = tube radius, and

K = Reynolds number (a constant equal to about 1000 for vascular fluids).

Note in this equation that tube radius is in the denominator. Thus the smaller the tube, the higher the transition velocity required from laminar to turbulent flow. Within most of the vascular system blood flow is almost always streamlined. Flow in the large aorta is turbulent during early systole, but other areas experience laminar flow, for which less energy is required.

Laplace Equation for Transmural Pressures

Poiseuille's equation describes flow in terms of pressure differences along the *length* of tubes. Pressure also has an effect across the wall of a tube (transmural), and this is particularly important in blood vessels that are elastic or where the tension in the wall is related to pressure.

A tube has only one radius of curvature, so Laplace's law (see Fig. 4.12) for tubes is

$$P = \frac{T}{R},$$

where

    P = pressure,

    T = tension, and

    R = radius of tube.

A small radius within the vascular system means pressure across the wall will be low at a given tension. Thus thin-walled capillaries can withstand capillary pressure because of their small dimensions. The thickness and elasticity of arteries and veins are partly due to the higher pressures that must be sustained across the walls of these larger tubes.

Laplace's relationship plays a role in differences observed between blood flow in capillaries and blood flow through small rigid tubes. Less force is required to produce flow within capillaries than in small rigid tubes of the same size. This difference can be due to one or both of two factors. One is a change in the dimensions of capillaries. If the radius increases because of the distensibility, or elasticity, caused by a given transmural pressure, R in Poiseuille's equation will increase, resulting in increased flow.

The other factor which can change and influence rate of flow through small capillaries is the viscosity term in Poiseuille's equation. There is evidence that blood can exhibit a lower viscosity in capillaries than in small rigid tubes. As a consequence of capillary dimension changes and/or changes in blood viscosity in capillaries, the energy required for flow is considerably less than in nonvascular tubes.

Changing Local Supplies when Local Demands Change

Not every organ of the body receives the same rate of blood flow nor should it, since demands for circulatory transport vary among tissues. In addition, demands for transport will change differentially among tissues. Figure 4.20 illustrates this for humans. At rest, blood flow to different tissues reflects their relative demands because of differences in the quantity of tissues (e.g., muscles versus skin) or differences in transport functions other than demands for oxygen (e.g., kidney filtration of blood; see Chapter 13). During heavy exercise the blood flow rate to some tissues, such as the contracting muscles, increases dramatically, while rate of flow to other tissues decreases (Fig. 4.20). How is this control over local blood distribution accomplished?

A major variable that influences blood flow is the radius of the arteriole tubes, and it changes as a consequence of either (1) *local* controls within different tissues or (2) *reflex* controls. Both control mechanisms involve contraction or relaxation of smooth muscle cells in the walls of the arterioles, referred to as *vasoconstriction* and *vasodilation*, respectively. This activity leads to localized shunting of blood flow from resistance differences within vascular tubes.

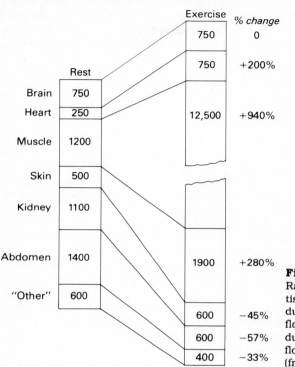

Figure 4.20
Rate of blood flow to different tissues in humans at rest and during exercise. Note that rate of flow to some tissues increases during exercise, while rate of flow to other tissues decreases (from Vander *et al.*, 1975).

Considerable local changes in blood flow occur within a tissue because of local changes in a variety of stimuli associated with demands for oxygen and/or metabolites. These controls operate entirely without nerve or hormonal components. For example, decreased P_{O_2}, increased P_{CO_2}, decreased pH, or increased temperature may all have a direct effect on the smooth muscles of arterioles, resulting in relaxation and leading to increased local rate of flow from increased tubular radius. Changes in these variables in the opposite direction lead to local vasoconstriction within tissues. Thus local negative-feedback mechanisms will result in changes of blood flow rates through tissues as well as in increased or decreased degrees of hemoglobin saturation at the same time (see Chapter 3).

This local control of supply is supplemented or overridden by reflex controls. Reflex controls involve nerves and/or hormones and are particularly important when transport supplies in blood are insufficient for the demands of *all* tissues. Soon we will consider an example when we discuss the "diving" reflex.

Reflex control of arteriolar radius occurs by means of sympathetic nerves and/or hormones. Increases or decreases in activity of sympathetic nerves in arteriole smooth muscle constrict or dilate the vessels, respectively, and epinephrine released from the adrenal medulla results in increased constriction of blood vessels in some tissues. Other hormones, such as angiotensin (see Chapter 13), also cause vasoconstriction. In general, reflex control of local blood flow occurs to maintain supplies of oxygen to tissues particularly sensitive to lack of oxygen (the brain and heart). This supply is obtained by restriction of flow to other tissues that use oxygen but that can also utilize other mechanisms for obtaining energy (see below).

Capillary Exchange

The walls of the capillaries are composed of a single layer of cells, and the distance between the center of a capillary and the cells where oxygen and substrates are used is small, allowing effective exchange to occur by diffusion. Furthermore, the density of capillaries is higher in tissues with higher demands for oxygen, so diffusion exchange will be more rapid in these tissues. This density difference has been studied in the muscles of small and large mammals. Small mammals have higher rates of oxygen use in their tissues, and they also have more capillaries passing through each section of muscle.

In addition to diffusion, some exchange occurs from bulk flow of fluid into and out of capillary tubes. The membranes of the capillaries are permeable to water, salts, and most substrates in solution in blood plasma. When blood first enters the capillaries, blood pressure is sufficiently high to *filter* plasma fluid out of the capillaries. Thus some fluid will flow from capillaries to cells as well as through the capillaries, effectively decreasing distances for diffusion exchange.

Not all of the chemicals in blood plasma filter through the capillary membrane. Relatively large blood plasma proteins stay in solution in the capillaries. They exert a pressure called *colloid osmotic pressure*. It is similar to any other type of osmotic pressure, getting its name from the collective naming of plasma proteins as "colloids." As fluid is filtered from the capillary tube by blood pressure, the plasma proteins become more concentrated in the capillaries. This concentration increases the osmotic pressure, forcing fluid to move back *into* the capillary tube (see Chapter 12 for a detailed discussion of osmotic pressure). Eventually, as blood pressure falls along the length of a capillary, the osmotic pressure acting to draw fluid back into the capillary will exceed blood pressure. Thus on the venous side, bulk flow will occur from the tissues *back* into the capillaries.

The average colloid osmotic pressure is equivalent to about 20 mmHg in mammals. Figure 4.21 summarizes the net effects of interactions between blood pressure and colloid osmotic pressure on bulk fluid exchange both under normal circumstances and when blood pressure along a capillary tube changes because of arteriole constriction or dilation. Depending on the relative placement of capillary tubes within a tissue, blood pressure filtration and colloid osmotic pressure movement can produce local channels of circulation between cells that contribute to some transport to and from cells.

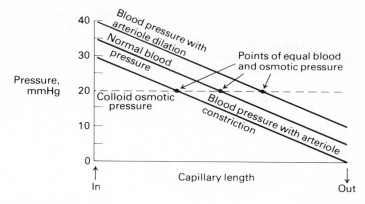

Figure 4.21

Changes in blood pressure along the length of a capillary of three
types: normal (middle line), dilated (right line), or constricted (left
line) arterioles. The "points of equal blood and osmotic pressure"
indicate the transition points between bulk flow *out* (filtration) of
capillaries because of blood pressure (to the left of the points) and
osmotic movement of fluid back *into* the capillaries (to the right of
the points).

ADAPTATIONS TO ASPHYXIA: THE DIVING
REFLEX AND ANAEROBIC VERSUS AEROBIC METABOLISM

At certain times all animals experience periods of low or no oxygen supply from
the external environment. However, in some species it is a common occurrence.
Among mammals and birds, for example, there are a variety of species that
"dive" to obtain food; i.e., they breathe air but frequently enter water, where no
external supply of oxygen is available for respiration. Beavers, muskrats, seals,
manatees, whales, penguins, loons, and ducks are examples. In addition, a
number of species of fishes, such as flyingfish, periodically enter air.

   All of these animals regularly face a problem of *asphyxia* (lack of oxygen
supply for respiration) while their demands for energy to remain active continue. The ability to tolerate asphyxia for various periods involves two major
sets of adaptations: (1) circulatory and respiratory adaptations for redistribution of blood flow to tissues with relatively high demands for oxygen, and
(2) switching from use of oxygen to nonoxidative means for generating energy
from chemicals in certain tissues. These adaptations are present in all animals,
but they are most obvious in those that normally dive for extended periods.

   One way to provide sufficient supplies of oxygen is for diving animals to
evolve respiratory modifications and circulation characteristics so that they can
take more oxygen with them when they dive. There are four possible modifications that can occur: (1) more air could be taken into larger lungs prior to a

dive; (2) the blood could carry more than the normal amounts of oxygen; (3) more oxygen could be associated with myoglobin in tissues; and/or (4) perhaps more oxygen could be dissolved in tissues.

Figure 4.22 compares these possibilities for humans, the sea lion, harbor seal, and the ribbon seal. Note that dissolved oxygen represents a small proportion of total internal oxygen supplies and does not vary between species. On first glance, the amount of oxygen carried in the lungs during a dive does not appear to be consistently higher in the diving animals as compared with humans, but this amount depends on the depth to which a species normally dives.

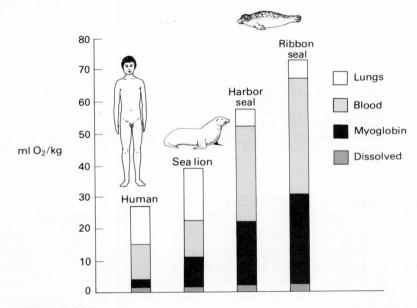

Figure 4.22
The quantity of oxygen stored in different locations in humans and three species of diving mammals. Note that lung oxygen storage in the harbor and ribbon seals reflects expiration prior to deep dives (modified from Hill, 1976).

The sea lion is a shallow diver and has a lung volume that is large for a mammal of its size. This is also the case for the sea otter (see Chapter 2). The harbor seal and ribbon seal dive to greater depths and *expire* prior to a dive; Fig. 4.22 shows a lower quantity of oxygen in their lungs. This reduction is important to prevent the "bends," or caisson disease. For every ten meters of depth, pressure on air in the lungs increases by one atmosphere, increasing the quantity of nitrogen that dissolves in the blood and tissues (see Chapter 1). If the pressure from a deep dive is released rapidly when the diver surfaces, the nitrogen will come out of solution in tissues and form bubbles which can block circulation through capillaries (gas embolisms). Thus animals that normally dive to depths that could present embolism problems decrease lung volumes prior to a dive, whereas shallow divers use increased storage of oxygen in lungs (Fig. 4.22).

The total amount of oxygen stored in blood is the product of carrying capacity (ml O_2/ml blood) and blood volume (total ml blood). Most diving mammals have a relatively high blood carrying capacity. This condition appears to be particularly true for species that are deep divers and that have lower lung storage of oxygen (Fig. 4.22). Carrying capacity for humans is about 18-20 ml O_2/100 ml blood. In the pygmy sperm whale it is 32 ml O_2/100 ml blood; in the harbor seal and ribbon seal the values are 26-29 and 34 ml O_2/100 ml blood, respectively. Many diving species also have a larger blood volume. For seals blood volumes are generally between 130-160 ml O_2/kg body weight, while in humans 60-80 ml O_2/kg is a typical value.

Myoglobin concentrations in muscles of diving birds and mammals are generally quite high as compared with nondiving species. Skeletal muscles in humans contain 4-5 mg of myoglobin per gram wet weight of muscle. Harbor seals have 55 mg/g, and skeletal muscles in ribbon seals contain 80 mg/g. Thus the muscles of diving animals have 10-20 times the supply of intracellular oxygen stores as nondivers (Fig. 4.22).

Despite these adaptations, internal oxygen supplies are still insufficient to meet demands for prolonged dives if rate of oxygen use continues at pre-dive levels. This fact was demonstrated for the hooded seal by Per Scholander. He measured each component of oxygen storage and calculated a total possible internal supply of 1520 ml O_2. The resting rate of oxygen use by seals breathing air was 250 ml/min, so the supply would suffice at this rate of use for a maximum of only 6 minutes if *all* the oxygen was used at the resting rate. The seal remained submerged for up to 18 minutes, leading to the conclusion that modifications must occur in the distribution of oxygen and its use during a dive.

The Diving Reflex

Two major changes in circulation occur within a few seconds of the start of a dive. The most dramatic event is an immediate, pronounced reduction in heart rate. A decrease in heart rate is called *bradycardia* (increased heart rate is *tachycardia*). The effect of diving on the heart rate of a hair seal is shown in Fig. 4.23. Prior to submergence heart rate is about 110 beats/min; within a few seconds of submergence it decreases to about 12 beats/min and remains low until emergence from the dive. Bradycardia is a universal response to asphyxia among vertebrates. It occurs in nondivers but is not as well developed. Moreover, bradycardia is not restricted to animals that breathe air. Figure 4.24 shows the bradycardia developed by a flyingfish when it "dives" into air.

Because stroke volume remains about the same, bradycardia results in a decrease in cardiac output. Cardiac output decreases by about 90% for the hair seal (Fig. 4.23). There is considerable redistribution of blood flow at the same time that heart rate decreases. Blood flow to most peripheral tissues is reduced, but blood flow to the heart and brain is maintained. Figure 4.25 compares the proportions of total blood flow to different tissues of a duck prior to and during a

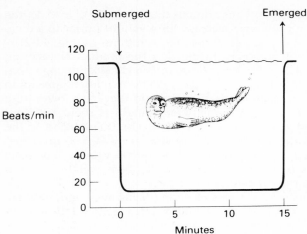

Figure 4.23
Almost immediately after submergence the heart rate of a hair seal decreases and remains low until emergence (modified from Gordon *et al.*, 1972).

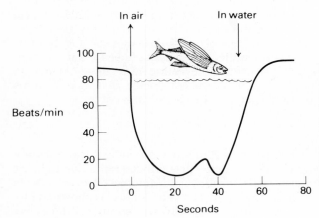

Figure 4.24
A flyingfish that "dives" out of water shows the same characteristic of rapid brady-cardia as air-breathing animals that dive into water.

prolonged dive. Note that blood flow to peripheral tissues such as the legs, intestine, and kidneys is reduced, while more of the temporarily lower cardiac output is distributed to the heart (coronary circulation) during a dive.

This redistribution is accomplished through rapid and sustained peripheral arteriole vasoconstriction. Figure 4.26 shows measurements of blood pressure in an arteriole of the flipper of a hooded seal. Within a few seconds of a dive arteriole constriction reduces arteriole blood pressure until it is close to venous pressure, while central blood pressure at the heart remains at a nearly normal level.

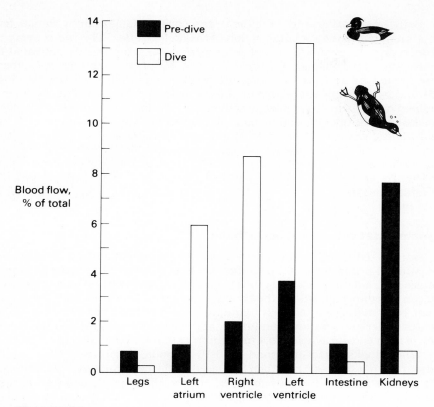

Figure 4.25
During diving by a duck, blood flow is redistributed so that the heart
receives a greater fraction of the total lower cardiac output while the flow to
peripheral tissues is decreased (modified from Johansen, 1964).

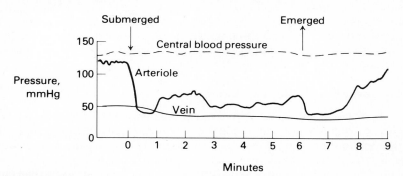

Figure 4.26
Constriction of peripheral arterioles in the flipper of a harbor seal
reduces blood pressure (and flow) during diving while central pressure
(and flow) is maintained.

Both bradycardia and peripheral vasoconstriction occur within a few seconds as a consequence of the action of the *diving reflex*. The vagus nerve to the heart and the sympathetic nerves to peripheral arterioles are activated to produce the rapid changes in heart rate and peripheral vascular resistance. These nerves are activated through the cardiac centers in the brain (see Fig. 4.15), and two types of stimuli are important for the production of the rapid and sustained changes in circulatory function. Stimulation of receptors on the face of ducks near the openings of the nasal passages by drops of water results in an initial slight decrease in heart rate. Within a minute or so of submergence the reflex is maintained and intensified from changes in blood chemistry detected in the carotid sinus. If a harbor seal is forced to dive and the carotid sinus is perfused with oxygen-rich blood from a heart-lung machine, the initial bradycardia is reversed and heart rate returns to pre-dive values, even while the animal is submerged.

Experimental versus Natural Diving

The extreme changes in circulatory function that occur during diving have been studied primarily in animals that are forced to dive. In order to monitor changes in circulatory function it has been necessary to restrain animals so that they can be connected with rather bulky equipment. However, with the recent advent of miniaturized telemetry equipment it has been possible to monitor changes under more natural conditions, and the results suggest that the diving reflex is not necessarily used to a maximum extent under all diving situations.

When ducks dived in a pond for 8-10 seconds, there was a slight initial bradycardia, but the heart rate then increased while the animals were submerged until it was similar to the rate observed prior to the dive. Heart rate was also higher during dives when the ducks were feeding.

These results suggest that earlier experimental studies may have observed only an extreme change, just one of a variety of possible responses to asphyxia. An animal in a natural setting may learn which dives are likely to be associated with varying degrees of asphyxia. Short, shallow dives may not require an extreme response but deep or prolonged dives may. Psychologists have shown that automatic behaviors, such as heart rate, can be conditioned to respond to environmental stimuli. For an animal whose head is pushed under water by a biologist an extreme response might be expected, particularly if the animal had little information about the length of the "dive."

Anaerobic versus Aerobic Metabolism

The redistribution of blood by the activation of the diving reflex conserves internal supplies of oxygen for the demands of those tissues (heart, brain) that cannot function without continuous supplies of oxygen. Other tissues of the body, particularly peripheral muscles, must function without an appreciable oxygen supply from blood during a prolonged dive. For a brief period, muscular contraction will be supported with oxygen from intracellular myoglobin stores, but even this supply will eventually be exhausted, after which metabolism continues by *anaerobic glycolysis*.

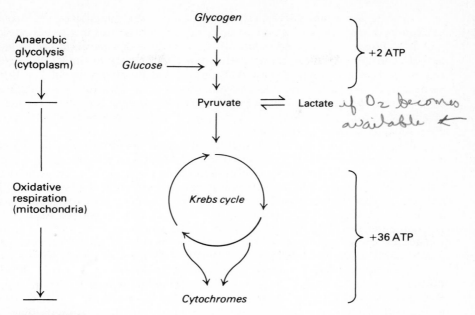

Figure 4.27
Schematic diagram of the pathways for anaerobic glycolysis and aerobic oxidation of glucose. The net quantities of ATP produced are with respect to molar amounts per mole of glucose. The relatively complicated steps occurring within mitochondria during the Krebs cycle take 1.8 times longer to produce ATP from glucose than ATP production takes during anaerobic glycolysis.

Figure 4.27 summarizes the intermediate metabolism of glucose (from glycogen), which produces adenosine triphosphate (ATP), used for the energy required for muscular contraction and other aspects of cellular function. The initial formation of pyruvate from glucose in the cytoplasm yields a net total of 2 molecules of ATP for each molecule of glucose transformed to pyruvate. *If* oxygen is available, pyruvate will enter the Krebs cycle in mitochondria and yield an additional 36 ATP for every initial molecule of glucose. If oxygen is not available, the initial anaerobic glycolysis formation of pyruvate results in lactic acid formation as an end product, with no additional formation of ATP.

This process illustrates the energy efficiency of oxidative metabolism. For each glucose molecule a total of 38 ATP molecules are formed by glycolysis plus mitochondrial oxidation; only 2 ATP molecules are formed by glycolysis alone. Thus we would expect a diving animal or any animal with access to oxygen supplies to utilize oxidative metabolism if and when it is feasible, since the glucose molecules in food generate *eighteen times* more ATP. This fact is the argument given for why a switch to anaerobic metabolism only occurs in animals after oxygen supplies in peripheral tissues have been exhausted.

Despite the lower energy efficiency of anaerobic glycolysis, the process does have a very important advantage, one which is due to the *speed* with which ATP can be produced to supply the demands of contracting muscles in an active animal. The complicated enzymatic machinery of the Krebs cycle is relatively slow if compared with ATP production just from glycolysis. In the same time it takes to produce 36 molecules of ATP from one molecule of glucose by oxidation, 64 molecules of ATP can be formed by glycolysis of 32 molecules of glucose. Thus glycolysis is 1.8 times faster, although it is only 1/18 as efficient for glucose (food) utilization.

Speed is very important to a diving animal (or an animal sprinting to avoid a predator; see Chapter 18) toward the end of a dive. Efficient extraction of energy from glucose is not as important as using a mechanism that will ensure survival by rapidly producing the energy sources to reach the surface and get oxygen. Moreover, once supplies of oxygen have been replenished, some of the lactic acid produced from anaerobic glycolysis can be converted to pyruvate and can enter the Krebs cycle (Fig. 4.27). If this process occurs, a diving animal will achieve some of the best of two possibilities: rapid ATP production from glycolysis, with subsequent delayed energy efficiency from oxidative metabolism.

Switches between aerobic and anaerobic energy production are not limited to diving animals. Asphyxia depends on the extent to which oxygen supplies to tissues match demands for aerobic metabolism, not just on environmental lack of oxygen. Among air- or water-breathing animals there are important differences in the extent to which increased demands for energy (such as during activity) are met by aerobic versus anaerobic processes. In general, anaerobic metabolism carries advantages for rapid energy production but is limited in duration. Aerobic metabolism carries advantages for sustained power production and immediate, effective substrate utilization. Differences between these modes of energy production will become apparent in the next part when we discuss demands for energy.

SUMMARY

Rate of blood flow together with concentrations of substances in blood determine total blood transport through circulation. Fluid circulation is efficient when the cost for producing sufficient flow to meet tissue demands is minimized. This efficiency is accomplished with different patterns of circulation, different pumping mechanisms, and differences in tubular-flow characteristics as demands change relative to supplies among and within animals.

As demands for circulatory transport increase (with increase in size, complexity, and/or activity in animals), circulatory transport patterns change. Internal flow in very small, inactive species occurs in an open system; more active and larger species (i.e., cephalopods, vertebrates) have blood flow restricted to tubes. With increases in demands in some species, separate pulmonary and systemic circuits have evolved with separate pumps for each circuit, which permit the maximizing of oxygen concentrations in blood delivered to tissues and the production of different pressures for flow in each circuit.

Although the hearts of some amphibians and reptiles may not be anatomically divided between pulmonary and systemic circuits, flow of blood to each can still be functionally separate. The pattern of circulation in these animals is not "primitive" with respect to circulation in other vertebrates but is adaptive to a variety of supply-demand changes by means of versatile shunting of flow through different circuits. Hypotheses concerning function that are based on taxonomy ("primitive" versus "advanced" characteristics) are usually logically inconsistent, and they are not necessary to explain observed differences between species.

Hearts produce the pressures to pump fluid through a circulatory system, and the highest flow rates relative to energy expended for pumping are produced by chambered hearts. In mammals, blood from systemic veins enters the right ventricle via the right auricle, and blood from pulmonary veins enters the left ventricle via the left auricle. Pacemaker cells in the right auricle (sinoatrial node) initiate contraction of the auricles, and synchronized contraction spreads to the ventricles via the atrioventricular node.

Blood enters the ventricles during relaxation (diastole), and a stroke volume is pumped during a periodic contraction (systole). The internal dimensions of ventricles (radii of curvature) determine the tension required for producing pressure for flow (law of Laplace). The cardiac output (ml/min) is the product of stroke volume (ml/beat) and heart rate (beats/min). Cardiac output varies between species, depending on differences in demands for oxygen delivery to tissues. Stroke volume (heart size) increases with increases in (body weight)$^{1.0}$, while heart rate decreases with increases in (body weight)$^{-0.25}$; thus cardiac output for resting mammals equals demands for resting rates of oxygen use, which are related to (body weight)$^{0.75}$.

Cardiac output changes in individual mammals by means of changes in stroke volume and (primarily) changes in heart rate. Stroke volume is determined from the difference between end-systolic and end-diastolic ventricular volumes. Stroke volume will increase if increased venous return produces more ventricular stretching (Starling's Law of the Heart) because the force of muscular contraction depends on the degree of stretch. Stroke volume will also increase from increased force of contraction, which reduces end-systolic volume. Heart rate in mammals is controlled by sympathetic (accelerating) and parasympathetic (decelerating) nerves and the sympathetic hormone, epinephrine. Receptors sensitive to pressure (stretch) and/or changes in blood chemistry (P_{O_2}, P_{CO_2}, pH) in the carotid sinuses and aortic arch influence heart rate through a negative-feedback system. The system involves integration of information in cardiac centers in the brain stem such that heart rate (and cardiac output) will change when demands for circulatory transport to tissues change.

Among fishes there are differences in controls over heart function and the contribution of heart rate and stroke volume to cardiac output. The heart of hagfish is aneural, and some variation in cardiac output occurs as a result of intrinsic control through changes in end-diastolic volume. Other fishes are known to have parasympathetic (inhibitory) neural control but no sympathetic (excitatory) control.

The rate of blood flow through the tubular vascular system is directly proportional to the difference in pressure across the system and inversely proportional to vascular peripheral resistance. Resistance to fluid flow through tubes is related directly to the length of the tube and fluid viscosity and inversely related to the fourth power of tube radius (Poiseuille's equation). Blood pressure changes most in arterioles and venules, both of which show the largest change in radius relative to length as compared with other vessels. Blood pressure may be low (or below zero) in the veins, but contraction of muscles serves as an auxiliary pump to force blood to the heart against gravity through one-way valves in veins.

Rate of flow in most of the vascular system is sufficiently low that flow is laminar rather than turbulent. According to the Reynolds equation, the rate of flow for transition from laminar to turbulent flow is inversely proportional to tube radius. The pressures across the walls of tubes (with a given tension) are inversely proportional to tube radius (law of Laplace). Thus larger vessels with thicker walls are required for higher pressures. Changes in tube radius from transmural pressure (pressure across a tube wall) can influence the rate of flow through capillaries as compared with the rate of flow through rigid tubes.

Vasoconstriction and vasodilation (contraction and relaxation of arteriole smooth muscle) will change local blood flow rates by means of changes in tube radius. Blood flow within a tissue will change, as demands change, from the direct effects of P_{O_2}, P_{CO_2}, pH, or temperature on arteriole smooth muscle. These local controls may be overridden by nerves and/or hormone-mediated reflex controls if internal supplies of oxygen in blood are insufficient to satisfy the demands of all tissues of the body.

Pressure changes within capillaries result in some bulk flow of fluid from the vascular system to cells and vice versa. Blood pressure forces fluid through the thin capillaries on the arteriole side, and osmotic pressure from unfiltered dissolved proteins in blood draws water back on the venule side. Changes in blood pressure from arteriole radius changes modify the extent of capillary fluid exchange. These changes, together with the density of capillaries, determine supplies of materials by means of diffusion exchange between the vascular system and cells.

Animals that dive experience relatively long periods of asphyxia, i.e., lack of environmental supply of oxygen. Some modifications of blood-transport characteristics that operate together with the diving reflex prolong the time for survival without access to oxygen. Shallow-diving species have larger lung volumes, which allow greater quantities of oxygen to be available during a dive. However, deep-diving species exhale prior to a dive to minimize problems with nitrogen that enters solution under pressure and forms bubbles (comes out of solution) during surface decompression (the "bends"). Moreover, the blood of divers usually has a higher total oxygen carrying capacity (ml O_2/ml blood) and is present in higher volumes, and many diving species have much larger concentrations of intracellular myoglobin in muscle.

Despite these adaptations, demands for energy during a long dive (or during activity in many species) can exceed the ability to provide energy entirely from oxidative metabolism. Two adjustments mediate this problem. First, the diving reflex results in redistribution of blood flow mostly to tissues with the highest demands for a continuous supply of oxygen. It happens by means of a drastic lowering of heart rate (bradycardia) together with peripheral vasocon- striction, so that reduced cardiac output is distributed primarily to the heart and brain. Second, peripheral tissues such as muscles generate ATP for energy- demanding muscle contraction from anaerobic glycolysis after intracellular oxygen supplies have been exhausted.

Although aerobic oxidation of glucose through the Krebs cycle is a more energy-efficient use of glucose (yielding 18 times more ATP per molecule of glucose), cytoplasmic glycolysis is more rapid (yielding ATP 1.8 times faster); thus the high energy demand of some tissues can be met during asphyxia. In addition, some lactic acid produced from anaerobic glycolysis may enter the aerobic Krebs cycle at a later time when oxygen is available and produce some (delayed) energy-efficient use of glucose substrates.

ANNOTATED REFERENCES

Chapter 1: Aquatic Respiration

Brown, C.E., and **B.S. Muir** (1970). Analysis of ram ventilation of fish gills with application to skipjack tuna (*Katsuwonus pelamis*). *Jour. Fish. Res. Bd. Canada* 27:1637-1652. (Examination of mouth-open ventilation of gills.)

Cameron, J.N., and **C.V. Batterton** (1978). Temperature and blood acid-base status in the blue crab, *Callinectes sapidus. Resp. Physiol.* 35:101-110. (Mechanisms of regulation of acid-base status in an aquatic ectotherm.)

Dejours, P. (1975). *Principles of Comparative Respiratory Physiology.* North Holland Pubs., Amsterdam. (Detailed accounts of all aspects of respiration.)

Grey, I.E. (1954). Comparative study of the gill area of marine fishes. *Biol. Bull.* 107:219-225. (Relationship between gill gas-exchange surface area and oxygen demands for different species.)

Hazelhoff, E.H., and **H.H. Evenhuis** (1952). Importance of the countercurrent principle for the oxygen uptake in fishes. *Nature* 169:77. (Important ex- perimental demonstration of the relative efficiency of countercurrent ver- sus concurrent flow for gas exchange in gills.)

Hughes, G.M. (1964). *Comparative Physiology of Vertebrate Respiration.* Harvard University Press, Cambridge. (Brief introduction to comparative physiology of gas exchange.)

Hughes, G.M., and **M. Morgan** (1973). The structure of fish gills in relation to their respiratory function. *Biol. Rev.* 48:419-475.

Hughes, G.M., and **G. Shelton** (1958). The mechanism of gill ventilation in three freshwater teleosts. *Jour. Exp. Biol.* 35:807-823. (Determination of water flow patterns from measurements of pressure gradients across gills.)

Hutchison, V.H., H.B. Haines, and **G. Engbretson** (1976). Aquatic life at high altitude: Respiratory adaptations in the Lake Titicaca frog, *Telmatobius culeus. Resp. Physiol.* 27:115-129. (Examination of a variety of respiratory and circulatory adaptations, including the importance of gas exchange through the skin.)

Johansen, K. (1971). Comparative physiology: Gas exchange and circulation in fishes. *Ann. Rev. Physiol.* 33:569-612. (Review of a great deal of information published in journals.)

Radford, E.P. (1964). The physics of gases. In W.O. Fenn and H. Rahn (eds.), *Handbook of Physiology*, Section 3, Vol. 1. American Physiological Society, Washington, D.C. (Application of gas laws to respiratory gas exchange.)

Rahn, H. (1967). Aquatic gas exchange theory. In N.B. Slonin and J.R. Chapin (eds.), *Respiration Physiology*, Vol. 1. C.V. Mosby, St. Louis. (Application of gas laws to aquatic situations.)

Reeves, R.B. (1977). The interaction of body temperature and acid-base balance in ectothermic vertebrates. *Ann. Rev. Physiol.* 39:559-586. (Review of information about temperature effects on pH for aquatic and air-breathing animals having changeable temperatures.)

Chapter 2: Air Respiration

Bretz, W.L., and **K. Schmidt-Nielsen** (1971). Bird respiration: Flow patterns in the duck lung. *Jour. Exp. Biol.* 54:103-118. (Description of air movement through the avian respiratory system.)

Davies, D.G. (1978). Temperature-induced changes in blood acid-base status in the alligator, *Alligator mississippiensis. Jour. Appl. Physiol.* 45:922-926. (Information on change in ventilation relative to metabolic rate for acid-base regulation when body temperature changes.)

Dejours, P. (1962). Chemoreflexes in breathing. *Physiol. Rev.* 42:335-358. (Negative-feedback mechanisms controlling ventilation that involve changes in blood chemistry.)

Donnelly, P.M., and **A.J. Woolcock** (1978). Stratification of inspired air in the elongated lungs of the carpet python, *Morelia spilotes variegata. Resp. Physiol.* 35:301-315. (Study of the matching between lung air and lung blood perfusion.)

Duncker, H.R. (1972). Structure of avian lungs. *Resp. Physiol.* 14:44-63. (Examination of the anatomy of air passages within the lungs of birds.)

Farber, J.P., and **G.N. Bedell** (1973). Responsiveness of breathing control centers to CO_2 and neurogenic stimuli. *Resp. Physiol.* 19:88-95. (Changes in ventilation from effects on inspiratory and expiratory centers in the brain stem.)

Gans, C., and **B.D. Clark** (1978). Air flow in reptilian ventilation. *Comp. Biochem. Physiol.* 60:453-457. (Analysis of inward and outward air flow patterns in respiratory cycles of reptiles.)

Gans, C., H.J. deJongh, and **J. Farber** (1969). Bullfrog (*Rana catesbeiana*) ventilation: How does the frog breathe? *Science* 163:1223-1225. (Description of positive pump ventilation.)

Hitzig, B.M., and **D.C. Jackson** (1978). Central chemical control of ventilation in the unanesthetized turtle. *Amer. Jour. Physiol.* 235:R257-R264. (Dissociation of the effects of P_{CO_2} versus pH on respiratory ventilation control.)

Howell, B.J., F.W. Baumgardner, K. Bandi, and **H. Rahn** (1970). Acid-base balance in cold-blooded vertebrates as a function of body temperature. *Amer. Jour. Physiol.* 218:600-606. (Examination of regulation of acid-base status with changing temperatures with regard to frogs, toads, and turtles.)

Johansen, K. (1968). Air-breathing fishes. *Sci. Amer.* 219:102-111. (Examination of a variety of mechanisms for gas exchange with air by fishes.)

Leith, D.E. (1976). Comparative mammalian respiratory mechanics. *Physiologist* 19:485-510. (Emphasis on differences in lung ventilation mechanisms among mammals, particularly diving species.)

Lenfant, C., K. Johansen, and **D. Hanson** (1970). Bimodal gas exchange and ventilation-perfusion relationship in lower vertebrates. *Fed. Proc.* (U.S.) 29:1124-1129. (Review of gas-exchange relationships for air-and-water breathers.)

Levy, R.I., and **H.A. Schneiderman** (1966a). Discontinuous respiration in insects III. The effect of temperature and ambient oxygen tension on the gaseous composition of the tracheal system of silkworm pupae. *Jour. Insect Physiol.* 12:105-121. (Information about effects of temperature on the ventilation mechanism of insects.)

Levy, R.I., and **H.A. Schneiderman** (1966b). Discontinuous respiration in insects IV. Changes in intratracheal pressure during the respiratory cycle of silkworm pupae. *Jour. Insect Physiol.* 12:465-492. (Synthesis of events that produce discontinuous ventilation of trachea.)

Malan, A. (1978). Intracellular acid-base state at a variable temperature in air-breathing vertebrates and its representation. *Resp. Physiol.* 33:115-119. (Examination of regulation of protein net negative charge with changes in temperature.)

McMahon, B.R. (1970). The relative efficiency of gaseous exchange across the lungs and gills of an African lungfish, *Protopterus aethiopicus. Jour. Exp. Biol.* 52:1-15. (Extent of gas exchange via lungs and gills with supplies from air, water, or both.)

Piiper, J., and **P. Schied** (1973). Gas exchange in avian lungs: Models and experimental evidence. In L. Bolis, K. Schmidt-Nielsen, and S.H.P. Maddrell (eds.), *Comparative Physiology: Locomotion, Respiration, Transport, and Blood.* American Elsevier, New York. (Discussion of uniform-pool, countercurrent, and crosscurrent models of gas exchange.)

Stahl, W.R. (1967). Scaling of respiratory variables in mammals. *Jour. Appl. Physiol.* 22:453-460. (Discussion of a series of power function equations relating respiratory variables to body size.)

Von Euler, C., F. Herreo, and **I. Wexler** (1970). Control mechanisms determining rate and depth of respiratory movements. *Resp. Physiol.* 10: 93-108.

Whitford, W.G., and **V.H. Hutchison** (1963). Cutaneous and pulmonary gas exchange in the spotted salamander, *Ambystoma maculatum. Biol. Bull.* 124:344-354. (Changes in gas exchange across lungs and skin as supplies and demands change.)

Chapter 3: Blood Gas Transport

Benesch, R., R.E. Benesch, and **C.I. Yu** (1968). Reciprocal binding of oxygen and diphosphoglycerate by human hemoglobin. *Proc. Natl. Acad. Sci.* (U.S.) 59:526-532. (Mechanism of the effect of DPG on hemoglobin-oxygen association.)

Bland, D.K., and **R.A.B. Holland** (1977). Oxygen affinity and 2, 3-diphosphoglycerate in blood of Australian marsupials of differing body size. *Resp. Physiol.* 31:279-290. (Oxygen affinity of hemoglobin decreases with decrease in body size from changes in DPG concentrations.)

Chiodi, H. (1971). Comparative study of the blood gas transport in high-altitude and sea-level Camelidae and goats. *Resp. Physiol.* 11:84-93. (Left shift in llamas and vicuñas as compared with closely related species from sea-level habitats.)

Grigg, G.C. (1969). Temperature-induced changes in the oxygen equilibrium curve of the blood of the brown bullhead, *Ictalurus nebulosus. Comp. Biochem. Physiol.* 28:1202-1223. ("Fine-tuned" shifts in dissociation curves that result from changing demands associated with temperature acclimation.)

Hall, F.G., and **F.H. McCutcheon** (1938). The affinity of hemoglobin for oxygen in marine fishes. *Jour. Cell. Comp. Physiol.* 11:205-212. (Right position of dissociation curves in active fishes and left position of curves in fishes from hypoxic environments.)

Hazard, E.S. III, and **V.H. Hutchison** (1978). Ontogenic changes in erythrocytic organic phosphates in the bullfrog, *Rana catesbeiana. Jour. Exp. Zool.* 206:109-118. (Air respiration in adults is accompanied by lower organic phosphate concentrations as well as a metamorphosis to adult hemoglobin types.)

Hill, R. (1936) The oxygen dissociation curve of muscle myoglobin. *Proc. Roy. Soc.* (London) B120:472-483. (Extreme left shift in myoglobin dissociation curve as compared with hemoglobin.)

Laver, M.B., E. Jackson, M. Scherperel, C. Tung, W. Tung, and **E.P. Radford** (1977). Hemoglobin-O_2 affinity regulation: DPG, monovalent ions, and hemoglobin concentration. *Jour. Appl. Physiol.* 43:632-642. (DPG-related shifts in the positions of oxygen dissociation curves.)

Lenfant, C., P. Ways, C. Aucutt, and **J. Cruz** (1969). Effect of chronic hypoxic hypoxia on the O_2-Hb dissociation curve and respiratory gas transport in man. *Resp. Physiol.* 7:7-29. (Effects of either high altitude or low environmental P_{O_2} on oxygen transport in humans.)

Magid, E. (1967). Activity of carbonic anhydrase in mammalian blood in relation to body size. *Comp. Biochem. Physiol.* 21:357-360. (Higher enzyme activity for carbonic anhydrase yields more carbonic acid formation in smaller mammals.)

Powers, D.A., H.J. Fyhn, U.E.H. Fyhn, J.P. Martin, R.L. Garlick, and **S.C. Wood** (1979). A comparative study of the oxygen equilibria of blood from 40 genera of Amazonian fishes. *Comp. Biochem. Physiol.* 62:67-85. (Relationship between oxygen affinity of blood and environmental supplies relative to demands in a large number of fishes.)

Riggs, A. (1951). The metamorphosis of hemoglobin in the bullfrog. *Jour. Gen. Physiol.* 35:23-44. (Right shift in the dissociation curve of adults is compared with the curve of tadpoles.)

Schmidt-Nielsen, K., and **J.L. Larimer** (1958). Oxygen dissociation curves of mammalian blood in relation to body size. *Amer. Jour. Physiol.* 195: 424-428. (Right shift, i.e., lower affinity of blood for oxygen, with a decrease in body size.)

Snyder, G.K. (1977). Blood corpuscles and blood hemoglobins: A possible example of coevolution. *Science* 195:412-413. (Optimal hemoglobin transport depends on the molecular weight of hemoglobin and has implications for the evolution of red blood cells.)

Surgenor, D.M. (1964). Transport of oxygen and carbon dioxide. In C. Bishop and D.M. Surgenor (eds.), *The Red Blood Cell*. Academic Press, New York. (Discussion of a variety of mechanisms of transport.)

Tyuma, I., and **K. Shimizu** (1970). Effect of organic phosphates on the difference in oxygen affinity between fetal and adult hemoglobin. *Fed. Proc.* 29:1112-1114. (Effects of hemoglobin type plus DPG concentrations on positions of adult and fetal oxygen dissociation curves.)

Weber, R.E., and **G. Lykkeboe** (1978). Respiratory adaptations in carp blood: Influences of hypoxia, red cell organic phosphates, divalent cations and CO_2 on hemoglobin-oxygen affinity. *Jour. Comp. Physiol.* 128:127-137. (Variation in the position of dissociation curves for a fish with differing environmental supplies, and the mechanism for the shifts.)

Chapter 4: Circulation

Berger, P.J., and **N. Heisler** (1977). Estimation of shunting, systemic and pulmonary output of the heart, and regional blood flow distribution in unanesthetized lizards (*Varanus exanthematicus*) by injection of radioactively labeled microspheres. *Jour. Exp. Biol.* 71:111-121. (Examination of the diversity of circulation types achieved by shunting mechanisms.)

Burton, A.C. (1972). *Physiology and Biophysics of the Circulation: An Introductory Text*. Year Book Medical Pubs., Chicago. (Emphasis on human circulation, with application of various equations involved in tubular fluid flow.)

Butler, P.J., and **A.J. Woakes** (1979). Changes in heart rate and respiratory frequency during natural behaviour of ducks, with particular reference to diving. *Jour. Exp. Biol.* 79:283-300.

de Burgh Daly, M., R. Elsner, and **J.E. Angell-James** (1977). Cardio-respiratory control by carotid chemoreceptors during experimental dives in the seal. *Amer. Jour. Physiol.* 232:H508-H516. (Experimental manipulation of receptors during a dive by perfusion of the circulatory system with blood from a heart-lung machine.)

Doi, Y. (1920). Studies on muscular contraction. *Jour. Physiol.* (London) 54: 218-226. (Starling's Law of the Heart for frog ventricular cardiac output.)

Folklow, B. (1955). Nervous control of blood vessels. *Physiol. Rev.* 35:629-663. (Review of information on reflex control of blood vessel dimensions.)

Johansen, K. (1964). Regional distribution of circulatory blood during submersion asphyxia in the duck. *Acta Physiol. Scand.* 62:1-9. (Description of changes in distribution of blood flow during a dive.)

Johansen, K., and **D. Hanson** (1968). Functional anatomy of the hearts of lung-fishes and amphibians. *Amer. Zool.* 8:191-210. (Emphasis on adaptive shifts in circulatory patterns.)

Johansen, K., C. Lenfant, and **D. Hanson** (1970). Phylogenetic development of pulmonary circulation. *Fed. Proc.* 29:1135-1140. (Despite the taxonomic orientation, a variety of differences are discussed concerning systemic and pulmonary circulations among vertebrates that reflect changes in demands for oxygen.)

Johansen, K., and **A.W. Martin** (1962). Circulation in the cephalopod, *Octopus dofleini. Comp. Biochem. Physiol.* 5:161-176. (Description of the closed circulation in this relatively active invertebrate.)

Johansen, K., and **A.W. Martin** (1965). Comparative aspects of cardiovascular function in vertebrates. In W.F. Hamilton and P. Dow (eds.), *Handbook of Physiology,* Section 2, Vol. 3, *Circulation.* American Physiological Society, Washington, D.C. (Detailed review of circulatory mechanisms and controls in a variety of vertebrates.)

Martin, A.W. (1974). Circulation in invertebrates. *Ann. Rev. Physiol.* 36:171-186. (Review of patterns of circulation in a diversity of arthropod species, including insects.)

Paulev, P. (1965). Decompression sickness following repeated breath-holding dives. *Jour. Appl. Physiol.* 20:1028-1031. (The problem of the "bends" is related to nitrogen solubility at high pressures during a dive.)

Randall, D.J. (1968). Functional morphology of the heart of fishes. *Amer. Zool.* 8:179-189. (Relative roles of heart rate and stroke volume for cardiac output in fishes.)

Rushmer, R.F. (1960). Control of cardiac output. In T.C. Ruch and H.D. Patton (eds.), *Physiology and Biophysics,* 18th ed. Saunders, Philadelphia. (Emphasis on heart rate and stroke-volume determinants in humans.)

Satchell, G.H. (1970). A functional appraisal of the fish heart. *Fed. Proc.* 29:1120-1123. (Pressure and flow characteristics during a heart contraction cycle in sharks.)

Scholander, P.F. (1940). Experimental investigations on the respiratory function in diving mammals and birds. *Hvalradets Skrifter* No. 22:1-131. (Lengthy but classic study of oxygen use in diving mammals.)

Scholander, P.F. (1963). The master switch of life. *Sci. Amer.* 209:92-107. (General description of the diving reflex.)

Tazawa, H., M. Mochizuki, and **J. Piiper** (1979). Respiratory gas transport by the incompletely separated double circulation in the bullfrog, *Rana catesbeiana. Resp. Physiol.* 36:77-95. (Demonstration of functional separation of pulmonary and systemic circulations using measurements of blood chemistry.)

White, F.N. (1968). Functional anatomy of the heart of reptiles. *Amer. Zool.* 8:211-220. (Basis for functional separation of systemic and pulmonary blood flow in reptiles.)

Zweifach, B.W., and **M. Intaglietta** (1966). Fluid exchange across the blood capillary interface. *Fed. Proc.* 25:1784-1788. (Blood pressure- and osmotic pressure-related movements of fluid into and out of capillaries.)

Demands for Energy

Part II

Energy is crucial for survival and reproductive success. It is used for respiration, circulation, digestion, excretion, growth, movements, and reproduction. The efficiency with which energy is used forms the basis on which some individuals succeed and survive to reproduce and others fail and are selected against.

The principle of economically efficient use of food energy was important in Charles Darwin's formulation of the theory of evolution through natural selection. Food (energy) as a resource may be limited in availability to many animals. If an individual obtains only a limited amount because of competition with others, it will be more likely to survive if its rate of energy use is as low as possible. The level of minimum energy expenditure depends on characteristics evolved through the interaction of animals and their environments (see below). If demands for energy exceed supplies from food over sufficient time intervals, an individual will starve. Those individuals that have evolved characteristics that minimize rates of energy expenditure relative to supplies (within certain limits) are more likely to survive and produce offspring with similar adaptations.

The chapters in this part of the book differ somewhat from most of the others. Here we will be concerned primarily with determinants of energy expenditure rates, i.e., *demands* for energy. The next part of the book will deal with energy regulation and control, that is, the mechanisms animals have evolved to match energy supplies with demands by means of feeding.

There are two reasons for this approach. First, biologists have produced a great deal of information on the demands of animals for energy independently of the effect of those demands on energy supplies through feeding. To examine energy as a resource with any degree of completeness, we must treat energy demands independently of supplies.

The second reason is related to the first. One of the reasons demands for energy are measured independently of energy supplies is because the principle that determines efficiency of energy use depends on the *difference* between energy intake (from food supplies) and energy use (from demands). Within certain limits, the greater the difference (intakes − expenditures), the more energy an animal will have to ensure its survival and reproductive success.

In many cases the difference between intake and expenditures of energy is maximized simply by *minimizing the demands*. This condition is particularly true when an animal is resting after obtaining a certain amount of energy from feeding. Whatever amount of food an animal has been able to obtain will be most effectively utilized when the rate of energy expenditure is the least possible while it is resting and not feeding. The extent to which rate of energy expenditure is minimized for a resting animal depends on its size (Chapter 5), its temperature (Chapters 6-8), and a variety of other characteristics found in different animals and their environments that influence energy exchange.

Animals that are active (Chapter 9) expend energy at higher rates for different reasons. One of the functions of activity is to cover distance. Again, if an animal has a certain amount of energy it will use it most effectively in this situation if it travels a distance with a *minimum possible expenditure of energy*. In Chapter 9 we will examine the characteristics of movements involved in running, flying, and swimming to determine how and if animals minimize expenditures of energy for these types of movement.

Many animals are also active when they feed. In this situation they should not necessarily minimize energy expenditures, but they should minimize expenditures *relative to* energy intakes so that the difference between energy intakes and expenditures is the maximum possible for a given environmental condition. This principle forms the basis for part of our discussion of feeding in the third part of the book.

Energy Use At Minimum Rates

5

The rate at which an animal expends energy can vary considerably. When an animal is inactive and exposed to a particular set of environmental conditions, its rate of energy use will be the lowest possible for the set of characteristics it has evolved. In general, this occurs when it has not fed for some period of time (it is not assimilating food, which requires energy expenditure) and is exposed to a particular phase of a photoperiod while it is inactive (see below).

Temperature also influences the rate of energy expenditure, which we will examine in detail in the next three chapters. For birds and mammals (endotherms) there will be a particular environmental temperature where energy demands for body temperature regulation are minimum (sometimes called the *basal* metabolic rate). However, other animals (ectotherms) have rates of energy expenditure that change continuously with temperature; it is not possible for them to experience an environment that absolutely minimizes rates of energy expenditure. Nevertheless, it is possible to *standardize* the environmental conditions influencing metabolic rates so that comparisons can be made between species. For example, a variety of animals can be exposed to the same temperature under conditions of photoperiod phase and food availability that will minimize added energy expenditures.

The metabolic rates that are compared under these conditions are called *standard metabolic rates*. Comparison of standard metabolic rates permits us to examine major determinants of demands for energy independently of major environmental factors that also influence demands for energy. One determinant is the size of animals, and we will examine comparisons of standard metabolic rates for animals of different sizes after a discussion of the ways to measure energy and rates of energy expenditure.

MEASURING ENERGY

Energy exists in several forms (thermal, electrical, chemical, mechanical, etc.) that are interconvertible; that is, one form can be transformed to another. However, only the conversion of other energy forms to heat is complete, a consequence of the laws of thermodynamics. An engine (such as your body) that consumes fuel to do work is never completely efficient. The energy that does not appear as work is dissipated as heat. Moreover, the work that is done is ultimately dissipated as heat. For example, the turning of a generator arm converts mechanical energy to electrical energy, but part of the mechanical energy occurs as heat (produced mostly by friction) rather than electricity. The amount of heat energy produced relative to the electricity produced determines the physical efficiency of conversion from one energy form to the other. In addition, the electricity produced is ultimately dissipated as heat.

Because all forms of energy are ultimately converted to heat, energy is measured with the unit of thermal energy, the *calorie*. A calorie (cal) is defined as the amount of heat energy needed to raise the temperature of one gram of water from 14.5 to 15.5 °C at an atmospheric pressure of 760 mmHg. The *joule* is the unit of energy in the International System of units (SI; see Appendix I), and one calorie is equivalent to 4.184 joules. *Power* is the rate of energy expenditure. We will use units of cal/(time) to measure power. The SI unit for power is the watt (W), and one watt is one J/s, so one cal/s is 4.184 watts (see Appendix I).

Animals use chemicals in food as sources of energy. One way to obtain energy from chemicals is from oxidation, such as in the Krebs cycle for oxidation of glucose (see Fig. 4.27). Some of the energy from a mole of glucose will be transformed to other forms of energy, such as heat or the chemical energy in ATP molecules, but the complete oxidation of one mole of glucose will ultimately be manifested as heat energy when work is done. The complete oxidation of glucose to carbon dioxide and water will produce 673,000 calories of heat when one mole of glucose is burned in an oxygen atmosphere:

$$C_6H_{12}O_6 + 6O_2 \longrightarrow 6CO_2 + 6H_2O + \text{Heat (673,000 calories)}.$$

When glucose is oxidized in a cell, 38 ATP molecules per glucose molecule form in the intermediate step of the chemical conversion of energy:

$$C_6H_{12}O_6 + 6O_2 + 38ADP + 38H_3PO_4 \longrightarrow 6CO_2 + 38ATP + 44H_2O + \text{Heat (382,000 calories)}.$$

In this case only 382,000 cal of the total 673,000 cal are given off as heat. The remaining 291,000 cal of chemical energy are transferred from the glucose bonds to the phosphate bonds when adenosine diphosphate (ADP) forms ATP. This energy is used within cells for a variety of purposes. Chemicals are moved across membranes, electrons are transported within the cell, or, if a muscle cell is involved, contraction may occur. All of these activities require energy that has been transferred to the ATP molecules. When the energy is used for work, the remaining 291,000 cal will be given off as heat, with the *net* result that one mole of glucose *always* provides 673,000 calories of heat energy, regardless of whether it is oxidized in cells or in test tubes.

MEASURING METABOLIC RATE (ENERGY USE/TIME)

We wish to know how much energy an animal uses in a period of time, its internal (metabolic) power production. The process of oxidative metabolism can be written in a simplified way as:

$$\text{Chemical substrate} + O_2 \longrightarrow CO_2 + H_2O + \text{``Other waste''} + \text{Heat.}$$

In this equation CO_2, H_2O, and "Other waste" represent products that yield no additional energy. From the point of view of oxidation, they are "waste" products (although they are useful in other contexts), and some energy must be expended to transport and/or excrete them (see Part I and Part IV). "Other waste" products are formed in energy-yielding reactions such as protein oxidation, which involves nitrogen excretion. Lactic acid represents a "waste" product from anaerobic glycolysis (see Chapter 4), although it will not be a waste product if oxygen becomes available so that it can be used in the Krebs cycle.

The most direct method to measure rate of energy use is to measure the rate of heat production by an animal. Another way is to measure the energy value of chemicals in food and to subtract "waste" energy in the products of metabolism to indirectly estimate rate of heat production. Another indirect way is to measure the amounts of gases (O_2 and CO_2) consumed and/or produced when certain chemicals are oxidized.

Direct Calorimetry

Measuring energy use from the heat given off by an animal to its environment is difficult to do for air-breathing animals because special equipment is needed to detect heat flow from animals to air. However, for aquatic animals or diving animals direct calorimetry is relatively simple (Fig. 5.1). The animal is placed in water in an insulated container and after a certain time period the water temperature is compared with the temperature of only water in an identical container. Differences in water temperature between the containers over time give a direct measure of the rate of heat production by the animal, since every gram of water that increases in temperature by 1 °C represents one calorie of heat production by the animal.

Indirect Calorimetry: Measuring Substrates and "Waste" Products

There are two ways to estimate the amount of heat produced by an animal without actually measuring heat production. The first makes use of knowledge

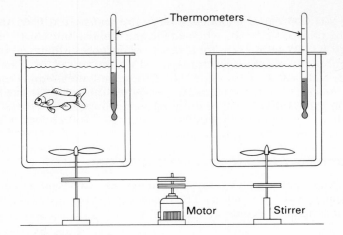

Figure 5.1
For aquatic animals, direct calorimetry involves measuring
the difference in water temperature between two containers
that are identical except for the presence of an animal in
one container. One calorie of heat will be produced
whenever one gram of water increases in temperature by
one degree Celsius.

about the ultimate energy content of consumed food. This value, minus the
energy value of "waste" products produced and excreted, is a measure of the
net energy value to the animal of what it has eaten.

Food components fall into three substrate categories: carbohydrates, proteins, and fats (neglecting for present purposes the nutritional values of food).
Table 5.1 shows the heat produced when one gram of these different substrates
is completely oxidized with a surplus amount of oxygen. If the composition of
ingested food is known, the total energy intake can be calculated. If the energy
lost from incomplete assimilation by the digestive tract (energy in feces) and
from nitrogen excretion (from protein oxidation) is subtracted, the *net* energy
gain by the animal from one feeding to the next can be calculated. Under appropriate (steady state) circumstances, net energy gain over time gives information on the rate of energy use.

Table 5.1   Energy from complete oxidation of food
components (from Kleiber, 1961)

COMPONENT	CALORIES/GRAM
Carbohydrate	4200
Protein	4300 (minus urea)
Fat	9400

There is a disadvantage to this method. It assumes that an animal will use exactly what it has consumed in net energy over the time interval between feedings. This may be true in some cases, but many animals store some ingested energy as fat or use stored fat reserves to supplement their requirements for energy (see Chapter 10). This situation can be monitored by weighing the animals. Any increase or decrease in weight for a grown animal not accounted for from food in the digestive tract or from water gains or losses will represent either storage of surplus energy or the use of energy from internal fat storage. Keeping track of all these variables is very complicated, although they provide some very interesting information on how animals deal with energy supplies (food) over various time periods (see Chapter 10).

Note that fat as a substrate provides more than twice the energy for every gram oxidized than does either carbohydrate or protein (Table 5.1). Fat is "energy-dense." This fact is important because animals can store energy in fat by adding less weight than they would if the energy were stored as another substrate. Moreover, storage of energy as carbohydrate by production of glycogen in cells requires water, which adds additional weight to an animal.

The weight of an animal has an important influence on how much energy must be expended for various activities. This limitation, together with the high energy density of fat, provides an explanation for its use as an important internal energy-storage tissue. Migratory birds and some mammals that hibernate, for example, add large quantities of fat prior to migration or hibernation (up to 100% of body weight before fat deposition for some species). The fat provides energy for periods when demands are high (migration) and/or supplies of food are low (hibernation) (see Chapters 8 and 9). Use of other substrates for internal energy storage would result in higher expenditures of energy relative to supplies.

Indirect Calorimetry: Measuring Respiratory Gases

The second procedure for indirectly estimating rate of energy use from oxidative metabolism is to measure the amount of oxygen consumed and the amount of carbon dioxide produced over a period of time. Oxygen consumption and carbon dioxide production are continuous, and there is little storage of oxygen within the body. Carbon dioxide is "stored" in appreciable amounts as bicarbonate in blood (see Chapter 3), and the amounts change under some circumstances, such as when body temperature changes. Thus measurements of oxygen uptake provide the most precise way to estimate rates of energy use indirectly, since oxygen consumption and energy expenditure are directly related under a variety of circumstances. However, note that measurements of oxygen consumption neglect possible short-term contributions to energy demands by anaerobic metabolism. If these are appreciable, measurements of lactic acid production (a "waste" product from anaerobic glycolysis) are necessary.

When substrates are used to estimate aerobic energy requirements, the heat resulting from complete oxidation of one gram of the substrate is measured (Table 5.1). The measurement is usually done with a "bomb" calorimeter. A gram of carbohydrate, protein, fat, or a known mixture is placed in a sealed container with an excess amount of oxygen. The contents are ignited, and the heat changes the temperature of a certain amount of surrounding water.

When gases are used to estimate energy requirements, measurements are made of the heat that results when a constant but limited amount of oxygen (e.g., one milliliter) is used to oxidize as much as possible of an excess amount of substrate; the amount of carbon dioxide produced from the oxidation is also measured. The amount of heat produced from the use of one ml of oxygen and from the production of one ml of carbon dioxide is shown for different substrates in Table 5.2.

Table 5.2   Heat produced from oxidation of excess substrate with one ml of oxygen or when one ml of carbon dioxide is produced, and the respiratory quotient for different substrates (from Kleiber, 1961)

SUBSTRATE	CALORIES/ml O_2	CALORIES/ml CO_2	ml CO_2 PRODUCED / ml O_2 CONSUMED
Carbohydrate	5.0	5.0	1.0
Protein	4.5	5.6	0.8
Fat	4.7	6.6	0.7

Note from Table 5.2 that there is not much difference in the quantity of heat produced from oxidizing different substrates with a constant amount of oxygen. The maximum difference, which is between carbohydrate and protein, is only 0.5 cal for every ml of oxygen. If the substrates used by an animal are unknown but oxygen consumption is known, there is a maximum error of only 5% if 4.75 cal/ml O_2 is used for the caloric value of oxygen consumption.

Note that for carbohydrates the same number of calories are produced for each ml of oxygen used and carbon dioxide produced, but there are different values for the two gases if various other substrates are oxidized (Table 5.2) because they contain different amounts of carbon. For carbohydrates the number of carbon atoms exactly equals the number of oxygen molecules required for oxidation, whereas for proteins and fat there are fewer carbon atoms relative to oxygen molecules.

These differences can be used to determine what substrate an animal is oxidizing. The ratio of ml of CO_2 produced to ml O_2 consumed is the *respiratory quotient* (R.Q.; see Chapter 2), which is different for different substrates (Table 5.2). If an R.Q. of 1.0 is measured an animal is probably using carbohydrate as a substrate, and an R.Q. of 0.7 indicates oxidation of fat. An intermediate R.Q. is difficult to interpret, because the oxidation of some mixture of carbohydrate and fat as well as the exclusive oxidation of proteins will give an R.Q. of 0.8.

STANDARD METABOLIC RATES: THE MINIMUM COSTS FOR EXISTENCE

We would like to know how demands for energy differ among animals expending energy at standard rates. In previous chapters we have discussed the respiratory and circulatory supply-demand relationships for animals that differ in *size*. Body size is a striking difference between species. What is its impact on rate of energy expenditure? How do we compare rates of metabolism among animals of different sizes so that differences in size are accounted for?

To answer these questions, we can measure the rates of energy use by animals of different sizes (weights) that are all deprived of food long enough to start using fat as a substrate (R.Q. = 0.7). At the same time the rates of energy use can be measured when these animals are inactive at a standardized temperature and photoperiod, telling us what the *minimum* energy costs are for existence of animals of different sizes.

Figure 5.2 shows how standardized rates of energy expenditure change with changes in body weight for some terrestrial vertebrates. There are two important features to these relationships. First, note that animals from different major groups have different minimum rates of energy expenditure at a given body weight. Passerine birds (perching song birds) have the highest values, while reptiles at temperatures of 37 °C have lower values (Fig. 5.2). Second, note that the lines in Fig. 5.2 are all approximately parallel; that is, they all have about the same slope, meaning that a change in body weight results in about the same change in minimum rate of energy expenditure by animals from different groups.

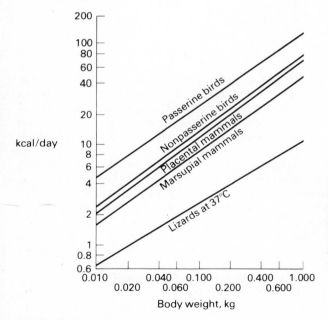

Reprinted with permission of Macmillan Publishing Co., Inc. from *Animal Physiology: Principles and Adaptations* (2nd ed.) by M.S. Gordon, G.A. Bartholomew, A.D. Grinnell, C.B. Jorgensen, and F.N. White. Copyright © 1972 Malcolm S. Gordon.

Figure 5.2
The relationship between the daily rate of minimum standard heat production and body weight for terrestrial vertebrates shows that the magnitude of minimum rate of energy expenditure is different for animals from different major groups, but the change in rate of expenditure with a change in weight (the slopes of the lines) are similar.

Why do we see these two major features? Table 5.3 summarizes the values for minimum expenditure rates of the different groups of vertebrates at a similar size (at one kilogram). Some of the differences are accounted for on the basis of differences in body temperatures. For example, passerine birds have body temperatures close to 42°C, nonpasserine birds are at about 40-41°C, placental mammals are at 38°C, and marsupial mammals are at about 35-37°C. Reptiles have body temperatures that change if environmental temperatures change (see Chapter 6), but those measured for Fig. 5.2 were exposed to an environment producing a body temperature of 37°C so that they could be compared with animals from other groups.

Table 5.3 Metabolic rate of animals from different major groups of vertebrates weighing one kilogram (from Gordon *et al.*, 1972)

GROUP	KCAL/DAY (WEIGHT = 1 kg)
Passerine birds	129
Nonpasserine birds	78
Placental mammals	70
Marsupial mammals	49
Reptiles (at body temperature = 37°C)	10

endotherms

ectotherms

Differences in body temperatures influence the rate of energy use (see Chapter 6 for details), but it is possible to adjust the lines in Fig. 5.2 to reflect metabolic rates at the same body temperature (37°C) between groups. When this is done the nonpasserine birds and the placental and marsupial mammals all converge to one line, while the lines for passerine birds and reptiles remain separate.

The differences between birds and mammals (endotherms; see Chapter 7) and reptiles (ectotherms; see Chapter 6) are striking. The reptiles shown in Fig. 5.2 are typical also of other vertebrates such as fishes and amphibians. Differences in minimum maintenance costs between these major groups is related to the extent to which total metabolic rate is supported by aerobic versus anaerobic metabolism (see Chapter 4). In reptiles, amphibians, and fishes, aerobic metabolism is lower, and thus demands for oxygen and energy substrates are lower.

Differences between these major groups are also related to the extent to which aerobic and anaerobic processes contribute to metabolic demands for activity. Reptiles, amphibians, and fishes rely more extensively on short-term, rapid anaerobic processes to increase power, but birds and mammals rely more on aerobic (sustainable) processes (Chapter 18). Thus the higher resting energy requirements of birds and mammals reflect basic differences in mechanisms used for generating power at rest and during activity.

The differences between groups of animals shown in Fig. 5.2 have also been thought to reflect taxonomic differences in degree of evolutionary "development." Thus the lower rate of metabolism for a reptile has been viewed as a reflection of its "less advanced," i.e., "more primitive," degree of progress in an evolutionary scheme. However, there are great difficulties with this type of historical hypothesis. Is it really "primitive" to have a low requirement for energy? According to the principle of efficient use of energy at minimum rates, it would be a distinct advantage.

The relative advantages and disadvantages between these major groups most likely involve costs and benefits associated with the degree of anaerobic versus aerobic metabolism. Low aerobic metabolism reduces demands for oxygen and energy, but reliance on anaerobic processes sets limits on long-term endurance for activity (Chapter 18). High aerobic metabolism produces high costs for maintenance but also permits a higher degree of sustainable activity.

PHOTOPERIOD AND STANDARD METABOLIC RATES

The minimum rate of energy expenditure depends on the time period when an animal is normally active and the period when it is normally resting or inactive. For many species active and inactive time is determined by the photoperiod. Some species are diurnal, that is, normally active during the light phase, while others are nocturnal. Birds are primarily diurnal, and Fig. 5.3 shows values of minimum rates of energy expenditure for passerine and nonpasserine birds in either the light phase (α) or the dark phase (ϱ) of a photoperiod. Note that the minimum rate of energy use is less by about a constant amount with respect to weight for the birds during the dark phase of the photoperiod, when they would normally be resting without food.

These changes illustrate the difficulty of defining standard metabolic rates as the absolute minimum rates of energy expenditure. Obviously, the minimum rate changes when the environment changes. As photoperiods change there will be a daily (*circadian*) rhythm of changes in standard metabolic rates.

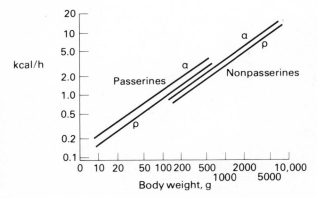

Figure 5.3
For passerine (perching songbirds) and nonpasserine birds of given weights, the level of minimum metabolic rate is lower during the dark phase (ϱ) of a photoperiod than it is during the light phase (α) (from Aschoff and Pohl, 1970).

Standard metabolic rates also change seasonally; thus an animal exposed to a particular photoperiod may have a different level of minimum energy expenditure in the fall months than it has in the spring months. Strictly speaking, comparisons between species differing in size should take these differences into account by attempting to standardize measurements with respect to photoperiod and seasonal effects.

THE 0.75 SLOPE AND BODY SIZE

Both the x and the y axes in Figs. 5.2 and 5.3 are logarithmic scales. Thus the Log of y is a linear function of the Log of x, and for rate of energy expenditure and body weight we can write (see Appendix II)

$$\text{Log (kcal/day)} = K' + (b)\text{Log(kg)}.$$

In this way it is put in the form of a general linear equation

$$y = a + bx,$$

where

a = intercept (value of y when $x = 0$), and

b = slope = $(y - a)/x$.

A logarithm of a number is the exponent to which 10 is raised to equal that number (see Appendix II for a review of logarithms). The first equation above can be written as a power function:

$$\text{kcal/day} = K(\text{kg})^b,$$

where

K' = Log K,

K = value of y when kg = 1.0 (since Log 1.0 = 0 or $10°$ = 1.0), and

b = slope.

Another way of stating these relationships is to say that the parallel lines in Figs. 5.2 and 5.3 all have a slope that is about the same but the actual values of K are different for different major groups of animals. Table 5.3 gives values of K for the lines in Fig. 5.2. The slope for the lines is 0.75.

The slope of 0.75 for the equations relating standard rate of energy expenditure and body weight is a property of most living organisms (some invertebrate groups have a slope of 1.0). Figure 5.4 shows minimum rates of energy expenditure versus weight for single cells and for multicellular organisms (both ectotherms and endotherms). Although the levels of metabolic rate are different for these different groups (different values of K), the lines are all parallel with a slope of about 0.75, that is, intermediate between a slope of 0.67 and 1.0 (Fig. 5.4). It is seldom that so many different organisms show such similarity in a characteristic; therefore, it deserves considerable discussion.

A slope of 0.75 has very important implications for interpreting demands for energy by animals of different sizes. As weight increases, demands for energy to meet minimum requirements increase, but they do not increase in

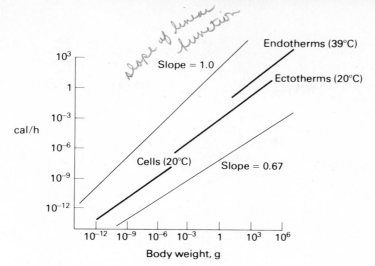

slope of linear function

Figure 5.4
For both unicellular and multicellular organisms, rate of minimum energy expenditure is related to weight with a slope of 0.75 (after Hemmingsen, 1960).

direct proportion; otherwise the slope would be 1.0 (a power function with a slope of 1.0 is a linear function). For example, a placental mammal that weighs 12 kg has a standard rate of energy expenditure of 451 kcal/day. A placental mammal that weighs 24 kg expends 759 kcal/day. Doubling size does not double minimum energy expenditure rates for these animals.

Weight-specific Metabolic Rate

There is another way to view standard metabolic rates. Instead of looking at the rate of energy use by the entire animal, we can look at the average rate of energy use by *one gram* of the animal. This rate is found mathematically by simply dividing the total metabolic rate by weight:

$$\frac{\text{cal/h}}{\text{g}} = \frac{K(g)^{0.75}}{(g)^{1.0}},$$

or

$$\text{cal/g} \times \text{h} = K(g)^{(0.75 - 1.00)}$$
$$= K(g)^{-0.25}.$$

Standard metabolic rate per gram increases with *decreasing* body size (negative slope) in an exponential manner, illustrated for mammals in Fig. 5.5.

Some caution is needed to interpret this view of standard metabolic rate and the standard metabolic rate of the entire animal, because the relationships are opposite with respect to weight (one slope is positive while the other is negative). The high metabolic rate per gram for small animals indicates that their cells and tissues (or what constitutes a *gram* of animal) use energy at a

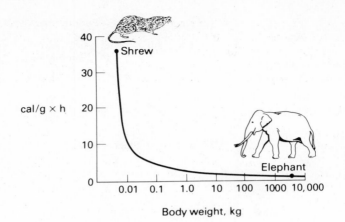

Figure 5.5
When minimum metabolic rate of mammals is expressed per gram body weight and plotted on semilogarithmic coordinates, energy expenditure rates per gram increase exponentially with decreasing body weight (after Schmidt-Nielsen, 1970).

high rate compared with larger animals. We have discussed some adaptations in function of the circulatory and respiratory systems that serve to supply oxygen to the tissues of small animals at relatively high rates (see Part I). For example, recall that respiratory rate and heart rate are related to $(weight)^{-0.25}$, while tidal volumes and stroke volumes are related to $(weight)^{1.0}$. Moreover, oxygen dissociation curves for small animals are shifted to the right, which means that oxygen is delivered to their tissues at a higher rate. However, the *total* quantity of energy required to meet the minimum requirements of small animals [(cal/g × h) × (g) = cal/h] is less than it is for larger animals (Fig. 5.2).

In general, you can keep these different ways of expressing rates of energy use straight by remembering that *per gram* measurements refer to demands within *tissues* of an animal, and energy requirements per gram have an impact on *how* total requirements are met. For example, the faster breathing and heart rates of smaller mammals as compared with larger mammals involve mechanisms to match supplies with *tissue demands* on rate of oxygen use. However, whether an animal survives depends on its *total* demands for energy. Individuals have evolved particular weights, and unless *all* demands for energy are met they will not survive. Natural selection acts on the total individual (cal/h); demands per gram (cal/g × h) determine how supplies must be provided to meet total tissue demands.

WHY 0.75 AND NOT SOME OTHER SLOPE?

Several hypotheses have been proposed to explain why the exponent for total minimum energy requirements is 0.75 and not some other number; they involve different relationships between areas and size. Most hypotheses involve mathematical manipulation of exponents, and it may be helpful to review the basic rules (see Appendix II). If we can provide an explanation for why total minimum rates of expenditure have an exponent of 0.75, we will also have an explanation for why minimum rates of expenditure per gram have an exponent of -0.25 and vice versa, since the two are mathematically equivalent.

Surface Area Hypothesis

Animals lose heat to their environments through their surfaces. The fact that small animals require more energy per gram was thought to be due to their greater surface areas relative to their volumes. Recall that the surface area of a sphere is related to the square of the radius ($4\pi r^2$), while volume is related to the cube of the radius ($4/3\pi r^3$) (see Fig. 1.3). When the linear dimensions of a spherical object become smaller (r decreases), volume decreases more than surface area; thus a decrease in size results in an increased surface area relative to volume. A similar situation occurs for other geometrically similar objects.

Because the surface areas of small animals are larger relative to their volumes, they will lose heat to an environment at a faster rate per unit area of surface. However, the surface area of animals is related to the 0.67 power of their volume (size) and *not* the 0.75 power (Fig. 5.6). Surface area is proportional to (characteristic length)^2 while volume is proportional to (characteristic length)^3. Thus length $\propto$ (surface area)$^{0.5}$ and length $\propto$ (volume)$^{0.33}$. Since length = length, (surface area)$^{0.5}$ $\propto$ (volume)$^{0.33}$ and surface area $\propto$ (volume)$^{0.67}$. Although heat is certainly lost from the surfaces of small animals at a higher rate per unit area, this loss cannot explain the exponent of 0.75.

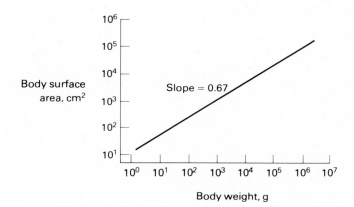

Figure 5.6
For vertebrates ranging in weight from about five grams to more than a ton, body surface area is related to weight with a slope of 0.67 and not 0.75 (after Hemmingsen, 1960).

Cross-sectional Area and the Structural Support Hypothesis

The supporting structures of animals are subject to forces that act to collapse them, and these structures have evolved so that their dimensions overcome this problem. For example, the leg of a vertebrate will have a bone structure whose length is sufficient for the requirements of movement and whose diameter (thickness) is sufficient to withstand forces acting to break the leg. At the same time there is economy in limb construction. A limb is not thicker than that which is necessary to withstand the forces acting on it; otherwise, more material would have to be obtained for its construction and more energy would be needed to move it.

As the length of a limb increases, the diameter must increase for structural support. The relationship between "limb" (trunk or branch) length and diameter has been studied for trees. The constraints for growth in trees are similar to those operating on the limbs of animals. As the height (length) of a tree increases, diameter increases so that forces acting on the tree (from gravity and winds) do not break the supporting structure. However, a tree has evolved with an economy of thickness that permits upward growth and the opportunity to spread leaves in the form of a canopy to compete for available sunlight.

As the height of trees increase, diameters increase with an exponent of two thirds (0.67; Fig. 5.7). For any cylinder, there is a "critical length" at a particular diameter beyond which the cylinder will buckle or collapse when a bend-

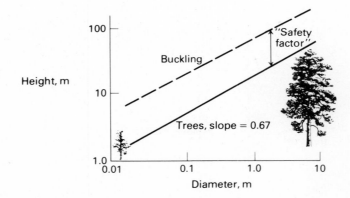

Figure 5.7
The relationship between the average heights of trees and their average diameters (solid line) shows height increasing as the 0.67 power of the diameter. A similar relationship exists for the "critical height" (dashed line), the height at a given diameter where bending forces would cause collapse (after McMahon, 1973).

ing force is applied. The critical length is also related to the 0.67 power of diameter:

$$l_{cr} = K(d)^{0.67},$$

where

l_{cr} = critical length of a cylinder,

K = constant determined by the elastic properties of the material used to construct the cylinder, and

d = diameter of the cylinder.

The critical length of trees is calculated from information on the elastic properties of green wood (the value of K). Trees have a length at a given diameter that is less, by about a constant amount, than the critical length where collapse would occur from bending forces (Fig. 5.7). The difference represents a safety factor.

A similar argument can be applied to the structure of animals. Their supporting structures (bone, muscle, etc.) are constructed such that length is proportional to the 0.67 power of diameter:

$$l \propto d^{0.67}.$$

Weight is proportional to volume if density remains the same. Since volume = length × cross-sectional area, or

$$\text{volume} = l\,(0.25\,\pi\,d^2),$$

and

$$\text{volume} \propto \text{weight},$$

then

$$\text{weight} \propto ld^2 \text{ (neglecting the constant } 0.25\,\pi\,). \tag{5.1}$$

Because

$$l \propto d^{0.67},$$

then

$$d^2 \propto l^3.$$

Substituting l^3 for d^2 in Eq. (5.1) gives

$$\text{weight} \propto l(l^3) \propto l^4$$

or

$$l \propto (\text{weight})^{0.25.}$$

Substituting $(\text{weight})^{0.25}$ for l in Eq. (5.1) gives

$$\text{weight} \propto (\text{weight})^{0.25}\, d^2$$

or

$$d^2 \propto (\text{weight})^{0.75}.$$

Cross-sectional area $= 0.25 \pi d^2$, suggesting a relationship between cross-sectional area and $(weight)^{0.75}$. Thus the physical limitations imposed by structural support and the resulting relationships between length and diameter could generate a dependence on the 0.75 exponent of weight. For example, the maximum force a muscle can produce is related to its cross-sectional area (see Chapter 18), and taking structural support limitations into account, maximum power should be proportional to $(weight)^{0.75}$. If maximum power (energy/time) is a constant multiple of minimum power, this factor can produce a relationship between cross-sectional area of animals and their minimum rates of energy expenditure related to $(weight)^{0.75}$.

Area in a Fourth-dimension Hypothesis

The effects of buckling and bending forces on structures could provide a rationale for complex terrestrial organisms such as trees and vertebrates subject to wind and gravitational forces, but the exponent of 0.75 also applies to unicellular and aquatic organisms that may not be subject to the same physical limitations (see Fig. 5.4). Is there any other relationship between areas and volumes (size or weight) that might be more general and apply to *all* living organisms?

We are used to thinking in only three dimensions, but there is a general relationship between surface area and volume for a space with any number of dimensions equal to or greater than three:

$$\text{surface area} \propto (\text{volume})^{(n-1)/n},$$

where n = number of dimensions. For three-dimensional space ($n = 3$) it gives

$$\text{surface area} \propto (\text{volume})^{0.67};$$

(see Fig. 5.6). For a four-dimensional space,

$$\text{surface area} \propto (\text{volume})^{0.75}.$$

Since volume and weight are proportional if density is the same, a relationship is produced between surface area in four dimensions and body weight with the exponent of 0.75.

What could provide a fourth dimension? One suggestion that has been made is time. Among mammals, for example, total life span is related to $(weight)^{0.25}$. Thus total minimum metabolic rate (total cal/h $= K (kg)^{0.75}$) times longevity (total h $= K' (kg)^{0.25}$), the total calories expended in a lifetime, is proportional to $(weight)^{1.0}$. This relationship suggests that mammals possess a finite total capacity for metabolism over their lifetimes that depends on their weight, and the rate at which metabolism occurs determines lifespan. This raises more perplexing problems. What determines lifetime capacity for metabolism? What is cause and what is effect?

By now you should be impressed with the extent to which exponents can be manipulated to examine ideas. At present the "mystery" of the 0.75 slope has not been solved, and it will take some imaginative experiments to resolve the problem. When the information that provides an unambiguous explanation for the 0.75 slope is available, considerable progress will have been made in understanding an important common characteristic of most living organisms.

SUMMARY

For inactive animals energy is used effectively when rate of use is minimized. The minimum rate of energy expenditure for birds and mammals is called the "basal" metabolic rate, and it occurs under certain environmental conditions (with respect to temperatures and photoperiods) and depends strongly on body size (weight). Since rate of energy expenditure in ectotherms changes continuously with body temperature, comparisons between species must be standardized. Standard metabolic rate refers to energy expenditure rates that are measured when major environmental variables influencing rates are equalized.

All forms of energy are ultimately transformable to heat, and the calorie (or joule) is the measure of heat energy. One calorie of heat is required to increase the temperature of one gram of water from 14.5 to 15.5 °C at 760 mmHg atmospheric pressure. Power is the rate of energy expenditure (cal/time, or watts). The chemical substrates of animals (carbohydrates, proteins, and fats) are ultimately transformed to standard amounts of heat energy, and the rate of heat production can be measured in three ways.

Direct calorimetry involves measuring the heat given off by animals to their environments as a consequence of their metabolism. Heat production can be estimated indirectly in two ways. Knowledge of the ultimate energy value of assimilated chemical substrates (assimilated intake − excretory losses) that are metabolized can be used to provide information on energy use over time if an animal is at a steady state (no changes in substrates stored versus those assimilated) from one feeding to the next. A less complicated indirect way to estimate rate of energy use is from information on the caloric equivalents of oxygen consumed and/or carbon dioxide produced. Carbohydrates, proteins, and fats that are oxidized differ in carbon structure, and thus in the quantity of CO_2 produced relative to the quantity of O_2 consumed (the respiratory quotient, or R.Q.). These differences can provide information about which substrates are oxidized.

Fats can provide more than twice the energy from oxidation than can either proteins or carbohydrates. Thus more energy can be stored internally in fat with less weight increase. Since an animal's weight has an important impact on its energy expenditures, this difference in energy storage provides an explanation for the evolution of fat as a major internal energy-storage tissue.

When the standardized metabolic rates of a large number of animals are measured, it can be seen that the minimum rate of energy expenditure increases with body weight in an exponential manner. Animals from different major groups (e.g., unicellular versus reptiles versus mammals versus passerine birds) have different levels (intensities) of minimum metabolic rates, but for most living organisms minimum metabolic rate increases with weight raised to an exponent (slope in a graph) of 0.75. It is equivalent to rate of energy expenditure *per gram* increasing with decreasing body weight related to $(weight)^{-0.25}$. The major difference between these ways of expressing minimum metabolic rates depends on whether *total* requirements of an animal for energy are considered or whether the mechanisms by which *tissues* (grams) of an animal are supplied with energy are considered.

Several hypotheses have been proposed to explain the exponent (0.75 or −0.25) of body weight for minimum rates of energy use. Although small animals lose heat at a higher rate per gram (or per unit surface area) because of their higher surface areas relative to their volumes, it does not explain the slope because surface area and volume (size) are related by a slope of 0.67 and not 0.75.

The lengths of supporting structures of animals are related to the 0.67 power of their diameters because of physical limitations imposed by bending or buckling forces, which leads to a relationship between cross-sectional area and $(weight)^{0.75}$. Thus supporting structures developed by animals that depend on cross-sectional area could lead to a relationship dependent on weight to the 0.75 power.

A third hypothesis suggests that surface and volume may be appropriately related if a fourth dimension is important. For four-dimensional space, surface area is related to $(weight)^{0.75}$. Perhaps time could provide a fourth dimension, since the life span among mammals is related to $(weight)^{0.25}$; thus rate of expenditure of energy (cal/h) times total time (h), that is, total metabolic capacity (cal), depends directly on weight. At present there is no unambiguous explanation for the 0.75 slope, and more experiments are required to allow us to determine which hypothesis could explain this important characteristic of living organisms.

Temperature Regulation in Ectotherms

6

The rate of energy use by an animal is influenced by its temperature. Temperature is a measure of heat, i.e., energy content, and chemical reactions occur at different rates at different temperatures. For example, many cellular chemical reactions are governed by the activity of enzymes interacting with substrates to yield important products. The kinetic properties of enzymes that determine the rates of reactions are highly sensitive to temperature, with increases in temperature usually accelerating the rates of enzymatic reactions up to a limit (see below). In addition, the two- and three-dimensional structure of an enzyme depends on its immediate energy environment. The chain of amino acids composing an enzyme folds on itself in a manner that influences ability to catalyze reactions. This folding is stabilized by weak bonds (hydrogen bonds, hydrophobic interactions, van der Waals interactions, and electrostatic interactions) that are influenced by temperature.

Figure 6.1 summarizes the importance of temperature for enzyme-mediated reaction rates. In general, because of the effects of temperature on enzyme function, there is a narrow range of temperatures where enzyme activity

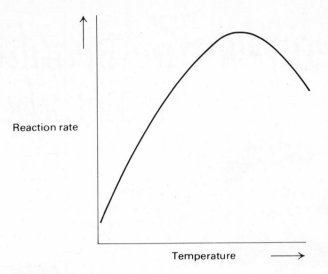

Figure 6.1
A general relationship between an enzyme-mediated
reaction rate and temperature. Increases in temperature
usually increase reaction rates up to a limit above which
thermal destruction of enzymes decreases reaction rates.

is "optimal." At an optimum temperature, enzymes interact most effectively
with their substrates to yield the required products of chemical reactions. The
optimum temperature does not have to be the temperature for maximum reaction rates. It can be any temperature that results in the formation of required
products at rates sufficient to meet demands.

The efficiency of energy use with respect to temperature depends not only
on enzymes but also on the presence of substrates with which enzymes interact
to provide required products. The substrates come from food, and if their supply
is reduced, continued high reaction rates at certain temperatures will exhaust
substrate supplies relatively quickly. Decreases in temperature will reduce rates
of substrate use (Fig. 6.1), so in this case an "optimal" temperature would be a
low temperature. We will examine this aspect of temperature effects on rates of
energy use in detail in Chapter 8, when we discuss torpor and hibernation.

When there are sufficient quantities of substrates from food, the interaction between enzymes and substrates are most effective when products are
formed at rates that match demands. Ability to meet product demands depends
on both catalytic efficiency and ability to change (regulate) product formation
when demands change. There is considerable flexibility in the mechanisms of
enzyme function that contribute to regulation of product formation. For example, the concentrations of substrates in cells are usually less than the concentrations that yield maximum reaction rates (catalytic activity). To meet different
demands, reaction rates can be controlled, i.e., changed (increased or decreased),

by means of adjustments in intracellular substrate concentrations. Other enzyme-substrate regulatory mechanisms also exist, but a discussion of them is beyond the scope of this book. However, despite a variety of mechanisms for regulation of enzyme-substrate interactions, the effects of temperature changes on reaction rates (Fig. 6.1) can override regulatory adjustments. Unless temperature variation is confined to certain limits, regulation of reaction rates cannot occur with maximum efficiency.

This knowledge leads us to predict that, when possible, animals should maintain (regulate) their temperatures at values conducive to most effective functioning. Both whether it is possible and the particular temperatures that should be maintained depend on a variety of characteristics of animals and their environments. One important determinant for temperature regulation is the *source* of heat utilized for regulation; this consideration forms the basis for the separation of our discussion of temperature regulation into two chapters, which deal with ectotherms and endotherms, respectively.

In these chapters concerning temperature regulation we will deal primarily with animal and environmental characteristics that dictate *gains* of heat by animals, where possible, to balance *losses* of heat to their environments. This situation involves a "cold" environment. The problems involved in temperature regulation where heat must be lost by animals ("hot" environments) involve important interactions with the regulation of water as a resource, and we will consider these in detail in Chapter 14 after we have developed the principles governing water regulation and control.

ECTOTHERMS AND ENDOTHERMS

All animals can be divided into two groups, depending on the primary source of the energy that determines their temperature. *Endotherms* ("internal heat") obtain heat to regulate body temperature primarily from internal aerobic combustion of fuels. We will be concerned with their function in the next chapter. *Ectotherms* ("external heat") obtain most of the heat energy for their temperature regulation from the external environment and rely less on aerobic processes for heat production. A common source of heat for some ectotherms is the sun, and animals that use this direct source are called *heliotherms*. Those that obtain heat from the surface of the earth (indirectly from the sun) are called *thigmotherms*. Endotherms include birds, mammals, and some fishes, insects, and plants (see Chapter 7). There is no clear taxonomic boundary between ectotherms and endotherms, although most unicellular organisms, plants, invertebrates, reptiles, fishes, and amphibians obtain a major source of heat from some part of their external environment.

HEAT EXCHANGE

Like other physical variables, heat moves from an area of high "concentration" to an area of low "concentration," but heat is different from other physical variables. It has no mass, length, volume, or density. Physical substances with these characteristics are called *extensive*. Chemicals, for example, can be characterized by their mass (atomic weights, molecular weights, concentrations, etc.); heat cannot and is called an *intensive* physical property. Heat is measured from temperature. An object with a high temperature has increased heat energy, and heat energy will move from objects having higher temperatures to objects having lower temperatures in several ways.

Radiation of heat energy is quantitatively an important avenue for heat exchange. With the exception of minor geological sources, most heat energy reaches the surface of the earth on a daily basis from the sun by radiation of the intense heat of nuclear fusion occurring on this star. This radiant heat is divided into three components: ultraviolet, visible, and infrared. Figure 6.2 shows the distribution of radiant energy from the sun measured at sea level. Note that the quantity reaching the surface varies considerably, depending on local conditions

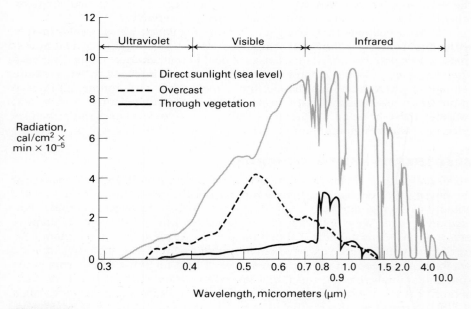

Figure 6.2
Distribution of radiant energy from the sun at sea level as direct sunlight or as influenced by local environmental conditions (modified from Gates, 1965).

associated with clouds or vegetation that intercept radiant energy. On a daily basis, about 655 calories hit one square centimeter of surface as light visible to man, while another 691 cal are received as invisible ultraviolet and infrared light.

The amount of heat energy that exchanges by radiation from one object to another depends on the difference in temperature between the objects. The equations that describe the exchange are complicated and depend on several physical characteristics of the objects as well as the wavelength of radiation. For infrared radiation, radiant heat exchange (H_r) is described by

$$H_r = \sigma\, e_1 e_2 A(T_1^4 - T_2^4),$$

where

σ = a physical constant (Stefan-Boltzmann's constant);

e_1 = "emissivity" of the warmer object, determined by physical properties of the object;

e_2 = "emissivity" of the cooler object;

A = area for radiation exchange;

T_1 = temperature (in °K) of the warmer object; and

T_2 = temperature (in °K) of the cooler object.

Heat also exchanges over short distances by *conduction* between two objects of different temperature and from warmer parts of an object to cooler areas. Although heat flows from an area of higher temperature to an area of lower temperature, there must be physical contact for the conduction exchange of heat. The equations that describe this exchange are also complicated and depend on the physical characteristics of the objects. For example, for an "infinite slab with parallel surfaces," heat exchange by conduction is described by

$$H_c = KA\left(\frac{T_1 - T_2}{l}\right),$$

where

K = thermal "conductivity", determined by the physical properties of objects;

A = area for conduction of heat;

l = length for conduction of heat;

T_1 = temperature of the warmer object; and

T_2 = temperature of the cooler object.

A special type of conduction is called *convection*. The density of air (or water) changes when its temperature changes; air or water is less dense hot than it is cold. This fact forms the basis for winds and currents on a global scale, and on a much smaller scale local currents of air or water are established around an object that has a temperature different from its environment. Global and local air and water currents vary considerably and have an effect on the exchange of heat between animals and their environments.

It is very difficult to determine the values of the variety of physical characteristics of animals and their environments that are needed in specific equations to determine heat exchange by radiation, conduction, and convection. The values of K, A, l, and e change within and between animals and from one environmental situation to another. However, the equations all depend on a *difference in temperature*, and biologists have described physical heat exchange for animals with a simplified equation that reflects the difference:

$$\begin{matrix} \text{Heat exchange} \\ \text{(conduction, radiation,} \\ \text{and convection)} \end{matrix} \quad = C(T_1 - T_2),$$

where

C = thermal "conductance,"
T_1 = temperature of warmer object (animal or environment), and
T_2 = temperature of cooler object (animal or environment).

Thermal conduct*ance* (C; note that it is not conduc*tion*) represents a combined value of physical characteristics of animals and environments influencing heat exchange by radiation, conduction, and convection. Since the value of C changes when animals or environments change, it is highly dependent on particular conditions associated with a set of measurements.

In addition to radiation, conduction, and convection, heat exchanges whenever water evaporates or condenses, changing state from a liquid to a vapor or back. Energy is required to change the state of water from a liquid to a gas; it is called the latent heat of vaporization of water. We will consider the importance of latent heat in some detail in Chapter 14 when we discuss interactions between water regulation and temperature regulation in hot environments.

For a living animal, heat is also generated internally from metabolism. This amount must be added to any net exchange of heat between an animal and its environment. For endotherms, internal aerobic heat production is a major component of temperature regulation (see Chapter 7); for ectotherms it is not.

All of the factors influencing heat exchange for an animal can be summarized for a situation where the animal has a temperature that is not changing (steady-state condition). If its temperature remains constant, heat loss must equal heat gain, and

$$\text{Heat loss} = \text{Heat gain} = H_m \pm H_r \pm H_c \pm H_w,$$

where

H_m = heat production from metabolism; *always pos*
H_r = heat gain or loss from radiation;
H_c = heat gain or loss from conduction and convection; and *maybe pos or neg*
H_w = heat gain or loss from changes in state of water.

The value of H_m is always positive, but the values for the other components may be positive or negative, depending on differences in temperature and factors influencing water loss and gain (although all of the components cannot be positive if temperature remains constant).

The types and avenues of heat exchange are also summarized in Fig. 6.3 for a typical complicated microenvironment. Heat is gained from metabolism and a variety of objects warmer than the animal through several components of radiation (direct from sun, reflected, etc.), conduction, and convection. Heat is lost to cooler objects by the same avenues, and heat will be lost from evaporation of water in most terrestrial environments, since animals usually have more water than their surroundings.

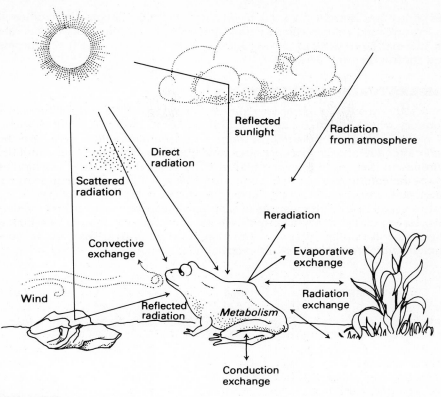

Figure 6.3
Schematic diagram of the avenues of heat exchange between an ectotherm and a variety of physical aspects of its microenvironment (modified from Gates and Porter, 1970).

In order to understand the importance of each avenue of heat exchange for an animal we would need to know a great deal about the physical characteristics of its environment, as well as properties of the animal that influence its metabolic rate (H_m) and rate of evaporative heat exchange (H_w). The determination of these factors is a very complicated task because animals and environments change considerably in ways that influence all avenues of heat exchange. Nevertheless, it is clear that sources of heat from the external environment influence the temperature of an animal through conduction, radiation, and convection.

For ectotherms, body temperatures are determined primarily by these external physical-environmental characteristics. H_m for ectotherms is usually low (see Fig. 5.4). Most importantly for the *regulation* of body temperature, some ectothermic animals can *move* within their environments to balance the exchange of heat by various avenues and achieve a degree of *control* over their temperatures by using heat-exchange aspects of their external environments.

This ability has not always been recognized, and early laboratory studies of temperature regulation in ectotherms emphasized the lack of ability to generate heat internally rather than looking for alternative ways in which ectotherms could control their temperatures.

TEMPERATURE PREFERENCES

If an ectotherm such as a lizard is exposed to a variety of air temperatures in a laboratory where external sources for heat exchange are limited, the temperature of the animal and the uniform thermal environment will in most cases be similar after a relatively short period of time (Fig. 6.4). If the temperature of the environment is changed, the temperature of the animal will change in a similar way. Some ectotherms have an ability to resist a temperature change over a brief period. For example, when the marine iguana is exposed to different temperatures its temperature changes at different rates when it heats and when it cools. However, for many ectotherms body temperature and environmental temperature are similar after an hour or two if the animals are exposed to environments with limited sources for external heat exchange.

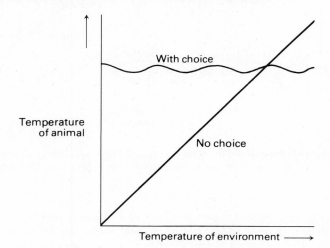

Figure 6.4
With no choice for heat exchange the body temperature of an ectotherm depends on the temperature of the environment. Under natural conditions (with a choice for heat exchange), many ectotherms will select a preferred body temperature.

Based on these and similar observations, ectotherms have been called *poikilotherms* ("changing heat") because of a limited ability to counteract the effects of an imposed temperature change in the external environment. In contrast, endothermic *homeotherms* ("same heat") have the same body temperatures over long periods during environmental temperature changes (from changes in H_m; see Chapter 7). However, the change in body temperature for many ectotherms that is seen in the laboratory is often a consequence of not giving them a *choice* of alternative sources of heat from the external environment or of not recognizing their temperature relationships under natural conditions, where they may have a choice of heat-exchange methods in a thermally diverse environment.

If some ectotherms are given a choice of several temperatures in the laboratory, they will select particular temperatures (Fig. 6.4). This behavior is called a temperature *preference*, and the selected temperature is called an "eccritic" or "preferred" temperature. The temperatures of ectotherms in a natural environment can also show a high degree of constancy (Fig. 6.5). The degree of precision in temperature regulation varies from one species to another and also depends on the heat-exchange characteristics of an animal's environment. In a number of cases the average preferred temperature in nature and in the laboratory are similar. When the behavior of these animals is taken into account, it is difficult to distinguish them from endotherms that maintain a relatively constant body temperature by internal mechanisms.

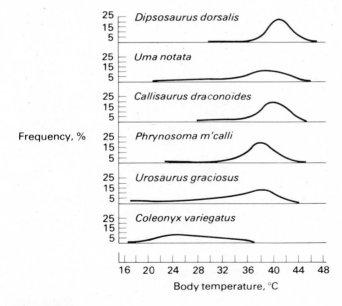

Figure 6.5
Frequency of occurrence of different body temperatures observed for several species of reptiles captured in the field. Note that most temperatures occur near a maximum, with more observations below the maximum than above the maximum (after Mayhew, 1968).

Some ectotherms use a variety of behaviors to achieve relatively precise control of their body temperatures. A heliotherm may bask in the sun in the morning, orienting its body so that a maximum surface area is exposed for gain of heat from solar radiation. Later in the day it will retreat to cooler, shaded areas or shuttle back and forth between sun and shade. Figure 6.6 shows the relationships between a series of behaviors of an ectothermic reptile and its body temperature. Behaviors such as "low-temperature burrowing" (to warmer substrates) and "orientation" (to the sun) lead to increases in body temperature from heat gains. Behaviors such as "shade seeking" and "high-temperature burrowing" (to cooler substrates) reduce body temperatures by changing environmental radiation and conduction heat-exchange characteristics.

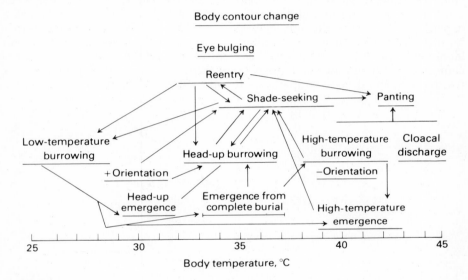

Figure 6.6
Relationships between a variety of behaviors for horned lizards (*P. coronatum*) that occur at or that influence different body temperatures (modified from Heath, 1965).

The preferred temperature resulting from behavior and heat exchange varies from one species to another as a consequence of differences in the availability of heat for exchange in different environments (Table 6.1). Terrestrial ectotherms usually have preferred temperatures close to 35-40 °C if environmental conditions permit, values similar to the temperatures maintained by many endothermic animals. We will examine some possible reasons for this similarity in the next chapter. Aquatic or amphibious ectotherms generally have lower preferred temperatures (Table 6.1) since aquatic environments are usually cooler and offer fewer alternatives for heat exchange.

Table 6.1 Preferred temperatures of some ectotherms from aquatic and terrestrial environments (from Brett, 1971, Dawson, 1975, Javaid and Anderson, 1967, Lillywhite *et al.*, 1973, and Precht *et al.*, 1970)

ANIMAL	PREFERRED TEMPERATURES
Sockeye salmon (*Oncorhynchus nerka*)	15 °C
Rainbow trout (*Salmo gairdneri*)	22 °C
Brook trout (*Salvelinus fontinalis*)	18 °C
Atlantic salmon (*Salmo salar*)	18 °C
Western toad (*Bufo boreas*)	27 °C
Turtle (*Pseudemys scripta*)	25-29 °C
Alligator (*Alligator mississippiensis*)	32-35 °C
Desert iguana (*Dipsosaurus dorsalis*)	38 °C
Collared lizard (*Crotaphytus collaris*)	38 °C
Anolid lizard (*Anolis carolinensis*)	30-33 °C
Beetle (*Ips typographus*)	27-33 °C
Fly larva (*Musca domestica*)	30-37 °C

Why is this precision and relative constancy of body temperature maintained? The answer relates to the effects of temperature on metabolic rate processes and enzyme functions influencing those rates (see above). One way to appreciate the importance of temperature regulation for effective function is to examine the consequences for ectotherms that cannot or do not regulate their temperatures.

RATE PROCESSES AND Q_{10}

If an ectotherm is exposed to different uniform thermal environments without being permitted a choice, the rates of its functions depend on its temperature. The animal is essentially a chemical system in which reaction rates depend on the heat energy of the cells. Many reaction rates are influenced by temperature in a similar way; thus a general description serves to illustrate the effects of temperature on rate processes.

As temperature increases the rate of a process increases, since more heat energy is added to the chemical system. However, if we start at different temperatures, the reaction rate does not change at the same rate as does the temperature (Fig. 6.7). If we start with a temperature of 10°C the rate may be 2. At a temperature of 20°C the rate may be 4, while at 30°C the rate may be 8. Thus in this example, every time the temperature changes by 10°C the rate changes by a factor of 2. The factor by which a rate changes when the temperature changes by 10° C is called the *temperature coefficient*, or Q_{10}. The Q_{10} for many biological processes is close to 2, although it may be higher or lower. Figure 6.7 shows how a reaction rate changes if the Q_{10} is 3.

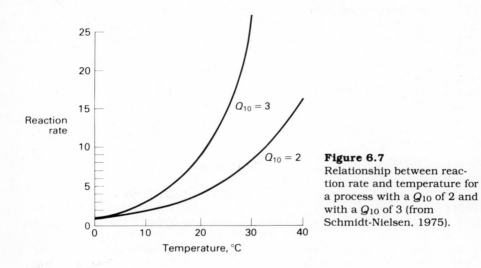

Figure 6.7
Relationship between reaction rate and temperature for a process with a Q_{10} of 2 and with a Q_{10} of 3 (from Schmidt-Nielsen, 1975).

The effects of temperature on rate processes are exponential in nature. To predict the rate at a new temperature we need to know the rate at another temperature and the value of the temperature coefficient. The new rate can then be calculated from the following exponential equation:

$$R_2 = R_1 Q_{10}^{\frac{(T_2 - T_1)}{10}},$$

where

R_2 = new rate at temperature T_2,

R_1 = known rate at temperature T_1, and

Q_{10} = temperature coefficient for the reaction.

Because exponential equations are so common and important in biology and physiology, their solution is discussed in Appendix II, where an example is given for the solution of temperature rate equations.

Q_{10} has been used by biologists to describe the effects of temperature on virtually every kind of rate process. However, Q_{10} is not a constant for any given reaction but instead varies slightly with temperature. The physicist Arrhenius formulated the expression to relate rates and temperatures; it depends on temperature as a variable and the constants R (the universal gas constant), e (the base of the natural logarithm; see Appendix II), and, for a given reaction, μ (the activation energy for a reaction). Q_{10} cannot be described entirely from the constants R, e, and μ, and will therefore change when temperatures change. However, changes in Q_{10} with temperature are usually small within the temperature range of interest to biologists (Q_{10} changes by about 10% within a 30°C range of temperature); it can thus be considered a useful "characteristic" of rates under most biological conditions.

The importance to an animal of regulation of body temperature can be appreciated because of these exponential temperature-dependent rate effects. Without an ability to keep the temperature of its body from changing, the rates at which an animal's functions proceed will be dictated by the thermal characteristics of its environment. If environmental heat exchange produces a high body temperature, the animal will use its chemical substrates at very high rates, perhaps faster than they can be replenished. If environmental heat exchange produces a low body temperature, rates will decrease perhaps to a level at which products from reactions are not formed at rates sufficient for most effective function. Function will be most effective when rates are "optimal" over a relatively narrow range of body temperatures; this fact is the basis for the evolution of body-temperature regulation.

It is important to note that the precision of temperature regulation by ectotherms depends not only on the benefits associated with maintaining particular body temperatures but also on the costs to achieve those benefits. For example, ectotherms in environments either with few sources for heat exchange or where the costs to achieve temperature regulation are high would be expected to have more variable temperatures. Lizard species in forested tropical habitats where available sources for heat exchange from solar radiation are few usually have lower and more variable body temperatures than species from more open habitats, where the costs for heat exchange are lower. In general, then, whether or not a physiologically "optimum" body temperature is maintained depends on the *net benefit* to the animal, and this net benefit can change considerably as a result of environmental changes affecting animals dependent on environmental sources for heat exchange.

Q_{10} **AND TEMPERATURE PREFERENCES**

Some of the finer details of ectotherm temperature preferences that determine the precision of temperature regulation can be interpreted relative to Q_{10} effects on rate processes. Note in Fig. 6.5 that the temperatures selected by many ectothermic reptiles are not distributed evenly around the maximum. There are usually more observations of lower temperatures below the maximum than there are observations of higher temperatures above the maximum. This type of distribution, with a long tail to the left, is said to be "negatively skewed" (skewed to the left on a graph). A number of ectotherms have frequency distributions of preferred temperatures that are negatively skewed, and an explanation can be found by examining temperature preferences with respect to the exponential effects of temperature on rate processes (see Fig. 6.8).

Temperatures may be selected by ectotherms to control rates. It is likely that there is some *error* in the ability of an ectotherm to determine a single optimal rate for a process. Since there will be some error, occasionally temperatures will be selected that result in values higher or lower than the optimum, so the frequency distribution for *rates* will be evenly (or normally) distributed around the optimum rate (left axis in Fig. 6.8).

When the normal frequency distribution for rates is translated into a frequency distribution for *temperature preferences* through an exponential Q_{10} curve (dashed arrows in Fig. 6.8), the resulting distribution for preferred temperatures is negatively skewed (x axis in Fig. 6.8). Thus the observed negatively skewed temperature-preference distribution for an ectotherm is expected from errors in determining optimum rates, strongly suggesting that temperatures are selected to regulate rates near optimum levels when possible.

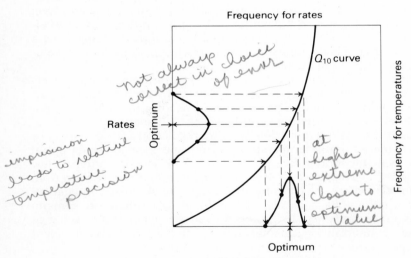

Figure 6.8
The distribution of preferred temperatures for an ectotherm (curve below the Q_{10} line) will be negatively skewed as a consequence of Q_{10} effects if the error associated with determining an optimum rate process (curve to the left of the Q_{10} line) is distributed evenly around the optimum rate (modified from Dewitt and Friedman, 1979).

ACCLIMATION

Temperature preferences produce rates close to optimal levels for ectotherms that have a choice and can select among different thermal environments with little cost. However, a number of ectotherms are not mobile, or their environments do not offer suitable choices for heat exchange at all times (the costs of temperature regulation are high). Plants are relatively immobile, and some animals, such as aquatic ectotherms, have few choices of sources for heat exchange. Even when different temperatures are available at some times, daily and seasonal changes modify the choices and the costs of achieving regulation. A heliotherm cannot bask at night, and ectotherms that do not migrate from temperate areas must deal with seasonal changes in climate. For these animals and situations the only alternative is to modify internal rate processes when temperatures change.

This internal modification is called *acclimation*. As an example, the rate of oxygen consumption for a frog previously at a temperature of 25 °C for a long time is about 35 μl O_2/g × h at a "new" temperature of 10 °C (Fig. 6.9). However, if the frog has been at a temperature of 5 °C for a long time (i.e., it is acclimated to that temperature), its rate of oxygen consumption at 10 °C is higher—about 80 μl O_2/g × h. As a result of previous exposure to either of two different body temperatures for prolonged periods the frog can have two different rates for the same process at the same body temperature. Some internal modifications have resulted from acclimation; the rate processes have been *compensated* after exposure to one temperature by being higher or lower than would be predicted just from a temperature coefficient.

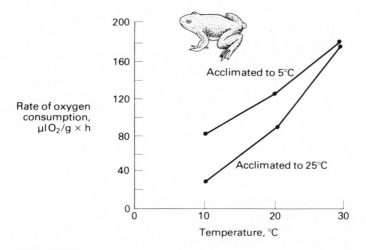

Figure 6.9
The rate of oxygen consumption of frogs (*Rana pipiens*) at a given temperature depends on the temperature of acclimation (from Rieck *et al.*, 1960).

It is important to be careful about removing an animal from its natural environment to study acclimation. If changes in internal rate processes occur seasonally, they may not be noticed if the animal is exposed to a particular temperature for only a few weeks or months. For example, the metabolic rate of the frog in Fig. 6.9 may be different if it is acclimated to 5 °C during the winter as compared with the summer. Other characteristics in an animal's environment will influence acclimation—the changes that occur when only one environmental variable is studied. For example, the degree of hydration or the seasonal reproductive condition of an animal can influence its ability to compensate reaction rates with respect to temperature. The term *acclimatization* is used to describe the interactions occurring under natural conditions that influence acclimation.

Compensation for temperature changes by means of acclimation can be examined relative to Q_{10} by studying Fig. 6.10. A rate of K is observed at a temperature T. If $Q_{10} = 2.0$ and the temperature is increased or decreased by 10 °C, an animal that showed no compensation will have a rate described by the line connecting $K/2$ and $2K$. If *perfect* compensation occurs, the rate will not change when the temperature changes. *Partial* compensation occurs between these extremes. Perfect compensation through acclimation is very difficult, since it requires a chemical system with an unusual response to temperature; that is, one that does not follow the Q_{10} response typical of most rate processes.

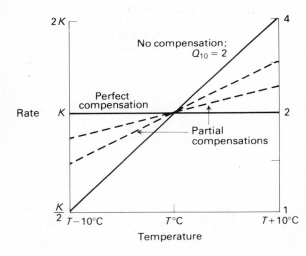

Figure 6.10
For a reaction rate with a Q_{10} of 2, no compensation from acclimation will result in the predictable doubling of the reaction rate when temperature increases 10 °C. Perfect compensation will yield no change in reaction rate when temperature changes. Acclimation usually results in an intermediate (partial) compensation (modified from Precht *et al.*, 1973).

The Mechanisms for Acclimation Compensation

What could explain acclimation or internal compensation for temperature changes? The answers should be sought from changes that occur within the cells of animals, particularly in those species with few alternatives for dealing with the problems of variation in their temperatures (where the costs of temperature regulation are high). The ability of enzymes to catalyze reactions depends on temperature (Fig. 6.1), and most hypotheses to explain acclimation concern mechanisms to *change* enzyme function when temperatures change.

For animals that select or maintain a particular body temperature, the optimal temperature for some enzymes and the preferred or maintained temperature can be similar. If a change in enzyme function occurs as the result of acclimation, the optimum temperature can change. This possibility is shown graphically in Fig. 6.11, and it has been described for several ectotherms. For example, the temperature at which the muscle enzyme phosphofructokinase has a maximum rate of reaction in trout depends on acclimation temperature. The maximum occurs at 10 °C for fish acclimated to 18 °C, and the maximum occurs at 5 °C for trout acclimated to 2 °C.

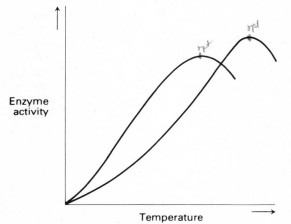

Enzyme activity

Temperature

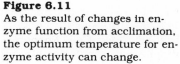

Figure 6.11
As the result of changes in enzyme function from acclimation, the optimum temperature for enzyme activity can change.

Enzyme function can change in several ways with different consequences of producing compensation. The amount of an enzyme present relative to substrate concentrations can change. For example, an increase in substrate concentration at a relatively low temperature can increase the overall rate of a reaction if the reaction rate is normally limited by substrate concentration. Alternatively, the three-dimensional configuration of an enzyme can change when temperature changes, influencing the effectiveness with which the enzyme interacts with its substrate. Some compensation will occur if more than one temperature results in effective enzyme structures, but there are limits over a wide range of temperature change. This type of change in enzyme structure with temperature is called *enzyme modulation*. The change in pH with a change in body temperature in ectotherms (Figs. 1.16 and 2.19) is one mechanism to produce enzyme modulation by the maintenance of a relatively constant protein net charge as temperatures vary.

A third mechanism for acclimation involves production of different enzymes by different genetic units within a cell when temperature changes. Enzymes produced by different genes that catalyze the same reaction are called *isozymes*. Isozyme production may be the most effective mechanism for achieving compensation when temperatures change, since different isozymes can have different optimal temperatures (Fig. 6.11). However, the limitation here is in the ability of animals to contain the diversity of genetic material needed to produce different enzymes when environments change. For example, several species that acclimate by producing isozymes when temperatures change are polyploid: i.e., they have more than the usual amount of genetic material. Perhaps because of genetic limitations, it appears that most compensation from changes in enzyme function results in only a few (one or two) separate optimum temperatures, and the optimum temperatures usually differ in absolute efficiency (Fig. 6.11).

UPPER AND LOWER LIMITS FOR ECTOTHERMS

Acclimation results in some changes in the internal response of an individual as its environment changes and represents one way for animals to deal with some variation in environmental heat-exchange characteristics. Temperature selection from a part of the environment represents another. However, there are upper and lower limits to temperatures for ectotherms. Although some plants and microorganisms are found in virtually every thermal environment, most multicellular ectotherms are found in environments where the temperature extremes are within the 0-45 °C range.

We will consider some possible reasons for the upper limit in the next chapter when we discuss temperature regulation in endotherms and the problem of why their body temperatures are maintained between 30° and 45°C. The lower limit to body temperature in ectotherms is set by the freezing point of cellular water. Cells are composed of 80-90% water, and the chemicals within them must remain in solution at particular concentrations to most effectively function. Even with this limitation, there are some exceptions in certain animals that have evolved to live in environments where the temperature is below the freezing point of cellular water.

Several species of fishes and invertebrates live in the cold waters of the Arctic and Antarctic regions. The water temperature below a depth of about ten meters remains at − 1.7 °C all year. Near the surface the water temperature increases to 5 °C in summer and decreases to − 1.7 °C in winter when ice forms on the surface. Some species of fishes and invertebrates live only in deep areas. If they are caught and brought to the surface in winter, they will freeze when they contact water containing ice crystals. Normally they never come near the surface but live in a continuous "supercooled" state. As long as they never contact ice crystals, there will be no focus for the formation of ice in their tissues. Species that live near the surface either migrate seasonally or produce chemicals that act as antifreeze by lowering the freezing point of their tissues below that of their environment in winter.

ACCLIMATION AND TEMPERATURE PREFERENCES

By exposing an ectotherm to different temperatures in the laboratory, it is possible to define the extremes of the temperature range within which survival is possible for a given species. It is usually done by exposing a group of animals to high and low temperatures and determining the upper and lower limits of body temperature at which 50% of the animals survive. These two points define the limit of temperature tolerance for a species. There are differences in tolerance among species, with those from colder, less variable environments usually having smaller ranges of tolerable temperatures.

The limits for temperature tolerance also depend on acclimation temperatures for ectotherms. An example is shown in Fig. 6.12 for bluegill sunfish. When the fish are acclimated to 10°C for a long time, they will survive within the limits of 0 to 25°C. When they are acclimated to 25°C, the limits are 8 to 33°C. The acclimation of ectotherms to different temperatures is shown graphically by a polygon describing the tolerance limits. For any ectotherm there will be an absolute upper acclimation limit above which survival is not possible. For the bluegill sunfish this is about 35-36°C (Fig. 6.12).

When ectotherms are acclimated to different temperatures and then permitted to select temperatures, they initially show a preference that depends on the acclimation temperature. This preference pattern is shown for the bluegill sunfish by the dashed arrows within the tolerance polygon in Fig. 6.12. However, within a short time individuals will begin to select different temperatures and move toward a "unique preferred temperature" for that species (shaded square in Fig. 6.12), suggesting that the process of acclimation can occur in a relatively short period of time.

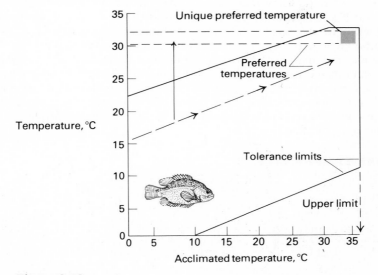

Figure 6.12
When ectotherms are acclimated to different temperatures, their ability to tolerate temperature extremes are described by a "tolerance polygon" with an upper limit. This example is for bluegill sunfish. Fish placed in a temperature-selection apparatus gradually shift preferences along the dashed line to a "unique preferred temperature" (shaded square) within a few hours. A gradual shift is important, since immediate selection of a high preferred temperature would result in death if acclimation to low temperatures has occurred (modified from Reynolds and Casterlin, 1979).

It is important that there is a gradual shift in preferred temperatures. If an ectotherm moves immediately to a high unique preferred temperature, it can end up outside the zone of temperature tolerance (vertical arrow in Fig. 6.12). A few individuals may do this, but since they die the behavior is genetically selected against. Such "abnormal" behavior is a consequence of individual ectotherms normally being acclimated to temperature changes that occur over long time periods, meaning that they have not evolved to deal with the short-term choices found in the laboratory.

The short-term shifts in preferred temperatures for acclimation in ectotherms can be interpreted relative to the underlying mechanisms for acclimation. The gradual shift toward a unique preferred temperature may be the consequence of substrate concentration changes or temperature effects on enzyme structure during acclimation. With these two mechanisms, an ectotherm acclimated to a low temperature that immediately selected a high (unique preferred) temperature would utilize substrates at excessive rates. The third mechanism for acclimation, the production of a separate set of enzymes, may require a certain amount of time for the induction of isozymes as an ectotherm shifts gradually toward a unique preferred temperature.

The unique preferred temperature for a species has been called its "ultimate" or "final" preferred temperature. It is achieved relatively soon after acclimation if suitable choices are available (Fig. 6.12), and it represents the range of temperatures where function is presumed to be "optimal," i.e., most effective for a given species. This temperature is the one selected to regulate processes, and it has also been called the "optimal" temperature. Although we will find that changes in an animal's environment will change which temperature range is optimal (from changes in benefits or costs of regulation), it is of interest to examine some of the characteristics associated with these temperatures and some hypotheses for why animals have evolved to select certain unique preferred temperatures.

OPTIMAL TEMPERATURES

If an ectotherm has a choice and the costs of choices are low, which temperature should it select? There is likely to be one or two temperatures of those available that provide the most effective function for a given ectotherm, and these are called the optimal temperatures. To study optimal temperature selection, it is important to examine rate processes that have a direct influence on the survival or reproductive ability of animals, since these processes are the ultimate measures of efficiency for living animals.

A number of rate processes can be measured that are directly important for survival and reproductive success. For example, the rate at which the desert iguana (*Dipsosaurus dorsalis*) assimilates energy from ingested food is maximal at its preferred temperature of about 38 °C. Rate of growth is also temperature-dependent, and animals that grow rapidly are more likely to survive and reach reproductive age sooner than those that grow more slowly, so this rate process is particularly instructive to examine with respect to temperature selection by ectotherms.

Rate of growth is a function of several processes. The food an animal consumes must be assimilated; some of the assimilated energy must be used for maintenance and the remainder can be used to add new tissue. Both assimilation and maintenance rates are temperature-dependent, with the net result that an animal will grow most rapidly at a particular body temperature (Fig. 6.13). Above the optimal temperature for growth the maintenance costs will be higher becuase of Q_{10} effects on rates of use of assimilated energy, while below the optimal temperature the rate of assimilation of food will be less than maximum.

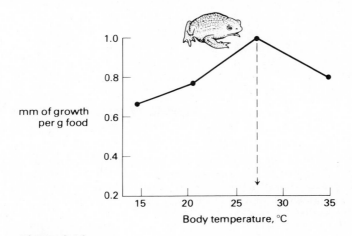

Figure 6.13
Increase in linear growth per gram of food consumed by western toads (*Bufo boreas*) given all the food they can eat and exposed to different environmental temperatures shows that growth efficiency is maximum at a body temperature of 27 °C (modified from Lillywhite *et al.*, 1973).

Because rate of growth is important for survival and reproduction and because it also depends on temperature, we can predict that ectotherms should select those temperatures that are optimal for growth rate if they have a choice. Only a few observations have been made concerning this hypothesis. For example, the toads used to measure the temperature for maximum growth rate shown in Fig. 6.13 selected a temperature of 27 °C when they were given access to all the food they could eat. Sockeye salmon selected a temperature of 15 °C between the periods of feeding at dusk and dawn (Fig. 6.14); 15 °C is the temperature for their maximum growth rate when they have sufficient food (Fig. 6.15).

Ectotherms do not always select temperatures optimal for growth rate. For example, during the day the sockeye salmon migrate to the bottom of the lake where temperatures are lower (Fig. 6.14). Daily migration patterns similar to this are also seen in a number of other aquatic vertebrates and invertebrates.

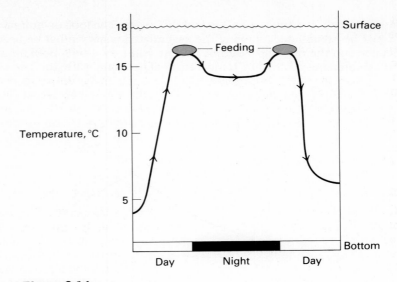

Figure 6.14
The daily movement patterns of salmon in a lake. At dusk and dawn
they feed near the surface. They remain in warm water near the surface at night, but they move to cooler water near the bottom during
the day (modified from Brett, 1971).

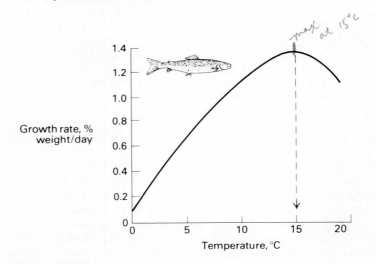

Figure 6.15
Rate of growth of salmon given all the food they can eat and at
different body temperatures (from being exposed to different
environments). Maximum growth rate occurs when body
temperature is 15 °C, and it is the temperature selected by
salmon at night (see Fig. 6.14) (modified from Brett, 1971).

Several additional factors explain this change in temperature preference. For example, if an ectotherm obtains less food, selecting a lower temperature will reduce maintenance costs because of Q_{10} effects. We will consider this possibility in the chapter dealing with torpor and hibernation (Chapter 8).

Predators can also force an ectotherm into a less than optimal temperature environment with respect to growth. Light intensity is relatively low at the bottom of a lake even during the day, possibly making it more difficult for predators to find salmon (Fig. 6.14). Characteristics of the natural environment such as predators will influence temperature selection by ectothermic animals by influencing the costs of temperature regulation, but the net effect should be to optimize function whenever possible within these limitations.

CONTROLS FOR TEMPERATURE SELECTION IN ECTOTHERMS

The muscles responsible for movement are effectors for temperature regulation in ectotherms that select different microenvironments for heat exchange. Locomotion for temperature regulation is controlled in vertebrates by means of information on body temperature processed in the central nervous system. This control can be demonstrated by heating or cooling a part of the brain called the anterior hypothalamus. It is located near the spinal cord, and we will discuss its importance for control of body temperature for endothermic vertebrates (see Chapter 7).

The anterior hypothalamus can be experimentally heated or cooled selectively by running hot or cold water through small tubes placed in this part of the brain. When this temperature alteration is performed on blue-tongued lizards (*Tiliqua scincoides*) that are allowed to choose between a high-temperature (45 °C) and low-temperature (15 °C) environment, the lizards remain in the hot environment longer when the hypothalamus is cooled and shift to the cooler environment soon after the hypothalamus is warmed (Fig. 6.16). Only a small amount of heat is added or removed by heating or cooling just this part of the brain, and when the hypothalamus is cooled the temperature of the rest of the body increases more than normal since the lizards remain in the warm environment longer. When the hypothalamus is heated, the lizards remain in the cooler environment, causing their temperatures to decrease more than normal (Fig. 6.16).

These results suggest that the anterior hypothalamus of the lizard functions in a manner similar to a thermostat with a set point. If the thermostat is cooled, the animal heats more than normal. This situation is similar to the thermostatic control of body temperature in endotherms (see Chapter 7). The major difference lies in the use of the muscle effectors for changing body temperature; either movement of the animal (ectotherms) or internal heat production caused by shivering (endotherms) is effected.

The activation of the effectors normally occurs when body temperature is displaced relative to some "reference" or "set point" temperature in the hypothalamus (see Introduction). If we heat or cool the hypothalamus the lizard is "tricked" and will behave as if its entire body was warm or cold. This result also suggests that the receptors for detecting changes in body temperature as determined from reference values are located in the anterior hypothalamus.

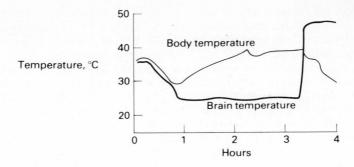

Figure 6.16
When the temperature of the brain of a blue-tongued lizard
is cooled, body temperature increases because the lizard
stays in a warm environment longer. When the brain is
heated, the lizard shifts to a cooler environment and body
temperature decreases (adapted from Hammel *et al.*, 1967).

TEMPERATURE REGULATION IN DINOSAURS?

Recently there has been considerable interest in studying the physiology of dinosaurs. Since they are extinct, we are forced to study the closest living relatives
available, the reptiles. Also, by careful study of the structures of fossil remains
some characteristics of function can be hypothesized.

Since many scientists initially studied ectothermic reptiles in the laboratory where the animals had few choices for heat exchange, they were unim-
pressed with the abilities of reptiles to regulate body temperatures, and reptiles
were labeled "poikilotherms." By analogy, dinosaurs were also believed to be
"primitive" with respect to temperature regulation, or "cold-blooded." But we
now know that some reptiles are quite sophisticated with respect to temperature regulation under appropriate conditions. Although we cannot observe temperature preferences of dinosaurs, we can argue that perhaps some were equally
effective in dealing with the heat-exchange characteristics of their environments. This point of view provides a framework to examine some other evidence suggesting that dinosaurs may have had regulated body temperatures.

One possibility is that some large dinosaurs may have had relatively constant temperatures simply as a consequence of their size. The specific heat of
animal tissue is about 0.9 cal/g × °C; that is, it requires 0.9 calorie to change one
gram of tissue one degree Celsius. A three-ton dinosaur would have a mass of
2,721,540 grams. A change in its temperature by only one degree Celsius
would require 2,449,386 calories, an appreciable amount of energy. If the
dinosaur had a minimum standard metabolic rate equal to that of modern
lizards with a temperature of 37 °C (Fig. 5.2), this amount of energy would be
about 70% of its total daily minimum rate of energy expenditure. Because of its
large size a rapid change in body temperature by a large amount would be
unlikely to happen, so body temperature could have been relatively constant for
large dinosaurs on a daily basis.

Other arguments for temperature regulation in dinosaurs are based on the morphology of fossil structures that are analogous to similar structures in endothermic birds and mammals. The cellular structure of some dinosaur bones suggests a high degree of vascularization similar to the bones of most endotherms but unlike the bones of many modern reptiles. In addition, reconstructions of the skeletal systems of some dinosaurs suggest they may have had a form of locomotion with an upright posture and gait similar to mammals (Fig. 6.17) but unlike the sprawling posture and gait of many modern reptiles.

Figure 6.17
Reconstructions of the fossil remains of some dinosaurs suggest that they could have had an erect posture and a rapid gait (adapted from Ostrom, 1978).

Further suggestive evidence for temperature regulation in some dinosaurs comes from examination of their external features. Dinosaurs such as *Brontosaurus* and *Brachiosaurus* had very long necks, so the distance from the heart to the brain may have required a four-chambered heart to produce arterial pressures sufficient to support a long column of fluid and to force blood up to the brain. This kind of circulation could have led to a circulatory system similar to birds and mammals but unlike most other reptiles.

The structure and arrangement of the large dorsal fins of *Stegosaurus* (Fig. 6.18) may have functioned as a method of heat loss as well as being a defense against flesh-eating predators. The fins appear to have been well vascularized and were staggered in two rows, meaning that when *Stegosaurus* oriented itself perpendicular to the wind the fins would present a maximum surface area for heat exchange. Experiments with scale models of *Stegosaurus* in wind tunnels indicate that this situation could have been the case. If the fins

Figure 6.18
The large dorsal plates of *Stegosaurus* were arranged alternately in two
rows and may have served as heat-loss radiators (adapted from Ostrom,
1978; also see Farlow *et al.*, 1976).

did in fact function as a heat-loss mechanism, it would suggest that *Stegosaurus*
may have regulated its body temperature.

   If some dinosaurs regulated their temperatures it could help us to understand their extinction. The earth underwent drastic climatic changes at the end
of the Cretaceous period. With cooler temperatures the demands of dinosaurs for
food (which in large dinosaurs were substantial) would have increased if
temperatures were regulated by endothermic mechanisms (see Chapter 7). At
the same time the supply of food could have decreased if plants were also influenced by the climatic changes. Extinction could have been related to increased
demands for energy with insufficient supplies of food. In addition, ectothermic

dinosaurs that may have regulated temperatures behaviorally could have been exposed to environments without the appropriate conditions for achieving optimal body temperatures for survival and reproduction. The result was the evolution of birds, mammals, and modern reptiles adapted to the heat-exchange characteristics of their environments.

Alternative Hypotheses

It seems reasonable to suggest that some dinosaurs may have had regulated temperatures, based on current understanding of temperature regulation in ectotherms. However, a major controversy concerns whether the source of heat for the regulation was internal or external (ectothermy versus endothermy). On this point there is great dispute and lively debate.

High internal heat production is usually associated with high oxidative metabolism (see Chapter 7). However, modern reptiles rely to a large extent on anaerobic glycolysis for short-term high demands for energy. Recall that anaerobic glycolysis has the advantage of very rapid ATP production, with the subsequent possibility for longer-term oxidative efficiency (see Chapter 4).

Most of the evidence discussed above for dinosaurs can apply to animals without high oxidative demands, that is, animals where the sources of heat for regulation are primarily external. For example, the long neck of a dinosaur might not require a four-chambered heart if demands for oxygen delivery to the brain were low. Perhaps blood simply could have been pumped when the head was low if anaerobic metabolism was sufficient to meet demands when the head was high. Posture, bone structure, and gait also would not necessarily be directly dependent on high internal rates of heat production, nor would heat loss structures necessarily depend on the source of lost heat.

Some form of temperature regulation in dinosaurs seems highly probable. The question of the source of heat for regulation cannot be resolved, since we do not have access to living dinosaurs. However, the arguments provide an interesting forum in which to discuss factors important for temperature regulation.

THE ADAPTIVE VALUE OF FEVER

Since many ectotherms demonstrate a preference for a body temperature that can be optimal, we can use this behavior to ask some interesting questions about some aspects of temperature regulation. It is a common observation that certain infections lead to a fever, which is an increase in body temperature above normal, controlled values. It is usually treated with chemicals to reduce body temperature, but perhaps an increase in body temperature is of advantage to an animal with an infection. This hypothesis can be tested experimentally by studying the temperature preferences of ectotherms that are infected with fever-producing bacteria.

Normally, desert iguanas will select a body temperature of about 38 °C and maintain this temperature by their behavior. If they are injected with a fever-producing bacterium species they will select a higher body temperature (40-42 °C). The preference for a behavioral fever suggests that the higher body temperature is optimal to combat the infection. A higher body temperature may make it more difficult for the infective bacterium species to survive in the host, since it too will have an optimal temperature for growth and reproduction.

If a fever has survival value, those iguanas that select higher body temperatures should be more likely to survive the infection. Since most of the iguanas will select a high body temperature after infection if they have a choice, the value of fever for survival can be tested by exposing infected lizards to environments producing different body temperatures without allowing the animals a choice for heat exchange. Note that ectotherms are ideal subjects for this type of experiment, since the body temperatures of endotherms (including humans) cannot be changed easily from changes in environmental heat-exchange characteristics.

Those iguanas exposed to environments in which higher body temperatures of 40-42 °C can be selected are more likely to survive than those exposed to environments producing a "normal" body temperature of 38 °C or lower (Table 6.2); it suggests that the fever helps combat the infection. Extreme fevers can be very dangerous because death can occur if body temperature increases too much. However, a relatively small increase in body temperature is not of great concern and may be beneficial.

Table 6.2 Survival of *Dipsosaurus dorsalis* infected with *Aeromonas hydrophila* bacteria and exposed to different body temperatures (from Kluger, 1978)

BODY TEMPERATURES	PERCENT SURVIVING AFTER FIVE DAYS
42 °C	85%
40 °C	70%
38 °C	25%
34 °C	0%

Further experiments with lizards and bacteria suggest a mechanism for the survival value of behavioral fever. When the fever-producing bacterium *Aeromonas hydrophila* is grown in culture, it will grow equally well at 38 °C and at 42 °C. However, when the level of iron in the growth medium is reduced, bacterial growth is inhibited at 42 °C but not at lower temperatures. The iron is an important nutritional component for the bacteria (see Chapter 11) and may form part of an enzyme needed for normal growth or metabolism. The rate of use of the iron is higher at a higher temperature because of Q_{10} effects; thus it limits growth and reproduction at higher temperatures.

When lizards were infected with *A. hydrophila* bacteria, the concentration of iron in their blood decreased at all temperatures, but a body temperature of 42 °C was required to inhibit bacterial growth because of a reduction in blood iron concentration. Moreover, experimental treatments that *increased* iron concentrations in the blood of infected lizards at temperatures of 42 °C caused an increase in the death rate of the lizards. These results confirm the hypothesis that behavioral fever is optimal for survival because of the production of a thermal environment that is not optimal for survival of the infective agent.

SUMMARY

Ectotherms obtain heat for regulation of body temperatures primarily from the external environment, while endotherms regulate body temperatures primarily from internal heat production. For any object, temperature is determined by the amount of internal heat production and the net exchange of heat through radiation, conduction (including convection), and changes in state of water. Some ectotherms exhibit a preference or choice of different microenvironments for temperature regulation through heat exchange, particularly via radiation, conduction, and convection, where exchange is determined by the difference in temperature between an animal and its environment.

Ectotherms that have no choice of body temperatures because of inability to move or lack of environmental sources for heat exchange are more subject to temperature effects on rate processes. The temperature coefficient, or Q_{10}, describes the effect of temperature on rates. The relationship is exponential, usually with a Q_{10} of 2, meaning that rates double for every 10°C increase in temperature. The Q_{10} effects on rates can be modified somewhat through acclimation. Exposure to a body temperature can produce compensation in rates that make them different from rates predicted just from Q_{10}. Partial compensation for temperature variation can occur from modification by means of substrate concentrations, structural changes in enzymes, or production of different enzymes (isozymes) at different temperatures. However, there are cost limits on the extent to which ectotherms can modify internal rates when their temperatures change.

Ectotherms that can move and have a choice of microenvironments would be expected to select conditions where body temperatures are optimal for their survival and reproduction. Experimental evidence suggests that some ectotherms do select temperatures that maximize rates of food assimilation and rates of growth, but other environmental factors, such as food availability and location of predators, also influence the selected body temperatures. Control of body temperature through movement in vertebrate ectotherms involves body temperature receptors in the anterior hypothalamus of the brain. Body temperature is controlled in a manner similar to a thermostat by comparison of body temperature with a reference temperature.

Dinosaurs used to be thought of as "cold-blooded," but recent recognition of temperature regulation in some present-day ectotherms suggests the possibility that they could have regulated their temperatures. Interpretations of the structure of fossil remains and the external morphology of dinosaurs also provides some suggestive evidence for temperature regulation. Extinction of dinosaurs may have been related to either increased demands for food accompanied by decreased supplies of food resources or the lack of suitable heat exchange characteristics from external environments when climate changes occurred at the end of the Cretaceous period.

The preference of some ectotherms for an optimal temperature leads us to examine the adaptive value of fever. When desert iguanas are infected with a species of fever-producing bacterium they select a higher body temperature. This behavioral fever results in a greater probability of surviving the infection, since a higher body temperature results in a nonoptimal temperature environment for the infective organism.

Temperature Regulation in Endotherms

7

The source of heat for body temperatures in endotherms comes primarily from their aerobic metabolism rather than from the external environment. As a consequence, when the heat-exchange characteristics of a cold environment change, body temperatures will remain relatively constant if rate of energy expenditure changes. Recall (Chapter 6) that for a constant temperature,

$$\text{Heat loss} = \text{Heat gain} = H_\text{m} \pm H_\text{r} \pm H_\text{c} \pm H_\text{w},$$

where

H_m = heat production (gain) from metabolism;

H_r = heat gain or loss from radiation;

H_c = heat gain or loss from conduction and convection; and

H_w = heat gain or loss from changes in state of water.

For an endotherm in a "cold" environment, H_r, H_c, and H_w can all be negative; that is, heat *loss* can occur by all of these avenues of heat exchange, or

$$\text{Heat loss} = H_\text{r} + H_\text{c} + H_\text{w},$$

but heat *gain* will be provided entirely from metabolism, or

$$\text{Heat gain} = H_m = \text{heat loss} = H_r + H_c + H_w$$

when body temperature remains constant in a "cold" environment.

Homeotherms are animals that maintain relatively constant body temperatures when their environmental temperatures change. We have seen that some ectotherms are relatively homeothermic through their behavior. We now wish to understand the consequences and mechanisms of homeothermy for endothermic animals.

One consequence of endothermy is relative *precision* of temperature control. Ectotherms depend on external sources of heat that vary daily and seasonally. Endotherms function more independently of external environmental heat-exchange characteristics (H_r, H_c, and H_w can all be negative, but body temperature will still remain constant when H_m changes). However, this precision and independence carry considerable *costs* in energy expenditures from aerobic metabolism. Energy substrates must be used at high rates if internal heat production is to contribute to temperature regulation. For an endotherm, the efficiency of this energy use is maximized when the demands for energy expenditures for temperature regulation (from changes in H_m) are minimized as much as possible (see introductory paragraphs for this part of the book).

Another consequence for endothermy is the extent to which metabolism depends on sustainable aerobic processes. Ectotherms that rely more on rapid, anaerobic processes for increasing energy expenditures are usually limited in the time in which anaerobic metabolism can supply increased demands for muscular activity (Chapter 18). Although aerobic endothermy carries a high cost for maintenance (Chapter 5) and activity (Chapter 9), an additional advantage to precision of temperature regulation is increased endurance—the ability to sustain increases in metabolic rates for relatively long periods.

The extent to which it is possible for energy expenditures for temperature regulation in endotherms to be minimized during inactivity depends on two types of factors: (1) environmental factors, such as sources for gains and losses of heat by radiation, conduction, and convection; and (2) biological factors, such as body size and animal insulation characteristics. We will discuss these factors shortly. First, however, it is important to examine some of the characteristics of endotherms and endothermy. Which types of organisms are endotherms? What temperatures do they maintain? When are they maintained and why? Answers to these questions allow us to interpret the importance of endothermy and the importance of environmental and biological factors that can influence effective energy use for temperature regulation.

ENDOTHERMIC ORGANISMS

Just as physiologists used to think that all ectotherms were "poikilotherms," they also used to think that the only endotherms were birds and mammals. This is not the case. Consider the skunk cabbage (*Symplocarpus foetidus*). The flowering part of this plant remains at a relatively constant temperature (between 15 and 20°C) even when the environmental air temperature changes from −15 to +15°C. Similarly, the inflorescence (floral cluster) of the plant *Philodendron selloum* remains at a temperature between 38 and 44°C even when the air temperature changes from 0 to 40°C (Fig. 7.1). Most plants are immobile ectotherms that do not regulate their temperatures, but the inflorescences of these plants are endothermic. The maintenance of a high temperature in the flowering parts of these plants may be important for either rapid flower development or the production and release of volatile chemicals to attract pollinators such as bees.

There are other examples of endotherms that maintain relatively constant temperatures in certain parts of their bodies when environmental temperatures change (Fig. 7.1), e.g., the bumblebee (*Bombus vagans*) and the sphinx moth (*Hyles euphorbia*). Each keeps the temperature of its thorax above the temperature of the environment when it is active. Some of this heat comes from flight muscles located in the thorax. Similarly, tunas maintain the temperature of swimming muscles and some other organs at relatively constant temperatures when environmental temperatures change. Brooding Indian pythons (*Python molurus*) produce heat from shivering to keep their body temperatures and the temperatures of their eggs relatively constant when air temperatures decrease. Thus there are examples of endothermy shown by a taxonomically wide variety of multicellular organisms.

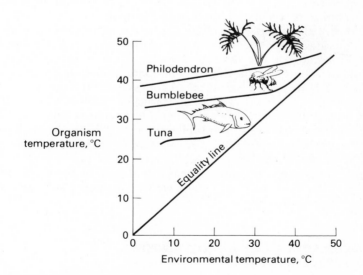

Figure 7.1
Temperatures of the flowers of philodendron, the thorax of bumblebees, and the swimming muscles of tunas remain relatively constant when environmental temperatures change (after Nagy *et al.*, 1972; Heinrich, 1972; Carey and Teal, 1969).

CHARACTERISTICS OF ENDOTHERMY

Endothermy is a matter of degree in both space and time. No animal maintains the temperature of all parts of its body at exactly the same temperature. For example, we will find that the appendages of birds and mammals are maintained at different temperatures than those of the rest of the body. Although the thorax of an active bee or sphinx moth is maintained at a relatively constant temperature some of the time, the head and abdomen will vary in temperature by a greater amount. In addition, philodendron and skunk cabbage do not regulate the temperatures of roots.

Endothermy is also a matter of timing. Even when part of the body is maintained at a relatively constant temperature from internal heat production, it does not necessarily occur all the time or to the same degree at different times. Bumblebees and sphinx moths, for example, only regulate the temperature of the thorax when they are active, and philodendron and skunk cabbage maintain a relatively constant temperature in an inflorescence only when flowering occurs.

Differences in the location or timing of temperature regulation have been described as *heterothermy*. A heterotherm could be any animal that did not maintain the same temperature in all parts of the body (regional heterothermy) all of the time (temporal heterothermy). All animals could be heterotherms by strict application of this criterion, and it is more useful to understand temperature regulation if less emphasis is placed on terms that serve to describe variation in temperature regulation than it is to attempt to explain whatever variation is observed in terms of its functional significance.

Functional significance involves the *costs* of temperature regulation (measured by rates of energy expenditure) relative to the *benefits* derived from regulated temperatures. If costs are excessive when compared with benefits there should be variation in the degree of temperature regulation in time and/or space. Whether costs are excessive when compared with benefits depends on the impact of costs (from energy expenditures) on survival and the ability to reproduce as compared with the advantages temperature regulation can produce for these two necessities. In this chapter we will examine costs and benefits for animals that maintain relatively constant temperatures for some period of time in some part of the body through internal heat production. In the next chapter we will examine the consequences of and some explanations for a regulated *change* in body temperature when we discuss torpor and hibernation in ectotherms and endotherms.

THE TEMPERATURES OF ENDOTHERMS: SOME POSSIBLE BENEFITS

Before we examine the costs and mechanisms involved in endothermy, we should examine the temperatures maintained by a variety of endotherms to see if there is any consistent *pattern* that provides a clue as to why certain temperatures are maintained. If there is a consistent pattern, it can provide a basis for understanding benefits related to certain temperatures.

The temperatures of a variety of endotherms are shown in Table 7.1. Although some have relatively low temperatures (such as parts of the skunk cabbage and the tuna), many endothermic temperatures are between 30 and 45 °C (Table 7.1). Recall that some ectotherms from terrestrial environments select body temperatures that are in the same range (see Table 6.1). Some differences in the temperatures of both ectotherms and endotherms are related to the heat-exchange characteristics of their environments that influence the costs of maintaining particular temperatures. However, several species from very diverse groups have temperatures in a similar range.

Table 7.1 Temperatures maintained by organisms through endothermy when environmental temperatures change

ORGANISMS	ENDOTHERMIC TEMPERATURE RANGES
Placental mammals	36-38 °C
Nonplacental mammals	34-36 °C
Birds	39-42 °C
Flower of skunk cabbage	15-20 °C
Flower of philodendron	38-43 °C
Bumblebee thorax	32-36 °C
Sphinx moth thorax	38-43 °C
Tuna muscle	18-23 °C
Indian python incubating eggs	29-34 °C

The ability to keep body temperature relatively constant when environmental temperatures change achieves the benefits of nearly perfect compensation for Q_{10} effects on rate processes (see Fig. 6.10) and permits the evolution of enzyme systems and optimal function (assimilation, growth, etc.) at the regulated temperature. But what is the benefit associated with this characteristic range of temperatures observed in a variety of animals? Why is a body temperature in the 30-45 °C range important? Several sets of hypotheses have been proposed.

The Historical Hypotheses

Two hypotheses based on the evolutionary history of animals have been proposed. One historical hypothesis claims that a temperature in the 37-42 °C range is a taxonomically "advanced" characteristic, because birds and mammals have these temperatures (and are homeothermic), but other more taxonomically "primitive" animals do not. We now know that species from many taxonomic groups have similar endothermic temperatures (Table 7.1); therefore this hypothesis (which also does not explain why the 37-42 °C range is "advanced" in any functional way) is not correct. In order to believe this hypothesis we would have to construct a large series of explanations for all the exceptions to the "rule." Nothing inherent in the taxonomic "hierarchy" of animals explains the importance of their temperatures.

Another historical hypothesis claims that the reason temperatures of 30-45 °C are observed is because this used to be the temperature of the earth at the time the ancestors of endotherms evolved, and the temperatures have simply been maintained at those values ever since. In this case it is the temperature rather than the animal that is considered "primitive."

This hypothesis is a poor one for at least two reasons. First, there is little reason to expect an animal to evolve the same temperature as its environment; there are several reasons why this should not occur. For example, all the avenues for heat exchange by radiation, conduction, and convection depend on a difference in temperature. If temperatures were the same, heat exchange by these mechanisms could not occur and the loss of water for temperature regulation (via H_w) would be excessive for terrestrial endotherms. Second, this hypothesis has great difficulty with the diversity of endotherms and their different evolutionary histories. With ancestors evolving at times when the temperature of the earth was very different, why have temperatures in the 30-45 °C range become so prevalent in different endotherms?

The "Maxi-therm" Hypotheses

These hypotheses have two parts. They claim that we observe a body temperature in the 30-45 °C range because (1) if temperatures were higher the animals could not survive, and (2) there are possible advantages to having a body temperature close to an upper limit and higher than the temperature of the environment.

The first part of these hypotheses is used to explain why body temperatures are not higher than 30-45 °C. The proteins and enzymes of many animals with body temperatures in this range do not function effectively at higher temperatures (Fig. 7.2; also see Fig. 6.1). The proteins "denature" (break down), and the temperature at which this occurs is usually just a few degrees above the normal body temperature (Fig. 7.2). However, if proteins and enzymes have evolved to function optimally at the normal 30 to 45 °C body temperatures of these animals, we would expect them to operate less effectively at higher or lower temperatures (see Fig. 6.1).

Rather than ask what happens to proteins and enzymes in these species, we should ask whether it has been possible for any other species to evolve so that their proteins function effectively at temperatures higher than 45 °C. Some unicellular algae and bacteria live and reproduce in hot springs in which temperatures can exceed 70-80 °C, and they have evolved proteins that function normally at these high temperatures. However, no multicellular animals have been found that survive or grow at these temperatures, so it is quite possible that certain tissues of metazoans may have real limitations that dictate a temperature of 45 °C or lower.

The second part of the maxi-therm hypotheses depend on the first. If there were an upper temperature limit to metabolic functions just above 45 °C, animals would have temperatures near this upper limit because of advantages that result from a body temperature as high as possible compared with the

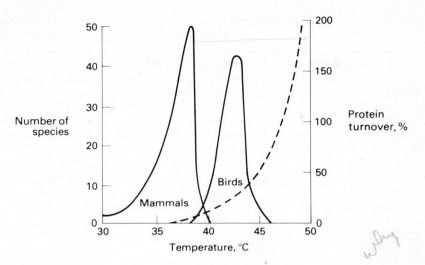

Figure 7.2
The distribution of body temperatures of many species of mammals and birds are just below the temperatures where their proteins would begin to denature at a high rate (after Calder and King, 1974).

temperature of the environment. For example, terrestrial animals will benefit by saving water if their body temperatures are not close to environmental temperature because heat loss by radiation, conduction, and convection will result in temperature regulation without increased heat loss from evaporation of water (H_w). Moreover, animals can lose heat by radiation, conduction, and convection more effectively when they are active if their body temperatures are high relative to environmental temperatures.

These benefits are based on *some* difference in temperatures, but they do not necessarily depend on a particular difference dictated by a specific temperature in the 30-45 °C range. (If there was no limit dictating a body temperature of 45 °C as the upper extreme, these benefits could be even greater.) However, not all organisms that maintain a high temperature from endothermy will realize these benefits. The skunk cabbage and philodendron, for example, do not maintain the temperature of flowers for these reasons.

The Thermodynamic Efficiency Hypothesis

Is there a principle that applies to all organisms to explain the prevalence of body temperatures of 30-45 °C? The logarithmic natures of many thermodynamic properties of water within cells have been suggested as a possibility (see Appendix II for a discussion of logarithmic functions). The reactions that occur in cells take place in water, and the thermodynamic characteristics of water (such as ionization characteristics, osmosis, viscosity, etc.) can influence the efficiency of cellular functions.

The broadest possible range of temperature for function in liquid water is from 0 to 100 °C. Below and above this range water changes state to a solid and a vapor, respectively. Water also shows characteristics of ice or vapor at temperatures between the 0 and 100 °C extremes; that is, the transition of water between states is not instantaneous at the extremes. Between the extremes the thermodynamic characteristics of water change in a logarithmic way. Ionization constants, osmosis, viscosity, and equilibrium constants all involve *natural logarithms* in the equations that characterize their dependence on temperature (Table 7.2); thus water will show characteristics of being like a solid or a vapor over the 0-100 °C range in a nonlinear way.

Table 7.2 Thermodynamic characteristics for water and the equations that describe them all involve natural logarithms (ln); in all of these equations T is temperature (from Calloway, 1976)

Ionization constants (K)	$d \ln K/dT = \triangle H/RT^2$
Osmosis $(\overline{V})$	$-\overline{V}\pi = RT \ln x$
Viscosity (μ)	$\ln \mu = (B/T) + A$
Equilibrium constant (k)	$d \ln k/dT = \triangle H/RT^2$

If thermodynamic properties of water were not logarithmic but depended on temperature in a *linear* way, and if there were advantages to having a process occur in water halfway between the 0 and 100 °C extremes when water is *least* like either a solid or a vapor, we would predict the optimal temperature to be 50 °C (Fig. 7.3). However, because thermodynamic characteristics of water are described by curvilinear, logarithmic functions (Fig. 7.3), the predicted midpoint will be the 100 °C range divided by the base of the natural logarithm (e = 2.71828), or 36.8 °C (upper logarithmic curve in Fig. 7.3). Thus thermodynamic characteristics of water that depend on a midpoint of water between ice and vapor will have optimal values at temperatures in the 30-45 °C range. If efficiency of cellular function depends on these optima it would provide an explanation for the prevalence of body temperatures in the 30-45 °C range.

There are also difficulties with this hypothesis. It predicts that animals living at high altitudes should have lower body temperatures because the predicted thermodynamic optimal midpoint for water temperature decreases with a decrease in pressure. Humans and cattle living at high altitudes do not have body temperatures lower than those at lower elevations. Ectotherms (reptiles—sceloporine lizards) that have evolved at high altitudes do have lower average body temperatures, but the temperatures are somewhat lower than what would be expected on a thermodynamic basis. Furthermore, even though the optimal point for the thermodynamic characteristics of water is near 38 °C at sea level, the change in the characteristics over the full 0-100 °C range can be slight; thus the change in relative value of these characteristics to an organism whose body temperature is near 38 °C can be slight.

It would be interesting to test some of these hypotheses experimentally. Is it possible to select for a multicellular animal that will survive and reproduce at

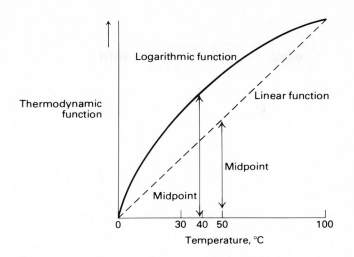

Figure 7.3
The midpoint between ice and vapor for water on a linear
scale is 50°C, but on a natural logarithmic scale (with the
base "*e*") it lies in the 30-45°C range.

body temperatures significantly higher than 45°C? If so, what is the impact on
its reproductive success compared with animals having lower temperatures?
Ectotherms with a short generation time, such as *Drosophila*, may be the best
animals to use.

THE MECHANISMS OF ENDOTHERMY: THERMOSTATIC HEAT PRODUCTION

Although the costs of temperature regulation differ from one species to another,
the general effect of environmental temperature on rate of energy use is similar
for all endotherms. We will examine the general characteristics of heat production for temperature regulation in endotherms and then examine some adaptations that have evolved that minimize the expenditure cost. Finally, we will examine some characteristics of the internal proportional control mechanisms for
maintaining a constant body temperature in some endotherms.

Figure 7.4 shows a typical rate of energy expenditure and body-temperature response of an endotherm to changes in environmental air temperature.
At low environmental temperatures, rate of energy expenditure (H_m) increases
as a consequence of loss of heat from the animal to its environment from radiation, conduction, convection, and the evaporation of water. At low environmental temperatures H_r, H_c, and H_w are all negative, which means that H_m must increase if heat loss increases and body temperature remains constant. Because
heat production in endotherms is analogous to thermoregulation in a building,
this response is called *thermostatic* heat production.

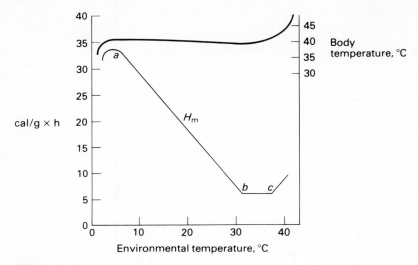

Figure 7.4
Thermostatic heat production by endotherms occurs between the lower
critical temperature (*b*) and the maximum possible rate of heat production (*a*). At environmental temperatures above the lower critical
temperature, metabolic rate is constant in the thermoneutral zone (to *c*).
Above the thermoneutral zone, metabolic rate increases when body
temperature increases.

If the environmental temperature decreases too much, an animal will
reach an upper limit in its ability to generate heat (*a* in Fig. 7.4). At environmental temperatures below this point, heat loss exceeds capacity for heat production from metabolism and body temperature decreases. When environmental temperature increases from this point (*a*), metabolic rate decreases until a
minimum value is reached (*b* in Fig. 7.4). This minimum rate of energy expenditure represents "basal" metabolic rate under appropriate conditions (e.g., for
inactive animals in a certain phase of a photoperiod; see Chapter 5), and the
lowest environmental temperature at which the minimum occurs is called the
lower critical temperature.

Below the lower critical temperature, endothermic animals experience thermoregulatory costs because their rates of energy use must increase for maintenance of a constant body temperature. Above the lower critical temperature
there can be a range of environmental temperatures over which rate of energy
expenditure does not change (from *b* to *c* in Fig. 7.4), called the *zone of thermoneutrality.* At environmental temperatures above this zone, rate of energy
expenditure increases when body temperature increases because of Q_{10} effects.
This situation represents "heat stress" for an animal because its rate of energy
expenditure must increase as a consequence of increased body temperature
(see Chapter 14).

ADAPTATIONS TO MINIMIZE HEAT PRODUCTION COSTS IN ENDOTHERMS

There are a variety of possibilities to increase effective energy use by minimizing the rate of energy expenditure for body-temperature regulation in endotherms. These can be examined from the simplified equation developed by biologists that incorporates the various characteristics of heat exchange by radiation, conduction, convection, and evaporation of water. For constant temperature,

$$\text{Heat gain} = H_m = \text{heat loss} = C(T_B - T_A) + H_w.$$

In this equation the "temperature" of the animal is written as its body temperature (T_B), and the "temperature" of the environment is written as T_A, or "ambient" temperature. C is thermal conductance, which incorporates the variety of physical heat-exchange coefficients for radiation, conduction, and convection (K, e; see Chapter 6). If the units on H_m are cal/g × h, the units on C are cal/g × h × °C. Thus thermal conductance is a measure of the rate at which heat is lost (and must be balanced by gain) from an endotherm for every °C change in environmental temperature below the lower critical temperature (when body temperature remains constant). For the general linear relationship in Fig. 7.4, C is the slope of the straight line between b and a (where the line extrapolates to zero metabolic rate at T_B).

To minimize H_m for temperature regulation, each of the terms, C, $(T_B - T_A)$, and H_w could possibly be minimized for an endotherm that is resting with a certain amount of energy that can be expended. Minimizing as many of these as possible will result in a certain quantity of energy being used most effectively.

Minimizing ($T_B - T_A$)

There are two ways to reduce the value of this term: (1) reduce T_B at a low T_A, or (2) increase T_A at a constant T_B (for a "cold" environment) so that the rate of energy expenditure occurs in the zone of thermoneutrality or as close as possible to the lower critical environmental temperature. Both of these adjustments are possible, but they depend on very different mechanisms. The first is called torpor or hibernation, which we will examine in the next chapter. Usually T_B is not reduced substantially for endotherms when there are sufficient energy substrates available from food to keep body temperature at "normal" regulated values.

Animals change "T_A" from changes in their behavior. The mechanisms are entirely analogous to the ways ectotherms select certain microenvironments to optimize heat-exchange characteristics. "T_A" in this case represents *any* environmental factors that will influence the rate of heat loss by radiation, conduction, or convection; within any environment endotherms have choices that influence these factors.

An example of an effect of "T_A" on heat loss is the influence of wind speed. Wind speed influences convective heat loss. When the rate of heat production by snowy owls (*Nyctea scandiaca*) was measured at an air temperature of −20 °C, rate of heat production increased exponentially with wind speed (Fig.

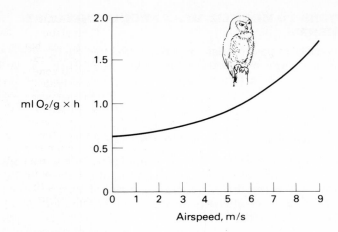

Figure 7.5
When snowy owls are exposed to an air temperature of
$-20\,°C$, their rate of energy expenditure increases exponentially as wind speed increases (after Gessaman,
1973).

7.5). When endotherms have a choice, it seems that they should select environments that minimize this type of "T_A" effect on H_m, and there is evidence
that this selection occurs. For example, chickadees spend their time in vegetation sheltered from winds when air temperatures are low and wind speeds are
high. In addition, long distance migration, which occurs in a number of species
of endotherms, represents a mechanism to modify characteristics of "T_A" on a
seasonal time scale (although this and other choices require energy and can
also modify access to food energy).

Minimizing C

The higher the value of C, the higher the slope of the line between b and a in Fig.
7.4 and the greater the rate of heat production required to keep body
temperature constant at a given difference between body and environmental
temperature. Thus an important consideration for endotherms in cold environments is to minimize C.

Minimizing C is equivalent to *maximizing* $1/C$. This inverse of C is a
measure of *insulation* between the heat production of an animal and heat loss
to the external environment. The better insulated an animal is, the less heat
production is required to keep body temperature constant when environmental
temperature decreases (Fig. 7.6).

Note in the simplified equation where $H_m = C(T_B - T_A)$ that $T_B = T_A$ when
$H_m = 0$. Mathematically this means that the thermostatic heat production lines
should intersect the x axis at the body temperature of the animal (if C remains
constant) (Fig. 7.6). If the standard metabolic rates and body temperatures of
two animals are the same (e.g., they may be endotherms of the same size), the

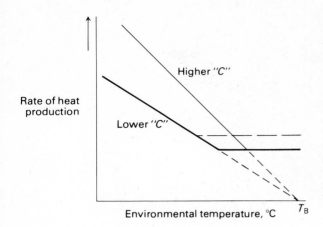

Figure 7.6
If insulation increases (*C* decreases) but standard metabolic rate in the thermoneutral zone is the same, the lower critical temperature will decrease because thermostatic heat production extrapolates to body temperature on the *x* axis. If insulation stays the same and standard metabolic rate in the zone of thermoneutrality increases, the lower critical temperature also decreases.

lower critical temperature will be less for the animal with a lower value of *C*. If insulation is the same but one animal has a higher standard metabolic rate in the zone of thermoneutrality (e.g., it may be smaller), it will also have a *lower*, lower critical temperature (Fig. 7.6).

Insulation Thickness and Color

One obvious way to change insulation is to change the quality or quantity of material on the surface of an animal. Terrestrial endotherms from cold environments have evolved thick coats of fur (Fig. 7.7), resulting in a required increase in heat production only at air temperatures lower than those requiring heat production if the insulating coats were thinner. The white fox, for example, has a lower critical temperature of −40°C as compared with values above +20°C for more poorly insulated species of foxes from more tropical areas. Terrestrial endotherms from cold environments must expend more energy to produce the thicker coats, but the investment is returned in benefits of lower energy expenditures for heat production requirements of temperature regulation.

Biologists have been puzzled for a number of years by the colors of the insulating coverings of some animals. Dark colors absorb more incident radiant energy and light colors reflect more incident radiant energy (because of differences in "emissivity"). Thus for an insulating coat of a given thickness, animals from cold environments might be expected to have dark coats and those from hot environments might be expected to have lightly colored coats. However, there are exceptions to this prediction. Many species of Arctic animals are white in color and some species from hot desert environments are black. What can explain this apparent paradox?

A white color is adaptive for an animal that must move across snow, camouflaging it from potential predators and prey. Whether an animal will survive predation or obtain food can be as important as saving energy by means of heat-exchange characteristics. Moreover, some measurements of heat ex-

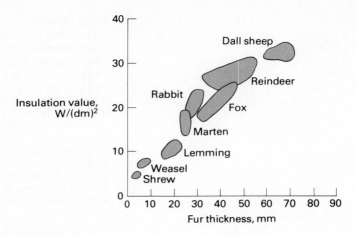

Figure 7.7
Insulation depends on fur thickness in many terrestrial
mammals and also increases with body size (modified from
Scholander *et al.*, 1950).

change suggest that a white coat can offer some advantages in saving energy.
Radiant heat penetrates *into* a coat of some thickness to a greater degree if it is
white. For an endotherm with relatively thick white insulation, more incident
radiant energy can be reflected down to the skin.

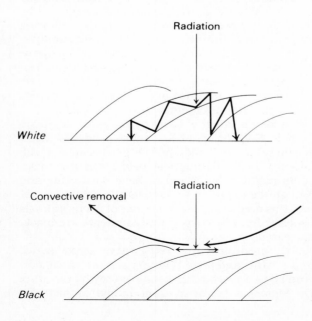

Figure 7.8
Schematic diagram of pro-
posed differences resulting
from different insulation colors on net gain of heat from
solar radiation. When white,
more heat could penetrate to
reach the skin. When black,
more heat could be absorbed
at the surface and lost by
convection if wind speed was
strong enough (based on
Walsberg *et al.*, 1978).

For darkly colored insulation, incident radiation is largely absorbed at the surface of the insulating layer. In this situation, convective loss from winds can remove an appreciable amount of the absorbed heat. Measurements indicate that black insulation results in less heat gain from radiation than does white under comparable conditions when wind speed is greater than about three m/s. These differences are summarized schematically in Fig. 7.8 and provide an explanation for why some animals from different environments have evolved different insulation colors.

Insulation and Size

Small endotherms from cold environments are not as well insulated as are larger endotherms. For example, the lemmings of the Arctic do not have as thick a coat of fur as do Arctic foxes or reindeer (Fig. 7.7). If the thermal conductance of a large number of birds and mammals is measured as a function of their size, the conductance of small animals is usually more than that of large animals (Fig. 7.9), a reflection of the relationship between cal/g × h for metabolic rate and $(weight)^{-0.25}$ (see Chapter 5). Recall that although a higher rate of heat loss per gram (less insulation at small size) dictates supplying energy at a higher rate *per gram*, the *total* energy requirements for small animals are still lower.

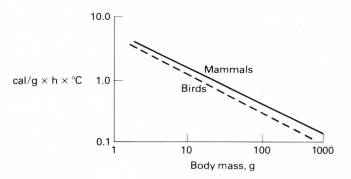

Figure 7.9
Thermal conductance increases with decreasing body size for birds and mammals (modified from Herreid and Kessel, 1967).

The relative value of insulation (reflected in the metabolic rate per gram) versus total energy requirements is important to the interpretation of the observation that animals with similar characteristics are larger in colder climates (*Bergmann's rule*); deer, fox, and other species that have a wide geographic distribution are usually larger in colder areas. At one time, the lower per-gram costs for metabolic rate (or the higher insulation values) for larger animals were thought to provide an explanation for the advantage of being large in a cold environment. However, the advantage cannot be that *total* energy expenditures per animal are less, because the total requirements for large animals are higher.

The advantages of large size in cold climates must be based on factors influencing survival. The advantages will be determined by total energy requirements for an animal *relative to* supplies of energy, such as from food availability, or perhaps internal energy-storage ability. For example, a large animal that can store more energy for its requirements for temperature regulation will have to feed less frequently than will a smaller animal able to store less energy relative to its thermoregulatory demands. Demands for energy as compared with supplies can provide explanations for advantages to large size in the cold, but demands alone or insulation values alone do not.

Insulation and Shape

The geometry of an animal has an effect on heat retention. Consider the weasel (*Mustela frenata*), which is long and thin. Comparison of its thermal conductance with other mammals of similar weight but more compact shapes shows that the weasel is not as well insulated (Fig. 7.10). The weasel has a larger surface area for heat loss than do more compact mammals of the same weight.

It does not mean that weasels are necessarily less effective than other mammals are in dealing with energy as a resource. Again, efficiency (and survival) depends on energy demands (costs) compared with supplies (benefits). Weasels have been successful partly because they have evolved a shape that permits them to pursue prey into long, thin burrows. This benefit (food, or energy availability) apparently compensates for an increased thermoregulatory cost associated with its shape; thus the supply of energy compared with use is sufficient.

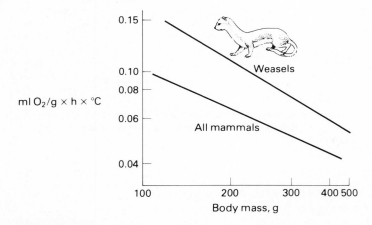

Figure 7.10
The thermal conductance for weasels is higher than it is for mammals of the same weight but having more compact shapes (modified from Brown and Lasiewski, 1972).

Insulation and Extremities

Not all of the body of an endotherm is covered with insulation, even for animals that have evolved in cold environments. Some extremities or parts thereof, such as the nose, the tips of ears, and the bottoms of feet, may not have any insulating covering or may have less insulation than is on other parts of the body.

For extremities of very small size there is a physical reason for lack of insulation. There is a "critical radius" for a cylinder below which the addition of any insulation will increase rather than decrease heat flow from the cylinder to the environment. The critical radius depends on the ratio of the thermal conductivity of the insulation (K, or physical characteristics other than temperatures influencing conductive heat exchange) and the convective heat transfer coefficient (h, determined by the nature of the materials transferring heat from convection). When the radius of a cylinder plus an insulating layer is greater than the critical value (K/h), adding insulation has the desired effect of decreasing heat loss, but when the cylinder is small enough so that the radius is less than K/h (usually in the order of centimeters or less), adding insulation *increases* heat loss (Fig. 7.11). In effect, adding material to a very small cylinder increases the surface area for heat loss rather than reducing heat loss. This relationship has been used to explain why some small newborn young are naked and only develop insulation later when body size is larger. It may also help to explain why some small extremities and some very small animals lack insulation.

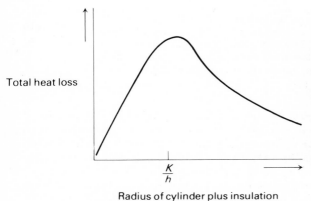

Total heat loss

$\dfrac{K}{h}$

Radius of cylinder plus insulation

Figure 7.11
For very small cylinders, heat loss will increase if the radius is increased by the addition of insulation up to a "critical radius" equal to the value of K/h. Above the critical radius size heat loss will decrease with added insulation (from Balmer and Strobusch, 1977).

Animals from cold climates usually have shorter extremities than animals with otherwise similar characteristics from warmer environments (*Allen's rule*). If the insulating covering on smaller extremities is not that effective, shorter length would reduce total heat loss. However, extremities are necessary, and adaptations within the circulatory system provide a means to minimize extremity heat loss in resting animals (Fig. 7.12).

Warm blood is pumped from the body in centrally placed arteries. If environmental temperature is low for an inactive animal, the veins near the sur-

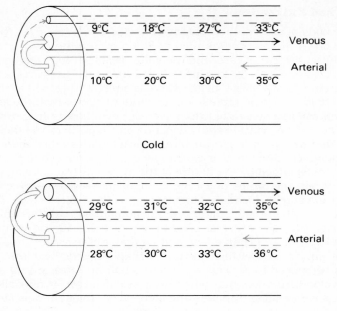

Figure 7.12
In cold environments, surface veins constrict and countercurrent heat exchange between centrally placed arteries
and veins returns heat to the body core. In warm environments, heat loss from the surface is increased when
surface veins dilate (modified from Schmidt-Nielsen,
1964a).

face will constrict and those near the arteries will dilate (see Chapter 4). The
warm arterial blood will lose heat to the cooler venous blood, and heat is returned
to the body rather than being pumped out to the tips of extremities. It effectively
increases insulation because it increases the distance between returning blood
and the surface of the extremity where heat is lost to a cold environment. This
countercurrent heat exchange in the vascular system results in much lower
temperatures at the surface of extremities in the cold, and thus results in less
heat loss. In "hot" environments, where heat loss is important for temperature
regulation, the central venous flow is constricted, and veins near the surface
dilate, increasing transfer of heat from the animal to the environment through
its extremities (Fig. 7.12).

There are limits to the operations of vascular countercurrent heat exchange for minimizing heat loss. The temperature of the extremities cannot go
below the freezing point of tissues or cellular damage will occur (frostbite), and
there are built-in safeguards in some Arctic species to prevent frostbite. If the
paw of an adult Arctic gray wolf is immersed in an alcohol-and-dry-ice bath hav-

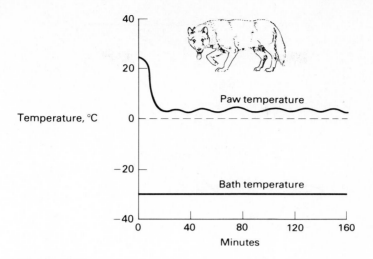

Figure 7.13
When the paw of an Arctic wolf is immersed in a bath at a
temperature of − 35 °C, the temperature of the paw oscillates
just above the freezing point as warm arterial blood is
periodically shunted through capillaries in the paw (modified
from Henshaw *et al.*, 1972).

ing a temperature of − 35 °C, the paw rapidly cools to 2-3 °C (Fig. 7.13). Further
cooling is prevented by a periodic increased flow of warm arterial blood through
the capillary network of the paw. Flow of warm blood is increased or decreased
by means of local changes in the diameter of blood vessels, so the temperature
oscillates just above the freezing point. It minimizes heat loss within the limitations set by avoidance of tissue damage from freezing.

Respiratory Heat Exchange and Minimizing H$_w$

Heat and water must be added to the air that terrestrial animals inhale before it
reaches the lungs. During expiration, a considerable amount of heat can be lost
to the environment as a consequence of loss of warmed air and from loss of
water evaporated to humidify the air. For example, humans lose up to 20% of
total resting heat production just from expiring warmed, humidified air when
inspired air is dry at a temperature of 0 °C. If the air that is expired is cooled
before it leaves the body, some of this heat can be recovered, thereby reducing
the required rate of energy expenditure.

A number of animals cool expired air, and some are more effective at it
than others (Fig. 7.14). Humans expire air at a relatively high temperature, but
small birds and mammals expire cooler air. The extent to which expired air is
cooled depends on the dimensions inside the nasal passages. Inspired air is
humidified when water evaporates from the surface of the nasal passages, cooling the surface, because about 580 calories are required for each milliliter of

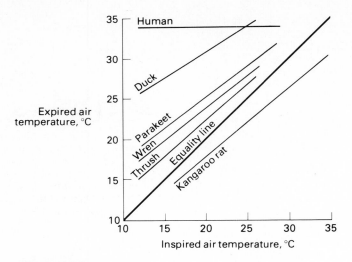

Figure 7.14
Humans do not cool expired air appreciably, but
respiratory countercurrent heat exchange in small birds
and mammals results in greater cooling of expired air (from
Schmidt-Nielsen *et al.*, 1970).

water evaporated (the latent heat of vaporization of water; see Chapter 14). In
addition, heat is transferred from the blood passing through nasal tissues to the
tissues and ultimately to the cooler inspired air by conduction and convection.
The shorter the distance from the center of the airstream to the nasal tissue, the
more readily heat and water exchange, and the cooler the nasal tissues will
become during inspiration.

During expiration, the heat from the warmed air will pass back to the
cooler nasal tissues. Again, the smaller the distance from the center of the
airstream to the tissue, the more rewarming of tissue and cooling of air will take
place. Moreover, when the humidified air is cooled water will condense on the
surface of the nasal passages because air contains less water vapor when it is
saturated at lower temperatures (see Chapter 13). We will examine this important principle when we consider adaptations for water conservation in terrestrial animals, but even in very cold environments heat is recovered when
water recondenses on the nasal tissues.

The process of heat recovery in the respiratory passages is similar to heat
recovery in the vascular system by countercurrent heat exchange (Fig. 7.12). In
the vascular system, heat exchange occurs across *space* from one channel to
another at the same time. In the respiratory passages, heat exchange occurs
across *time* (between inspiration and expiration) but in the same channel or
space (Fig. 7.15).

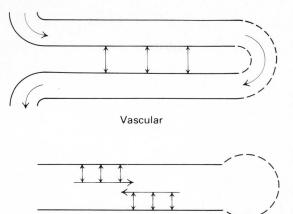

Vascular

Respiratory

Figure 7.15
Heat exchange in the vascular and respiratory systems is similar. In the vascular system, exchange occurs at the same time between two channels. In the respiratory system, exchange occurs at different times in the same channel (from Schmidt-Nielsen *et al.*, 1970).

If the efficiency of respiratory heat exchange depends on the size of the respiratory passages, is efficiency of heat recovery during respiration dependent just on body size, or can it represent an adapation to minimize respiratory heat loss in cold environments? We could predict that small mammals and birds with small nasal passages should efficiently minimize respiratory heat loss in any environment just because of their small noses. Are there large animals (with large noses) from cold environments that are also effective in cooling expired air? If so, the mechanisms for cooling expired air would represent adaptations to minimize heat loss.

When reindeer (*Rangifer tarandus*) are active, they produce a considerable amount of heat. Because they are well insulated (Fig. 7.7), the heat produced from their activity in excess of requirements for temperature regulation must be lost; otherwise their body temperatures would increase. However, when they are resting, they should minimize heat loss to their cold environment. Heat exchange in the respiratory passages accomplishes both functions (Fig. 7.16).

There is a flap of skin at the openings to the nasal passages of reindeer (and other relatively large mammals) that reduces each opening to a relatively small slit when they are inactive. When they are resting, expired air temperature is lower than body temperature (Fig. 7.16). When they are exercising, the nostrils flare open, increasing the distance for heat exchange. In addition, the rate of respiration increases during exercise, which means that a volume of air in the nasal passages has less time for heat exchange. Moreover, some air may be expired through the mouth where dimensions are even larger (see Chapter 14). The results of dimension and time changes produce an expired air temperature that is higher when heat production exceeds requirements for temperature regulation (Fig. 7.16). Thus respiratory heat exchange in some large Arctic species represents an adaptation for minimizing heat loss at rest as well as increasing heat loss during activity.

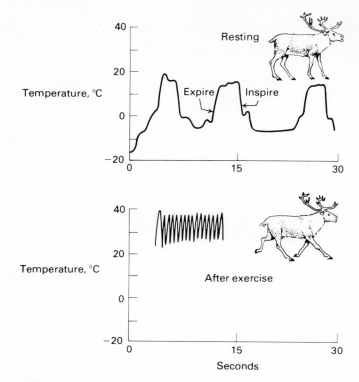

Figure 7.16
While reindeer are resting they expire air slowly and cool the air more than it is cooled immediately after exercise, when increased respiratory rates and nasal passage dimensions result in increased heat loss to the environment (modified from Hammel, 1962).

CONTROL MECHANISMS FOR TEMPERATURE REGULATION IN ENDOTHERMS

An endotherm that responds to a change in heat loss by increasing heat production must have a control system to produce an appropriate response for balancing losses with gains. As we have discussed elsewhere (see Introduction), the control system will involve receptors (detectors), information processing (integration), and effectors to produce the response. There is great diversity among endotherms, which include plants, insects, and vertebrates, and the control systems differ in some details. However, a model of proportionally controlled thermostatic operation applies in general to all endotherms. We will examine some details in mammals and some aspects of the temperature control of an insect endotherm.

Temperature Controls in Mammals

In mammals a change in temperature is detected in two locations: at the skin and in the central nervous system. The part of the nervous system involved in temperature control is the anterior hypothalamus (Fig. 7.17). If this area is removed, mammals lose the ability to generate heat in the cold. Thus the area is necessary for the activation of effectors for heat production. In some birds the upper part of the spinal cord is important for body-temperature control.

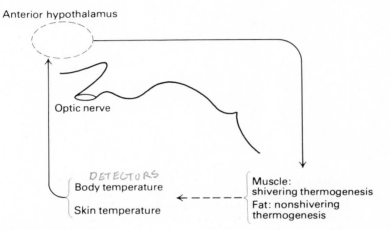

Figure 7.17
Diagrammatic relationship between major components of the negative-feedback control system for thermostatic temperature regulation in mammals.

It is possible to place small tubes in the anterior hypothalamus and circulate cold water through them, causing the metabolic rate to increase even though only a small area is cooled (Fig. 7.18). Cooling other areas of the hypothalamus does not produce this response. It is similar to the control over temperature regulation in ectotherms (except for the nature of the response) and suggests that this area of the brain contains detectors for body temperature.

The nerve cells in the anterior hypothalamus operate by comparing a change in body temperature with some reference, or "set point," temperature. If body temperature drops below the set point, heat production increases to counteract the drop and return temperature to the normal reference value (Fig. 7.17). Moreover, the response is proportional to the difference between the set point and body temperature. The larger the difference, the greater the rate of heat production. This can be seen from the results of experiments in which the hypothalamus of a rabbit was cooled (Fig. 7.18).

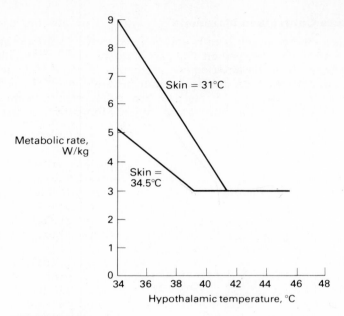

Figure 7.18
Rate of heat production in rabbits depends on the variation
of hypothalamic temperature from the set point
temperature as well as skin temperature. The magnitude of
heat production is proportional to the variation (modified
from Hardy and Stitt, 1976).

Information from temperature receptors in the skin is integrated (combined)
with information on body temperature in the hypothalamus to determine the
heat production response. If the hypothalamus is cooled and skin temperature
is changed, the rate of heat production is higher for a lower skin temperature
(Fig. 7.18). Because a relatively large amount of heat loss occurs from the skin,
information on both body and skin temperatures provides more precise control
for body temperature regulation when heat loss changes. The integration of information on skin temperature with hypothalamic temperature produces an
anticipatory type of response to changes in environmental temperature. Thus
detection of a low skin temperature can lead to an increased rate of heat production without an appreciable drop in hypothalamic temperature.

Depending on skin temperature, there is a linear dependence of heat production (increase from basal values) on the difference between the temperature
of the hypothalamus and some reference or set point temperature (Fig. 7.18).
Therefore,

Response (rate of heat production) = $\alpha (T_{hyp} - T_{set})$.

In this equation α represents the slope of the line describing the proportional
relationship (Fig. 7.18).

The response that is observed depends on two variables that can change. First, note in Fig. 7.18 that the slope for the proportional relationships differ depending on skin temperature (α is lower for a higher skin temperature). Second, note that an increase in rate of heat production begins at different hypothalamic temperatures for different skin temperatures (a lower hypothalamic temperature at a higher skin temperature), indicating that the set point for a response changes. We will examine other cases of a change in set point for control of body temperatures in the next chapter where we will examine hibernation and the mechansims involved in body temperature regulation when the temperature that is controlled decreases.

How is the rate of heat production in mammals increased when body and/or skin temperatures decrease? There are two mechanisms for mammals: shivering thermogenesis and nonshivering thermogenesis (Fig. 7.17). Shivering produces heat from contraction of muscles. Substrates are oxidized to yield the ATP required for muscle contraction (see Chapter 18). Oxidation and the use of ATP energy increase heat production in a way similar to that of increased exercise.

Mammals can also increase heat production without shivering. This fact was demonstrated through the use of a drug called *curare*, which acts to paralyze skeletal muscles. If it is given in small amounts, it only paralyzes skeletal muscles that are involved in heat production by means of shivering. Mammals treated in this way still increase the rate of heat production when they are exposed to cold.

The mechanism for nonshivering heat production involves release of fats from adipose tissue. The fats enter cells, where they are oxidized to generate heat. Oxidation also occurs directly within certain types of fat cells, which we will consider when we discuss hibernation. The increase in oxidation of fats in the cold is controlled in part with a neural hormone (norepinephrine) that is released at some fat cells as a consequence of the integration process in the anterior hypothalamus (also, see Chapter 8).

Temperature Control in the Sphinx Moth

Heat is generated in the thorax of the sphinx moth during contraction of the flight muscles, and the thorax is covered with an insulating layer of scales. If heat is applied to the thorax of a restrained moth, the thoracic temperature increases (Fig. 7.19). When the temperature reaches 38-39 °C the rate of increase of temperature slows; thus thoracic temperature is regulated and kept close to these temperatures. At the same time, the temperature of the abdomen begins to increase (Fig. 7.19). Excess heat in the thorax is shunted back to the abdomen (which is not as well insulated) through blood vessels; thus the temperature of the thorax is controlled, once the insect has warmed, by the rate of movement of fluid in the circulatory system between the thorax and abdomen.

Presently, it is not known where the receptors are in the thorax, if they provide information that is integrated with respect to a specific reference temperature at a certain location, or if other effector mechanisms (such as fat utilization or respiratory ventilation) are involved in control of heat production and heat loss. However, the thermostatic proportional control model of temper-

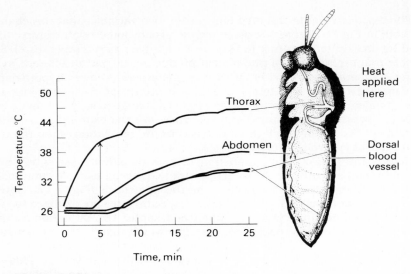

Figure 7.19
When heat is applied to the thorax of a sphinx moth, thoracic temperature (upper line) increases. When thoracic temperature reaches 38-39°C, heat is shunted via the circulatory system back to the abdomen, which is not as well insulated, and abdominal temperature increases (from Heinrich, 1974).

ature regulation will provide a good basis for further studies of other aspects of the negative-feedback systems for temperature control in these and other endotherms.

ENVIRONMENTAL TEMPERATURE AND DEVELOPMENT

All endotherms change in size from small to large during development, and there are varying degrees to which the young are influenced by their thermal environments after they are born. For birds, mammals, viviparous fishes, and reptiles, the developing young remain in the body of an adult up to a certain point in development. After birth they must deal with their thermal environment while development continues.

There are two kinds of endotherm young at birth; they can be classified according to their ability to generate heat internally. *Precocial* young are born with the ability to control their temperatures without appreciable assistance. *Altricial* young lack the ability to generate metabolic heat for body temperature regulation when they are born, but they develop the capacity at a later time.

Most precocial young are relatively large at birth, are born covered with insulation, and have the ability to move in their environments to obtain food. Altricial young are smaller, are usually born naked, and lack the ability to move. They are born at an earlier stage or development. Many altricial-birth animals are small and do not have the ability to allow development inside the mother to continue to the same extent as precocial-birth animals. Because altricial animals are "born early," they give us the opportunity to examine the development of temperature regulation and some of the interactions between requirements for development and an animal's thermal environment.

Altricial young are initially helpless and cannot easily avoid predators. This situation places a premium on rapid growth; they should attain the ability to move and maintain themselves as fast as possible. Their rate of growth depends partly on the amount of energy the parents supply, which is affected by the number of young the parents have to care for. The thermal environment of the young and their ability to generate heat influence the rate at which they use the energy provided by the parents.

Because altricial young initially do not generate appreciable metabolic heat for body temperature regulation, their body temperatures will change if they are exposed to different environmental temperatures (Fig. 7.20). In this respect they are similar to ectotherms, and physiologists used to consider altricial young as "poikilotherms" that had "primitive" physiological abilities. However, the inability to regulate body temperature provides altricial young with a distinct advantage in maximizing their growth rate early in development. The energy they would otherwise have to use to maintain their body

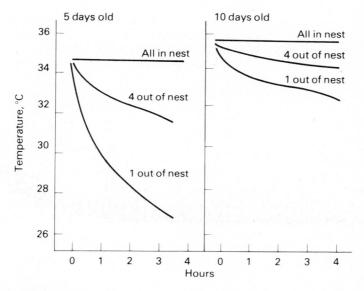

Figure 7.20
Five-day-old and ten-day-old rat pups are maintained at a relatively constant temperature in the nest by parental brooding (upper lines). When four pups huddle together out of the nest they will cool; one pup by itself will cool more rapidly. By ten days after birth, some ability to generate heat internally slows the rate of cooling (modified from Alberts, 1978).

temperatures can be used to add new tissue instead. Using as much as possible for growth and as little as possible for maintenance of body temperature will decrease the time they are subject to predation particularly if there is a limited amount of food they can process.

Despite this, there is an advantage to growth occurring at a regulated temperature. There is an optimum temperature for growth of newborn altricial young, as there is for growth of ectotherms (see Chapter 6) and for growth of precocial young that remain within the parent at a similar stage of development. Above or below the optimum temperature the growth rate will be less because of Q_{10} effects on rates of energy use or assimilation of food.

The temperature environment for altricial young is provided by the parents. A nest is built that serves to insulate the young and hide them from predators. Nest temperature is maintained by the incubating and brooding behavior of the parents. A parent will sit on eggs or young and heat will be transferred from the adult to the young. In some cases there is a special patch of skin on the abdomen of breeding adult birds (the brood patch) that becomes highly vascularized just prior to incubation. It is brought into contact with the young to facilitate external heat transfer from the adults.

Incubating parents are very sensitive to the temperature requirements of the young. If an artificial egg is placed in the nest of an incubating herring gull and cold water is circulated through it, the parent will increase its heat production by shivering. Conversely, if hot water is circulated through the egg, the parent will stop incubating, allowing the egg to cool.

The energy for regulation of the nest temperature ultimately comes from food consumed by the parents. In addition to using food energy to maintain their own temperatures and the temperature of the nest, the parents must provide the growing young with food. The more mouths to be fed, the more food must be obtained, particularly as the young rapidly increase in size and their total demands for energy increase. Depending on availability of food, the demands for energy by the adults will place a premium on the time that must be used to search for food. Time used to search for food cannot be used to maintain the temperature of the nest.

The temperature and energy demands for reproduction lead to several adjustments in breeding animals (see Chapter 15 for a further discussion of resource use during reproduction). Many species time the birth of their young to occur when food availability is high. In several species, the parents share the care of young to varying degrees, depending partly on the ability of one parent to supply all of the energy for raising young. The number of young can also be adjusted, although it should be kept as high as possible within the constraint of survival. The average number of young born at one time (the clutch size) is influenced by energy demands for reproduction, and if food availability is low the number of young can be reduced by selective starvation so that the supply of energy matches the demands of the remaining young. This process is called brood reduction, and it has been observed in several species of birds.

The requirements of the young for energy and an optimal thermal environment influence these adjustments in clutch size, time of breeding, and extent of parental care. The maintenance demands of the young are initially low but require the adults to expend time brooding that cannot be used to search for food. Later, when the young have developed the ability to regulate their temperatures, the adults have to spend less time brooding but more time searching for food to supply larger young. An interesting dependence of the energy and temperature demands met by parents on the number of young in the nest influence the number of young that can be produced at one time.

When the adult leaves the nest the altricial young will cool if they have not developed the ability to maintain their temperatures. The rate of cooling depends on the number of young (Fig. 7.20). The more young there are in a nest, the lower will be the total surface area for heat loss as compared with the total volume of young. If there are several young, an adult can spend more time off the nest gathering food before the nest cools and requires more heat from brooding.

This advantage means that young can be added to a brood where the cost of adding an additional young is less per individual the more young there are (Fig. 7.21). Later in development, when the young can control their temperatures, less heat production by an individual is required if there are others in the nest because the other young add heat and insulation. The relative saving in cost for adding additional young has to be balanced by the increasing *total* cost of providing for more young (Fig. 7.21), but the interactions between (1) parental timing for brooding and feeding, (2) nest insulation, (3) food availability, and (4) brood size will generate an optimum number of young relative to the supply of energy resources.

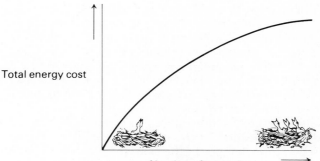

Total energy cost

Number of young in nest

Figure 7.21
As more young are added to a nest, the additional young increase the effective insulation because total surface area for heat loss decreases relative to the total volume of young. Although the total overall energy cost increases with the production of more young, each individual that is added does not increase cost to the same extent as do the initial young.

SUMMARY

Endotherms maintain relatively constant body temperatures by changing internal heat production when heat loss increases in cold environments. There are endothermic animals from several taxonomic groups of multicellular organisms, and a wide variety of terrestrial endotherms maintain temperatures between 30 and 45 °C. Because the characteristic pattern of body temperatures is not related to taxonomy, some other explanation based on advantages to animals must explain the observed temperatures. The "maxi-therm" hypotheses claim an inherent upper limit of about 45 °C, with advantages of a body temperature close to this limit and different from environmental temperatures. The thermodynamic efficiency hypothesis claims that an optimum temperature for metabolic functions in cellular water would occur between 30 and 45 °C because thermodynamic functions are logarithmic with midpoints between 0 and 100 °C (when water is least like either vapor or ice) in the 30-45 °C range.

Endotherms increase heat production (H_m) thermostatically when heat loss increases below a lower critical environmental temperature. The magnitude of heat production required to keep body temperature constant depends primarily on the rate of heat loss from radiation, conduction, convection, and the evaporation of water. Heat loss by radiation, conduction, and convection is minimized by maximizing insulation. Adaptations to cold involve the insulative covering of animals, and large endotherms are better insulated than small endotherms. Although large animals lose less heat per gram, they still have higher total energy expenditures in the cold; therefore, the advantage of large size in cold climates (Bergmann's rule) has to be due to a favorable rate of energy use relative to energy availability.

The effectiveness of insulation for minimizing heat loss also depends on the geometry of animals. Weasels are long and thin and not as well insulated as more compact mammals of the same weight. Long and thin extremities are also not well insulated, although countercurrent heat exchange in the vascular system will minimize heat loss within limitations of tissue damage from freezing. Heat loss from the respiratory passages is minimized by countercurrent heat exchange over time, and some large Arctic mammals conserve heat at rest and increase heat loss during activity by changing the timing and the dimensions for heat exchange in the respiratory passages.

The control of body temperatures in mammals involves receptors for detecting body temperatures in the hypothalamus (part of the brain) and for detecting temperature changes in the skin. Information from both areas is integrated in the brain to determine the heat production response, which is proportional to the difference between the integrated brain and skin temperatures and some reference temperature, or "set point." Heat production in mammals is increased by shivering thermogenesis, involving skeletal muscle contraction, and by nonshivering thermogenesis, involving mobilization and utilization of fats for oxidative metabolism.

Endotherms must deal with their thermal environments in varying degrees during development. Precocial young have the ability to regulate their temperatures at birth, whereas altricial young do not. The parents of altricial young provide both the heat for their development in a nest environment and the food for their growth. The interrelationships of time needed for parental feeding, the requirements of the young, and food availability produce several adjustments in patterns of reproduction. One factor influencing parental demands for energy is the number of young in the nest. As the number of young are increased, the effective insulation per individual increases; thus the total cost of producing young does not increase in an additive way as young are added to a nest. This insulation factor makes possible an optimum number of young as compared with supplies of energy resources.

Torpor and Hibernation

8

The rate of heat production by an endotherm in a cold environment (below thermoneutrality) depends on the difference between its temperature (T_B) and the temperature of the environment (T_A). Effective insulation will minimize the rate of heat loss, but there are situations where survival depends on saving even more energy. This savings is accomplished in some inactive animals by a reduction in body temperature so that ($T_B - T_A$) is minimized (within limits) when T_A is low. Animals that reduce their body temperatures in the cold are said to enter a state of *torpor* because they are lethargic, and functions occur at rates much slower than normal when body temperature is reduced.

Different terms have been used to describe time periods when animals reduce body temperatures. *Hibernation* ("winter sleep") refers to a decrease in body temperature in the winter, usually for relatively long periods. *Estivation* ("summer sleep") refers to a decrease in body temperature in the summer. *Daily torpor* refers to a decrease in body temperature on a daily basis rather than during a particular season.

These terms serve to classify the timing of a decrease in body temperature. We will try to understand the functional significance of a reduced body temperature in terms of the relative advantages (benefits) and the costs associated with torpor at different times. We will then examine some interesting aspects of physiological control mechanisms associated with torpor and with the increase of heat production needed to raise body temperature when animals come out of torpor to resume a "normal" body temperature.

WHY ABANDON AN OPTIMAL TEMPERATURE?

In the previous two chapters we have developed a number of arguments suggesting that animals should maintain particular body temperatures if possible. Now we are suggesting that temperatures should change. If there is an advantage (benefit) associated with a body temperature in the 30-45 °C range, why should an animal change from this "optimal" condition? Similarly, if the benefit associated with a particular body temperature is effective enzyme function, why should this be compromised through a decrease in body temperature?

As is so often the case, what is "optimal" will change because of environmental changes. Whether it is *possible* to maintain a particular temperature depends not only on the benefits associated with that temperature but also on the costs to realize those benefits. For example, the maintenance of a relatively high body temperature for an endotherm in a cold environment involves a cost in increased rate of energy expenditure. For an ectotherm the cost may involve movement for heat exchange in an environment such that other functions (such as feeding) become more restricted or exposure to predators becomes more probable. These costs must be taken into account to determine which body temperatures are "optimal" for survival when environments change.

When both benefits and costs depend on temperature, an animal should evolve to maximize the *difference* between the two. This point is illustrated graphically in Fig. 8.1 for a situation in which costs increase exponentially with temperature and where benefits also change. If an animal can realize benefit 1, it should have an optimal temperature at T_1 (at higher cost) because the *net* benefit would be maximum at this temperature. If the benefit that can be realized changes to benefit 2, the optimal temperature for a maximum difference changes.

A change in what is an optimal temperature for an animal can be visualized in less theoretical terms by considering an extreme situation with respect to energy as a benefit and a cost. If an animal can obtain *no* food, its survival depends on how much energy it has been able to store internally up to that time and the rate at which the internal reserves are being used. The higher the rate of energy use for the fixed supply, the less time for survival. Because decreasing body temperature will decrease the rate of energy use, it can be optimal for survival by prolonging the time to starvation. The more time an animal has, the more likely it is that it will obtain food to replenish internal energy reserves.

Whether an animal decreases body temperature to minimize demands for energy depends on the availability of a sufficient supply (benefit) of energy in

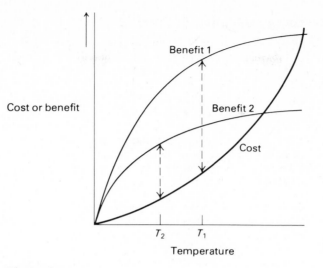

Figure 8.1
Maximizing the difference between benefits and costs
will generate different optimal temperatures (T_1 and T_2)
if benefits change.

the form of food. Animals (species) that predictably achieve a more or less continuous access to food (energy) supplies are not subject to the evolutionary selection pressure of having to reduce demands by changing body temperatures. Animals that are periodically subjected to situations in which energy availability is not sufficient to meet demands at a high body temperature have evolved to reduce their demands for energy by decreasing body temperature.

Note that a decrease in body temperature (torpor) is opposite to fever (see Chapter 6), but it occurs for similar reasons. Fever is optimal if a short-term increase in body temperature promotes survival by combating infection. A decrease in body temperature occurs under other conditions, but it will also promote survival because of a change in an animal's environment.

TORPOR IN ECTOTHERMS

As was the case in the experiments dealing with behavioral fever, ectotherms will demonstrate a preference for lower body temperatures under appropriate circumstances. Some ectotherms are subjected to lower body temperatures because they have no choice with respect to heat exchange in certain environments. Others select lower body temperatures if they have a choice when benefits and/or costs change. The benefits and costs can involve energy and other aspects of an animal's environment.

Torpor and Energy Supplies and Demands

When they have a choice, ectotherms should select a lower body temperature if energy supplies from food are reduced. The usual procedure by which we investigate these changes is to "starve" ectotherms (completely remove their access to food) and examine their temperature preferences over a period of time. For example, if certain species of salmonid fishes are given all the food they can eat, they will select a relatively high body temperature, but if they are starved, they will select a lower body temperature (Fig. 8.2). After they are fed again, they select the higher temperature. Similarly, western toads select a low temperature within a thermal gradient when they are starved, and they select a higher temperature that is optimal for maximizing their growth rate after they are fed (see Fig. 6.13).

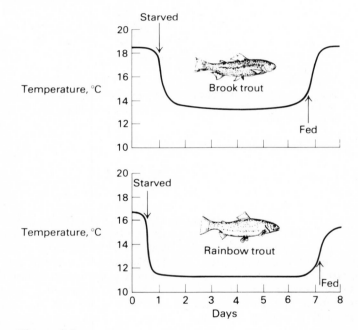

Figure 8.2
When brook trout and rainbow trout are starved, they select lower body temperatures. The selected temperatures remain the same from day to day and are different between the species. After they are fed they select a higher body temperature (adapted from Javaid and Anderson, 1967).

Note in Fig. 8.2 that the fishes selected a *particular* low body temperature when they were starved, even though they could have selected a body temperature lower still. This selection is very important because it suggests that the animals were not selecting temperatures that reduced their rates of

energy expenditure to the minimum possible values. We will find that endotherms entering torpor also change body temperatures to specific lower values
that are not the lowest possible for absolute minimization of rates of energy expenditure (see below). The explanation may be related to a second, lower "optimal" temperature for enzyme function (see Fig. 6.11). Although a reduced
body temperature will lower demands for energy, the reduction incorporates
other factors (such as effective enzyme function) that may be important.

The interaction between body temperatures and food consumption is
shown for ectotherms by means of measurements of their rates of growth when
they are given different amounts of food and are exposed to different thermal
environments with no choice for body temperature selection. Results for
sockeye salmon are shown in Fig. 8.3. When an "excess" amount of food is
available to the fish, growth rate is maximum at 15 °C. Below this body temperature less energy is assimilated, and above this temperature more energy must
be used for maintenance (see Chapter 6). When the amount of available food is
reduced, the body temperature for the maximum rate of growth is also lower
(Fig. 8.3). Rate of use of energy for maintenance is less at the lower temperature,
so a greater proportion of the lower amount of energy can be used for growth at
a lower body temperature.

Note that the rate of growth is always less at low temperatures (Fig. 8.3). If
maximizing the rate of growth provides a survival advantage, ectotherms

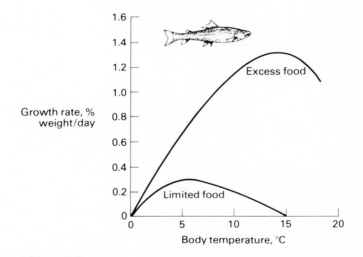

Figure 8.3
With all the food they can eat but with no choice for body
temperatures, sockeye salmon grow best at a body
temperature of 15 °C. When their food intake is reduced to
a small fraction of maximum daily intake at 15 °C, growth
rate occurs best at a lower body temperature (adapted from
Brett, 1971).

should select higher temperatures when sufficient food is available because it always produces higher growth rates (15 °C for the salmon in Fig. 8.3). Lower temperatures should only be selected when food benefits are reduced or when other aspects of an animal's environment force it to select lower temperatures (see below).

Even though a low body temperature will reduce the rate of energy use, there are advantages for ectotherms in the selection of higher temperatures to process whatever amounts of food they have been able to consume. Digestive enzyme function and assimilation is usually maximum at a relatively high body temperature. For example, when western toads are kept at a temperature of 14 °C with all the food they can eat, they will grow more slowly than if they are kept at their preferred temperature of 27 °C (Fig. 8.4). If they are kept at 14 °C and are given access to a heat source for five hours a day, they will grow faster (Fig. 8.4), permitting the food in the digestive tract to be assimilated more completely. Other observations also support the importance of relatively high temperatures for food processing. If a long boa constrictor is fed and placed in a cold cage with a heat lamp, the snake will move such that the lump of food within the snake will stay beneath the heat lamp as it passes through the long digestive tract.

If either a high or a low body temperature can be optimal depending on digestion and maintenance costs, we would expect an ectotherm to move back and forth between heat-exchange locations depending on when it fed and how much it consumed relative to demands. Consumption of relatively large quantities of food would lead to an animal staying at a high temperature for a long time, while consumption of small quantities would result in a switch to a lower temperature sooner after feeding.

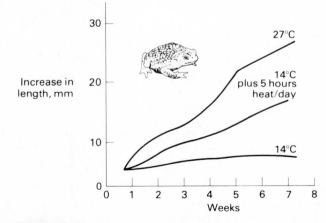

Figure 8.4
Western toads grow most rapidly with excess amounts of food at a body temperature of 27 °C. If they are kept at 14 °C and given access to a heat source for 5 hours per day, rate of growth is higher than when they are kept at 14 °C without access to a heat source (adapted from Lillywhite et al., 1973).

   Unfortunately, no experiments have been done to examine patterns of body-temperature preferences in ectotherms as a function of the amount of food consumed. Most observations of temperature-selection behavior have been based on animals that either were given all the food they needed or were completely starved. However, there are some observations suggesting that ectotherms can time their temperature preferences. If lizards (*Klauberina riversiana*) are starved for a few days, they will select a lower temperature at night when they normally do not feed. During the day they will select higher temperatures (Fig. 8.5). It would be interesting to examine this pattern of temperature preference when benefits from food availability are varied.

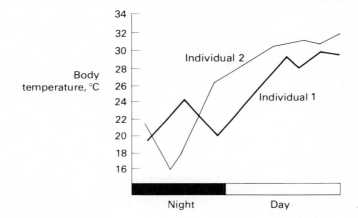

Figure 8.5
When two lizards (*Klauberina riversiana*) are kept in a thermal selection apparatus without food, they pattern their temperature selections so that body temperatures decrease at night when they normally do not feed (adapted from Regal, 1967).

Other Costs and Benefits of Torpor

The optimal balancing of energy expenditures with energy gains through temperature selection is only one possible explanation for torpor in ectotherms. Other aspects of an animal's environment will also dictate selection of low or high temperatures at different times when it may appear that the animals have a choice. For example, recall that lake salmon selected a temperature at night that was optimal for food processing and growth, but they selected a lower temperature near the bottom of the lake during the day (see Fig. 6.14). Selection of the lower temperature during the day could be due to less food consumption relative to expenditures, to the presence of predators (a changed cost for survival) in warm-temperature areas during the day, or to a combination of these circumstances.

   Interactions between animals other than predation also result in different patterns of temperature selection. For example, bluegill sunfish (*Lepomis macrochirus*) are very aggressive toward one another. If small juvenile sunfish are given all the food they can eat, they will select a temperature of 31 °C. If a large adult sunfish is added to the temperature selection apparatus, it will chase the juveniles away from the location of the preferred temperature. In this situation, an ectotherm with an apparent ability to make a choice is restricted by its social environment to heat-exchange areas other than what would be predicted just from normal temperature selection for maximizing the efficiency of energy use.

TORPOR IN ENDOTHERMS

Most hypotheses concerning torpor in endotherms deal with the effective use of energy as a resource, which is in part due to the higher demands of endotherms for energy (see Chapter 5). Thus the implications for changes in demands as energy supplies change are more obvious in endotherms than in ectotherms. The major classifications of torpor in endotherms, according to when it occurs (daily versus seasonal), provide a basis to interpret changes in benefits and costs.

Daily Torpor

The pattern of torpor in endotherms occurs on either a daily or a seasonal basis. Endotherms that decrease body temperatures daily are all small, including a number of species of small mice and the smallest birds, the hummingbirds. The metabolic rates of two species of hummingbirds when they are "homeothermic" and torpid at different environmental temperatures are shown in Fig. 8.6.

Small endotherms have high rates of energy expenditure per gram. Moreover, small endotherms are not able to store large amounts of energy internally to supply their relatively high per-gram rates of metabolism. By decreasing body temperature, the rate of energy use per gram can be reduced by as much as 90% (Fig. 8.6); thus a small endotherm with energy reserves that are limited when compared with demands can survive for longer periods without feeding if it enters torpor.

As we will see, entering torpor does carry a cost for the "arousal" that raises body temperature, but daily torpor is still advantageous for saving energy even with this cost. For example, the amount of energy expended by a pocket mouse (*Perognathus californicus*) when it enters torpor and then immediately expends energy to raise its body temperature back to "normal" values is less than the amount of energy expended to keep body temperature high for the same period of time.

Most endotherms feed either during the day or at night, and when torpor occurs it is usually during periods when the animals do not normally feed. Not all small endotherms enter torpor on a daily basis. The small shrews, for example, may maintain a high temperature, but to accomplish this they must search for and consume food almost continuously. Therefore, they do not normally experience long periods in which there is no feeding.

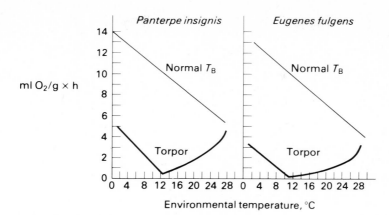

Figure 8.6
The rate of oxygen consumption for two tropical hummingbird species decreases when they are torpid until body temperatures reach 12 °C. At environmental temperatures below 12 °C body temperatures of these torpid hummingbirds are controlled by means of thermostatic heat production (adapted from Wolf and Hainsworth, 1972).

If small endotherms reduce energy-use rates by lowering body temperature, we would expect them to always enter torpor at times when they do not normally feed, if saving energy is the only important aspect of entering torpor. This situation does not occur. When small endotherms are inactive, they will continue to expend energy at relatively high rates and maintain a high body temperature until they have expended all but a fraction of their energy reserves. When the energy expenditures and intakes of hummingbirds were measured for several days, they were observed to enter torpor at night only when their energy reserves fell below some relatively low value (Fig. 8.7). They did not always enter torpor at night, a time in which they normally do not feed. The reason is due to some other cost for torpor, i.e., some risk associated with being torpid, that results in the use of torpor only in an "energy emergency." The risk could be due to a greater probability of being caught by a predator while in a lethargic state. Torpor occurs when risks involving death from starvation outweigh other risks.

Note in Fig. 8.6 that the hummingbirds increase their metabolic rates when they are torpid if the environmental temperature drops below about 12 °C. This characteristic is seen in other animals that are torpid, although the temperature where rate of energy expenditure increases varies from one species to another. When the rate of energy expenditure during torpor increases, the animals are controlling their body temperatures but at reduced levels. In Fig. 8.6 the lines for thermostatic heat production at "normal" and at torpid body temperatures are parallel; these animals have all the capabilities for regulating their temperatures when they are torpid. The only difference is that body temperature is maintained at a lower "set point" during torpor.

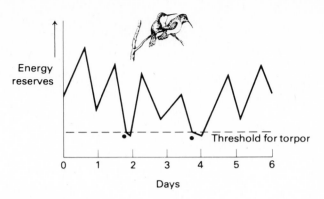

Figure 8.7
Hummingbirds do not enter torpor (indicated by •) at
night (when energy reserves are decreasing) unless they
reach some lower "threshold" of energy reserves (from
Hainsworth *et al.*, 1977).

This observation is important because physiologists used to consider
torpor in endotherms as a "primitive" characteristic. Endotherms that entered
torpor were believed to have "lost" the ability to control body temperature and
were considered to be essentially "poikilothermic." We now know that this is
not the case, and historical hypotheses cannot explain the occurrence of torpor.

Endotherms maintain specific low temperatures in torpor if environmental conditions permit. Hummingbirds that have evolved in cold environments
maintain low body temperatures in torpor, whereas those that have evolved in
warmer environments maintain a higher body temperature in torpor. Recall
that the maintenance of a specific body temperature in torpor also occurs in
some ectotherms (see Fig. 8.2). The temperature that is maintained in torpor by
internal heat production or by behavior involving heat exchange is one that is
normally available in the environment of the animal. The maintenance of a
specific low body temperature may be due to an advantage associated with having a second, predictable temperature for regulation. Perhaps a set of enzymes
functions most effectively at a particular lower temperature. For example, the
affinity of the enzyme phosphoenolpyruvate for its substrate differs depending
on the temperature to which rainbow trout have been exposed (Fig. 8.8). Having two optimal temperatures for enzyme function can generate two distinct
selected temperatures when environmental conditions change.

Seasonal Torpor

Although similar principles of costs and benefits relative to energy expenditure
for body temperature regulation apply to seasonal hibernators, the time scale
over which expenditures and availabilities of energy change are much longer.

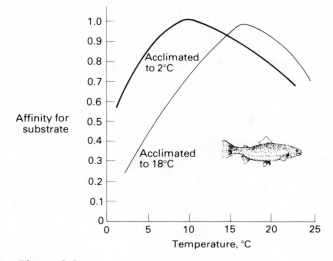

Figure 8.8
The affinity of the enzyme phosphoenolpyruvate for its
substrate is maximum at two temperatures that depend
on the temperatures of acclimation (2 °C or 18 °C)
(modified from Somero and Hochachka, 1971).

This situation leads to some differences in the ways by which seasonal hibernators deal with energy and changes in body temperature. The longer time
periods make it more difficult to quantitatively determine the relationships between energy availability and energy use, but a great deal of information has
been gathered on the characteristics of seasonal hibernators, which can be interpreted in terms of changes in costs and/or benefits with respect to energy
use.

Most hibernators are exposed to cold environments for several months of
the year, when the availability of food is reduced. Prior to the winter season they
store large energy reserves internally by means of fat deposits (Fig. 8.9). These
reserves are used at reduced rates during torpor until spring. This process is an
example of an "anticipatory" response. Energy is stored prior to its expenditure, and the severity of the thermal environment over the hibernating period
will determine whether storage is sufficient to meet demands.

With the balance between supply and demand operating over long time
periods, many seasonal hibernators have evolved "built-in" mechanisms to
deal with long-term patterns of energy use. These mechanisms were studied by
an experiment that involved measuring the body weight (an index of fat reserves)
of golden-mantled ground squirrels (*Citellus lateralis*) that were kept in a
laboratory at an environmental temperature of 35 °C for a year or more with a
constant photoperiod of 12 hours light and 12 hours dark (Fig. 8.10). At this environmental temperature torpor was not possible because body temperature
cannot drop below environmental temperature. Furthermore, there was no ex-

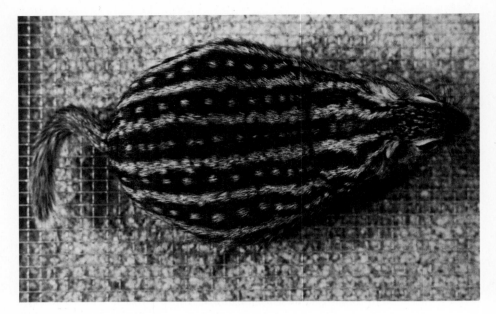

Figure 8.9
A "fat" thirteen-lined ground squirrel (*Citellus tridecemlineatus*) just prior to the on-set of the winter hibernating season. Photo courtesy of Nicholas Mrosovsky, University of Toronto.

ternal, environmental cue to changing seasons because photoperiod was kept constant. Nevertheless, the squirrels still increased their weight by about 100% by consuming food in excess of their short-term requirements in the "autumn" months. They also ate less and lost weight during the normal hibernating season (Fig. 8.10). Because this built-in rhythm is known to occur over approx-imately one year of time, it is called a *circannual* (about one year) rhythm.

Despite the built-in rhythm for energy storage and use, some hibernators will obtain more or less food depending on its availability in the late summer and fall. In addition, an extremely severe or prolonged winter season will place extra demands on stored energy reserves. This type of variation in supply and/or demand has to be compensated for by the length of time a hibernator spends in torpor during the winter together with the time it spends out of torpor attempting to supplement energy reserves by getting food. Many seasonal hibernators do not spend the entire winter in torpor. They undergo "periodic arousals" from torpor (Fig. 8.11), and less time is spent in torpor in the fall and spring compared with the winter. Even during the winter they occasionally undergo arousal from torpor.

Arousal from torpor requires energy to increase body temperature, so it should not occur unless the benefits from arousal exceed its cost. One possible benefit is to replenish energy reserves. Several species of hibernators occa-

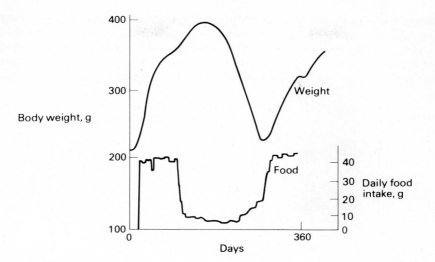

Figure 8.10
The body weight (an index to energy storage in fat) and food-consumption
pattern for ground squirrels (*Citellus lateralis*) show an annual rhythm
even when body temperature cannot decrease and photoperiod is held
constant (from Pengelley and Fisher, 1963).

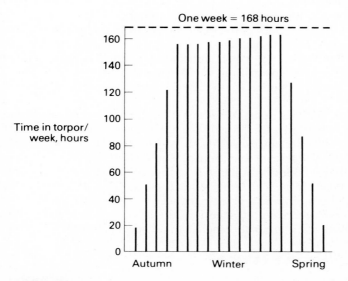

Figure 8.11
Schematic illustration of the pattern of periods of torpor for
a ground squirrel throughout an entire hibernating season
(from Torke and Twente, 1977).

sionally consume food during the hibernating season, and food is more likely to be available in fall and spring when periodic arousals are more frequent. Periodic arousals occur in species of hibernators that are relatively small (Table 8.1), and these species are not able to store as much energy internally relative to their rates of expenditure as do larger species of hibernators.

Table 8.1 Occurrence of feeding during the hibernating season for a number of species of hibernators (from Fisher and Manery, 1967)

SPECIES	WEIGHT, g	OBSERVED TO FEED
Marmot	2000	No
Hedgehog	1000	No
Columbian ground squirrel	700	No
European hamster	300	Yes
Dormouse	120	Yes
Golden hamster	100	Yes
Chipmunk	60	Yes

If hibernators consume food to adjust their energy reserves, do they always add a maximum amount of fat at all times during a season? Partway through the season a hibernator requires less energy to provide for expenditures until spring than it does at the beginning of the season, and the amount of food it consumes when it has been deprived is less partway through the season than at the beginning, which indicates that they have a "sliding set point" for proportionally replenishing energy reserves. Their requirements for energy change during the winter, as do their responses to energy supplies when they feed.

There are other hypotheses to explain the function of periodic arousals. Not all species of hibernators feed during the hibernating season (Table 8.1), yet some of them exhibit periodic arousals. Most hibernators urinate soon after arousal from torpor, suggesting that arousal may be necessary to excrete waste products, particularly if kidney function is suppressed during torpor. However, the evidence for this is equivocal. Some studies suggest that kidney function occurs during torpor although at a greatly reduced rate. Other studies, in which metabolites such as urea were injected into the blood of hibernating animals, did not show a consistent arousal response in all species tested.

SIZE LIMITS AND HIBERNATION

Very small endotherms can enter torpor on a daily basis, and mammals of intermediate size can exhibit seasonal hibernation (Table 8.1), but there are no very large true hibernators. Even though a bear may sleep for long periods in winter, very careful measurements of the temperatures of "hibernating" bears show that they do not enter torpor. Body temperature decreases a few degrees, but a bear sleeping in winter can be easily awakened.

The reasons for the absence of torpor in large endotherms involve limitations and advantages imposed by their size. For example, the amount of heat (the energy cost) required to raise the temperature of a large endotherm can be considerable. The specific heat of animal tissue is about 0.9 cal/g × °C. A grizzly bear that weighs 386 kg requires 347,400 calories to raise its temperature one degree. If body temperature has to increase from 5 to 37 °C for arousal from torpor, the heat requirement is 11,116,800 calories. The basal metabolic rate of a bear this size is about 6,100,000 calories per day. Although the rate of heat production increases above basal levels during arousal, the production of large amounts of heat in a short (daily) time period may not be feasible. Moreover, since large animals are able to store more energy internally relative to their rates of energy use, a decrease in body temperature is less important for them. Because large size permits internal storage of energy supplies sufficient to meet demands for temperature regulation, torpor is neither necessary nor desirable.

CONTROL MECHANISMS DURING TORPOR

The torpor cycle can be divided into three parts for the purpose of examining how function changes: entry, torpor itself, and arousal from torpor.

Entry into Torpor

When endotherms enter torpor they appear to reduce their rate of heat production to a minimum value. When this happens, heat is lost to a cold environment faster than it is being produced, and body temperature decreases (Fig. 8.12). There is also some information suggesting that heat loss increases during entry in mice, because measured values for the rate of body-temperature decrease fall between curves that assume minimum and maximum heat loss (Fig. 8.12). Increased heat loss can occur by means of a change in effective insulation, such as from redistribution of blood flow to the surface of the animal.

Experiments in which the hypothalamus of ground squirrels was cooled while they entered torpor suggest that reduced heat production during entry is accompanied by a gradual decrease of the hypothalamic set point. Thus a ground squirrel entering torpor will still respond to hypothalamic cooling by increasing heat production during the period when its temperature is decreasing, but lower and lower hypothalamic temperatures are required to produce heat production as entry continues.

Function during Torpor

The drastically reduced body temperature during torpor is accompanied by adjustments in function that reflect changed demands for oxygen uptake and delivery. Because the temperature of the animal is less, demands for oxygen are less. This condition is reflected by decreased lung ventilation, primarily from a drastic reduction in the rate of ventilation. For an animal in hibernation, breaths are separated by minutes instead of seconds.

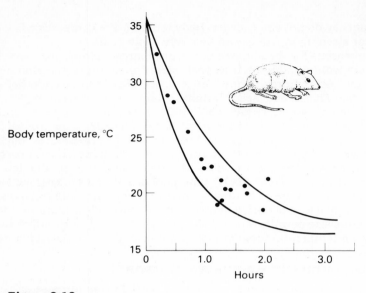

Figure 8.12
The body temperature of the mouse *Perognathus californicus*
entering torpor decreases at a rate intermediate between
minimum heat loss (upper curve) and maximum heat loss (lower
curve) with minimum rate of heat production (from Tucker,
1965).

Torpor also involves a decrease in circulatory transport function as compared with animals with higher body temperatures (Fig. 8.13). The circulation
to heart muscle and special fat deposits called "brown fat" is maintained at a
higher level than it is to other tissues, but the total blood flow is generally reduced to 10% or less of blood flow at higher body temperatures.

This reduction is accomplished, in part, from a drastic decrease in heart
rate (bradycardia; see Chapter 4). For example, the brown bat (*Eptesicus
fuscus*) reduces heart rate from about 700 beats/min at a body temperature of
37 °C to 12 beats/min at a body temperature of 5-6 °C. As a consequence of the
lower heart rate, cardiac output is also drastically reduced. For ground squirrels
(*Citellus tridecemlineatus*), cardiac output is 69 ml/min at a body temperature
of 37 °C but only 1 ml/min at a body temperature of 7-8 °C. Stroke volume remains about the same, and blood volume is redistributed by local changes in
blood vessel diameters. In general, the changes that occur in circulatory function in hibernating endotherms are strikingly similar to the changes that occur
during diving (see Chapter 4).

Despite the reduced intensity of function, hibernators still maintain precise control over their reduced body temperatures by mechanisms similar to
those that control higher body temperatures. If the anterior hypothalamus of a
hibernating ground squirrel (*Citellus lateralis*) is cooled, heat production is increased proportionally when very low hypothalamic temperatures are reached

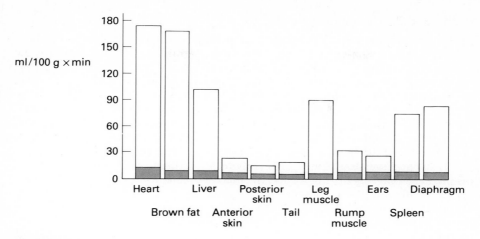

Figure 8.13
The rate of blood circulation to tissues in the dormouse, *Glis glis*, is drastically reduced during torpor (shaded areas) as compared with higher body temperature (open areas) (modified from Wells, 1971).

(Fig. 8.14). The set point (the reference value) has shifted to a much lower value in torpor, and it is different for different species, but heat production is still proportional to the difference between the temperature of the brain and the new set point temperature (Fig. 8.14). This type of mechanism results in the thermostatic heat production during torpor that is observed in a number of endotherms known to be torpid at low environmental temperatures (see Fig. 8.6).

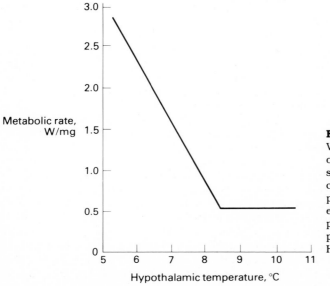

Figure 8.14
When the hypothalamus of a hibernating ground squirrel (*C. lateralis*) is cooled, heat production is proportional to the difference between brain temperature and a low set point temperature (from Heller and Colliver, 1974).

Acid-base Status during Torpor

Recall that the regulation of respiration for terrestrial animals involves the acid-base balance of blood (see Chapter 2). Air-breathing ectotherms regulate the degree of alkalinity of blood when their temperatures change by adjusting pH from changes in ventilation relative to CO_2 production so that the same enzyme net negative charge is maintained. When body temperature decreases, pH increases; thus the level of alkalinity stays the same (see Fig. 2.19). This adjustment is important for maintaining optimal enzyme function, but what happens to the pH of the fluids of hibernators when their body temperatures decrease?

The pH of the blood of several hibernators is shown in Table 8.2 for high (36-37 °C) and low (less than 10 °C) body temperatures. Note that although the pH increases slightly during torpor, the increase is not as much as it should be if degree of alkalinity, or enzyme net negative charge, was regulated. For example, in the marmot (*Marmota monax*) the observed pH at a body temperature of 8 °C is 7.60 (the $\triangle$pH/ °C = 0.005 expected only from temperature effects on buffers; see Chapter 3). The pH should be 7.95 ($\triangle$pH/ °C = 0.017) if enzyme net negative charge is to remain constant when body temperature decreases 29 °C. Thus these hibernators are relatively more *acidotic* than expected when their temperatures decrease.

Table 8.2 The pH of arterial blood for some hibernating animals and for the same animals with a "normal" high (36-37 °C) body temperature (from Malan *et al.*, 1973)

SPECIES	pH AT 36-37 °C	pH AT $T_B < 10$ °C
Ground squirrel (*Citellus tridecemlineatus*)	7.385	7.390
Marmot (*Marmota monax*)	7.45	7.60
Hamster (*Cricetus cricetus*)	7.40	7.57

What is the role of an acid concentration higher than predicted for hibernators in torpor? One possibility could be more dissociation of oxygen from hemoglobin because of the Bohr shift. Although this shift may occur, the demand for oxygen during torpor is reduced relative to environmental supplies of oxygen, so little advantage would be gained from an increased dissociation supply other than a somewhat reduced circulatory demand. Another possible advantage of acidity is the inhibition of rate processes. By changing the acidity, enzyme function might be suppressed and contribute to a general reduction of metabolism. At the time of arousal, respiratory hyperventilation occurs in hibernators, serving to reduce carbon dioxide concentration in the blood. As a consequence, pH shifts toward a more optimal (alkaline) condition for enzyme function as body temperature increases (see Fig. 3.20).

Arousal from Torpor

Hibernators must generate large amounts of heat to raise their temperatures within relatively short periods when they arouse. Mammalian hibernators use nonshivering thermogenesis to a large degree; mammals have specialized collections of fat cells, called *brown fat*, to produce large quantities of heat in short time periods.

The color of brown fat comes from a large concentration of oxidative enzymes located in the mitochondria of these fat cells, where a large proportion of nonshivering thermogenesis takes place (Fig. 8.15). No other tissue has more mitochondria per cell, and each mitochondrion is connected with a separate fat droplet, its own fuel supply. The mitochondria in brown-fat cells are also extensively filled with membranes where oxidation takes place (Fig. 8.15); thus this tissue is admirably suited to generate large amounts of heat from oxidation in a short time.

Figure 8.15
Electron micrograph of mitochondria from a brown-fat cell. Note the highly developed internal mitochondrial membranes, where oxidation of substrates takes place (from Hayward and Lyman, 1967).

Brown fat is found in several places in the body, but there are large deposits near the spinal cord and between the shoulder blades, called the interscapular brown fat. When bats (*E. fuscus*) arouse from torpor, this tissue is much warmer than the rest of the body (Fig. 8.16), but how does heat get from the interscapular brown fat to the rest of the body?

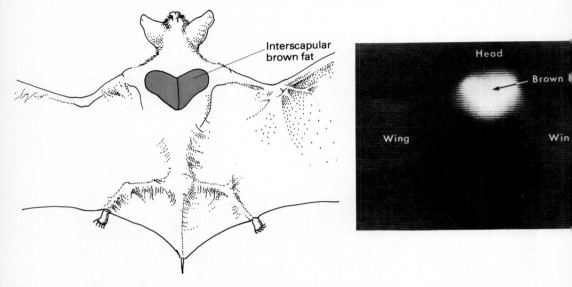

Figure 8.16
Photos taken with heat-sensitive film indicate that the temperature of interscapular brown fat in the bat *E. fuscus* (drawing) is high during arousal (from Hayward and Lyman, 1967).

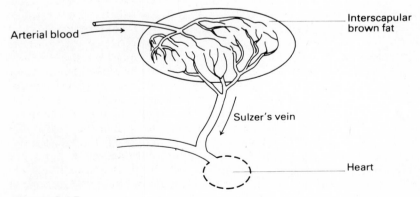

Figure 8.17
The arrangement of the circulation to and from interscapular brown fat permits passage of warmed venous blood to the heart with a minimum of heat loss via countercurrent heat exchange (adapted from Smith and Horwitz, 1969).

The vein that carries blood from the brown-fat deposit to the heart dilates to carry relatively large amounts of warmed blood that has passed through this tissue. In addition, this vein (called Sulzer's vein) is isolated from arteries that carry blood to the interscapular brown fat (Fig. 8.17). This isolation means that the warm venous blood reaches the heart *without* losing heat to cooler arterial blood by countercurrent heat exchange.

The Mechanisms of Mitochondrial Heat Production

Brown fat is highly suited to the study of the mechanisms by which oxidation produces relatively large amounts of heat for temperature regulation by non-shivering thermogenesis. The mechanisms are best studied in hibernators, but all mammals have deposits of brown fat; thus the mechanisms of increased oxidative heat production will be found in all mammals.

The production of heat in brown-fat cells is initiated when the chemical norepinephrine is released from nerves of the sympathetic nervous system that innervate brown-fat deposits (see Chapter 17 for details of general control mechanisms involving the sympathetic nervous system). The norepinephrine is believed to influence one or both of two processes that can change the amount of heat produced from oxidative events within mitochondria.

The events involved in oxidation of substrates within a mitochondrion are summarized in Fig. 8.18. Oxidation involves a substrate (such as NADH) accepting a proton (H^+). This event occurs on the inner surface of internal mitochondrial membranes and results in movement of the protons across the membrane to the outside. For increased heat production through substrate oxidation, the protons must be transported from the outside back inside across the inner mitochondrial membrane (Fig. 8.18).

A proton on the outside of the membrane could follow one of two paths from outside to inside, where it is required for oxidation of the substrate. In *ATP-coupled oxidation* (upper path in Fig. 8.18), the combination of ADP with phosphorus (P_i) to yield ATP results in proton movement into the inner surface. Thus any use of ATP to generate ADP, making ADP available for phosphorylation, will increase heat production from the increased movement of protons from outside to inside (for combination with substrates). In ATP-coupled oxidation, increased heat production occurs if norepinephrine increases demands for ATP by means of increased use of ATP in the cell. It is analogous to the mechanism for increased heat production by shivering thermogenesis in which ATP use in muscular contraction results in increased oxidative heat production.

Another way to increase heat production is to *uncouple* the across-the-membrane movement of protons from a dependence on ATP production and use (middle path in Fig. 8.18). For example, the membrane could develop a "leak" for protons, perhaps as a consequence of the presence of norepinephrine, so that heat would be produced from oxidation without a requirement for increased ATP production and utilization.

Which one of these mechanisms occurs? There is evidence that both are important for the increased heat production observed in brown-fat nonshivering thermogenesis. ATP-dependent processes within brown-fat cells (such as

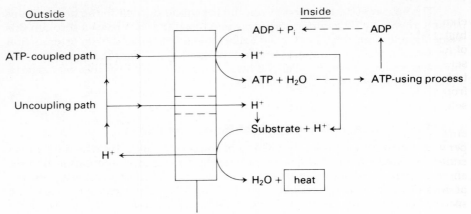

Figure 8.18
Oxidation on the inner surface of the internal mitochondrial membranes involves proton (H^+) acceptance by substrates to produce heat. This can be coupled to ATP production through proton transport across the membrane (upper path) so that cellular demands for ATP and oxidation are linked, or heat production may be uncoupled from ATP production if "leaks" allow proton movement across the membranes independently of phosphorylation (middle path) (adapted from Horwitz, 1978).

ion movements across cellular membranes; see Chapters 12 and 13) increase in rate somewhat (by 50-70% in some cells) but do not account for all the heat generated. The remainder could come from *some* uncoupling. For this reason, oxidation in the mitochondria of brown-fat cells is described as "loosely coupled"; that is, there is a degree of uncoupling for heat production independent of cellular demands for ATP supplies.

All mammals have deposits of brown fat, but the amount is larger in hibernating mammals. In addition, brown fat is important for heat production in some newborn mammals that are poorly insulated at birth. As was the case with the diving reflex (see Chapter 4), brown-fat thermogenesis represents a common mechanism of function in a large number of animals, but it has been modified through natural selection to provide for a different *degree* of function in certain groups of animals in which supplies and demands for energy as a resource are different.

SUMMARY

Although a relatively high body temperature is most effective under some circumstances, a lower body temperature will be optimal if environmental conditions change such that the costs for maintaining a high temperature become prohibitive. Costs can involve the rate of energy use, and reduced energy availability can lead to the evolution of torpor to reduce energy costs.

Some ectotherms select lower environmental temperatures for heat exchange if food availability is reduced, but there is an advantage in selecting a high body temperature with which to process (assimilate) even limited amounts of food with maximum efficiency. It should result in a pattern of temperature selection timed with respect to food consumption. Other aspects of an ectotherm's behavior with respect to its environment, such as periodic shelter from predators or aggressive individuals, will also produce torpor by forcing selection of a lower body temperature.

Torpor in endotherms occurs daily or seasonally. Some small endotherms enter torpor daily; this action is related to their high rates of energy expenditure per gram as compared with their lower energy-storage abilities. However, small endotherms do not enter torpor unless they have used all but a fraction of their energy reserves because torpor involves risks. When torpor occurs, body temperatures are regulated at lower values characteristic of the environment of the species, and this may permit relatively effective function at a lower body temperature.

Seasonal torpor involves supply and demand for energy over longer time periods. It has led to the evolution of anticipatory, built-in, circannual rhythms of energy storage in late summer and fall and energy use in winter. The energy demands of seasonal hibernators change during the hibernating season and so do their responses through feeding. By periodically arousing during the season they may be able to proportionally supplement energy reserves or excrete waste products. There are no very large hibernators because it is not feasible for large animals to generate large amounts of heat for arousal in a short time. Large animals are also able to store more energy relative to their requirements; thus torpor is not necessary.

When endotherms enter torpor, heat production is reduced and heat loss can increase, meaning that body temperature will decrease. During torpor, respiration and heart rates are reduced, and cardiac output is less along with a redistribution of blood flow so that heart and brown fat receive a larger proportion of the reduced flow. The brain is still important for body-temperature control but has a reduced set point during torpor. The pH of blood is lower than it would be if endothermic hibernators were regulating the relative alkalinity of blood during torpor. This acidosis during torpor may aid in suppressing some functions. Hyperventilation during arousal shifts pH to more optimal (alkaline) values with respect to temperature.

Arousal from torpor involves production of large amounts of heat in short periods. Nonshivering thermogenesis in mammals caused by highly oxidative brown-fat deposits produces much of the heat, which is shunted to the heart through veins positioned so that countercurrent heat exchange with cooler arterial blood is minimized. The mechanisms for increased heat production by nonshivering thermogenesis involve increased use of ATP; also involved is increased oxidation from uncoupled proton transport within mitochondria, meaning that heat production is not entirely tied to changing cellular demands for ATP.

Locomotion

9

Animals increase energy expenditures when they move. Although the amount of time spent in activity can be small, the increase in metabolic rate can be substantial. Effective energy use while animals are active involves minimizing the required increase in energy expenditures to achieve the functions of activity.

There are several functions of activity. Animals move to feed, to mate, to escape predators, or simply to cover distance such as during migration. Movements for feeding involve increasing energy expenditure to obtain more energy, and economic efficiency for this kind of activity involves maximizing energy intakes (gain) in relation to expenditures (see Chapter 10). Movements for predator escape involve maximizing the distance between predators and prey (for the prey), so prey survival is related to the extent of increase in expenditures for movement as compared with the location of predators. Detection, maximum speed, and maneuverability are important components of activity for predator-prey interactions over short distances (see Part VI).

In this chapter we will be concerned with the biological and environmental characteristics that influence effective movement in which the function of activity is to cover distance without the constraints imposed by feeding or predator avoidance. We will examine movement by modes of locomotion: running, flying, and swimming. Each one involves different environmental characteristics; that is, movement on the ground, in the air, and in water. For each locomotion mode we will examine relationships between energy expenditures, speed, and distance. We are particularly interested in defining the aspects of locomotion that contribute to effective energy use by minimizing the energy costs to cover distance (the costs of transportation).

The discussion of locomotion deals primarily with endothermic vertebrates that support increased power requirements mainly from aerobic metabolism (increased oxygen consumption). Most detailed information has come from studies of these animals in which aerobic processes can produce sustained changes in activity. Other animals that rely more on anaerobic metabolism during activity (reptiles, amphibians, and fishes among vertebrates) are subject to similar physical factors that influence energy expenditure but over shorter times. Anaerobic processes provide a high initial "scope" (potential increase) for muscular activity, but they are capable of sustaining activity only for short periods (see Chapter 18). Thus stamina or endurance varies considerably among animals depending on major processes that provide the energy for activity, but within these limits similar physical factors influence the expenditure of energy for locomotion.

TERRESTRIAL LOCOMOTION

When mammals of different sizes run on a treadmill while their rates of oxygen consumption are being measured, their rates of energy expenditure per gram increase linearly with speed but with a much steeper slope for small species (Fig. 9.1). Why are these relationships linear? What are the implications for the costs of covering distances? Some answers can be found by examining the forces operating on a terrestrial animal as it moves.

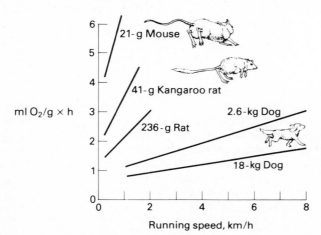

Figure 9.1
The per-gram rate of oxygen consumption increases linearly as a function of running speed for mammals of different sizes, with the steepest increases for smaller species (after Taylor *et al.*, 1970).

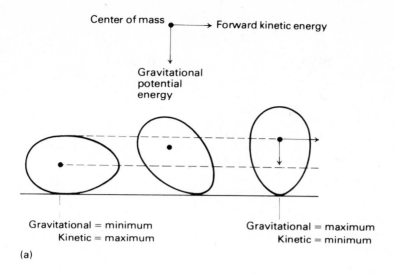

(a)

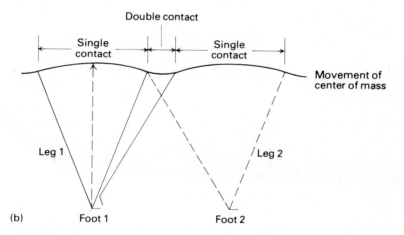

(b)

Figure 9.2

Walking is analogous to a rolling egg (a) and an inverted pendulum
(b). For the egg, forward kinetic energy lifts the center of mass to a
maximum gravitational potential energy that is then translated back
to kinetic energy. For the inverted pendulum, the center of mass is
at its greatest height (maximum gravitational potential energy) during single-foot contact (dashed arrow). Kinetic energy can be at a
maximum when the center of mass is lowest, which is during
double-foot contact (b based on Cavagna *et al.*, 1977).

The Mechanics of Terrestrial Locomotion

The forces operating on an animal during walking are analogous to those operating on a rolling egg (Fig. 9.2a). If the *forward kinetic energy* is sufficient, the egg will move and the center of mass will rise, increasing the *gravitational potential energy*. At the fulcrum (when the egg is on its apex), the gravitational energy is translated into forward kinetic energy. The two forms of energy are out of phase; i.e., as one increases the other decreases, and the egg falls forward more rapidly from its apex (where speed is instantaneously zero) to its side.

As you walk, your body initially rises, increasing the gravitational potential energy. You then "fall" forward, and gravitational potential energy is translated into forward kinetic energy. At the start of the next stride, the kinetic energy from your forward movement serves to raise your center of mass again.

The two forms of energy oscillate back and forth as you walk. When they are in exactly opposite phases during a stride, the additional energy needed from internal muscular contraction is *minimum*. If kinetic and gravitational energy become more in phase, the translation of one to the other is less than maximum, and more internal power must be generated to maintain forward movement.

Walking is also analogous to an inverted pendulum (Fig. 9.2b). Halfway through a stride, when one leg is in contact with the ground, the center of mass is displaced upward by a maximum amount. For the part of a stride when both feet contact the ground, the center of mass is at its lowest point while forward movement is maximum. A similar relationship holds for animals with more than two legs. Animals with four legs walk as if they were two individuals (each with two legs) walking one behind the other.

Figure 9.3 shows the phase relationships between kinetic and potential energy for a turkey walking at two speeds. Similar relationships hold for other walking animals. At a speed of 3.4 km/h kinetic and gravitational energy are maximally out of phase during a stride, while at the higher speed of 4.6 km/h kinetic and potential energy are less out of (i.e., more in) phase.

One way to assess the impact of these changes in phase relationships that occur as speed changes is to determine how much power (cal/kg × min) *would* be required for walking if there were *no* exchange between kinetic and potential energy during a stride, and compare this hypothetical power with the *actual* power requirements resulting from some exchange between the two energy forms. If no exchange can occur, all the power for movement has to come from internal muscle contraction.

The extent to which external exchange of energy reduces internal power requirements is shown in Fig. 9.4 for the turkey. About 70% of the total power that would have to be generated without any exchange is "recovered" (does not have to be generated from muscle contraction) at a walking speed between 3 and 4 km/h. At lower and higher speeds, more internal power must be generated because of changes in the phase relationships between kinetic and gravitational energy during a stride (Fig. 9.3).

Note in Fig. 9.4 that the percent recovery for a walking turkey reaches zero when speed increases to 5-6 km/h. It is because of the change in phase relationship; kinetic and potential energy are completely in phase and no exchange can occur between them. It occurs also in animals that run, trot, or hop; for example, the phase relationships for a running turkey are shown in the lower part of Fig. 9.3.

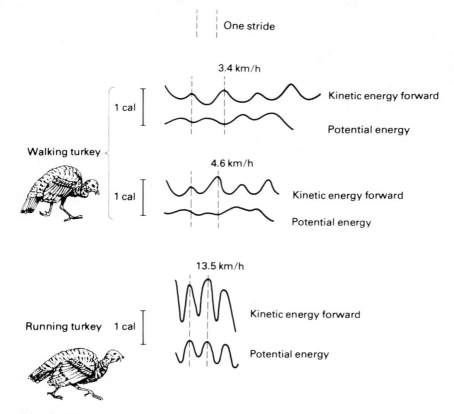

Figure 9.3
When a turkey walks at a speed of 3.4 km/h, kinetic and potential energy exchange maximally because they are maximally out of phase during a stride. When forward kinetic energy is maximum, potential energy is minimum. At higher walking speeds (4.6 km/h) or during running (13.5 km/h), the two energy forms are less out of phase (more in phase); less exchange occurs between them during a stride (adapted from Cavagna et al., 1977).

Since kinetic and potential energy do not exchange during running, power must be generated to: (1) lift the center of mass of a running animal, and (2) propel it forward. These two power requirements must be added because they cannot exchange for a running animal. Figure 9.5 shows the power of each component for the turkey. Note that the power to lift the center of mass is independent of speed whereas the power for forward movement increases with increasing running speed, as does the total. This situation produces the general shape of

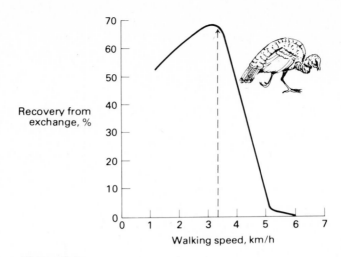

Figure 9.4
For the turkey, the maximum exchange between kinetic
and gravitational energy when they are out of phase at
3.4 km/h (dashed arrow) equals about 70% of the energy
required for walking. Changes in phase between these
energy components when speed changes reduces exchange between the two at other speeds, increasing required internal power for movement (adapted from
Cavagna *et al.*, 1977).

the linear curves in Fig. 9.1 relating rate of oxygen consumption (internal power)
and running speed.

There is another form of energy that contributes to the efficiency of
locomotion of terrestrial animals that is quite important in running, trotting,
and hopping. During part of a stride some energy is stored in the elastic
elements of muscles and tendons. This storage is analogous to the winding of a
spring, and the kangaroo is a good biological example. It is difficult to measure
the energy stored and released in the elastic elements, but its importance can be
estimated by measuring the physical efficiency of doing external work.

The physical efficiency of doing work is calculated by dividing the internal
power produced (from the rate of oxygen consumption) into the power required
for doing external mechanical work on the environment. For muscles, the efficiency of transformation of internal chemical energy to external mechanical
work is no greater than about 25% (that is, only 25% of energy from oxidation
of substrates ends up as "useful" work in terms of movement of a mass). Thus if
the ratio of external mechanical power to internal metabolic power exceeds
0.25, some other energy must be contributing to the higher-than-predicted efficiency.

Figure 9.6 shows measurements for the efficiency of doing external work
(power required for mechanical work/power measured from metabolic rate) for

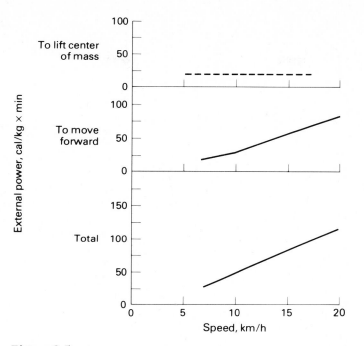

Figure 9.5
For a turkey, the total external power required to lift the center of
mass remains constant as speed changes, but the power required
for forward movement and the total increase linearly with increasing running speed (adapted from Cavagna *et al.*, 1977).

the turkey and for the kangaroo. Note that the kangaroo's efficiency for doing
external work when it is hopping exceeds what is expected just from the efficiency of muscular contraction because of the elastic storage and release of
energy in its large tendons when it is hopping.

In summary, there are two basic mechanisms that operate to minimize
power rquirements for terrestrial locomotion: (1) moving at speeds for walking
where gravitational and kinetic energy are in opposite phases, and (2) using
elastic storage and recovery during running, trotting, and hopping when the
first mechanism cannot operate.

The use of these mechanisms depends on the gait utilized for terrestrial
locomotion. Running, trotting, and hopping all involve the utilization of elastic
energy storage and recovery because kinetic and potential energy are in phase.
Galloping involves a combination of the two mechanisms. When animals gallop
at low speeds, up to 30% of energy can exchange between kinetic and gravitational modes. At high galloping speeds elastic storage and recovery become
more important as animals alternately "bounce" off their back feet and front
feet.

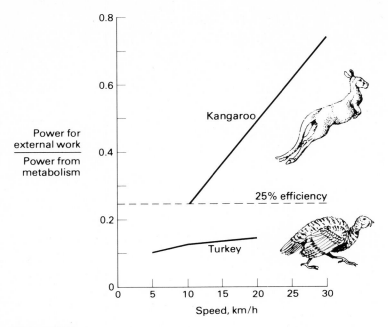

Figure 9.6
The efficiency of transformation of internal power for muscular contraction to external work should not exceed about 25%. The kangaroo achieves greater efficiency because of storage and recovery of elastic energy when it hops (adapted from Cavagna *et al.*, 1977).

Minimizing Costs for Terrestrial Transportation

The locomotion function we are concerned with is covering distance. The cost to cover distance is easily calculated by dividing the cost per unit time by the speed at which that cost is measured (cal/g × h ÷ km/h = cal/g × km). The result is called the *cost of transport* (Fig. 9.7). It is similar to the inverse of fuel use as miles per gallon because food as fuel supplies the energy to cover distance by locomotion.

Note that the cost of transport per gram for running mammals decreases as speed increases and eventually reaches a constant minimum value (Fig. 9.7). This relationship is most pronounced for animals of small size. If minimizing the energy expended to cover distance contributes to energy efficiency, animals should run at *minimum-cost-of-transport speeds* (where the curves in Fig. 9.7 are flat). For large animals, such as dogs, this is accomplished by running at any speed because the cost of transport remains essentially constant as speed changes. Smaller animals should run only at high speeds to minimize the cost of transport. Very small animals, like mice, cannot run at speeds that produce an absolute minimum transport cost, but their minimum costs can be estimated by extending the lines in Fig. 9.1 beyond measured values (producing the dashed lines in Fig. 9.7).

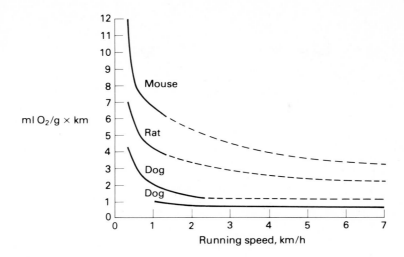

Figure 9.7

The per-gram cost of transport (the cost of running) for one kilometer as
a function of running speed. The dashed lines represent theoretical
values because the animals, while on a treadmill, only ran as fast as indicated by the solid lines (adapted from Taylor *et al.*, 1970).

Although the minimum cost of transport *per gram* is higher for smaller
animals, the *total* cost for transport at minimum-transport-cost speeds is higher
for larger animals because they must move a larger mass. Because mice may
not be able to move fast enough to reach minimum-transport-cost speeds, they
may incur some loss of efficiency to cover distance as compared with larger
animals that can move at sufficient speeds to achieve minimum transport costs.
The importance to mice depends on the impact of having to move at speeds at
which energy supplies may be used at a higher rate. Otherwise the general interpretation of minimum transport costs per gram are similar to the interpretation of size effects on standard rates of minimum energy use, which show a
similar (negative-slope) relationship with size (see Chapter 5).

Is there any evidence that animals are responsive to these costs for
locomotion? Does observed activity match what would be predicted based on
cost minimization? Costs are likely to be important when animals must migrate
over relatively long distances (with fixed fuel supplies), and the constancy of
minimum transport costs for running over a wide range of speeds makes the
velocity of running a poor indicator of movement effectiveness for large animals.
Small species are predicted to move at maximum speeds to cover distances
(Fig. 9.7). Your impression of movement in mice may be that it occurs at high
speed, but it may be because of their close proximity to you (predator avoidance) rather than because of the minimization of transport costs. It would be interesting to examine speeds of movement for small animals in detail under conditions in which the function of movement is to cover distance without other
constraints.

Cost minimization seems important to the pattern of migratory locomotion for some groups of animals. Spiny lobsters (*Panulirus argus*) migrate for considerable distances at certain seasons by walking single file in relatively large groups (queues). This behavior results in minimizing locomotion costs because individuals toward the rear experience less drag, i.e., resistance to movement, as they move through the water along the sea bottom (Fig. 9.8). Leadership often changes, meaning that individuals alternate in the position that is equivalent to high-cost individual locomotion.

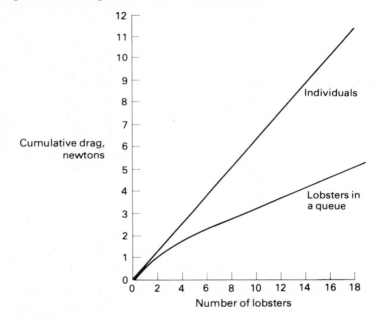

Figure 9.8
The total drag for each lobster is less if it joins a straight line formation than if it moves individually (modified from Bill and Herrnkind, 1976).

There are other advantages that are possible for this type of group movement, such as reduced predation or advantages of individual orientation to reach a destination. However, the minimization of locomotion costs could occur as well. In addition, we will find similar possible advantages to group movements by some animals that fly or swim in which individual movement effectiveness can be enhanced from utilization of forces that act between individuals (see below).

FLYING

Humans have been fascinated with flight for ages, but it was not until relatively recently that animal flight could be studied in detail. A major advance in the

Figure 9.9
A parakeet flying in a wind tunnel can be equipped with a mask to measure oxygen consumption rates for flight at different speeds (from Tucker, 1968).

study of flight physiology occurred when biologists trained birds to fly in wind tunnels where they could be studied as they flew in "stationary flight" (Fig. 9.9). Small masks placed on the birds permitted measurements of oxygen consumption, and the speed of flight could be controlled simply by adjusting the motor speed on the wind-tunnel fan.

In flight, the change in power (rate of oxygen consumption) with speed is called a *performance curve*. The performance curve for a parakeet flying level in a wind tunnel is shown in Fig. 9.10. At low airspeeds rate of energy expenditure is high; power decreases with increasing speed until a minimum value is reached at about 35 km/h; above this speed the energy expenditure rate increases again. The performance curves for different birds have different absolute values, but they are usually U-shaped with a particular speed for minimum power (denoted V_{mp} in Fig. 9.10). Why is this the case? Again, we can find some answers by examining forces operating on animals as they fly.

The Mechanics of Flight

It is possible to consider animals as aircraft to examine the basic physical forces operating in flight (Fig. 9.11). Lift must overcome weight for an animal to remain airborne, and thrust must overcome drag for it to move forward. Power is required for both lift and thrust.

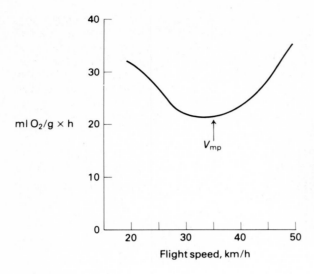

Figure 9.10
The rate of oxygen consumption for a 35-gram flying parakeet is high at low speeds, decreases to a minimum (V_{mp}) as speed increases, and increases at higher speeds (after Tucker, 1968).

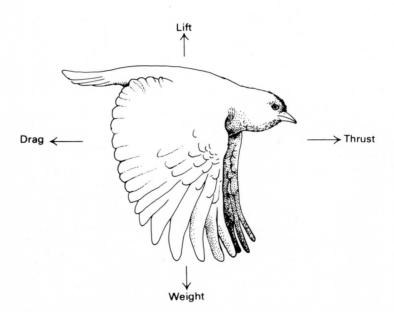

Figure 9.11
Flight involves power to generate lift (to overcome weight) and thrust (to overcome drag).

The power for lift is called *induced power* (P_i). In animals induced power comes from the movement of air over the wings (Fig. 9.12). The wings are *air-foils*; rapid flow of air over the wings produces a partial vacuum beneath the wings that provides an upward lifting force. The faster the air moves, the lower the induced power required to balance weight. Thus the faster an animal moves through the air, the lower the requirement for power to produce lift because the speed of movement of air over the wings increases. Figure 9.13 shows the decrease in P_i that occurs for a flying animal as flight speed increases.

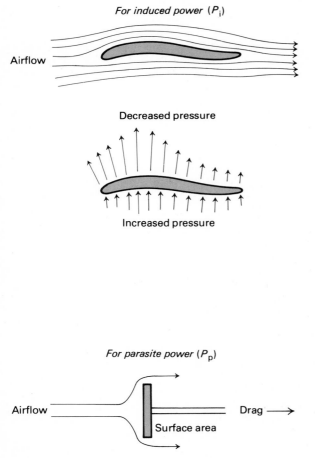

For induced power (P_i)

Airflow

Decreased pressure

Increased pressure

For parasite power (P_p)

Airflow

Surface area

Drag $\longrightarrow$

Figure 9.12
Differential airflow over wings produces induced power for lift, and the cross-sectional area of the body that hits against air determines drag, which must be overcome by thrust.

Induced power also depends on the morphology of wings, which differs among species. When wings beat, they move through an arc called the *wing disc*. As the wing disc area increases, induced power decreases because air is moving over a larger lifting surface. The size of the wing disc area is related to wing length ($\frac{1}{4} \pi b^2$, where b = wing length). Animals that have short wings compared with their body weights are said to have high *wing disc loading* because the load (weight) that must be supported by the sweep of wings through a wing disc is high. An increase in weight among species is usually accompanied by an increase in wing disc area to provide a larger lifting surface (up to a given weight limit). However, different groups of flying animals have different degrees of wing disc loading. Geese, for example, have high wing disc loading as compared with other birds, so they must expend more power during wing beats to produce sufficient lift. These birds spend large amounts of time on water and relatively little time flying other than during migration, which is reflected in the allocation of resources to the structures that influence power requirements for flight.

The power needed for thrust to overcome drag (Fig. 9.11) is usually divided into two parts: power to overcome drag on the body (fuselage), or *parasite power* (P_p), and power to overcome drag on the wings, or *profile power* (P_o). Wing drag is very difficult to measure because it can change during the beat of the wings. For simplicity we will consider it constant with flight speed (Fig. 9.13; P_o), although it usually increases somewhat with flight speed, particularly for small passerine birds with relatively long wings that beat their wings rapidly (such as finches, warblers, and sparrows).

Drag on the body of an animal is determined by the surface area that hits against air (Fig. 9.12). Highly streamlined objects have a low drag because surface area is minimized. Moreover, the power required to overcome drag on the body increases with increasing flight speed. Fig. 9.13 shows the increase in P_p that occurs for a flying animal as flight speed increases when surface area influencing drag remains constant.

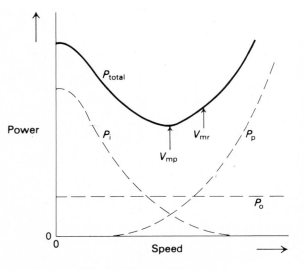

Figure 9.13
Total power for flight (P_{total}) is the sum of induced power (P_l), profile power (P_o), and parasite power (P_p). For this simplified figure P_o is considered constant, although it usually increases as flight speed increases and may contribute to the upsweep of the U-shaped performance curve. Minimum power speed (V_{mp}) is less than speed for maximum range (V_{mr}) (adapted from Pennycuick, 1969).

If the three power terms in Fig. 9.13 are added together, we can construct a performance curve based on the forces operating in flight. The U shape is a consequence of changes in the relative importance of power for lift (decreasing with increasing speed) and power to overcome drag on the body (increasing with increasing speed). Drag on the wings (P_o) will also contribute to an upswing of the U-shaped performance curve at high speeds for some species. This analysis of flight differs from direct measurements of rates of oxygen consumption. The total mechanical work done on the environment for lift and drag is called *power output*, and the power required for an animal to do that work is called *power input*. These amounts are related by a physical efficiency term.

Minimizing Costs for Flight Transportation

As a consequence of changes in power for drag and lift, there is a flight speed at which the *rate* of energy expenditure is least (V_{mp} in Fig. 9.13), but it is not the flight speed that results in a minimum energy expenditure over a *distance*. If the power in flight at different speeds for the parakeet in Fig. 9.10 is divided by flight speed, minimum cost of transport occurs at a flight speed of 42 km/h (Fig. 9.14). This speed is also called the speed for *maximum range* (V_{mr}) because the parakeet will fly a maximum distance with a given supply of fuel (food or fat) when it moves at this speed.

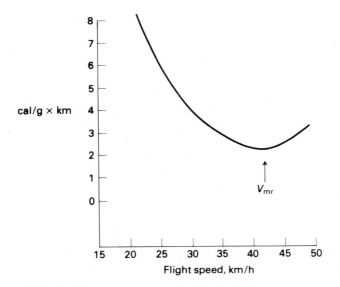

Figure 9.14
For the parakeet, minimum cost of transport (at maximum range speed, V_{mr}) occurs at a flight speed of 42 km/h, but note that the curve is relatively flat for a variety of flight speeds close to V_{mr} (modified from Tucker, 1968).

Note that the cost-of-transport curve (Fig. 9.14) is relatively flat above speeds of 30 km/h for the parakeet. Although the cost is least at about 42 km/h, a parakeet will not incur greatly increased costs for transport when it moves at somewhat different speeds. Thus like locomotion by running, speed for flight is a relatively poor indicator of whether efficiency of energy use is important for distance movement.

The minimum cost of transport is less for flying than for running. For an animal that weighs 1 kg, flight at minimum cost of transport speed requires (costs) about 0.9 cal/g × km. A 1-kg runner expends about 3 cal/g × km. The relationships are similar for animals of other sizes (see below). However, the expense relationships are opposite for *rates* of energy use. Flight involves increase in the rate of energy expenditure by 6-10 times for flight at maximum-range speeds compared with resting values, whereas running involves an increase in rates of energy expenditure by 2-3 times compared with resting values for movement at minimum-transport-cost speeds (compare the y axes of Figs. 9.1 and 9.10 for comparable sizes, using the same units for rates of expenditure). These differences exist because of the interaction between power and speed for the different locomotion modes. Although flight requires a higher rate of energy expenditure, speeds are much higher; thus distance is covered with less cost.

Hovering Flight: A Special Case

Some insects and birds can fly at zero forward speed for prolonged periods. The mechanisms for hovering flight in animals are analogous to the mechanisms utilized for flight in helicopters, illustrated in Fig. 9.15 for hovering flight by a hummingbird.

Many birds generate power for lift at low speeds only during the downstroke of the wings. At the bottom of a downstroke the wings are "feathered" and brought quickly up to the start of the next downstroke. Hovering in hummingbirds involves generation of power for lift on both downstroke and upstroke (Fig. 9.15). If the speed of wing movement (up to 80 beats in one second in very small hummingbirds) is "slowed" by high-speed filming, the wings are seen to pass through a "figure-eight" arc (Fig. 9.15). At the end of a downstroke (which is actually tilted forward), the wings are quickly flipped over so that the upper sides of the wings beat down as they move backward and up. This operation is analogous to the pivoting by the wings (blades) of a helicopter, and it propels a stream of air downward through the wing disc. It requires the pivot in each wing of a hummingbird to be very flexible. The wings are pivoted at the "wrist" rather than at the elbow or shoulder; you can simulate the figure-eight pattern by a movement of your hand while keeping your arm stationary.

Flight at zero speed involves no lift from forward movement, meaning that movement of air over the wings depends entirely on the movement of the wings. Thus hovering flight is energetically demanding. Note in Fig. 9.13 that the curve for P_{total} intercepts the power axis at zero speed at a high value. The rate of oxygen consumption by hovering hummingbirds has been measured for seven species. The results show some of the highest rates of oxygen consumption per gram for vertebrates, about 43 ml O_2/g × h. Resting rate of oxygen consumption for these birds is about 3-4 ml O_2/g × h, so hovering involves an in-

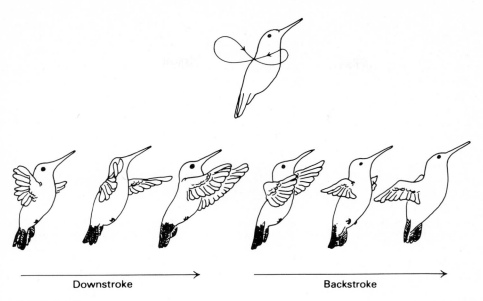

Downstroke Backstroke

Figure 9.15
The movement of the wings of a hummingbird describes a "figure eight" pattern during hovering. During a backstroke the wings are "flipped over" so that the upper surfaces beat down, generating lift.

crease in cost of about 10-15 times the minimum resting values. These increases can be sustained as long as the hummingbirds obtain sufficient energy from the nectar of flowers from which they feed while hovering.

Minimizing Flight Costs through Individual and Group Behavior

Because each species has a predictable speed for movement to minimize locomotion costs, it might be possible to measure flight speeds to see if animals move to minimize their costs of covering distance. Again, animals that are migrating are best to observe because they must fly long distances with fixed amounts of fuel.

Unfortunately, there is little precise information on flight speeds for migrating animals. Sometimes it is difficult to tell if an animal is migrating or just moving a short distance. It is also difficult to account for winds. Birds can use tail winds or fly into head winds, both of which will change the observed flight speeds. In addition, recall that the cost-of-transport curve for flight by individuals is relatively flat at high speeds, meaning that individuals could be relatively insensitive to the absolute minimum costs that occur at a specific speed.

Recent studies of bird flight in wind tunnels suggest that the performance curve may be adjustable by individuals to some extent at higher speeds. Measurements of starlings showed a relatively constant rate of oxygen consumption at speeds above V_{mp}, where drag effects would be expected to produce the upward U effect. The constancy of rate of energy expenditure at high speeds is thought to be due to adjustments of drag by the birds; that is, they "trim" body contours to reduce drag as flight speed increases. If it occurs, it would have an important effect on the cost of transport and would allow movement at speeds somewhat higher than predicted, assuming drag remained constant.

Another individual behavior that is characteristic of flight in some birds is "bounding flight." In this flight mode, the wings beat for a portion of flight time and are then folded against the body; it produces an up-and-down movement during flight. Bounding flight is observed in small passerine birds with relatively long wings and rapid wing beats. These birds have an appreciable increase in profile or wing drag as speed increases. Folding the wings for part of flight reduces the impact of increasing wing drag on the upsweep of the U-shaped performance curve and lowers the cost of transport.

Some birds fly in groups when they migrate, which may reflect a mechanism to minimize costs for transport similar to the pattern of movement of lobsters, and perhaps bird groups are more easily studied to test ideas about cost minimization. The classic example of group migration is the V formation of Canada geese (*Branta canadensis*). Geese are also interesting to study because they have relatively high wing disc loads, and any mechanism to decrease power for flight would contribute substantially to their movement effectiveness.

When the wing of one bird sweeps downward it produces an upwash of air past the tip of the wing, and the magnitude of the upwash is greatest near the tip (Fig. 9.16). If another bird moves close enough, the upwash will reduce the amount of its induced power required for lift. Similarly, the upwash from the second bird will benefit the first bird. The upwash also trails behind a bird as a vortex, much like eddies of water; birds displaced from a straight line will also receive a benefit in reduced requirement for induced power if they are positioned appropriately with respect to the vortex.

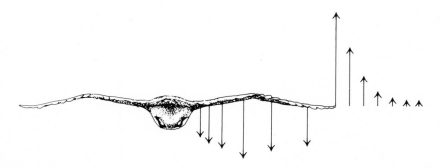

Figure 9.16
The beat of the wings induces air flow downward beneath the wings and upward beyond the tip of the wings. The magnitude of the upwash is greatest near the tip of the wings (from Lissaman and Shollenberger, 1970).

The potential energy savings from flying in formation depend on how much upwash is produced by the wings. It depends on wing length (b in Fig. 9.17) and the spacing between wing tips of adjacent birds (s in Fig. 9.17). For one species in which wing length is essentially constant, the ratio $b/(b + s)$ is 1.0 when s is 0 and it approaches 0 when s is very large.

As s decreases, the power required for lift in formation flight as compared with power required for lift in solo flight should decrease (Fig. 9.17). The savings in power depend somewhat on the number of birds in the formation, but even for small formations of 3-9 birds, individuals could save 40-50% of the energy required for lift by flying in formation. It is shown in Fig. 9.17 that when $s \rightarrow 0.0$ (distance between bird wing tips is minimum), formation flight involves only 50-60% of lift power as compared with solo flight.

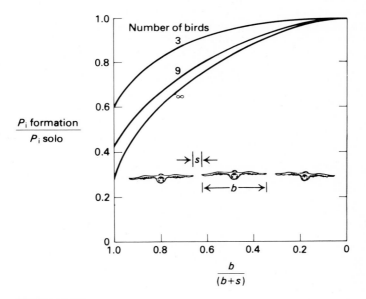

Figure 9.17
The potential savings in lift power for an individual flying in formation as compared with solo flight increases as spacing between wing tips decreases ($s \rightarrow 0$) and increases somewhat with the number of birds flying in formation (modified from Lissaman and Shollenberger, 1970).

Why should formation flight in geese involve a V shape? If the formation is a straight line with equal spacing between all birds, the birds in the middle will gain a greater lift than the birds at each end, possibly leading to continuous competition among the birds in formation for the center position (the one with the most advantage). A V shape with different spacing from the apex of the V to the tails will distribute lift equally among the members of the formation. Putting a lead bird farther out reduces its lift savings, but the savings can be equalized by having the second birds in the V close to the lead bird. Trailing birds near the ends of the V will receive more lift from the backwash of birds in front of them, which can be equalized by increasing distance to adjacent birds. The theoretical formation that results in equal savings for each individual is a swept V (Fig. 9.18).

Since induced power will be less in a formation, the speed for maximum range should also be less. The curve for P_i in Fig. 9.13 will be shifted downward relative to the curve for P_p. Where they intersect determines V_{mp} and V_{mr}.

All of these arguments are based on theoretical considerations of power output that predict why a formation might save energy and modify speeds for migration. The important question now is whether Canada geese actually behave as predicted. Some recent measurements of formation flight in geese indicate that wing tip spacing is such that energy savings do occur. Measurements of s for geese formations indicate an average savings of 10% with a maximum of 35% savings; it should have an important impact on the effectiveness of energy use by geese and their ability to move long distances with minimum expense.

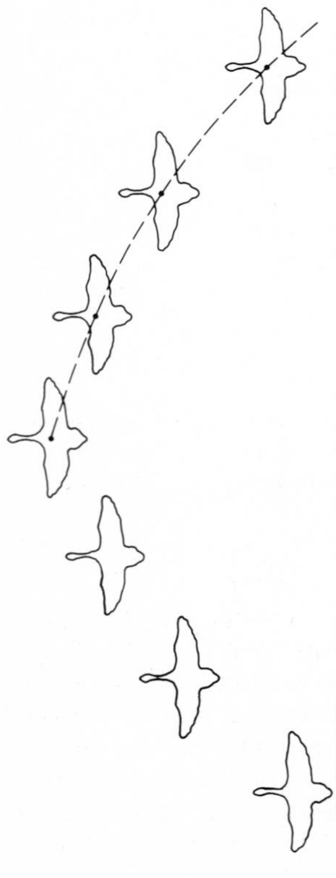

Figure 9.18
The optimal formation geometry for equalizing savings among individuals is a swept V in which individuals farther out on a leg of the V are positioned farther behind the bird in front of them (from Lissaman and Shollenberger, 1970).

SWIMMING

When the rate of oxygen consumption is measured for trout swimming at different speeds, the power increases with speed in a nonlinear way (Fig. 9.19). The shape of the curve is the same as the shape of the curve that describes power required to overcome drag on the body for flying animals (Fig. 9.13, P_p), thus power for swimming animals primarily involves overcoming drag as they move through water. It certainly has an important impact on minimizing movement costs. Why is there no measurable cost for lift in swimming fishes?

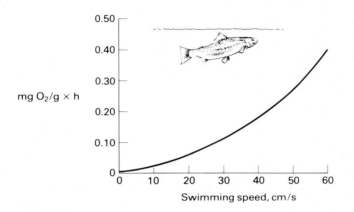

Figure 9.19
The rate of oxygen consumption for a fish increases
nonlinearly as swimming speed increases in a manner
similar to the increase in the power requirement to overcome drag during flight in birds (see P_p in Fig. 9.13)
(modified from Webb, 1971).

Buoyancy Adjustments for Lift

A swimming animal does not have to expend energy for lift if its weight is adjusted so that it just equals the weight of the water it displaces. If the animal is heavier than water it will sink and will have to expend energy to achieve and maintain lift.

Weight relative to volume (g/ml) is specific gravity. The specific gravity of fresh water is 1.000 g/ml, and the specific gravity of seawater is 1.023 g/ml (because of heavier salts dissolved in seawater). The total specific gravity of an animal is determined by the specific gravities of its components. Because the skeletal structures of many animals have a specific gravity greater than water (usually 1.2-3.2 g/ml due to the use of calcium for structural support), they must be counterbalanced by other, lighter substances to adjust the total specific gravity downward to that of the environment. Because seawater has a higher specific gravity than fresh water, less downward adjustment is necessary to achieve neutral buoyancy in marine animals.

One way to adjust specific gravity downward is to minimize the use of heavy ions. If a solution is prepared with the same total number of ions as another solution but relatively light ions are used, the specific gravity of the one with light ions will be less. Seawater composed of NaCl, KCl, $CaCl_2$, $MgCl_2$, and Na_2SO_4 has a specific gravity of 1.023 g/ml. A solution composed of just NaCl and KCl (without the heavier Ca^{++}, Mg^{++}, and $SO_4^=$ ions) with the same total number of ions has a specific gravity of 1.020 g/ml. If heavy ions are excluded from a large enough portion of an animal, it will result in neutral buoyancy. However, since an adjustment of only 0.003 g/ml is achieved by this mechanism, a large proportion of the total volume of an animal must be adjusted. Nevertheless, several invertebrate species contain fluids with mostly light ions.

Another way to adjust specific gravity downward is to have a proportion of the body composed of (relatively) low-specific-gravity fats. Fats have specific gravities of from 0.85 to 0.90 g/ml. Since the change in specific gravity from the use of a ml of fat can be greater than the use of a ml of solution with light ions, this mechanism represents a way to adjust buoyancy with less of the total volume composed of the lighter substance. A number of species of plants, diatoms, and fishes have neutral buoyancies from low-specific-gravity fat deposits.

Gases have the lowest specific gravities. The specific gravity of oxygen is only 0.0014 g/ml, so a small volume of gas will produce neutral buoyancy. Many species of fishes and cephalopods (squid, cuttlefish) have gas-filled structures that produce lift. The chambered nautilus and cuttlefish have gas spaces enclosed in rigid chambers, and many species of fishes have gas spaces in a flexible swim bladder (Fig. 9.20). The proportion of the volume of the animal required to produce neutral buoyancy when filled with gas is only about 5-7%. The specific gravity of fishes and cephalopods without gas spaces is about 1.07 g/ml; this amount must be adjusted downward by 0.07 g/ml (7%) in freshwater and 0.05 g/ml (5%) in seawater.

Gases expand or contract when pressure changes, and pressure in water changes by one atmosphere for every ten meters of change in depth. Without some mechanism to keep gas volume constant when depth changes, buoyancy will change. For example, if the gas space expands when a fish ascends, buoyancy will increase and increase lift. Without an adjustment back to the original volume, energy would have to be expended to keep from rising in water. The rigid structure surrounding the gas spaces in cephalopods prevents expansion of the gas when the animals ascend, and some adjustments in the volume of the gas space can be produced by adding or removing liquid.

The volume of the flexible swim bladder of fishes is kept constant by adding or removing gas from the bladder when pressure changes. Some species of fishes have a connection between the swim bladder and the esophagus by which gas volume is adjusted by swallowing and/or burping air. However, some deep-sea fishes never surface. They maintain neutral buoyancy by adding and removing gas via the circulatory system. A part of the wall of the swim bladder is highly vascularized and separated from the gas space by a ring of muscle. This "oval chamber" (Fig. 9.20) opens to the gas space when pressure in the swim bladder increases as environmental pressure decreases during ascent.

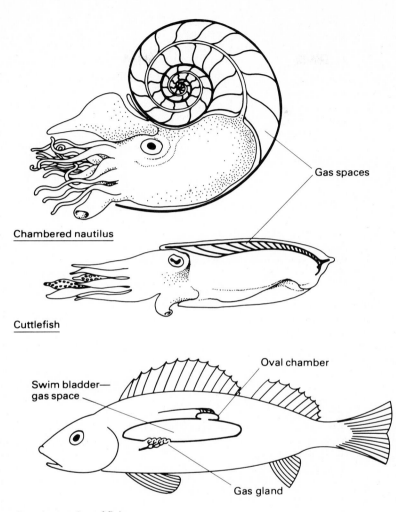

Chambered nautilus

Gas spaces

Cuttlefish

Swim bladder—
gas space

Oval chamber

Gas gland

Certain species of fishes

Figure 9.20
The chambered nautilus and cuttlefish adjust specific gravity with gas
spaces enclosed in rigid structures. Some fishes adjust specific gravity by
adding or removing gases from a flexible swimbladder, thereby keeping its
volume constant when pressure changes (adapted from Denton, 1961).

Because the partial pressure of gases in the bladder is greater than in the blood,
the volume decreases as gases diffuse to the blood.

When environmental pressure increases during descent, gas is secreted
into the swim bladder by a structure called the "gas gland" (Fig. 9.20). The gas

gland has two important structures (Fig. 9.21). A "hairpin" loop of capillaries supplies blood. The long, physical juxtaposition of arterial and venous capillary blood serves to maintain high pressures in the gas space. Gases such as oxygen and nitrogen enter the blood from the bladder because pressure in the bladder is higher than in the blood (as high as 100 atmospheres at a depth of 1,000 meters versus a total of 1.0 atmosphere for gases in solution in blood fluid). However, since arterial blood contains lower amounts of gases in solution, the gases diffuse back to the arterial side as they pass down the venous side of the loop (Fig. 9.21). Thus there is countercurrent exchange of gases, with the net effect that high pressures can be maintained in the swim bladder despite the large gradient for diffusion back to the blood.

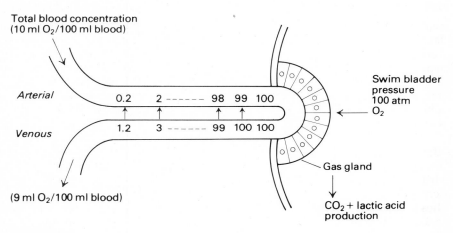

Figure 9.21
Arterial blood entering the countercurrent loop to the gas gland has less oxygen in solution but more oxygen associated with hemoglobin. Oxygen in solution diffuses to arterial blood from the venous side, and higher acidity at the gas gland dissociates oxygen from hemoglobin and reduces gas solubility. These conditions increase gas quantities in solution and produce a gradient for diffusion into the gas space (adapted from Schmidt-Nielsen, 1975).

How are gases added to the bladder? It partly involves total blood oxygen concentration, which is influenced by hemoglobin binding of oxygen (Fig. 9.21). Recall that oxygen dissociates from hemoglobin when blood becomes acidic (see Chapter 3). The blood of fishes is influenced by acidity to a large degree (the Root effect), and the cells of the gas gland produce carbon dioxide and lactic acid. The countercurrent flow of arterial and venous blood also keeps the acid concentration high in the region of the gas gland. Oxygen dissociates rapidly from hemoglobin in the acid region of the gas gland (called the Root "off" effect) and associates more slowly with hemoglobin when pH decreases as blood moves away from the gland (called the Root "on" effect). Thus oxygen is available for diffusion into the swim bladder at the gas gland when pH decreases. In addition, the addition of any substances to blood decreases the solubility of gases (such as nitrogen; see Chapter 1); gases are available for diffusion at the gas gland from solubility changes due to carbon dioxide and lactic acid production.

As a result of these mechanisms, the proportional production of CO_2 and lactic acid when a fish descends leads to secretion of gas into the swim bladder such that volume is adjusted and remains constant. The production of CO_2 and lactic acid require some energy; the cost to reduce lift is not zero. However, the expenditure is much less than would be required if the fish had to swim continuously to provide lift.

Minimizing Costs of Swimming Transport through Group Behavior

Despite the minimal energy expenditure required to maintain lift in water, the effects of drag require swimming animals to increase rates of energy expenditure with speed (Fig. 9.19). Although the cost of transport is minimum at a particular speed of swimming, other important factors in the environment of a fish (such as food or predators) will influence its speed. However, as was the case with geese and lobsters, migrating fishes associate in formations or "schools," which could reduce the energy required to overcome drag for an individual in the formation.

When a fish moves its tail back and forth to generate thrust, a vortex of water is generated that trails the fish (Fig. 9.22). The resulting water current directly behind the fish is in the opposite direction to its movement, and the swirling vortex produces a current in the same direction as its movement at a lateral position. The best way to reduce drag by using these currents is for fishes in alternate rows to be midway between each other (Fig. 9.22), generating the diamond pattern of spacing that is a characteristic of many fish schools. Being in school does not have to be a drag.

Direction
of
water
currents

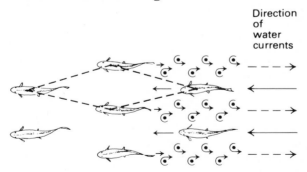

Figure 9.22
The beating tail of a fish creates trailing water currents behind the fish that are opposite to its direction of movement, but toward its direction of movement at a position lateral to the fish. A diamond geometry reduces minimum transport costs most effectively (from Weihs, 1973).

Do fishes behave as predicted in schools? According to hydromechanical theory, some individuals could save up to five times the energy that is required for swimming alone by appropriate positioning in schools. However, measurements on saithe, herring, and cod indicate that energy savings are much less. Other factors that influence positioning such as how benefits are distributed among individuals, abilities to maneuver, and abilities to maintain visual input have been hypothesized to outweigh the considerable advantages of greater energy economy.

LOCOMOTION FOR FEEDING AND ENERGY EFFICIENCY

Although we have examined characteristics of locomotion from the point of view of minimizing expenditures to cover distances, it is not the only function for locomotion. For long distance movements with a fixed fuel supply, we expect animals to minimize expenditures to maximize range. However, movements for obtaining energy (food) usually occur over short distances at a variety of speeds and costs. Nevertheless, we still expect an animal when feeding to move at speeds at which the *benefits* from capturing food are maximum relative to the *costs* to move and capture food. Animals that maximize the rate of *net* gain of energy (intakes/time – costs/time) will be most efficient in dealing with energy as a resource. We will examine this principle of economic efficiency in detail in the next chapter, but some studies with a certain species of fish are interesting at this point because they illustrate how the integration of movement costs and feeding benefits helps determine an "optimal" speed for locomotion while feeding.

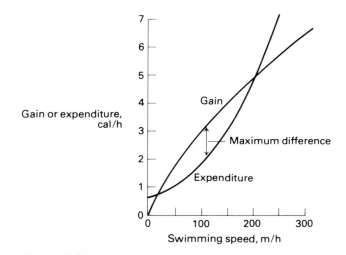

Figure 9.23
The rate of energy expenditure for swimming bleak increases as swimming speed increases, but so does the gain of energy from feeding. For the situation in this figure, the maximum *net* gain of energy would occur at a speed of 111 meters/hour; the fish swam at a speed of 107 meters/hour (modified from Ware, 1975).

Studies of individuals of the bleak (*Alburnus alburnus*) that were moving to feed permitted prediction of the swimming speed that would maximize the rate of net gain of energy. As we have seen, costs for swimming increase nonlinearly with swimming speed (Fig. 9.19; Fig. 9.23). For *Alburnus* the amount of food energy that could be captured also increased with swimming speed (Fig. 9.23) because the faster they moved, the more prey items they captured per unit time. The speed for the maximum *difference* between intakes and expenditures of energy (Fig. 9.23) was 111 meters/hour. The observed swimming speed for bleak was 107 meters/hour, which is very close to the rate of locomotion that would be most effective for maximizing the difference between intakes and expenditures of energy.

We will examine some other aspects of the interactions between energy expenditures and energy intakes in the next chapter when we consider feeding. In the present context it appears that expenditures for locomotion, by means of speed and possible interactions between individuals in some groups, are adjusted to contribute to a variety of functions that may be optimal, including long-distance movements as well as movement over short distances for feeding.

COMPARISON OF LOCOMOTION MODES

We can compare the energy expenditures for running, flying, and swimming by determining the increase above resting values in energy use rates at speeds for the minimum cost of transport for animals of different sizes. Thus we are comparing animals that run, fly, or swim, assuming that they are all moving at speeds for maximum range (Fig. 9.24).

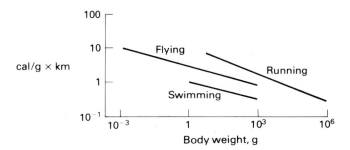

Figure 9.24
The minimum cost of transport per gram decreases with increasing size. It is also least for swimming animals (because of lower requirements for producing lift), higher for flying animals (moving at high speeds), and highest for running animals (from Schmidt-Nielsen, 1972).

Flying is less expensive for covering distance than is running, and swimming is the least expensive mode of locomotion because essentially no power is required for lift (Fig. 9.24). Although flight effectiveness is due to high rates of energy expenditure occurring at high speeds as compared with running, the effectiveness of swimming is due to the shift to an environment that permits low expenditures for lift by means of buoyancy adjustments. Note also that even though minimum transport costs per gram decrease with increasing size within each locomotion mode, the *total* minimum transport costs per animal will increase with size. Conclusions concerning expenditures for locomotion that contribute to efficiency of energy use will ultimately depend on the relationships between rates of use and rates of supply. The details of these relationships are explored in the next chapter.

SUMMARY

Activity involves high rates of energy expenditure, and energy is used most effectively when increased costs are minimized as compared with the function of activity. When the function of activity is to cover distance, efficiency of energy use involves minimizing expenditures for the distance traveled.

The rate of energy expenditure for running increases linearly with speed, with higher per-gram costs for smaller animals. Animals that walk will minimize internal power production by moving at speeds at which the transfer between kinetic and gravitational potential energy is maximum. No transfer between them is possible during running, trotting, and hopping, although energy is stored and recovered in elastic form. During running, the linear increase in power with speed is a consequence of a linear increase in the power required for acceleration, which is added to a constant power requirement for lifting the center of mass.

The energy cost to cover distance (cal/g × km, the cost of transport) by running is high at low speeds for small animals and decreases toward a constant value as speed increases. Small animals should run at maximum possible speeds to minimize costs of transport. Large animals can run at a variety of speeds and achieve maximum range, so speed is a poor indicator of movement effectiveness for them. However, group movement of lobsters in single-file formation by which resistance to movement can be reduced suggests that minimizing expenditures for locomotion is important.

Power for flight is high at low speeds, decreases to a minimum at intermediate speeds, and increases at higher speeds. This relationship is a consequence of high power requirements for lift at low speeds and high power requirements for thrust (to overcome drag) at high speeds, with intermediate requirements for both at intermediate speeds. Minimum cost of transport for flight occurs at a speed higher than that for minimum power. Birds flying in formation can reduce costs for flight and increase maximum range by using updraft air currents produced at the tips of the wings of other individuals. Theoretical considerations suggest up to 50% of the power for lift can be saved, and savings can be equalized to reduce competition among individuals in a formation when they distribute themselves in a swept V. Preliminary measurements of geese formations suggest that, by migrating in groups, birds save 10-35% of energy that would otherwise have to be expended in solo flight.

Power for swimming involves primarily thrust to overcome drag, because lift is not required if buoyancy is adjusted such that a swimming animal will displace its weight in water. Specific gravity is adjusted downward in aquatic animals by adjusting the proportion of light versus heavy ions, the proportion of the body composed of fats, or the use of a gas space. Fishes may have gases in a flexible swim bladder. In deep-sea fishes gas volume is adjusted when environmental pressure changes by adding or removing gases from the swim bladder. Gases are removed via the circulatory system and added and maintained through a countercurrent arterial-venous exchange system that involves acid production to dissociate oxygen from hemoglobin and to change the solubility of dissolved gases in blood.

Movement in formation by fishes in "schools" could reduce drag by the use of trailing water currents established by the strokes of the tails. The diamond geometry of fish schools would decrease costs for transport and increase range. Although costs for locomotion should be minimized when supply of energy is fixed, if energy availability changes with speed (such as during feeding for some animals), speed should be adjusted so that an animal maximizes energy benefits from food with regard to the costs for moving to obtain the food.

The minimum cost for transport is lowest for swimming animals because of the very low energy costs needed to generate lift in water. The minimum cost of transport for flight is lower than it is for running because flight occurs at much higher speeds. Higher speed of movement in flight counterbalances a higher power requirement for flight as compared with running.

ANNOTATED REFERENCES

Chapter 5: Energy Use at Minimum Rates

Aschoff, J., and **H. Pohl** (1970). Rhythmic variations in energy metabolism. *Fed. Proc.* 29:1541-1552. (Importance of photoperiod for standardization of minimum energy-use rates.)

Blum, J.J. (1977). On the geometry of four dimensions and the relationship between metabolism and body mass. *Jour. Theor. Biol.* 64:599-601. (Surface area in a fourth-dimension hypothesis.)

Boddington, M.J. (1978). An absolute metabolic scope for activity. *Jour. Theor. Biol.* 75:443-449. (Suggestion of time as a fourth dimension.)

Hemmingsen, A.M. (1960). Energy metabolism as related to body size and respiratory surfaces, and its evolution. *Rep. Steno. Mem. Hosp.* 9:1-110. (Detailed account of scaling of metabolic variables to body size.)

Kleiber, M. (1961). *The Fire of Life*. Wiley, New York. (Excellent discussion of mechanisms of measurements in animal metabolism, with some implications for animal production.)

McMahon, T. (1973). Size and shape in biology. *Science* 179:1201-1204. (Cross-sectional area and structural support hypothesis.)

Schmidt-Nielsen, K. (1970). Energy metabolism, body size, and problems of scaling. *Fed. Proc.* 29:1524-1532. (Discussion of mechanisms for oxygen delivery to match tissue demands that change with size.)

Chapter 6: Temperature Regulation in Ectotherms

Bakker, R.T. (1972). Anatomical and ecological evidence of endothermy in dinosaurs. *Nature* 238:81-85. (Postulation of possible endothermy based on a variety of pieces of information, including anatomy and community organization.)

Brett, J.R. (1971). Energetic response of salmon to temperature. A study of some thermal relations in the physiology and freshwater ecology of sockeye salmon (*Oncorhynchus nerka*). *Amer. Zool.* 11:99-114. (Summary of detailed digestive and growth studies that suggests an "optimal" temperature for growth with excess food.)

Dawson, W.R. (1975). On the physiological significance of the preferred body temperature of reptiles. In D.M. Gates and R.B. Schmerl (eds.), *Perspectives of Biophysical Ecology: Ecological Studies*, Vol. 12. Springer-Verlag, New York. (Review of data on temperature rate effects on physiological processes in reptiles.)

DeWitt, C.B., and **R.M. Friedman** (1979). Significance of skewness in ectotherm thermoregulation. *Amer. Zool.* 19:195-209. (Examination of frequency distributions of preferred temperatures with respect to Q_{10} for rate processes.)

Farlow, J.O., C.V. Thompson, and **D.E. Rosner** (1976). Plates of the dinosaur *Stegosaurus:* Forced convective heat loss fins? *Science* 192:1123-1125. (Anatomical and wind-tunnel studies suggesting possible heat loss mechanism.)

Gates, D.M., and **W.P. Porter** (1970). The energy budget of animals. In J.D. Hardy, A.P. Gagge, and J.A.J. Stolwijk (eds.), *Physiological and Behavioral Temperature Regulation.* C C Thomas, Springfield, Ill. (Detailed account of physical factors influencing microenvironmental heat exchange.)

Grieger, T.A., and **M.J. Kluger** (1978). Fever and survival: The role of serum iron. *Jour. Physiol.* 279:187-196. (Experimental analysis of importance of iron for optimal growth of *A. hydrophila* bacteria *in vivo* and *in vitro*.)

Hammel, H.T., F.T. Caldwell, Jr., and **R.M. Abrams** (1967). Regulation of body temperature in the blue-tongued lizard. *Science* 156:1260-1262. (Experimental manipulation of preferred temperatures by means of heating and cooling the brain.)

Hochachka, P.W., and **G. Somero** (1973). *Strategies of Biochemical Adaptations.* Saunders, Philadelphia. (Integration of discussion of enzyme-substrate interactions with other aspects of physiology.)

Kluger, M.J. (1978). The evolution and adaptive value of fever. *Amer. Sci.* 66:38-43. (Summary of evidence for role of fever as an "optimal" change in body temperature.)

Ostrom, J.H. (1978). A new look at dinosaurs. *National Geog.* 154:152-185. (Popular account of ideas for homeothermy in dinosaurs.)

Precht, H., J. Christophersen, H. Hensel, and **W. Larcher** (1973). *Temperature and Life.* Springer-Verlag, New York. (Encyclopedic account of factual data, with occasional rationale for observations.)

Reynolds, W.W., and **M.E. Casterlin** (1979). Behavioral thermoregulation and the "final preferendum" paradigm. *Amer. Zool.* 19:211-224. (Deals with a shift in preferred temperatures, depending on acclimation temperatures. This paper is the first in an extensive symposium dealing with temperature regulation in ectotherms.)

Somero, G.N. (1978). Temperature adaptation of enzymes: Biological optimization through structure function compromises. *Ann. Rev. Ecol. Syst.* 9:1-29. (Discussion of several mechanisms for regulation of enzyme-substrate interactions with respect to temperature.)

Chapter 7: Temperature Regulation in Endotherms

Alberts, J.R. (1978). Huddling by rat pups: Group behavioral mechanisms of temperature regulation and energy conservation. *Jour. Comp. Physiol. Psychol.* 92:231-245. (Interactions between number of young and rate of cooling for altricial young.)

Balmer, R.T., and **A.D. Strobusch** (1977). Critical size of newborn homeotherms. *Jour. Appl. Physiol.* 42:571-577. (Detailed mathematical development of the dependence of insulation on the size of a cylinder.)

Brock, T.D. (1967). Life at high temperatures. *Science* 158:1012-1019. (Accounts of adaptations of proteins to high temperatures in thermophilic bacteria and algae.)

Brown, J.H., and **R.C. Lasiewski** (1972). Metabolism of weasels: The cost of being long and thin. *Ecology* 53:939-943. (Effect of shape on insulation.)

Calloway, N.O. (1976). Body temperature: Thermodynamics of homeothermism. *Jour. Theor. Biol.* 57:331-344. (The thermodynamic efficiency hypothesis.)

Carey, F.G., and **J.M. Teal** (1969). Regulation of temperature by the bluefin tuna. *Comp. Biochem. Physiol.* 28:205-213. (Description of countercurrent heat exchange mechanisms for control of temperature in bluefin tuna.)

Hamilton, W.J. III (1973). *Life's Color Code*. McGraw-Hill, New York. (Summary of "maxi-therm" hypotheses for patterns of body temperatures.)

Hammel, H.T. (1962). Thermal and metabolic measurements on a reindeer at rest and in exercise. *Arctic Aeromedical Lab. Report* AAL-TDR-61-54. (Evidence for cooling of expired air in a large arctic animal.)

Heinrich, B. (1974). Thermoregulation in endothermic insects. *Science* 185: 747-756. (Summary of information for temperature regulation in some insects.)

Knutson, R.M. (1974). Heat production and temperature regulation in eastern skunk cabbage. *Science* 186:746-747. (Evidence for endothermy in this plant.)

Nagy, K.A., **D.K. Odell**, and **R.S. Seymour** (1972). Temperature regulation by the inflorescence of *Philodendron*. *Science* 178:1195-1197.

Schmidt-Nielsen, K., **F.R. Hainsworth**, and **D.E. Murrish** (1970). Countercurrent heat exchange in the respiratory passages: Effect on water and heat balance. *Resp. Physiol.* 9:263-276. (Mechanisms of respiratory heat conservation in small birds and mammals.)

Scholander, P.F., **V. Walters**, **R. Hock**, and **L. Irving** (1950). Body insulation of some arctic and tropical mammals and birds. *Biol. Bull.* 99:225-236. (Insulation thickness in relation to habitat and body size.)

Snyder, G. (1979). Thermodynamics and body temperature. *Jour. Theor. Biol.* 80:145-147. (Body temperatures of some reptiles at different altitudes compared with predictions of the thermodynamic efficiency hypothesis.)

Vinegar, A., V.H. Hutchison, and **H.G. Dowling** (1970). Metabolism and thermoregulation during brooding of snakes of the genus *Python* (Reptilia, Boidae). *Zoologica* 55:19-48. (Relationship of thermoregulation to habitat distribution in different species of *Python*.)

Walsberg, G.E., G.S. Cambell, and **J.R. King** (1978). Animal coat color and radiative heat gain: A re-evaluation. *Jour. Comp. Physiol.* 126:211-222. (Examination of interaction between coat color and convective heat loss.)

Chapter 8: Torpor and Hibernation

Brett, J.R. (1971). Energetic responses of salmon to temperature. A study of some thermal relations in the physiology and freshwater ecology of sockeye salmon (*Oncorhynchus nerka*). *Amer. Zool.* 11:99-114. (Summary of detailed digestive and growth studies that suggest a lower "optimal" temperature when food availability decreases.)

Fisher, K.C., and **J.F. Manery** (1967). Water and electrolyte metabolism in heterotherms. In K.C. Fisher, A.R. Dawe, C.P. Lyman, E. Schonbaum, and F.E. South, (eds.), *Mammalian Hibernation III*. American Elsevier, New York. (Hypotheses related to body size and excretion with regard to periodic arousals in seasonal hibernators.)

Hainsworth, F.R., B.G. Collins, and **L.L. Wolf** (1977). The function of torpor in hummingbirds. *Physiol. Zool.* 50:215-222. (Torpor involves some risks because it is not observed unless survival is threatened by starvation.)

Heller, H.C., and **G.W. Colliver** (1974). CNS regulation of body temperature during hibernation. *Amer. Jour. Physiol.* 227:583-589. (Proportional thermostatic control when hypothalamic temperature decreases below a lower set point in torpor.)

Horwitz, B.A. (1978). Neurohumoral regulation of nonshivering thermogenesis in mammals. In L.C.H. Wang and J. Hudson (eds.), *Strategies in Cold: Natural Torpidity and Thermogenesis*. Academic Press, New York. (Mechanisms of coupled and uncoupled mitochondrial heat production.)

Huey, R.B., and **M. Slatkin** (1976). Costs and benefits of lizard thermoregulation. *Quart. Rev. Biol.* 51:363-384. (Theoretical discussion of implications of changes in "optimal" conditions for temperature regulation.)

Lillywhite, H.B., P. Licht, and **P. Chelgren** (1973). The role of behavioral thermoregulation in the growth energetics of the toad, *Bufo boreas*. *Ecology* 54:375-383. (Interaction of food availability and temperature for maximum growth rate.)

Malan, A., H. Arens, and **A. Waechter** (1973). Pulmonary respiration and acid-base state in hibernating marmots and hamsters. *Resp. Physiol.* 17:45-61. (Comparative acidosis during torpor in hibernating mammals.)

Pengelley, E.T., and **K.C. Fisher** (1963). The effect of temperature and photoperiod on the yearly hibernating behavior of captive golden-mantled ground squirrels (*Citellus lateralis tescorum*). *Can. Jour. Zool.* 41:1103-1120. (Description of circannual rhythms for a seasonal hibernator.)

Popovic, V. (1964). Cardiac output in hibernating ground squirrels. *Amer. Jour. Physiol.* 207:1345-1348. (Changes in cardiac output that occur during torpor.)

Regal, P.J. (1967). Voluntary hypothermia in reptiles. *Science* 155:1551-1553. (Selection of lower body temperatures at night.)

Tucker, V.A. (1965). The relation between the torpor cycle and heat exchange in the California pocket mouse *Perognathus californicus. Jour. Cell. Comp. Physiol.* 65:405-414. (Quantitative measurements of heat exchange during entry and arousal from torpor.)

Wells, L.A. (1971). Circulatory patterns of hibernators. *Amer. Jour. Physiol.* 221:1517-1520. (Redistribution of reduced cardiac output during torpor.)

Wolf, L.L., and **F.R. Hainsworth** (1972). Environmental influence on regulated body temperature in torpid hummingbirds. *Comp. Biochem. Physiol.* 41:167-173. (Thermostatic heat production during torpor at characteristic lower body temperatures.)

Chapter 9: Locomotion

Bennett, A.F., and **J.A. Ruben** (1979). Endothermy and activity in vertebrates. *Science* 206:649-654. (Roles of anaerobic and aerobic metabolism in activity of vertebrates.)

Bill, R.G., and **W.F. Herrnkind** (1976). Drag reduction by formation movement in spiny lobsters. *Science* 193:1146-1148. (Description of queue mechanism for minimizing locomotion costs.)

Cavagna, G.A., **N.C. Heglund**, and **C.R. Taylor** (1977). Mechanical work in terrestrial locomotion: two basic mechanisms for minimizing energy expenditure. *Amer. Jour. Physiol.* 233:243-261. (Interactions between kinetic, gravitational, and elastic energy for walking, running, trotting, hopping, and galloping.)

Denton, E.J. (1961). The buoyancy of fish and cephalopods. *Prog. Biophys. and Biophys. Chem.* 11:177-234. (Review of mechanisms for reduction of specific gravity in aquatic animals.)

Lissaman, P.B.S., and **C.A. Shollenberger** (1970). Formation flight of birds. *Science* 168:1003-1005. (Major theoretical basis for energy savings from V formation flight.)

Partridge, B.L., and **T.J. Pitcher** (1979). Evidence against a hydrodynamic function for fish schools. *Nature* 279:418-419. (Observed patterns of geometry for some fish schools are not as predicted for high-energy savings.)

Pennycuick, C. (1972). *Animal Flight.* Edward Arnold, London. (Summary of major physical forces operating during flight and their impact on animals.)

Schmidt-Nielsen, K. (1972). Locomotion: Energy cost of swimming, flying, and running. *Science* 177:222-228. (Comparisons of minimum costs of transport.)

Taylor, C.R., **K. Schmidt-Nielsen**, and **J.L. Raab** (1970). Scaling of energetic cost of running to body size in mammals. *Amer. Jour. Physiol.*

219:1104-1107. (Linear dependence of cost on speed, with calculations of costs of transport.)

Torre-Bueno, J.R., and **J. Larochelle** (1978). The metabolic cost of flight in unrestrained birds. *Jour. Exp. Biol.* 75:223-229. (Effects of drag adjustments by birds on performance curves.)

Tucker, V.A. (1968). Respiratory exchange and evaporative water loss in the flying budgerigar. *Jour. Exp. Biol.* 48:67-87. (Power-output performance from wind-tunnel experiments.)

Ware, D.M. (1975). Growth, metabolism, and optimal swimming speed of a pelagic fish. *Jour. Fish. Res. Bd. Canada* 32:33-41. (Prediction and observation of optimal swimming speed based on rate of net energy gain.)

Weihs, D. (1973). Hydromechanics of fish schooling. *Nature* 241:290-291. (Hypothesis of school geometry based on minimizing individual drag.)

Energy and Nutrients: Meeting Demands through Feeding

Part III

Survival and reproduction depend on an effective supply of energy and nutrients, which for animals come from food. Chemicals must be found, ingested, and processed to provide the substrates for energy demands. In addition, effective metabolism depends on ingestion of a variety of other chemicals, such as vitamins and minerals, as well as avoidance of chemical poisons in certain foods. Our discussion of food supplies for meeting demands will deal first with energy (Chapter 10) and then with nutrients (Chapter 11). However, this separation is somewhat arbitrary because all required chemicals are mixed in food sources.

Animals are faced with a bewildering array of choices between food items. Which should they select? Why should certain types of food be selected? How can they be recognized and processed most effectively?

We have already discussed one aspect of feeding for energy regulation: the uses of energy by an animal influence its demands for energy from food. There must be a sufficient *difference* between energy intakes and expenditures to provide energy for maintenance, growth, reproduction, or periods when no food can be obtained. Those animals that select and process foods providing the greatest difference between expenditures and intakes over time will be more likely to survive and to produce offspring. Thus for any animal we would expect the *rate of net energy gain* (rate of intake − rate of use) to be maximum whenever possible when it feeds.

This principle of efficiency of energy resource use forms the basis for our examination of feeding for energy regulation. Now that we have an understanding of factors influencing demands for energy, we can examine both food intake and controls for food processing that serve to provide sufficient supplies as compared with demands.

In addition to energy substrates, unless food intake provides a sufficient supply of vitamins, minerals, and the required amino acids and fats, the metabolism of an animal will become "poisoned" because functions will be impaired. In addition, many foods (prey) contain chemicals that can directly poison the metabolism of predators. Again, the principle of net gain is important. Nutrients must be supplied at rates that match rates of use, and the net gain of poisons should be minimized.

A major problem for regulation and control of nutrient intake and avoidance of poisons is the wide variety of different chemicals that are required or that must be avoided. There are large numbers of nutrients and poisons but only three basic energy substrates. We will find that this situation has resulted in the evolution of two very different types of control systems for nutrients and poisons. One involves biochemical mechanisms for specializing on certain food types. The other involves adjustments in feeding from learning.

Our discussion of learning for nutrient and poison control involves principles of psychology and behavior not traditionally considered as a subject within animal physiology. However, our basic goal is to understand the ways animals survive. Learning basic principles of psychology and behavior can be as important for understanding survival in some cases as learning basic principles of chemistry and physics is for the understanding of others.

Feeding: Energy Regulation and Control

10

Most animals do not feed continuously. Food is ingested during meals and must provide for expenditures that occur over time periods when feeding does not or cannot occur. Some of the food energy must be stored internally for future use; the more energy that is stored (within limits) as compared with demands, the longer an animal can survive without feeding. In addition, the more energy that is stored, the more can be used for energy-demanding processes such as growth, reproduction, migration, hibernation, or maintenance costs due to changes in environmental temperatures.

For survival, the supply of energy from food must equal its biological demand over some period of time, or

$$\frac{\text{Energy assimilated}}{\text{Time}} = \frac{\text{Energy expended}}{\text{Time}}.$$

If these are exactly equal over a *short* time period, an animal cannot store energy for periods when expenditures may be high or food availability may be low. Thus we can divide the uses of energy into two components: use for "maintenance" over short time periods, and use for "storage" required to meet demands for longer time periods. For survival,

$$\frac{\text{Energy assimilated}}{\text{Time}} = \frac{\text{Rate of energy}}{\text{use for maintenance}} + \frac{\text{Rate of energy}}{\text{use for storage.}}$$

Biologists normally look at expenditures over relatively short periods of time. For example, the rate of energy use for maintenance is usually measured over periods of a few hours. Moreover, it is common to examine feeding over relatively short time periods because many animals feed during brief intervals several times a day. If we rearrange the equation and place both short-term maintenance and short-term feeding (assimilation) on the same side, we have for survival

$$\frac{\text{Rate of energy}}{\text{assimilated}} - \frac{\text{Rate of energy use}}{\text{for maintenance}} = \frac{\text{Rate of energy}}{\text{use for storage}}$$

or

$$\frac{\text{Rate of } net \text{ energy gain}}{\text{while feeding}} = \frac{\text{Rate of energy}}{\text{use for storage.}}$$

An animal will be most likely to survive if the rate of *net* energy gain over the short periods when it feeds is sufficient to equal demands over relatively long periods.

In order to understand rate of net energy gain from feeding (left side of the equation), it is important to understand what determines long-term rates of energy storage (right side of the equation). Furthermore, to know if an animal is efficient when it feeds we must know how much energy it consumes, how much is assimilated and kept in storage, and how much is used for shorter-term maintenance. All these requirements are difficult to determine because feeding and expenditures of energy for maintenance and storage are influenced by a number of variables in the environments of animals.

Figure 10.1 illustrates some of the components involved in energy regulation through feeding. Some internal change, called "hunger," results from the maintenance and storage uses of energy, and both are influenced by daily and seasonal characteristics of the environment, such as temperatures and photoperiods. "Search" in the environment is initiated and continued until prey is detected. The prey must then be pursued and captured. Ingestion follows successful capture, the food is digested, and part is assimilated to provide for further maintenance and storage.

Not all animals actively search for food in their environments. Some predators "sit and wait" for prey to come to them, including predators, such as spiders, that may spin a web and wait for food to be caught in it, or filter feeders that may be sedentary. In a way, even these predators must "search," detect, and capture prey items because they can sit and wait after moving to different areas of their environments, and they selectively consume different prey that they may capture. The relative expenses for actively searching predators and sit and wait predators are different, but the general sequence involved in "hunger" (Fig. 10.1) is the same. We will concentrate on information about

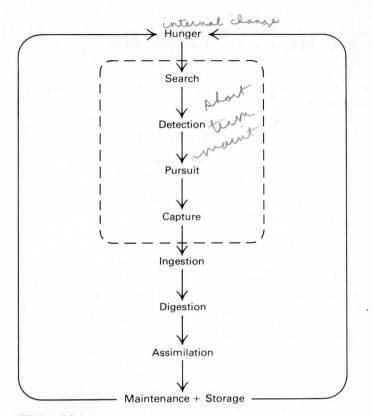

Figure 10.1
Feeding is divided into a sequence of events from "hunger" to
maintenance and storage uses of energy, which influence hunger.
Search, detection, pursuit, and capture are influenced primarily
by variables in the external environment. Ingestion, digestion,
and assimilation involve variation in the internal environment.
Hunger is influenced by both internal and external environments
by means of maintenance and storage uses for energy.

predators that are more active in their search for prey, because most experimental studies of feeding concern these types of animals.

Search, detection, pursuit, and capture (usually all considered part of
short-term maintenance costs) all require energy, and they all depend on
variable characteristics of an animal's environment. Food may be rare and may
be distributed unevenly. In addition, many prey have evolved mechanisms to
avoid detection and capture by predators (see Chapter 16). They may be cryptically hidden in their environments or they may be able to detect and avoid
these searching predators. The diet of a predator may be varied; quantification
of the amount of energy consumed from a large number of different food items

is difficult. Once food is captured, ingestion, digestion, and assimilation also require energy (other components of maintenance costs), and the amount expended varies depending on the type and amount of food in a diet.

Because feeding is so complex, most quantitative studies of energy regulation have been carried out in the laboratory, where the many components of the feeding process can be easily observed and where the food given to a predator can be controlled and composition and energy value can be measured. We will examine evidence for energy *regulation* by studying the rates of net energy gain for a few species in which *both* expenditures and intakes of energy have been measured over various time periods. We will then return to the components of feeding (Fig. 10.1) to ask how animals have evolved to deal effectively with energy supplies by means of feeding; we will examine some *control* systems that operate such that the rate of net energy gain is sufficient to ensure long-term survival and reproduction.

TIME PERIODS FOR ENERGY REGULATION

For most animals, energy supplies come from discrete meals that are consumed periodically. The pattern of energy regulation over relatively long time periods depends on when food is normally consumed and how much energy is expended at various times. Intake patterns and expenditure patterns will vary from species to species depending on whether the animals are nocturnal or diurnal, their patterns and extent of activity, and the availability of food. Some basic information on the *daily* patterns of both energy intakes and expenditures comes from studies of laboratory rats in which both energy intakes and energy expenditures have been measured.

The pattern of intake and expenditure shown in Fig. 10.2 is observed when the rate of energy intake of laboratory rats is measured at different times during a day and their rates of energy expenditure are measured for the same times. Rats are nocturnal and consume most of their food at night, even when

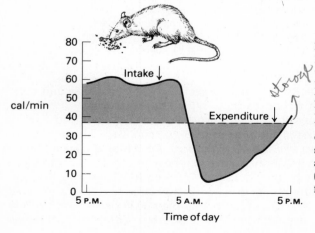

Figure 10.2
Patterns of rate of energy intake and expenditure for laboratory rats kept on a 12L:12D photoperiod with lights on at 5 A.M. Energy intakes in excess of expenditures (left shaded area) are stored and used when intakes are less than expenditures (right shaded area) (modified from LeMagnen *et al.*, 1973).

they have unlimited access to food for the entire light and dark period of a day. Moreover, under laboratory conditions there is very little change in the rate of energy expenditure during a day.

At night the rats consume energy at higher rates than it is expended, and during the day they expend energy at higher rates than it is consumed (Fig. 10.2). The excess, or net, energy consumed at night (shaded area in the left portion of Fig. 10.2) is stored internally and used during periods when expenditures exceed intakes (shaded area in the right portion of Fig. 10.2). Measurements of the respiratory quotient (R.Q.; see Chapter 5) indicate relatively high values at night, when food is assimilated and used as substrates, and relatively low values during the day, when more energy storage from fat is used (Fig. 10.3). For the example shown in Fig. 10.2, there is a slight excess amount of energy at the end of a day, so for these animals feeding results in a sufficient supply of energy to meet demands over a daily period, plus some additional energy for storage.

This daily pattern of energy regulation is similar to the seasonal pattern of energy storage and use by hibernators (see Fig. 8.10). The time scale is shorter, but some energy is stored in "anticipation" of future expenditures. The pattern occurs on a *circadian* (about one day) basis rather than on a circannual (yearly) basis, and excess energy stored at the end of a day can contribute to a seasonal net energy gain.

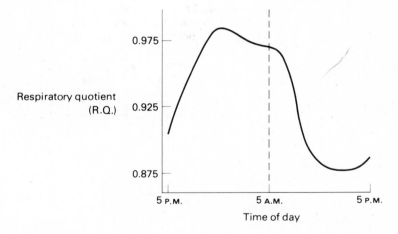

Figure 10.3
Pattern of the respiratory quotient for the rats shown in Fig. 10.2. Higher R.Q. is associated with energy storage; lower R.Q. is associated with use of stored energy (after LeMagnen *et al.*, 1973).

Meal Patterns

We can learn more about the regulation of energy by looking at both intakes and expenditures of energy over smaller time invervals. The smallest possible interval for energy regulation is from one meal to the next. The rate of energy intake over relatively long periods (hours or days) is the product of meal size and feeding frequency:

$$\text{Energy/Time} = \text{Meal energy} \times \text{Feeding frequency.}$$

Note that this equation for rate of supply is similar to other important supply rates in physiology. In respiration, minute volume is the product of tidal volume (breath volume) and respiratorty rate, and in circulation, cardiac output is the product of stroke volume (beat volume) and heart rate. For feeding, the time intervals for rates of energy gain are longer than those involved in respiration and circulation, but regulation of energy still involves the product of amount and rate, both of which can change if supplies and/or demands change (see below).

Does energy regulation occur from one meal to the next? Figure 10.4 presents a series of hypothetical meals. Each meal is preceded by a time interval (time since the last meal) and followed by an interval to the next meal. For example, meal 3 is preceded by interval "b" and followed by interval "c." Energy is expended continuously at various rates, and intervals between meals are usually longer than the time needed to consume a meal. We can ask if there is any relationship between the amount of energy consumed in a meal and either (1) the amount expended *prior to* that meal (in the interval since the last meal), or (2) the amount expended *after* that meal (until the next meal is initiated).

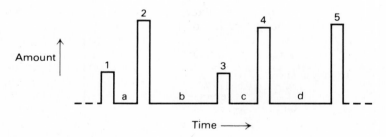

Figure 10.4
A series of hypothetical meals (1 → 5) separated by intervals (a → d) by which energy for each meal can be examined relative to energy expended in the *prior* interval or energy expended in the *following* interval. Energy expended to the *next* meal is related to meal size rather than meal size being dependent on energy expended prior to a meal.

For most animals that have been studied (laboratory rats, mice, flies, hummingbirds, cats), there is an excellent relationship between the amount of energy consumed in a meal and the energy expended to the *next* meal (Fig. 10.5). There is no relationship between the amount of energy expended since the last meal and the amount of energy consumed in the present meal.

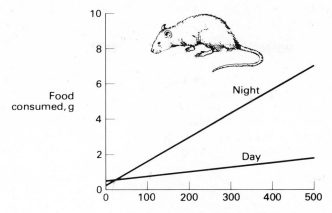

Figure 10.5
Time to the next meal for rats depends on meal size. Furthermore, more food is consumed at night relative to the
initiation of the next meal than is consumed during the
day. Thus more energy from a meal is stored at night by
rats (after LeMagnen *et al.*, 1973).

This pattern of energy regulation indicates that these animals expend
energy they have consumed rather than replenish energy they have expended
from one meal to the next. Thus supplies of energy are used for demands;
demands do not determine the required supplies (on this short time scale). It
does not mean that animals will not consume more energy if they are deprived
of normal access to food for a long time. Response to deprivation occurs in most
animals to adjust supplies to previous expenditures and is an important part of
energy regulation when normal feeding patterns are interrupted. However,
when access to food is not interrupted, many animals operate on energy intake
followed by expenditure rather than vice versa.

Not all the energy consumed in a meal is expended prior to the initiation of
the next meal. Note in Fig. 10.5 that more food is consumed by laboratory rats
relative to the interval to the next meal at night than is consumed during the
day (the slope is higher for the night line). The excess is stored for future use
and produces the pattern seen for laboratory rats in Fig. 10.2. These meal-to-
meal relationships are summarized in Fig. 10.6 from measurements of intakes
and expenditures of energy by hummingbirds. The straight line represents
storage of energy.

Changing Supplies Relative to Demands from Changes in Meal Size and/or Feeding Frequency

The regulation of energy supplies to meet demands from one meal to the next
produces the pattern of energy regulation over daily and seasonal periods, and

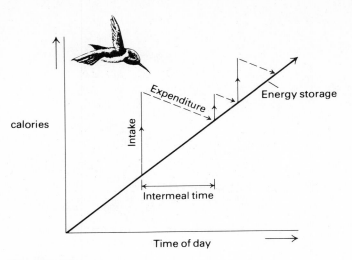

Figure 10.6
With part of energy intake used for storage, expenditure of the remaining intake (for maintenance) will generate the meal-to-meal pattern of feeding by hummingbirds (from Wolf and Hainsworth, 1977).

the maintenance and storage patterns can provide information about what is important for initiating feeding or producing the initial state of "hunger" in the sequence shown in Fig. 10.1. After food is consumed and assimilated, part is diverted for storage through control mechanisms that we will discuss shortly. The remaining assimilated energy is used for short-term maintenance, and when it is gone another meal is initiated. Thus changes in either the use of energy for short-term maintenance or the use of energy stored from one meal to the next should influence "hunger" by means of meal size and/or meal frequency.

If the energy expended for short-term maintenance is increased, feeding frequency and/or meal size should increase. If the use of stored energy is increased, meal frequency and/or meal size should also increase because a larger fraction of assimilated energy would be diverted for storage. For many animals it is not understood whether changes in meal frequency, size, or a combination of both factors are important for energy regulation. However, there is information showing that changes in demands for energy storage and maintenance influence feeding in appropriate ways for some animals.

The storage of energy for future use should change if expenditures change during periods when stored energy is normally used. One convenient period is when feeding does not normally occur (*i.e.*, overnight for animals that feed during the day). When house sparrows (*Passer domesticus*) are forced to be active at night (by making them run in a wheel) they respond to the increased use of stored energy by storing more energy the next day (Fig. 10.7). Hummingbirds also store more energy following a night when their expenditures are increased with cold environmental temperatures. For hummingbirds, increased demands

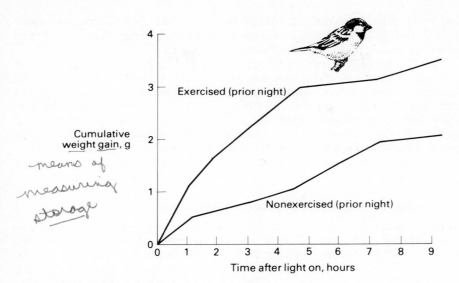

means of
measuring
storage

Figure 10.7
The weight gain (primarily from energy storage) during the day is greater
when house sparrows are forced to exercise during the prior night, using
greater amounts of stored energy (from Kendeigh *et al.*, 1969).

for energy storage result primarily in increases in meal size, whereas increases
in maintenance costs (from decreasing environmental temperatures during the
day) result in increases in meal frequency (Table 10.1).

Table 10.1   Changes in meal frequency or meal size when
maintenance or storage energy uses are increased; increased
use was produced by decreasing environmental temperature
for homeothermic hummingbirds (from Hainsworth, 1978)

	Increase in Meal Frequency in 17 Cases	Increase in Meal Size in 17 Cases
Increased maintenance (low temperature during the day)	13	
Increased storage (low temperature at night)		14

CONTROL SYSTEMS FOR ENERGY REGULATION

The ability of animals to meet demands for energy from the amount and/or frequency of energy intake depends on variation in the components of feeding

listed in Fig. 10.1. Each one can change over short time periods, and we expect animals to show adaptations to adjust the rates of net energy gain so that they are the maximum possible when the components vary.

We can divide the feeding components into two parts (Fig. 10.1). Those within the rectangle (search, detection, pursuit, and capture) depend primarily on external environmental variables, such as variation in prey quality, quantity, and distribution. Those outside the rectangle deal primarily with internal mechanisms for processing food that depend on variation in the quality and quantity of food ingested. We will discuss factors important to these major aspects of feeding and will then consider how information from both the external and internal environments is integrated for effective regulation of energy supplies to meet demands.

Controls for External Environmental Variation

Animals should search for food at those locations in their environments where the availability and nature of prey provide the greatest possible rate of net energy gain while they are feeding. This rate will depend on variation in *prey distribution*. If the same prey types (*i.e.*, same quality or energy value per prey item) are distributed in patches, a predator will initially have a high capture rate when it first enters a patch, but the rate of capture will decrease as prey items are removed and become harder to detect, pursue, and capture, as shown graphically in Fig. 10.8. In this situation a predator should leave a patch when the rate of net energy gain (the slope of the curve in Fig. 10.8) has decreased to a point where it would be more profitable for the animal to leave and find another patch.

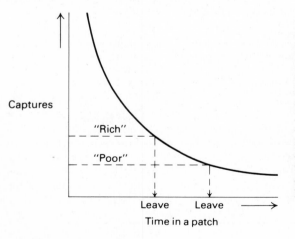

Figure 10.8
The rate of capture (slope of the line) will decrease with time spent in a patch as prey are removed and the remaining prey become harder to detect and capture. If an animal has been foraging in a "rich" environment (higher average capture rate), it should leave a patch sooner than if the average capture rate for the environment has been low ("poor" environment) (modified from Krebs *et al.*, 1974).

If the predator initially has no information about the relative quality of different patches it has not yet visited (prey may be cryptically hidden), it should leave a patch when its rate of net energy gain is equal to the average it could obtain in the entire set of patches it could visit. This situation can only be assessed from the nature of other patches it has already visited in its environment. A "rich" environment (one that has provided a high average capture rate) would lead to leaving a patch after *less* of a capture rate decrease than if the environment was "poor" or had resulted in a lower average capture rate prior to entering a patch of food (Fig. 10.8).

These ideas have been tested experimentally. When chickadees (*Parus atricapillus*) eat mealworms provided in a series of cups (or patches) in which they must search for the worms, they will leave a cup in a shorter period of time after the last capture in that cup if their average rate of capture for a whole series of cups has been high (Table 10-2). The period of time it takes to leave a patch after the last capture is called "giving-up time." A higher average capture rate for all cups (or richer environmental quality) should produce a shorter giving-up time if the chickadees searched so that they maximized their rate of net energy gain. The observation that giving-up time does change appropriately with environmental quality (prior experience with patches) suggests that these predators do make adjustments in their searching for prey so that the rate of net energy gain is the highest possible when the environment changes.

Table 10.2   Time to leave a patch after the last capture in that patch ("giving-up time") is lower for chickadees feeding in environments that have produced high average capture rates ("rich" environments) than in environments that have produced lower average capture rates ("poor" environments) (from Krebs *et al.*, 1974) ($t = 2.81$; $p < 0.025$)

	RICH	POOR
	Average capture rate = 3.5/min	Average capture rate = 2.8/min
Giving-up time (seconds)	11 ± 4.8	16 ± 5.4

If a predator must travel different distances from one patch to another, a greater distance will increase costs (as a result of more energy expended for locomotion) and will result in a lower overall rate of net energy gain (lower environmental quality). When great tits (*Parus major*) are forced to travel longer distances between identical patches of food in the laboratory, they remain in a patch longer so that rate of net energy gain falls to a lower value at the time they leave compared with less cost for travel between patches. Increased costs between patches produces a "poorer" environment (Fig. 10.8); this type of an adjustment is also expected if the rate of net energy gain is maintained at a maximum possible value.

Movement Patterns for Feeding

Part of effective patch exploitation involves the pattern of movement because the less energy expended in movement per prey item, the higher the rate of net energy gain. Effective movement involves minimizing the distance between prey items. If the prey items cannot be detected until they are encountered, distance between items in a patch can be minimized by changing *direction* of movement when a prey item is discovered. For example, the blowfly (*Phormia regina*) performs a "dance" when it encounters a small drop of food (sugar solution). The food is detected by cells at the base of hairs on its legs (see below). When a fly walks into a drop of food it will ingest the drop and then turn (Fig. 10.9).

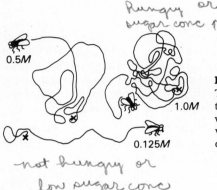

Figure 10.9
The "dance" of a hungry fly involves more tight turns and keeps a fly in a relatively small area when food (sugar) concentration is high (1.0*M* versus 0.5*M* versus 0.125*M*). **X** indicates sugar drop (after Dethier, 1976).

Turning is done in a tighter circle if the fly is "hungry" (has not fed) and the food has a high energy value (high sugar concentration). The fly is more likely to continue in a straight-line search path sooner if the food has a low energy value (Fig. 10.9; 0.125*M* sugar solution) or if the fly has fed recently and is not "hungry" (has experienced a "rich" environment). The "dancing" keeps the fly in patches of relatively high quality longer when it is "hungry" (environmental quality has been low) and takes it away from patches of relatively low quality sooner. As a result, time spent in a patch will be related to the rate of net energy gain. When the rate of net energy gain has decreased to a particular value (depending on average environmental quality; see above), the fly will stop dancing and continue on its way.

This pattern of movement has been called "area-restricted search," and it has been described for many actively searching predators that cannot detect prey items until they are physically encountered. Among these predators are included robins, ovenbirds, and a number of unicellular animals (e.g. *Paramecium*) that show "taxes," i.e., movement patterns tending to favor movement toward favorable environmental conditions. You would probably perform in a similar way, but it is likely that you would respond more to money than to food. If you found a penny in the grass you would probably take a brief look in the near vicinity and continue on your way. If you found a dollar bill you would probably spend a fair amount of time searching in the vicinity for more, and if you found a hundred-dollar bill, you would probably search long and hard for others. In addition, your degree of search for items of money of different quality (denominations) could depend on your economic condition, i.e., how "hungry" you were for the resource.*

Selection of Prey Quality

A searching predator can encounter prey items of different energy values, or *qualities* (such as different sugar concentrations in Fig. 10.9). To maximize the rate of net energy gain the predator should select high-quality prey whenever they are encountered, but if there are both high- and low-quality prey and high-quality prey become rare or hard to obtain, it is more profitable for the predator to *switch* to searching for and attempting to capture lower-quality prey, if it takes less energy to find and consume them with respect to the energy value gained. In essence, high-quality prey are not always the most profitable if they are rare.

This prediction has been studied by estimating the rate of net energy gain for flycatchers (birds) capturing flies of different sizes (Fig. 10.10). Although large flies contain more energy per item, the flycatchers should not search for them exclusively if the energy costs for search and capture produce the *net* result of a lower rate of net energy gain. For the example shown in Fig. 10.10 the flycatchers should—and do—select prey of intermediate sizes that require less energy to obtain with respect to their energy value.

The detection of prey quality can occur at a distance in some predators, but for others it occurs only after capture and prior to ingestion or rejection, depending on the signals given off by prey and the sensory abilities of predators. We will discuss the details of sensory filtering of important environmental information in Chapter 16. The blowfly depicted in Fig. 10.11 detects food quality from receptors located on its feet, but it can also orient toward food at a greater distance by using information from smelling, i.e., olfaction. Eyes, ears, and other receptors can be used for orientation to potential food sources. However, as a general rule the most important information concerning whether food is finally accepted or rejected is generated from receptors that detect chemical characteristics of food just prior to or immediately after the food enters an animal.

There is an interesting relationship between the ability of animals to detect the quality of captured chemicals and the energy value of the chemicals to the animals. For example, hummingbirds normally consume sugar in water, which is secreted as nectar by flowers they visit. If a hummingbird is given a choice of feeding from two food sources that differ in the concentration (energy value) of the sugar solutions, they can detect very small differences in concentration when the choices available are of low concentration (Table 10.3).

---

*I tried a few experiments of this kind on undergraduates (particularly "hungry" predators). To my surprise, a common response was to grab the money and run. Some searching in the vicinity was apparent on a short-term basis, but this searching was often followed by an "escape" response. I interpret it as a type of "predator avoidance behavior." Staying in the vicinity where unearned money is found may increase the probability that the owner will appear to claim it. This "lost wallet experiment" makes an interesting laboratory exercise.

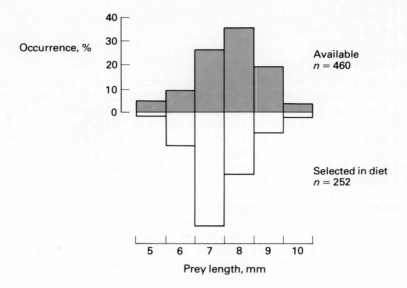

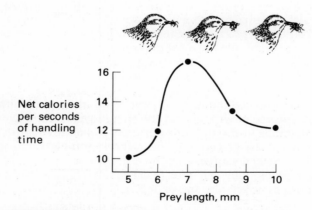

Figure 10.10
The selection of different size classes of flies by a flycatcher, the pied wagtail. The bar graphs show the actual available distribution of flies of different sizes and shows the selection by the birds of most prey in the 7-mm size class. The line graph shows the rate of net calories obtained as a function of prey size (from Davies, 1977).

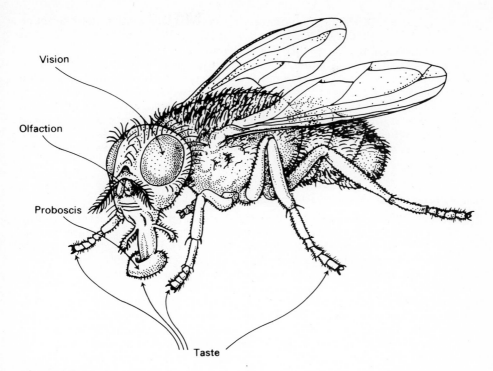

Vision

Olfaction

Proboscis

Taste

Figure 10.11
Location of the major external environmental receptors and the feeding apparatus (proboscis) for a blowfly (after Dethier, 1976).

Table 10.3   If hummingbirds are given sucrose solutions at the concentrations listed on the left, the concentration of an alternative food source must be greater by the amount shown on the right for the birds to treat them as different foods. Note that a greater difference is required as the quality of food is increased (from Hainsworth and Wolf, 1979)

SUGAR CONCENTRATION (MOLAR SUCROSE)	MINIMUM NECESSARY INCREASE IN CONCENTRATION (MOLAR SUCROSE)
0.25	0.05
0.50	0.10
0.80	0.20
1.20	> 0.20

However, if the choices are of higher concentration a small difference cannot be detected, and it takes a larger difference between them for the birds to treat them as foods of different quality (Table 10.3).

As we shall see (in Chapter 16), receptors are usually more sensitive to differences in the external environment if the level of environmental information is low. For a feeding animal this fact means that it will be more likely to detect a difference and treat foods as different if the environment signals low quality. In this case just a little more energy will make the difference between survival or starvation. When the environment signals higher quality, the same small difference is less important for an animal's survival because supplies of energy are more likely to be sufficient to meet demands.

Controls for Internal Food Processing

Digestion follows ingestion (Fig. 10.1). Food is pulverized and/or treated chemically at a variety of internal locations, and the morphology of the digestive system varies from one species to another (Fig. 10.12). However, many animals have a special organ where food is held for a time after ingestion. The structure may be very simple if little food treatment is required. For example, blowflies and hummingbirds, both of which feed primarily on liquids, hold food in a simple sack-like area called a crop. More complicated foods, such as grasses or proteins, require mechanical, acid, or bacterial treatment to break them down into simple chemicals in a stomach or rumen (see below). Once the food has been processed it is released to an intestine; chemical treatment also occurs there. A major function of the remainder of the digestive system is the assimilation of simple chemicals into the circulatory system from the gut.

The major digestive organs for a human are illustrated in Fig. 10.13. Masticated food enters the stomach, where it is treated by a variety of secretions. Hydrochloric acid destroys bacteria, breaks down cell walls, and provides an acid environment for other digestive enzymes. Pepsinogen breaks internal bonds in protein molecules at an optimal acidic pH. Amylases (from saliva) and lipases break up carbohydrates and fats, respectively. The pancreas and intestine also secrete enzymes that act on the chemicals in food (Fig. 10.13). Bicarbonate from the pancreas reduces the acidity of the fluid coming from the stomach. Like pepsin, trypsin and chymotrypsin break protein molecules in the middle; other enzymes, such as carboxypeptidases, break the ends of certain protein molecules. A series of other enzymes (maltase, lactase, sucrase) act on different sugar molecules, and nuclease acts on nucleic acids.

Figure 10.14 illustrates the extent of the surface area for digestive assimilation in the intestine. The comparison is made with a simple, smooth-walled tube of the same length as the intestine in a human. Relatively large folds on the internal surface (the folds of Kerkring) increase surface area threefold. Smaller projections from these folds, the villi, provide an additional tenfold increase in surface area, and the smallest projections (the microvilli) increase total surface area twenty times more.

Some animals have evolved complex digestive functions to deal with complexities of food composition. An interesting example is the function of the

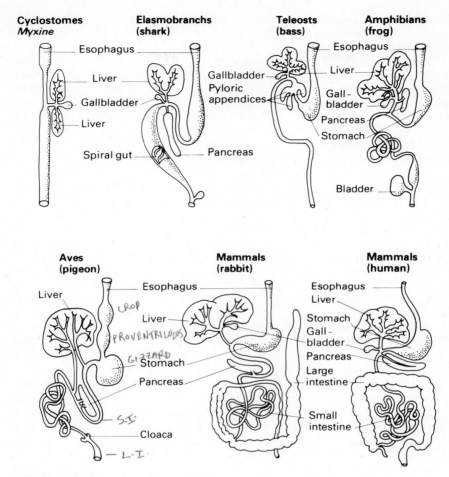

Figure 10.12
The digestive systems of different vertebrates all involve some organ for storage (stomach) and/or a gut for assimilation of processed food (from Florey, 1966).

rumen of some herbivores (Fig. 10.15). Plants contain a large amount of cellulose that cannot be digested by the enzymes of most vertebrates. However, some microorganisms produce enzymes (cellulases) that can break cellulose into glucose units.

The glucose is used by the symbiotic microbes as a substrate for fermentation in the anaerobic digestive system. Some of the products of fermentation (proprionic acid, acetic acid, and butyric acid) are assimilated by the host ruminant and used as energy substrates. In addition, some urea produced as a waste product of protein metabolism by the ruminant passes back to the digestive

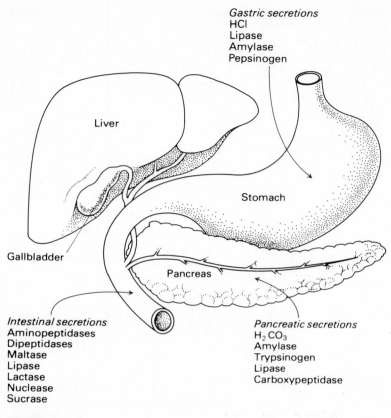

Gastric secretions
HCl
Lipase
Amylase
Pepsinogen

Liver

Stomach

Gallbladder

Pancreas

Intestinal secretions
Aminopeptidases
Dipeptidases
Maltase
Lipase
Lactase
Nuclease
Sucrase

Pancreatic secretions
H_2CO_3
Amylase
Trypsinogen
Lipase
Carboxypeptidase

Figure 10.13
The major digestive organs of a human, with some of the secretions involved in the chemical processing of foods (modified from Houpt, 1977).

system in saliva and from diffusion from the blood; it is then converted to ammonia by the microbes and used as a source of nitrogen for protein synthesis in microbial growth (Fig. 10.15). The protein produced by the microorganisms for their growth is, in turn, assimilated in the small intestine by the ruminant. The microorganisms also provide a source for both the B group of vitamins and vitamin C.

Changes in Digestion when Energy Supplies or Demands Change

The internal supply of food chemical energy from digestive processing should meet the demands of animals for energy. If either supplies or demands change, the digestive system should change in function. This point can be illustrated by considering the relatively simple digestive function of the fly when it is fed food

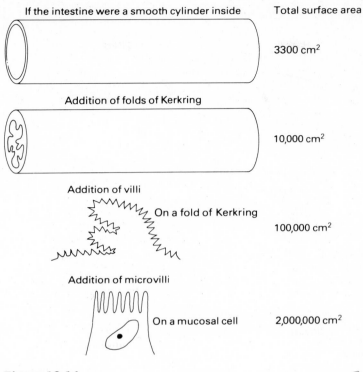

Figure 10.14
Comparisons of changes in the surface area of the small intestine
of a human that result from various folds and projections on the
inner surface (from Houpt in Goldstein).

From *Introduction to Comparative Physiology* edited by Leon Goldstein.
Copyright © 1977 by Holt, Rinehart and Winston. Reprinted by permission of Holt, Rinehart and Winston.

consisting of different sugar concentrations (i.e., the *supply* of energy is changed
while demands for energy stay relatively constant).

Food initially passes through an open valve to a crop (Fig. 10.16). The
valve closes after feeding stops, and food passes in small amounts from the crop
back up to the region of the closed valve and then to the gut, where the simple
sugars are digested and assimilated into the blood. If food of low sugar concentration is fed to a fly, the crop will empty faster than if the fly is fed food of high
concentration (Fig. 10.17). Thus *energy* is supplied from the crop at a rate that
meets demands because faster supply of food with less energy value provides
energy at the same rate as slower supply of food with higher energy value.

The rate of delivery of food from the crop of the fly is controlled by means
of detection of the osmotic pressure of blood. Osmotic pressure, a property of liquids, depends on the number of particles (molecules or ions) dissolved in the liq-

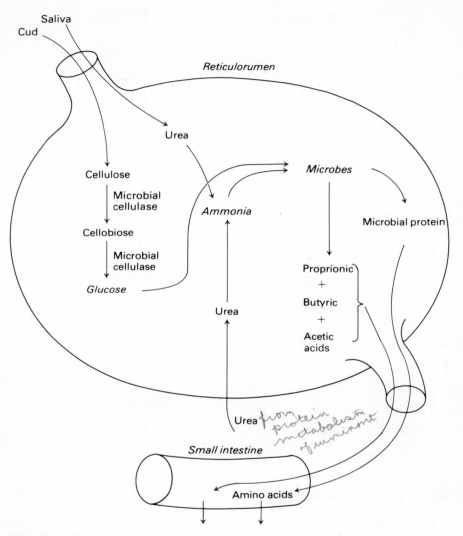

Figure 10.15
Summary of the major symbiotic relationships between a ruminant and its microbial flora that occur by means of digestion (modified from Houpt, 1977).

uid. The more particles there are, the higher the osmotic pressure is (see Chapter 12). But osmotic pressure does not necessarily depend on the types of particles or their energy values. We can demonstrate that the fly uses blood osmotic pressure to control the rate of supply of crop contents by injecting nonnutritive particles into its blood. If a solution of sodium chloride is injected into the blood

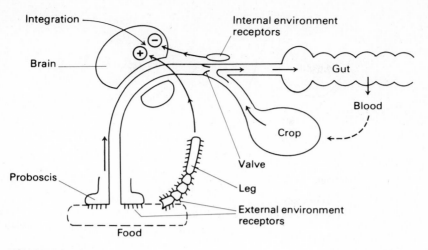

Figure 10.16
The digestive system of a blowfly, with the major internal and external environmental sensory control processes that are involved in determining a state of "hunger" or "no hunger" (based on Dethier, 1976).

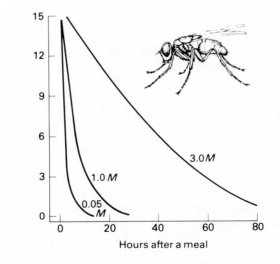

Figure 10.17
The contents of the crop of a blowfly are emptied more slowly when a meal has higher energy value, resulting in constancy of the internal supplies of energy to meet demands (adapted from Gelperin, 1966).

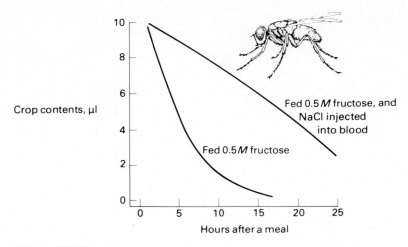

Figure 10.18
When a fly is fed 0.5M fructose the crop will empty in about 15 hours, but if the fly is fed the same meal and blood osmotic pressure is increased by injection of sodium chloride, the crop will empty more slowly (adapted from Gelperin, 1966).

it slows the rate of crop emptying, even though the fly cannot use sodium chloride for energy (Fig. 10.18). Normally, the sugar molecules that are assimilated into the blood cause a change in osmotic pressure, and a higher concentration of sugar means that more sugar molecules will reach the blood for each small quantity of food released from the crop. Under normal circumstances, this slows the rate of delivery when more sugar molecules are available as substrates for energy demands.

The rate of supply of food energy from the digestive organs of vertebrates operates in a similar manner, although there are more complex control mechanisms associated with more complicated foods. The control mechanisms for humans are illustrated in Fig. 10.19. The stomach empties at a rate proportional to its contents in a manner similar to the rate of crop emptying in the fly; however, the rate changes for different properties of the food released from the stomach to the intestine.

Pancreatic secretions of bicarbonate and enzymes are stimulated during entry of acidic material into the intestine from the stomach. These secretions are partly controlled by the hormone *gastrin*, which also stimulates the production of acid in the stomach. When fats enter the intestine, cells in the intestinal lining secrete a hormone called *cholecystokinin* (abbreviated CCK), which has the effect of decreasing rhythmic contractions of the smooth muscle of the stomach and intestine, slowing the rate of passage of energy-rich fats and thus proportionally adjusting the rate of processing to the demands for energy. Moreover, when molecules from digested protein or carbohydrate reach the intestine, the osmotic concentration they produce has an effect on stomach emptying in a manner similar to that of the fly; higher osmotic concentrations (more substrate molecules) slow emptying.

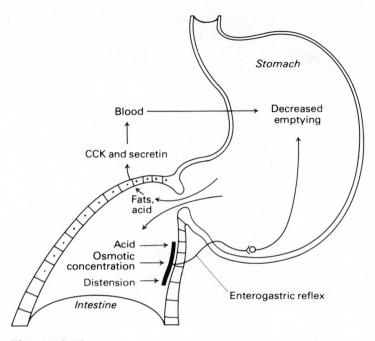

Figure 10.19
Schematic illustration of some feedback controls involved in regulation of the supply of food from the stomach to the intestine in a
human. Controls involve responses to fats, acid, distension, and
osmotic concentration of fluid reaching the intestine, all of which
"index" the energy value such that the rate of stomach emptying
will change when supplies reaching the intestine change.

In vertebrates, control of such emptying is also mediated partly through
the "enterogastric reflex," composed of local receptors to detect osmotic concentration and nerve connections from the intestine back to the stomach. This
nerve-mediated control mechanism also responds to distension of the intestine
by food by slowing passage of more food from the stomach. Finally, the
enterogastric reflex and the secretion of another hormone, called *secretin*, slows
rate of stomach emptying in response to increases in pH of the fluid in the intestine, which can also be an index of substrate quantity for assimilation processing.

The result of these control mechanisms is to proportionally control internal supplies of energy to meet demands. Higher-quality foods will be processed
more slowly and result in increased intervals between meals. Experiments in
which CCK and intestinal osmotic concentrations were increased decreased
food intake in rats. In addition, distention of the digestive tract, which simulates
internal energy availability, decreases short-term food intake in vertebrates.

The internal processing of food from the digestive tract after it has been
assimilated falls into two discrete categories: (1) processing of assimilated foods,

and (2) processing of stored energy provided from assimilated foods. Both can be illustrated for vertebrates, for which there is a great deal of information, and both categories involve the function of the liver. Note that the first category primarily concerns the use of energy for maintenance with some storage, and the second category involves the use of stored energy (Fig. 10.1).

Processing during Assimilation

Three general types of chemicals are assimilated into blood for use as energy substrates: carbohydrates (e.g., glucose), amino acids (from protein), and fats (Fig. 10.20). They are transported to the liver through a special circulatory connection between the veins of the gut and the liver called the hepatic portal system; most food chemicals go directly to the liver and not to it through the heart and lungs (see Fig. 4.5). Some of these chemicals are directly utilized as substrates for oxidation. Glucose is used by all the tissues of the body. In the liver, some glucose is used for energy by oxidation, some is stored as glycogen (by glucose molecules being recombined and packaged in cells), and the remainder not used for oxidation by other tissues is stored in adipose tissue after conversion to triglycerides (fats) (Fig. 10.20).

Assimilated amino acids are recombined into new protein or are used for energy either from direct oxidation or by conversion of amino acids to carbohydrate or lipid in the liver (Fig. 10.20). The latter occurs after the nitrogen in amino acids has been removed (see Chapter 13). Assimilated fats are utilized as immediate sources of energy by most tissues except the brain, and excess amounts are sequestered directly into adipose tissue (Fig. 10.20).

The liver plays a central role in processing assimilated chemicals from a meal. Maintenance use of energy is continuous and must be provided for from whatever supplies come to the liver; excess energy is stored for future use. If a meal is high in carbohydrates, a sufficient amount is supplied to all tissues for maintenance utilization, and the excess is stored as glycogen or (mainly) as fats in adipose tissue. If a meal consists primarily of amino acids, the excess (above maintenance requirements for structural uses as protein and immediate deamination and oxidation) is converted to glycogen and fats. If a meal is primarily fat, the excesses are diverted to adipose tissue for storage.

The fascinating details of the various biochemical pathways of intermediary metabolism involved in control of interconversions of fats, carbohydrates, and proteins are beyond the scope of this book. However, for our purposes it is important to realize that the interconversions provide considerable flexibility at the level of cellular function, depending on the supplies and demands of different substrates that may enter metabolic pathways via different routes. The metabolism of glucose may occur by anaerobic glycolysis or aerobic oxidation in the Krebs cycle (see Fig. 4.27). Fats are hydrolyzed to glycerol and fatty acids. Glycerol is converted to phosphoglyceraldehyde and enters the glycolytic pathway. Fatty acids are broken down to two-carbon fragments and enter the Krebs cycle as Acetyl-CoA. Amino acids are converted to a number of compounds that can enter the Krebs cycle (pyruvic acid, Acetyl-CoA).

The interconversions between various substrates are possible because of the reversible nature of many of the chemical reactions of cellular metabolism.

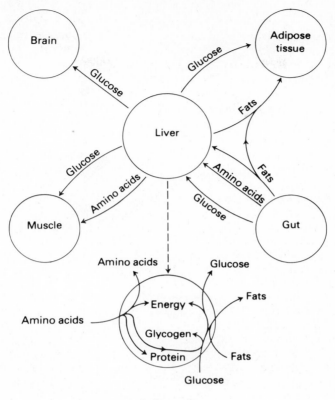

Figure 10.20
The major paths of energy substrate flow and conversion
during the assimilation phase (from Friedman and Stricker,
1976).

Therefore, depending on the relative concentrations of reactants and products
in glycolysis and oxidation, the substrates will either be used for production of
ATP (when demands from use of ATP are high) or be converted from one form
to another. The reversible nature of intermediary metabolic pathways exerts
considerable control over the ultimate flow of substrates.

Processing Stored Substrates

When an animal does not or cannot feed, its energy requirements must come
from what has been stored as excess (net gains) from assimilated energy (Fig.
10.21). No energy comes from the gut, but substrates are transported to the
liver from adipose tissue and muscle. Moreover, when fats in adipose tissue are
converted to fatty acids and glycerol, the fatty acids can be utilized directly by
tissues such as muscle (Fig. 10.21).

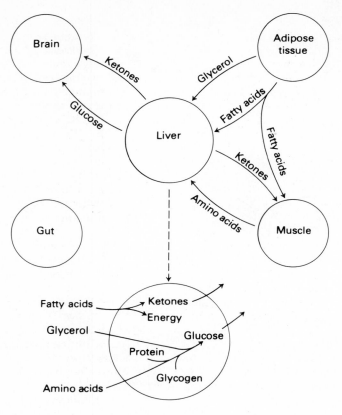

In liver cells

Figure 10.21
The major paths of energy substrate flow and conversion during
the use of stored energy (from Friedman and Stricker, 1976).

The brain of vertebrates cannot utilize fatty acids as substrates for energy
(Fig. 10.21), but the brain can utilize converted forms of fatty acids and amino
acids that are transformed in the liver. Fatty acids are converted to "ketone
bodies" in the liver (acetoacetate and betahydroxybutyrate), and ketone bodies
are used by the brain as substrates for energy. Glycerol, glycogen, proteins, and
amino acids are also transformed in the liver to glucose, which can be utilized
by the brain during starvation or post-assimilation energy processing (Fig.
10.21).

Hormonal Components of Substrate Processing

Two important hormones produced by the pancreas (Fig. 10.13) are involved in
the two aspects of internal food processing. *Insulin* is produced by the pancreas

during assimilation and facilitates processing of assimilated substrates. It stimulates glucose uptake by cells, glycogen synthesis, protein synthesis, and, indirectly, fat synthesis in adipose tissue. *Glucagon* is produced by a separate set of cells in the pancreas after assimilation of food ceases and facilitates processing of stored energy reserves. It stimulates breakdown of glycogen, production of glucose from fats and amino acids in the liver, and increased breakdown of triglycerides to glycerol and fatty acids. Other chemicals, such as norepinephrine, also influence the use of energy substrates (see Fig. 8.18), although they are traditionally not considered "digestive" hormones. We will discuss the mechanisms of cellular activation via hormones in Chapter 18 (see Fig. 18.1).

Individuals that exhibit a malfunction of the pancreas causing an insulin deficiency develop *diabetes mellitus*. It means "sweet urine," and individuals suffering from the disease have the transport of glucose into cells sufficiently impaired such that glucose concentrations in the blood increase. As a consequence, glucose concentrations in the urine formed in the kidney increase (see Chapter 13).

Some forms of diabetes mellitus are hereditary. Why is such a deleterious gene maintained in a population if it can result in inefficient use of glucose substrates and death? Some recent experiments with mice suggest a hypothesis.

Strains of mice exhibit diabetes having a genetic basis, and individuals that are homozygous recessive for the trait (db/db) develop the symptoms of insulin insufficiency, but those heterozygous (db/+) or "normal" homozygous dominant (+/+) do not. In experiments in which mice of different genotypes were starved, it was found that heterozygotes survived significantly longer than did normal dominant homozygotes, suggesting that the heterozygotes could be more efficient in using stored energy substrates. More effective function by a heterozygote produced heterozygotic dominance.

This condition has been called the "thrifty gene" hypothesis for diabetes. If individuals in a population were subjected to starvation such that those with the heterozygous (db/+) genotype survived, this situation would maintain the gene in a population at the cost of diabetes in those individuals inheriting the (db/db) genotype. The mechanism is similar to one described for the maintenance of the sickle cell trait by which individuals heterozygous for sickle cell anemia survive malaria to a greater degree than do normal dominant homozygotes.

Integration in the Brain

We have followed the sequence in Fig. 10.1 from searching for food in the environment to the internal maintenance and storage uses of energy. Information on changes in the external and internal environments is generated at each step in this sequence. Receptors for the external environment (olfaction, vision, taste) provide information on environmental quality concerning food, and internal receptors provide information on what kinds and how much substrate is available from ingested food and/or stored internal reserves (see below).

Internal and external information should be added together (or integrated) at any moment of time to determine if a state of "hunger" exists, requiring behavior to obtain food. The integration of external and internal information involves the brain, and the most completely understood systems are for the blow-fly and vertebrates.

Information from the receptor hairs on the feet of a blowfly passes directly to the brain (Fig. 10.16). If a fly has not fed recently and it steps in a drop of sugar, it will extend its proboscis (Fig. 10.11) and feed by sucking up the sugar solution. However, if the fly has fed recently it will *not* feed when it steps in the food. Therefore, as the result of having recently ingested food, the same information from the external environment produces different results, which we interpret as "hunger" or "no hunger."

The information that goes to the brain from the external environment in these two situations is the same. However, after feeding is completed, information from the internal digestive system *inhibits* (prevents) feeding. Receptors near the valve that is anterior to the entrance to the crop (Fig. 10.16) detect the passage of food from the crop to the gut. When food passes through this region the cells are stretched and send information along nerves to the brain to prevent a hunger response. If the connection between the brain and the digestive system is cut, a fly will feed continuously, become bloated with food to the extent that it cannot fly (Fig. 10.22), and may even burst. Cutting the connection between the brain and the area of the internal digestive process removes the negative (inhibitory) effect on response to the positive external environmental stimulation and disrupts the normal integration process in the brain.

Integration in vertebrates may occur in a similar way, but at present it is not clear where integration occurs or precisely what information is integrated. For many years the hypothalamus has been thought to be the specific site of integration. If the ventromedial part of the hypothalamus of a laboratory rat is removed, the rats will overeat and become obese. Thus this area was thought to be normally involved with inhibiting feeding when sufficient energy substrates were available internally, leading to the postulation of a series of "stasis" hypotheses for control of feeding in vertebrates. For example, the "glucostatic," "lipostatic," and "aminostatic" hypotheses each claimed that information from these substrates was involved in integration in the hypothalamus.

These ideas have recently had to be reevaluated. When the ventromedial hypothalamus is removed, integration may not be disrupted but some interaction with an effector may be impaired. For example, removal of the ventromedial hypothalamus results in increased insulin secretion, forcing increased storage of energy as fat without necessarily interfering with internal information on digestive processing.

A current theory of feeding control in vertebrates is called the "hepatostatic" theory. Proponents of this theory argue that the liver may play an important role in the integration of internal information for energy control because it is the site of most energy substrate processing. Information from the liver on various substrates may pass to the brain, where it could be integrated with information from the external environment. This theory has considerable appeal because it involves a variety of energy substrates. Future investigations of internal control mechanisms may find that each substrate can be important, but *when* they are important may be different. Timing of substrate use can be important for producing a sufficient rate of net energy gain to meet demands for energy.

Figure 10.22
Photograph of a "fat" fly. The nerves from the stretch receptors that normally inhibit feeding have been cut, with the result that this fly has continuously fed and has become bloated (from Dethier, 1976). Photo courtesy of V.G. Dethier, University of Massachusetts.

SUMMARY

Feeding efficiency involves both intakes and expenditures of energy. The rate of *net* gain of energy (intakes minus expenditures) over the relatively short periods when feeding occurs must provide for expenditures of energy when feeding does not or cannot occur (longer-term storage).

Daily and seasonal patterns of energy regulation are the result of the shorter-term patterns of regulation from one meal to the next. Rate of gains of energy are the product of meal size and feeding frequency. Most animals consume energy and then expend it rather than consume energy to replenish energy that has been expended. Part of the energy consumed and assimilated in a meal is diverted for longer-term storage (for daily periods when feeding cannot occur and/or for seasonally energy-demanding activities), and the rest is used for short-term maintenance until the next meal is initiated. Variation in either maintenance or storage uses for energy will influence feeding by means of meal size and/or feeding frequency.

"Hunger" leads to search, detection, pursuit, and capture of prey, and these activities depend primarily on external environmental variables such as prey quality, quantity, and distribution. Predators use mechanisms to maximize the rate of net energy gain while feeding when their environments change. Energy-storage rates are maximized by leaving a patch of food at different times as capture rate decreases depending on the overall quality of the environment with respect to obtaining energy. Movement patterns for area-restricted search activity are related to prey quality and distribution so that rate of net energy gain is maximized, and selection of prey of different quality will also maximize the rate of net energy gain. Predator ability to detect prey quality is also related to the value of prey for supplying energy demands.

After capture, food is ingested, digested, assimilated, and used for either short-term maintenance or longer-term storage. These processes depend primarily on internal environmental variation in supplies of substrates or demands for them. Digestive organs (crops, stomachs, rumens) supply energy for internal assimilation at different rates, and changes in energy value of foods (from changes in food quality) or changes in demands for energy lead to adjustments in rates of digestive processing.

Three types of energy substrates are assimilated: carbohydrates, amino acids, and fats (triglycerides). A meal that contains an excess amount of energy in any of these types can be processed in the liver so that net gains of energy are sequestered primarily in adipose tissue and secondarily in glycogen or protein. During nonfeeding periods the substrates are passed back to the liver, where fatty acids are converted to ketone bodies and amino acids are converted to glucose. Ketone bodies and glucose are utilized by the brain, and other tissues use ketone bodies and glucose as well as fatty acids.

The states of "hunger" and "no hunger" (satiation) depend on the integration of information from both the external and internal environments. If the internal environment signals that substrate availability is sufficient for maintenance and storage demands, the signals from the external environment will be inhibited and feeding will not occur. When the inhibition is removed (or when additional supplies are required for maintenance and storage requirements), hunger is manifested as a response to the external environmental signals related to food.

Control mechanisms to generate this regulation of feeding in blowflies involve inhibitory signals from the digestive tract related to rate of passage (by means of the stretch of cells) of food from the crop. The inhibitory information can prevent response to positive external environmental information produced by food. The control mechanisms in vertebrates may be similar, although it is not clear where the information is integrated or what stimuli are involved. The liver may play an important role in internal controls because each energy substrate is processed in the liver, during both assimilation and post-assimilation food processing.

Feeding: Nutrient Regulation and Control

II

Nutrients are chemicals important for effective overall function that are not sources of energy. They represent most of the composition of an animal (except energy substrates for the purposes of our discussion), and they are usually divided into the general categories of vitamins, minerals, and essential fats and amino acids. Their importance is demonstrated from symptoms of illness that occur when a diet is deficient in a required nutrient. For example, vitamins are, or form parts of, several enzymes that speed cellular reactions. If they are absent the reactions are impaired, and the animal will become ill and may die unless the nutritional deficiency is alleviated by the ingestion of a food containing the required nutrient.

Table 11.1 lists vitamins that must be supplied in the diets of humans and some symptoms that result from deficiency. Vitamins A_1 and A_2 are required for structural synthesis in epithelial cells. The aldehydes of these vitamins also form part of a visual pigment (rhodopsin) that is essential for the functioning of light receptors (see Chapter 16). Vitamin D is essential for the mechanism of calcium absorption in the small intestine; without it the skeletal system

Table 11.1 Some vitamins required by humans

VITAMIN	DEFICIENCY SYMPTOMS	FOOD SOURCES
Vitamins A_1 and A_2	Broken surface of skin; night blindness	Fruit, green and yellow vegetables, dairy and egg products, fish oil
Vitamin D	Rickets (soft and deformed bones, poor muscular development)	Egg and dairy products, fish oil
Vitamin E	Not established as a requirement in humans. Male sterility in rats; muscular dystrophy in some animals	Available in most foods
Vitamins K_1 and K_2	Impaired blood clotting	Green vegetables
Thiamine (B_1)	Beriberi (muscular atrophy, paralysis in severe cases)	Yeasts, nuts, whole grain cereals, liver, pork
Riboflavin (B_2)	Conjunctivitis; sores on lips and tongue	Dairy and egg products, yeasts, liver, leafy vegetables
Pyridoxine (B_6)	Dermatitis; convulsions	Eggs, grains, fresh vegetables
Nicotinamide (B vitamin)	Pellagra (dermatitis, diarrhea, abdominal pain)	Yeasts, meat, whole wheat
Pantothenic acid	Adrenal cortex insufficiency, impairment of antibody synthesis	Fresh vegetables, meat, eggs, whole grains
Biotin	Dermatitis; conjunctivitis	Available in most foods
Folic acid	Anemia	Leafy vegetables, liver
Cyanocobalamine (B_{12})	Pernicious anemia	Meats
Ascorbic acid (C)	Scurvy (bleeding gums, loose teeth, anemia, emaciation)	Citrus fruits, tomatoes

becomes deformed (Table 11.1). The K vitamins are required for synthesis of blood proteins, including those responsible for blood clotting. The B complex of vitamins (including nicotinamide) and pantothenic acid are essential parts of enzymes in cellular reactions; a common feature of vitamin B deficiencies is impairment of nerve function. Vitamin B_{12} (cyanocobalamine) and folic acid are essential for maturation of red blood cells in bone marrow, and vitamin C (ascorbic acid) plays a role in oxidative metabolism.

Minerals are also required for several important functions. Calcium is necessary for bone formation, nerve conduction, glandular secretion, muscular contraction, and blood coagulation. Minerals such as calcium, sodium, chloride, potassium, and phosphorus are usually present in the body in appreciable amounts. Other minerals are present in very small amounts, yet their absence also leads to symptoms of impaired function. Because their concentrations are usually low, they are referred to as "trace" elements.

Iron is a trace element that is an important component of hemoglobin, myoglobin, and intracellular enzymes called cytochromes. Copper is essential in small amounts for hemoglobin formation (it composes part of cyanocobalamine). Chromium is involved in insulin function. Iodine is a trace constituent of thyroid hormones involved in control of growth and metabolism in vertebrates (see Chapter 15). Cobalt is a trace component of vitamin B_{12}. Zinc is a constituent of enzymes such as carbonic anhydrase.

The enzymes or vitamins of which many trace elements are a part are usually present in low concentrations, meaning that relatively small amounts of the nutrients are needed for effective function. The exceptions are the minerals sodium, chloride, and potassium. These nutrients are present in relatively large amounts, and the next part of this book will be devoted to the details of their regulation and control.

There are about twenty amino acids found in animal proteins, and a few of them cannot be synthesized internally but must be consumed in food. In humans these "essential" (required in a diet) amino acids include arginine, valine, leucine, isoleucine, lysine, methionine, threonine, phenylalanine, and tryptophan. In addition, some fatty acids such as linoleic, linolenic, and arachidonic acids are required as structural components of cellular membranes.

The amino acids that are needed in a diet usually require more complicated metabolic machinery for their synthesis than do the "nonessential" amino acids synthesized internally. For example, arginine requires six or seven enzymes for its synthesis, whereas most nonessential amino acids require only one or two enzymes for synthesis. Thus if an amino acid can be obtained from food, less expense (cost) is involved in its acquisition with respect to the requirements for internal synthesis.

Although some amino acids and fatty acids can be obtained from food with less expense than by internal synthesis, the requirements of animals for them in food and for other nutrients pose severe problems of providing a diet that is balanced; i.e., containing sufficient quantities of a relatively large number of chemicals used for metabolism and growth. Protein and lipid components other than those synthesized internally and all minerals must be obtained from an animal's external environment in sufficient amounts to balance requirements for them.

There is considerable variation in the availability of nutrients in different foods. Not all foods have the same quality or quantity of nutrient supplies. Thus many animals face the problem of avoiding consumption of "junk" foods, that is, foods insufficient by themselves to meet all nutritional requirements. The major challenge of adequate nutrition faced by many animals is to insure a balanced supply of all the trace nutrients.

In addition to the problem of supply of many nutrients, animals must deal with the problems of chemical poisons in their food. Prey can promote their own survival and reproductive success by evolving to poison the metabolism of their predators. This defense mechanism is particularly common in nonmobile prey, and we are all familiar with certain plants that are toxic (e.g., hemlock, deadly nightshade).

The presence of a poison in food is similar to the lack of a required nutrient (Fig. 11.1). In both cases the animal's function is disrupted so that it no longer functions effectively, and it will become ill and die if either the poisoning or the nutritional deficiency is severe. Ultimately an animal will be "poisoned" when products cannot be formed at rates that match demands. Poisoning will occur either when certain chemicals (called poisons) prevent product formation when they are *present* or when other chemicals (called nutrients) prevent product formation when they are *absent*. We will find that some animals deal with the problems of adequate nutrition acquisition and poison avoidance in a similar way—learning which foods to avoid—and other animals have evolved to deal with nutrition and poisoning problems by biochemically specializing on certain food types.

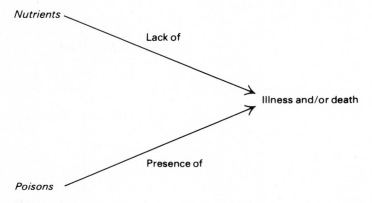

Figure 11.1
Lack of a nutrient or presence of a poison in food both produce the same result of illness and/or death.

REGULATION OF NUTRIENTS THROUGH SPECIFIC HUNGERS

We found in the previous chapter that animals demonstrate "hunger" for energy, involving detection and processing of information on three basic substrates in food so that energy supplies are sufficient for energy demands. Some animals also demonstrate "hungers" for specific nutrients, and the feeding occurs so that nutrient intake is sufficient for demands if suitable foods are available. This type of "hunger" is common among animals that normally consume varied foods, and the omnivorous laboratory rat has served as a model animal to study this behavior.

If a laboratory rat is made deficient in a nutrient (by feeding it a diet without the required chemical), it will preferentially select a food that contains the required nutrient if given access to that food. This behavior has been demonstrated many times for many different nutrients, and the degree of precision with which many animals select foods containing the appropriate required chemicals resulted in this feeding being described as a *specific hunger* for nutrients.

This behavior is believed to be typical of most animals that consume varied diets. The rat requires 30-40 separate nutrients in its food including water, 9 essential amino acids, 10 vitamins, and at least 13 minerals. If the rat is given its food such that certain nutrient components are available in separate containers, it will select an appropriate amount from each container for a nutritionally balanced diet. This study is called a "cafeteria" experiment, and the results from one study are shown in Fig. 11.2.

In this "cafeteria" (Fig. 11.2) carbohydrate was present as dextrose, protein as casein, and fats as cod-liver oil and olive oil. Yeast provided a source of nutrients, particularly the B vitamins. The rat selected some of each of these foods (as well as some from several other sources of specific minerals not shown in the figure), and it grew in a normal way.

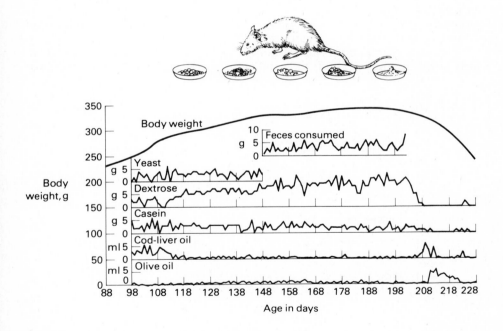

Figure 11.2
Result of a cafeteria experiment in which dietary components were selected from separate containers by a rat. Fecal consumption increased after removal of yeast, and fat consumption increased while sugar consumption decreased after prevention of fecal consumption. All of these changes in diet were adaptive for lack of B vitamins (modified from Richter and Rice, 1945).

The yeast was then removed from the cafeteria and growth rate decreased, although the animal remained healthy. It did so by increasing consumption of its own feces, from which it obtained some B vitamins. The microorganisms in the intestine (such as *E. coli*) provided a source of the vitamins, so the rat changed its diet specifically with respect to the nutritional deficiency. When it was prevented from eating feces it lost weight and became sick. However, even during this deficiency it changed its diet appropriately. The rat consumed more fat and less carbohydrate (Fig. 11.2). The B vitamins are important for enzymes that process carbohydrates, and the absence of the vitamins resulted in a switch to another substrate (fat), an action that could bypass some (but not all; see Fig. 10.21) of the deficiency. When the rat was offered a food containing B vitamins, it preferentially selected it and consumed an amount sufficient to restore normal function.

Specific hungers have been demonstrated for just about every nutrient. Sodium-specific hunger is particularly well known, and it has been observed in a variety of animals under natural and laboratory conditions. In mammals, a hormone is responsible for the retention of sodium in the body. The hormone is called *aldosterone*, and it is produced in the cortex of the adrenal gland near the kidneys (in Chapter 13 we will study this in more detail). If the adrenal cortex is removed or does not function properly to produce aldosterone, large amounts of sodium are lost from the body in urine.

There is only one way to counteract this loss—the consumption of relatively large amounts of salt (NaCl). Laboratory rats will do this when their salt

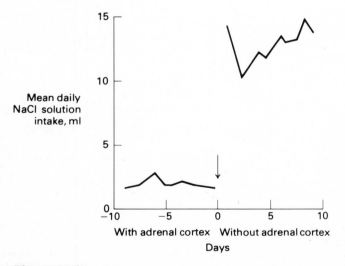

Figure 11.3
Following removal of the adrenal cortex and the development of a sodium deficiency (denoted by the arrow), rats immediately increase intake of salt solutions (after Epstein and Stellar, 1955).

loss is experimentally increased. After removal of the adrenal cortex they immediately increase consumption of salt solutions after a sodium deficiency develops (Fig. 11.3), and they maintain normal function by increased intake supply to meet the large demand resulting from urinary sodium loss.

Specific hunger for sodium also occurs in humans. A number of years ago a 3½-year-old boy was admitted to a hospital with the primary symptoms of overdevelopment of secondary sexual organs. While in the hospital, he ate very little and died a week later. The autopsy revealed that he died from lack of sufficient adrenal cortex function. He had previously eaten large amounts of salt, but the hospital diet did not provide enough sodium, and the physicians did not consider the boy's interest in salt to be important.

After the child's death, the following letter (in Wilkins and Richter, 1940) was written by the boy's parents to some biologists interested in sodium-specific hunger.

> When he was around a year old he started licking all the salt off the crackers and always asked for more. He didn't say any words at this time, but he had a certain sound for everything and a way of letting us know what he wanted. This was the first we had noticed his wanting the crackers for salt. Finally he started chewing the crackers; but he only chewed them until he got the salt off, then he would spit them out. He did the same with bacon, but he didn't swallow the pieces. When he was about sixteen months old, crackers were the first food he chewed and swallowed; but it was quite a while after that before he would chew up and eat a whole cracker. He would usually just make a mess of them eating the salt off.
>
> In an effort to try to find a food that he would like well enough to chew up and swallow, we gave him a taste of practically everything. So, one evening during supper, when he was about eighteen months old, we used some salt out of the shaker on some food. He wanted some too. We gave him just a few grains to taste, thinking he wouldn't like it; but he ate it and asked for more. This was the beginning of his showing that he really craved salt, because this one time was all it took for him to learn what was in the shaker. For a few days after that, when I would feed him his dinner alone at noon, he would keep crying for something that wasn't on the table and always pointed to the cupboard. I didn't think of the salt, so I held him up in front of the cupboard to see what he wanted. He picked out the salt at once; in order to see what he would do with it, I let him have it. He poured some out and ate it by dipping his finger in it. After this he wouldn't eat any food without having the salt too. I would purposely leave it off the table and even hide it from him until I could ask the doctor about it. For it seemed to us like he ate a terrible lot of plain salt. But when I asked Dr. _____ about it, he said, "Let him have it. It won't harm him." So we gave it to him and never tried to stop it altogether. After we gave it to him all the time he usually didn't ask for it with his dinner; but he wouldn't eat his breakfast or supper without it. He really cried for it and acted like he had to have it. Foods that he or-

dinarily wouldn't touch he would eat all right if I added more salt to them. He would take the shaker and pour some out on his plate and eat it with his finger, but we always tried to keep him from getting what we thought would be too much for him. He never did care for Zwieback, toast or bread or for cooked potatoes, but he did like raw potatoes, raw carrots, celery, tomatoes, lettuce and different other foods if he could dip them in salt. If I didn't give it to him, he always asked for it. At eighteen months he was just starting to say a few words, and salt was among the first ones. We had found that practically everything he liked really well was salty, such as crackers, pretzels, potato chips, olives, pickles, fresh fish, salt mackerel, crisp bacon and most foods and vegetables if I added more salt.

CONTROLS FOR NUTRIENT-SPECIFIC HUNGERS

Some animals selectively and precisely regulate the supplies of many different nutrients needed for survival so that intakes from feeding equal demands. How is this remarkably adaptive process accomplished?

The control of sufficient supplies of sodium involves a negative-feedback system. Although we will study this in detail in subsequent chapters, it can be said that control involves specific detection of sodium salts and water in an animal's environment. There are specific receptors that permit adjustments of supplies of these two chemicals as compared with demands. These control systems for water and sodium are similar in overall design to other negative-feedback controls for resource regulation that involve specific detection, integration, and proportional response.

The control of other nutrients does not involve specific detection of the required chemicals in separate negative-feedback control systems. We will examine the experimental evidence for this shortly, but for the moment consider what *would* be required if each and every nutrient had its own separate negative-feedback control system for regulation. Each of the 30-40 nutrients required by the rat from its feeding would have to have a separate receptor and integration mechanism, but most nutrients are required in only small amounts and would need to be detected only when a particular diet did not provide a sufficient supply. An animal might have a whole series of complicated control systems that would not even have to be used if there were no variations in diet resulting in nutritonal deficiencies. This type of solution to the problem of nutrient regulation is less effective than using a common feature of all nutrient deficiencies to control nutrient supplies.

All nutrient deficiencies cause *illness* (Fig. 11.1), and this common feature is used as a "stimulus" to produce control of nutrient supplies for many animals. Control can be demonstrated with a very simple experiment. Rats can be made deficient in thiamine (vitamin B_1), and they will show a clear preference for a diet that contains thiamine (Table 11.2, Group 1). However, if a group of rats is made deficient in thiamine and then injected with the vitamin so that they are no longer sick and have *no* dietary requirement for thiamine, they will *still* select the diet containing thiamine (Table 11.2, Group 2).

Table 11.2 Even following injections of thiamine that remove the illness and deficiency, rats still reject a deficient diet and prefer a diet with thiamine as a result of initial deprivation (from Rozin, 1965)

| | GRAMS EATEN OF EACH DIET | |
GROUP	WITH THIAMINE	WITHOUT THIAMINE
1) Thiamine deficiency	683	281
2) Thiamine deficiency followed by thiamine injections	1049	365

Rather than selecting a food that contains a specific required nutrient, the rats are *avoiding* a food that has previously resulted in illness. This behavior can be further demonstrated by producing a deficiency in a nutrient by feeding a rat a deficient diet and then offering a diet that is *different* but is still deficient. It could differ in texture or taste. The rat will prefer it to the diet that initially resulted in its illness, even though the new, different diet is still deficient in the required nutrient.

This control of nutrient supply involves learning. The rat learns to associate the negative stimuli related to its illness with the diet it has been eating; thus the deficient diet becomes *aversive*—it will be avoided if there are any alternatives available that can be detected as different in some way. The detection of suitable alternatives requires an amount of time that depends on the number of alternatives that could supply a nutrient to cure the illness (Fig. 11.4), and time after the ingestion of different foods is required in order to determine that the illness has been alleviated. Once an effective alternative has been found, the original diet that resulted in the illness will not be selected again (Fig. 11.4).

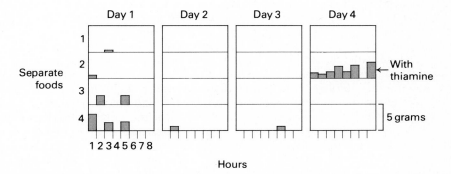

Figure 11.4
Meal patterns for a thiamine-deficient rat given three sources of food deficient in thiamine (foods 1, 3, and 4) and one that contained thiamine (food 2). Foods were sampled until information (feedback) on illness alleviation resulted in almost exclusive selection of the appropriate food (modified from Rozin, 1969).

The control system for sodium regulation does not appear to involve learning. The response to a deficiency of sodium is immediate (Fig. 11.3) and depends on information from taste. The "built-in" response to sodium deficiency is a consequence of a specific negative-feedback control system that involves detection of sodium in the environment; no learning is required (see Chapter 13).

CONTROLS FOR POISON AVOIDANCE

The presence of poisons in food produces the same result as does the lack of nutrients (Fig. 11.1), and many animals respond to poisons the same way they do to nutrient deficiency; that is, they learn to avoid foods that produce illness.

There is a special type of learning involved in behavioral controls for avoiding both poisons and foods without the required nutrients. Psychologists used to think that animals had to be reinforced very soon after a stimulus was presented in order to learn that a particular stimulus resulted in a specific consequence. A flashing light or a bell ring has to be followed soon by an electric shock for a rat to learn to avoid the light or sound. However, when food is ingested it may take several hours for a poison to be digested and assimilated into the blood before it produces illness. Thus for stimuli associated with food, there is normally a long delay between the stimuli and the consequences of eating foods.

Animals learn to associate *characteristics of food* (such as taste, texture, or color) with effects that occur much later, which they cannot do for other, nonfood characteristics of their environments. This association was shown in a classic experiment on poison control in which a group of rats was given flavored water to drink. Each time they took a drink a light flashed, a buzzer sounded, and they tasted the flavor in the water. Following this procedure, half the rats were poisoned to make them ill, and half were punished by electric shock to their feet. The next day the rats were tested to determine if they had developed an aversion to taste, light, and/or sound. Rats that had been poisoned would not drink flavored water, but they would drink other water when lights flashed and buzzers sounded. Rats that had been shocked avoided water when lights flashed and buzzers sounded, but they did not avoid the water with flavor.

These results forced psychologists to revise their ideas about learning. To understand these differences in learning it is necessary to consider and account for an animal's *biology* and *physiology*. Normally, light and sound in an animal's environment from nonfood sources are followed very soon by some consequence that is important for the survival of the animal. Predators can make noises and affect light intensity when they are close to their prey. To have survival value, learning should be most effective for these types of stimuli over very short periods. Potential prey should learn the advantages of a fast response to changes in light or sound stimuli from nonfood sources; a quick response to a snapping twig or flashing teeth can determine survival.

Because the consequences of poisons or nutritional deficiencies *cannot* be experienced immediately as a result of the time required for digestive and assimilative processing, the learning of food characteristics has evolved so that events over longer periods *can* be associated. Both short-term and long-term associations of stimuli with consequences become understandable when important aspects for survival and function are taken into account.

Most animals are not immediately killed by poisons. The effects of a toxic substance are less severe when the amount consumed is small, and a predator that is consuming a nutritionally adequate diet will not usually consume large amounts of a new food. Some "sampling" of new foods is important because a new food may be more effective for net energy gains, but the initial sampling is usually in small amounts. If an animal's experience with a new food results in some illness from poisoning or nutritional deficiency, the food will subsequently be avoided. If the experience results in no negative effects and an increase in the rate of net energy gain, the new food will be incorporated into the total diet in larger amounts.

The sampling, or "testing," of a new food for "consequences" has become apparent in attempts to eradicate populations of "wild" rats in urban areas where they pose a health problem. Even very toxic poisons have been ineffective because of the sampling behavior and learned aversions of these rats. To minimize aversions, rat exterminators follow a procedure called "prebaiting." The rats are first given a new food without any poison. After the rats "sample" or "test" the food, it will be incorporated in their diet in relatively large amounts over a period of time. The poison is then added. If the poison is so potent that it will kill the rat after ingestion of one meal and if the poison is not detected in the flavor or texture of the food (that is, it is not initially detected as a "new" food requiring "sampling"), it may result in the desired extermination.

Rats and humans are not the only animals that demonstrate food-associated learning aversions. A variety of other animals that consume diverse foods also show similar learning, including monkeys, coyotes, pigs, guinea pigs, hamsters, bats, chickens, quail, blue jays, frogs, garter snakes, and garden slugs, among others. An important prerequisite for whether an animal is expected to show food-associated learning is the normally varied nature of its food, i.e., its requirements for a nutritionally balanced diet from several sources.

Large grazing herbivores consume varied foods, and there are numerous reports of the poisoning of livestock that consume toxic plants, so there is some debate over whether large herbivores are influenced by learning with respect to food choice. Preliminary laboratory experiments indicate that cows can learn to associate food characteristics with illness. However, the problems of assessment of the nutritional quality of food by large herbivores under natural conditions are considerable. They consume large amounts of varied foods and spend a large part of a day feeding (5 to 9 hours for sheep and cattle). With consumption of varied foods over long periods, assessments of stimuli as related to consequences may be much more difficult. It has been suggested that the consumption of a diverse diet (with only a few poisonous items) and preferences for young plants (prior to development of poisons) or parts without poisons may minimize learning requirements for large herbivores.

DETOXIFICATION OF POISONS

Plants have evolved a large number of chemicals that can be toxic to animals consuming them. These chemicals include metabolic poisons such as cyanide (found in bitter almonds and the seeds of fruits) and compounds such as nicotine, rotenone, quinones, terpenoids, and other alkaloids. An advantage to the plants is that of reduced predation from the learned aversions of predators. However, there are some animals that do consume some of these chemicals with no ill effects. Most examples include insects or other arthropods that have evolved to consume one type of plant as food; that is, they have evolved to *specialize* in the ingestion and use of a certain set of chemicals. The common names for many insects convey this specialization: spruce budworms, tobacco hornworms, etc.

The insects have a general metabolism similar to other animals that cannot consume these foods, except that they can *detoxify* the poisons, the alternative mechanism for dealing with poisons. Rather than the animal avoiding the chemicals, they are ingested and processed so that they will not be harmful. Because there is such a large diversity of different poisons among plants (requiring very different mechanisms for detoxification), most detoxification mechanisms that have evolved involve specialization on specific chemical poisons by relatively small animals that obtain sufficient food from a single source (usually containing a limited number of poisons).

There are two general ways insects have evolved to detoxify harmful chemicals: (1) through biochemical transformations of the chemicals to harmless or useful products, and (2) through selective sequestration of toxic chemicals in storage sites within the body, where they will be harmless to the rest of the animal.

The biochemical detoxification mechanism for a chemical poison has been studied in larvae of the bruchid beetle, *Caryedes brasiliensis*, which consumes seeds of the legume, *Dioclea megacarpa*, as its sole food source during development. The seeds contain about 13 percent (dry weight) of L-canavanine, which is a potent neurotoxic amino acid lethal to other insects. A principal cause of canavanine toxicity is the production of functionally altered proteins in which the canavanine substitutes for other amino acids and changes the formed proteins so that normal function is disrupted.

In *Caryedes* larvae the transfer RNA responsible for protein synthesis does not interact with canavanine. Moreover, the canavanine is actually utilized as a food substrate; it is hydrolyzed to L-canaline and urea (Fig. 11.5). The urea is converted to CO_2 and ammonia, and the ammonia may then be involved in protein synthesis.

The L-canaline formed from L-canavanine (Fig. 11.5) is also a potent insecticidal toxin. It is a structural analog of the amino acid L-ornithine, and it has a paralyzing neurotoxic effect on many insects. In *Caryedes* larvae L-canaline is deaminated to form L-homoserine and ammonia (Fig. 11.5), which are used in the metabolism of the insect. Thus biochemical detoxification in this beetle involves two distinct steps to transform two poisons so that they not only are rendered harmless but also are changed to chemicals that may be used by the developing larvae.

How are toxic chemicals sequestered? One possible mechanism has been studied in the milkweed bug (*Oncopeltus fasciatus*), a brightly colored insect that consumes cardiac glycosides (poisons) from milkweed (*Asclepias curas-*

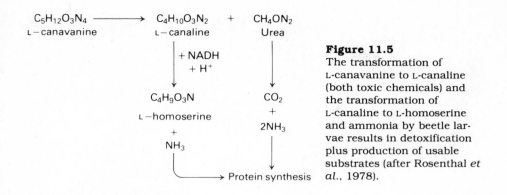

Figure 11.5
The transformation of L-canavanine to L-canaline (both toxic chemicals) and the transformation of L-canaline to L-homoserine and ammonia by beetle larvae results in detoxification plus production of usable substrates (after Rosenthal *et al.*, 1978).

savica) leaves and sequesters them in spaces within the thorax and abdomen. When the bugs are fed food with different concentrations of cardenoline poisons, the amount sequestered increases in direct proportion to the concentration of poisons in food. Moreover, the chemicals are sequestered even when the metabolism of the insect is inhibited (with an effective metabolic poison, 2,4-dinitrophenol). Examination of the fluid in the sequester spaces shows "emulsion particles" (droplets). The cardenoline poisons are sequestered in relatively high concentrations without large energy expense by removal from solution in blood and formation of droplets of fluid in the sequester spaces.

Sequestering toxic substances in an unaltered form produces an additional advantage for an animal; it makes it poisonous to *its* predators. This type of function has been studied extensively in a system in which a poison passes from a plant to its predator and then on to the next step in the food chain. Here learning results in specific aversions, producing advantages not only to the poisonous insect but also to other insect species that are not poisonous but look, sound, or taste like poisonous species.

CHEMICAL DEFENSE AND MIMICRY

The monarch butterfly lays its eggs on the milkweed plant. The milkweed produces toxic cardiac glycosides, but the larvae sequester the chemicals, and they are retained when the larvae undergo metamorphosis to become adult butterflies.

If a "naive" predator such as an inexperienced blue jay captures and eats an adult monarch butterfly, the cardiac glycosides cause it to vomit (Fig. 11.6). This is an unpleasant experience and causes the blue jay to lose whatever food was in its stomach. If it is given another monarch butterfly to eat the blue jay will reject it; it has learned to avoid a noxious prey item.

The importance of learning in this behavior can be demonstrated by feeding a naive blue jay monarch butterflies that have been raised on cabbage from which they can obtain no protective poisons. When nonpoisonous monarch butterflies are fed to inexperienced blue jays they are consumed with no ill effects and become a normal component of the blue jays' diet.

Figure 11.6
A blue jay will vomit after eating a monarch butterfly that contains cardiac glycosides.
Right photo from Brower *et al.*, 1968. Left photo courtesy of Lincoln Brower, Amherst
College.

The adult monarch butterfly is very colorful; it has a distinctive pattern of
orange and black on its wings. This pattern serves to "advertise" distastefulness to potential predators that have learned to avoid orange and black butterflies, which cause vomiting. Other distasteful or noxious prey also advertise
their unpalatability. A variety of species of bees with stingers are yellow and

black, easily recognizable to visually oriented predators. Bees also produce buzzing sounds (everyone has probably heard an "angry" bee), as do wasps and rattlesnakes. In addition, many biological poisons have a similar bitter taste.

The occurrence of a common sensory characteristic (visual, auditory, or chemical) in many different species that are noxious is called *Müllerian mimicry*. The advantages of similar environmental signals that warn of noxiousness are because of the generalization of learning by potential predators. Experience with one brightly colored, noisy, or bitter-tasting prey item that causes vomiting or illness or that stings is usually sufficient for a predator to avoid others with similar sensory characteristics.

Some potential prey species look, sound, or taste like other noxious prey species but are completely palatable, suitable food items. It is called *Batesian mimicry*. A classic example is the viceroy butterfly, which looks like the monarch but belongs to a completely different family of insects and is tasty to naive predators.

Batesian mimicry is a type of exploitation or aggression. There is a *cost* to a noxious prey that is associated with having the mechanisms that produce or sequester toxins responsible for the benefit of reduced predation. Batesian mimics gain the benefits of reduced predation by looking, sounding, or tasting like poisonous species but without having the additional costs of actual chemical defense. This mimicry exploits the noxious animals because a naive predator that samples a mimic will not be poisoned and will be more likely to capture another, similar-looking prey item; it could be an individual from the species that is being mimicked.

There is a cost to Batesian mimicry resulting from the "mistakes" (sampling behavior) of predators. This cost can be examined in the context of the advantage of being a mimic when the number of individuals that are being mimicked (the *models*) changes (Fig. 11.7). The advantage is that of not being eaten, but when the proportion of models falls below a certain value, predators will eat relatively large numbers of mimics (Fig. 11.7). When there are relatively few noxious items, the probability of a particular item being acceptable (mimic) increases. Predators sample (test) different prey items from time to time, particularly if they are hungry and there are few alternative food items. This behavior produces a cost of being a mimic if mimics are available in proportions high enough to be eaten. As a consequence, mimics are usually fewer in number than their models.

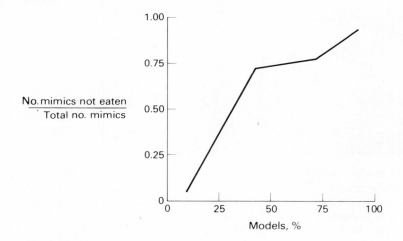

Figure 11.7
The number of mimics not eaten compared with the total number available is high only when there are relatively high proportions of models (actual noxious animals) in a population (modified from Brower, 1960).

There is another variation on the scheme of mimicry. For example, not all female monarch butterflies lay eggs on milkweed. Other plants may produce poisons but perhaps not to the same degree. This situation was studied by raising larvae on four plant species. To assess the potency of the poisons, the adults of the larvae grown on each species were ground into a powder and force-fed to adult blue jays in medicine capsules (naive blue jays are hard to find).

Not all blue jays vomit all the time when they are fed toxins, so an index was devised to measure the dose (concentration) of poisons in terms of the amount of adult butterfly that would make a blue jay vomit 50 percent of the time (the Vomit$_{50}$ index); the number of Vomit$_{50}$ units of poison sequestered by butterflies raised on different plant species was measured (Table 11.3). There was considerable variation in poison effectiveness as a result of the different host plant species. Thus some individuals among monarch butterflies were more palatable, mimicking other, more noxious individuals; it is called *automimicry* because it occurs within the same species. Batesian mimicry occurs between species.

Table 11.3 Variation in noxiousness occurs depending on which food plants female monarch butterflies lay their eggs (from Brower *et al.*, 1967).

FOOD PLANT	NUMBER OF VOMIT$_{50}$ UNITS PER ADULT BUTTERFLY
Asclepias	3.7
Calotropis	4.8
Gomphocarpus	0.8
Gonolobus	0.0

The constraints on automimics are similar to those on Batesian mimics. When predators "test" their potential diets for energy value and poisons, they can learn that orange and black butterflies are tasty if the proportion of automimics is high. However, there is also an advantage to some automimicry, i.e., the variation in the mimicry process within a species. If a predator eventually evolved a mechanism to detoxify a poison, and if *all* monarchs were exactly the same, they would all be suitable prey items and predation mortality would be high. For example, one way that blue jays could "detoxify" monarch butterflies would be to learn to eat only certain parts of a monarch, because most poisons are sequestered in the abdomen.

If female monarch butterflies evolve to lay *some* eggs on different plant species and these plants produce a poison that cannot be detoxified by a predator the way the most common poison is detoxified, those young could escape the new predation. Thus the female monarch butterfly should exhibit some "sampling" of host plants for predator-avoidance advantages. This sampling produces automimicry when host plants contain no poisons.

Moreover, if the young that develop have some variation in the locations of toxins sequestered in the body, the young would be more likely to survive *if* predators evolved to deal only with the most prominent defense mechanism (toxins in the abdomen). Evolution of some variation between individuals can produce advantages when the problems for survival change in the future.

"Tribute" versus "Nontribute" Defenses

Defenses usually influence nutrition because they can determine whether particular food items are consumed. There are other forms of defense besides the use of either chemical toxins or other obvious ways to inhibit prey consumption by predators (such as armor, spines, evasive behaviors, etc.). For example, the salamander *Batrachoseps attenuatus* exhibits tail autotomy. Part of the tail can be lost, without serious injury, to a predator trying to capture the fleeing salamander; the tail is also a region of large fat deposits rich in energy value. Also, the presence of toxins in some species may only occur at certain stages of development. In many plants, for example, young leaves are palatable to and are consumed by predators and only produce toxins when they mature. The use of palatable parts that "satiate" predators and thereby avoid death has been termed a "tribute" defense, distinguishing it from the "nontribute" use of chemical poisons, armor, mimicry, or running and hiding.

Why do different species use tribute or nontribute defenses? An economic hypothesis based on relative costs and benefits suggests that tribute defenses should be most common in numerous (highly productive) species within a community. Individuals of these species are more likely to be "sampled" by predators and to be subject to selective pressures that tend to circumvent nontribute defenses. If nontribute defenses are more likely to fail because of continued sampling pressure from predators, providing *some* energy to a predator will be more advantageous than the use of other types of defense even though the costs for the tribute may be relatively high.

The cost of a tribute defense is the amount of energy lost by a prey to a predator, but the cost of a nontribute defense is the energy use required to produce and maintain a chemical or other defense mechanism that prevents a predator from getting any energy (if effective). When the probability of a prey being sampled is relatively low because relatively few individuals of a prey species are in a community (lower productivity), the cost for nontribute defense should be lower than the cost for tribute defense for the benefit of predator protection, because the likelihood of the defense being circumvented is less.

This cost-benefit interpretation of different patterns of predator defense suggests some intriguing possibilities for further research. It would be very interesting to quantify the costs incurred by prey for different types of defense in a comparative context. For example, does tail autotomy cost more or less than production and maintenance of a potent chemical toxin in the skin? If it costs more, do the advantages of autotomy compensate for the additional cost? Information on these questions will permit tests of the economic hypothesis for predator defenses.

SUMMARY

Nutrients are chemicals not used as energy substrates but important for effective function. They include vitamins, minerals, and essential amino acids and fatty acids, the absence of any of which results in illness and/or death. Except for minerals such as sodium, calcium, chloride, and potassium, only small amounts of nutrients are required because they are usually components of enzymes normally present in low concentrations.

Animals not only must provide for supplies of a relatively wide variety of required nutrients, but also they must avoid a large number of toxic chemicals in their food. The presence of poisons and the absence of required nutrients both produce illness and inefficient function.

Nutrients are regulated in some animals by means of specific hungers so that supplies equal demands. Many animals select a variety of food items to obtain a nutritionally balanced diet. If a deficiency develops, a food component that contains the required nutrient is preferentially consumed. Specific hungers have been demonstrated for just about every nutrient. Specific hunger for sodium is well studied and involves an immediate response to increased demands for sodium.

Specific hunger for sodium involves a "built-in" negative-feedback control system. Controls involving specific hunger for other nutrients involve learning. Predators learn to avoid foods without suitable nutrients by associating the experience of illness with the characteristics of food. In this way the supply of a large number of chemicals are regulated with one basic mechanism.

Predators also learn to avoid foods that cause illness by poisoning. This learning is a special type that involves associating characteristics of food (such as taste, texture, or color) with consequences that occur after several hours of digestive and assimilation processing. The learning of consequences of other types of environmental stimuli (visual or auditory) not associated with food (e.g., danger from predators) cannot and should not occur over such long time periods.

Some predators have evolved to specialize on specific food types and deal with poisons by detoxifying them either by transforming the toxins chemically or by sequestering poisons internally. Sequestered toxic chemicals will make animals poisonous to their predators. The monarch butterfly stores cardiac glycosides, which makes blue jays vomit. The blue jays learn to avoid monarch butterflies and other individuals that look like them.

Noxious species that have evolved visual, auditory, or taste characteristics advertising distastefulness in a similar way are called *Müllerian mimics*. Palatable species that have advertisements similar to noxious species are called *Batesian mimics*. Individuals within a species that differ in degree of noxiousness are called *automimics*. Batesian mimics and automimics do not have to expend resources to produce or sequester chemicals for defense, but they are subject to the cost of relatively high predation if predators "sample" them. This cost is most important when the numbers of actual noxious prey items decrease or when the hunger of a predator increases. The consequences of predator behavior results in mimics being present in smaller numbers than their models.

Moreover, a cost-benefit (economic) hypothesis suggests that a tribute defense, i.e., one where a predator is "satiated" with a palatable part of a prey, should be most prevalent among species that are more numerous in a community because use of a nontribute (e.g., chemical) defense, although perhaps less costly, would be more likely to be circumvented by predators.

ANNOTATED REFERENCES

Chapter 10: Feeding—Energy Regulation and Control

Andersson, S. (1973). Secretion of gastrointestinal hormones. *Ann. Rev. Physiol.* 35:431-452. (Review of control mechanisms for gastric processing of foods.)

Coleman, D.L. (1979). Obesity genes: Beneficial effects in heterozygous mice. *Science* 203:663-665. (The "thrifty gene" hypothesis for the genetics of diabetes.)

Curio, E. (1976). The ethology of predation. In D.S. Farner (ed.), *Zoophysiology and Ecology*, Vol. 7. Springer-Verlag, New York. (Description of details of feeding controls in many species of animals.)

Davies, N.B. (1977). Prey selection and social behavior in wagtails (Aves: Motacillidae). *Jour. Anim. Ecol.* 46:37-57. (Selection of prey quality depends on rate of net energy gain.)

Dethier, V.G. (1976). *The Hungry Fly.* Harvard University Press, Cambridge. (Extensive summary of mechanisms responsible for hunger and food processing in the blowfly.)

Friedman, M.I., and **E.M. Stricker** (1976). The physiological psychology of hunger: A physiological perspective. *Psychol. Rev.* 83:409-431. (Detailed and critical review of current theories of feeding in vertebrates, stressing internal processing.)

Hunt, J.N., and **M.T. Knox** (1968). Regulation of gastric emptying. In C.F. Code (ed.), *Handbook of Physiology*, Section 6, *Alimentary Canal*, Vol. 4, *Motility.* American Physiological Society, Washington, D.C. (Summary of internal control mechanisms for stomach emptying.)

Kendeigh, S.C., **J.E. Kontogiannus**, **A. Malzac**, and **R.R. Roth** (1969). Environmental regulation of food intake by birds. *Comp. Biochem. Physiol.* 31:941-957. (Dependence of feeding on use of stored energy.)

Krebs, J.R. (1978). Optimal foraging: Decision rules for predators. In J.R. Krebs and N.B. Davies (eds.), *Behavioural Ecology: An Evolutionary Approach.* Sinauer Associates, Sunderland, Mass. (Summary of effects of variation of environmental factors on rates of net energy gain.)

Krebs, J.R., **J.C. Ryan**, and **E.L. Charnov** (1974). Hunting by expectation or optimal foraging? A study of patch use by chickadees. *Anim. Behav.* 22:953-964. (Dependence of giving-up time on prior rate of capture in patches.)

LeMagnen, J., **M. Devos**, **J.P.Gaudilliere**, **J. Louis-Sylvestre**, and **S. Tallon** (1973). Role of a lipostatic mechanism in regulation by feeding of energy balance in rats. *Jour. Comp. Physiol. Psychol.* 84:1-23. (Detailed energy-budget studies concerned with patterns of daily energy use in rats.)

Schoener, T.W. (1971). Theory of feeding strategies. *Ann. Rev. Ecol. System.* 2:369-404. (Review of the bases for different patterns of feeding, including "sit-and-wait" predation.)

Wolf, L.L., and **F.R. Hainsworth** (1977). Temporal patterning of feeding by hummingbirds. *Anim. Behav.* 25:976-989. (Dependence of meal size and feeding frequency on maintenance and storage uses of energy.)

Chapter 11: Feeding—Nutrient Regulation and Control

Brower, J.Z. (1960). Experimental studies of mimicry IV. The reactions of starlings to different proportions of models and mimics. *Amer. Nat.* 94:271-282. (Cost for Batesian mimics when proportion of models decreases.)

Brower, L.P., **P.B. McEvoy**, **K.L. Williamson**, and **M.A. Flannery** (1972). Variation in cardiac glycoside content of monarch butterflies from natural populations in eastern North America. *Science* 177:426-429. (Examination of automimicry.)

Denton, D.A. (1967). Salt appetite. In C.F. Code (ed.), *Handbook of Physiology*, Section 6, *Alimentary Canal*, Vol. 1, *Control of Food and Water Intake*. American Physiological Society, Washington, D.C. (Detailed accounts of sodium-specific hunger and its control.)

Duffy, S.S., **M.S. Blum**, **M.B. Isman**, and **G.G.E. Scudder** (1978). Cardiac glycosides: A physical system for their sequestration by the milkweed bug. *Jour. Insect Physiol.* 24:639-645. (A mechanism for toxin sequestration.)

Feeny, P.P. (1970). Seasonal changes in oak leaf tannins and nutrients as a cause of spring feeding by winter moth caterpillars. *Ecology* 51:656-681. (Pattern of feeding determined by nutrients and poisons in a specialized insect predator.)

Garcia, J., and **R.A. Koelling** (1966). Relation of cue to consequence in avoidance learning. *Psychonomic Sci.* 4:123-124. (Demonstration of basic difference in food-related long-delay learning.)

Maiorana, V.C. (1979). Nontoxic toxins: the energetics of coevolution. *Biol. Jour. Linn. Soc.* 11:387-396. (Discussion of factors influencing "tribute" and "nontribute" predator defenses.)

O'Dell, B.L., and **B.J. Campbell** (1971). Trace elements: Metabolism and metabolic function. *Comprehensive Biochem.* 21:179-266. (Role of many trace nutrients as enzyme components.)

Richter, C.P., and **K.K. Rice** (1945). Self-selection studies on coprophagy as a source of vitamin B complex. *Amer. Jour. Physiol.* 143:344-354. (Cafeteria experiments on specific hungers in rats.)

Rosenthal, G.A., **D.L. Dahlman**, and **D.H. Janzen** (1978). L-canaline detoxification: A seed predator's biochemical mechanism. *Science* 202: 528-529.

Rozin, P. (1965). Specific hunger for thiamine: Recovery from deficiency and thiamine preference. *Jour. Comp. Physiol. Psychol.* 59:98-101. (Experimental demonstration of avoidance of nutritionally inadequate foods.)

Rozin, P. (1976). The selection of foods by rats, humans, and other animals.

Adv. Study Behav. 6:21-76. (Review of nutritional and poison determinants of feeding for animals that learn.)

Wilkins, L., and **C.P. Richter** (1940). A great craving for salt by a child with corticoadrenal insufficiency. *Jour. Amer. Med. Assoc.* (AMA) 114:866-868. (Evidence for sodium-specific hunger in humans.)

Zahorik, D.M., and **K.A. Houpt** (1977). The concept of nutritional wisdom: Applicability of laboratory learning models to large herbivores. In L.M. Barker, M. Best, and M. Domjan (eds.), *Learning Mechanisms in Food Selection*. Baylor University Press, Waco, Texas. (Discussion of food-associated learning problems for large herbivores.)

Supplies and Demands for Water and Solutes

Part IV

The functions of all organisms ultimately take place in water. The chemicals within cells must be kept at certain concentrations for effective function, concentration being the amount of a solute in a volume of solvent (water in biological processes). Water is an excellent solvent because a large number of chemicals will dissolve in it without being chemically modified to a great degree. Life is thought to have originated in the sea, and the suitability of water as an environment for diverse cellular functions strengthens this hypothesis.

If environmental changes occur that influence the movement of water into or out of cells, mechanisms should exist to adjust the changes so that the concentration of chemicals within cells remains "optimal" for various functions. The extent to which an "optimal" condition occurs should depend on costs for regulation as compared with benefits derived from regulation. A variety of environmental changes lead to gains or losses of water by animals.

For animals in aquatic environments the problems of achieving gains or losses of water to balance losses or gains depend on the movement of water across semipermeable membranes by means of osmosis. We will examine these problems and the mechanisms (with costs) to counterbalance net water movements in marine (often hyperosmotic) and freshwater (hypoosmotic) environments (Chapter 12).

Terrestrial animals have a higher water concentration than do their environments; thus water is lost by evaporation. In addition, water is lost by excreting waste products of metabolism. These losses must be balanced by gains of water from metabolism, from food, and/or from drinking (all with some costs). The extent to which water is required and the extent to which it is available from these sources depend on variation in the characteristics of terrestrial environments that influence net benefits of regulation (Chapter 13).

Water is also an important resource for the control of body temperature in some situations. Heat is lost when water evaporates, and rate of water loss in terrestrial animals increases in hot environments or when heat production is high. These adjustments lead to some very interesting interactions between energy (heat) and water as resources (Chapter 14).

The regulation of water balance is intimately tied to the regulation of solutes. There are a few cases where pure water is lost or gained. However, solutes often accompany water as it exchanges in either direction, and the distribution of solutes within the body influences the distribution of water. A fish in fresh water is not only faced with the problem of large osmotic gains of water but also with the problem of replacing losses of solutes. Excretion involves not only losses of water but also losses of solutes contained in the excreted water. Each type of environment influences the amount of both water and solutes, and the mechanisms for balancing the exchange of one influence the mechanisms for balancing the exchange of the other.

In the following chapters we will consider a remarkable diversity of factors influencing the exchanges of water and solutes between animals and their environments. Despite this diversity, however, the uses of water and solutes as resources show many basic similarities with the uses of other important resources. The net change in amount of water and solutes over time within animals should remain within certain limits (or be balanced) for survival. Moreover, the precision of osmotic regulation (the extent of balance between losses and gains) varies among species and environments depending on costs of regulation as compared with benefits derived from regulation. The various control mechanisms we will consider all operate to achieve a degree of balance under a variety of circumstances in which the characteristics of regulation are influenced by the economics of resource exchange.

Osmoregulation and Control in Aquatic Environments

12

In any environment, it is important that the volume of water within cells, which influences the concentrations of cellular substrates, be kept within a certain range if costs of regulation are not prohibitive. This fact is illustrated for enzyme function in Fig. 12.1. As the concentration of a substrate is increased, the rate of an enzyme-mediated reaction will increase until the enzyme-substrate complex becomes "saturated." Under most circumstances the products from these reactions are formed at less than absolutely maximum rates, and the rates depend on substrate concentrations within cells being less than maximum (Fig. 12.1).

If the concentrations of substrates within cells increase or decrease, the rates at which products are formed will be more or less than the requirements for the products. The concentration of substrates within cells depends on intracellular water (Fig. 12.1). If cell water content increases, substrate concentrations will decrease, and if cell water content decreases, substrate concentrations will increase. For reaction rates to remain close to "optimal" values for effective function the water content of cells should remain at values that produce appropriate substrate concentrations.

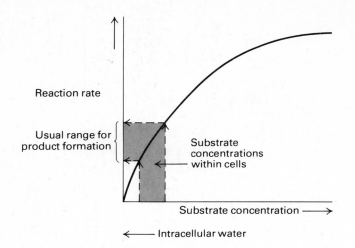

Figure 12.1
The rate for an enzyme-mediated reaction increases as substrate concentration increases until enzymes become
"saturated" with substrates. Products are usually formed
within cells at less than absolute maximum rates (shaded
area). If intracellular water content changes, substrate concentrations will change and produce different reaction
rates. Thus intracellular water content should normally be
maintained within certain limits (if costs permit). The solid
arrows indicate directions of increase for reaction rate,
substrate concentration, and intracellular water.

This pattern is subject to variation. For example, if demands for products
increase, one way to increase (regulate) product formation is by means of an increase in substrate concentration. However, under most circumstances we expect the intracellular water content to remain within relatively narrow limits if
possible (depending on costs versus benefits). Thus it is important to examine
the principal factors influencing movement of water into and out of cells. We
will consider the impact of changes in external environments on regulation of
water balance after a discussion of the principles influencing water exchange by
cells.

OSMOSIS AND OSMOTIC PRESSURE

A major function of cell membranes is to contain chemicals at particular concentrations for effective functions. The concentrations for some solutes are
higher inside a cell membrane than they are outside. Without the membrane,
chemicals would move by diffusion in water from regions of high concentration
to regions of low concentration. Membranes restrict movements by diffusion
and contribute in several ways to the maintenance of the intracellular chemical
environment.

Many biological membranes do not restrict the movement of water to the same extent. Water moves across membranes toward the region of highest *total* solute concentration. The movement of water across membranes because of differences in solute concentrations is called *osmosis*.

The process of osmosis is illustrated in Fig. 12.2. Two compartments are separated by a membrane that allows water to pass in either direction but is not permeable to a solute. At the start, the left compartment contains an amount of water equal to that in the right plus a solute. Over a period of time water moves from the right compartment to the left. Eventually the water movement stops, and at this point the volume of the left compartment has increased. The volume increase has in turn increased the *hydrostatic pressure* on the left, which forces water back to the right (dashed arrows) at a rate that equals the movement of water from right to left (solid arrows), meaning that equilibrium has been reached.

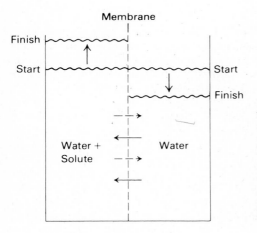

Figure 12.2
Osmosis and osmotic pressure illustrated for two compartments separated by a membrane permeable only to water. Movement of water by osmosis (solid arrows) results in increased volume on the left and increased hydrostatic pressure that forces an equal amount of water back to the right (dashed arrows) at equilibrium.

If more solute is added to the left the fluid level will increase and the pressure to produce a balance in water movement across the membrane will also increase. The magnitude of the equilibrium pressure change from osmosis depends on the *net difference* in solute concentration across the membrane. Thus if solutes were added to the right compartment in Fig. 12.2 the volume and pressure in the left compartment would decrease. If solute concentration in each compartment was identical there would be no net movement of water by osmosis.

The solute molecules in solution result in a hydrostatic pressure called *osmotic pressure* that depends on concentration, i.e., the number of moles of solute in a volume of water (n/V). At low concentrations solutes behave like gas molecules:

$$P = \frac{nRT}{V},$$

where

P = osmotic pressure,
n = moles of solute particles with osmotic effects across a membrane,
R = universal gas constant,
T = absolute temperature, and
V = volume.

Note that n is the number of solute *particles* with osmotic effects; it is not always equivalent to the number of moles of a chemical. For example, sodium chloride dissociates into Na^+ and Cl^- ions in water; one molecule of NaCl is equivalent to *two* solute particles in solution. Moreover, some particles may diffuse across membranes and others may not. The osmotic pressure (P) at equilibrium will be the result of water movement caused by the net difference in total concentration of dissolved particles across a membrane.

Measuring (n/V)

There are a variety of ways to measure the impact of solutes on osmotic pressure, but they all measure concentration, that is, (n/V). The most common measure of concentration is moles of solute/liter (mol/l). A mole is a gram-molecular weight. NaCl has a molecular weight of 58.5; thus 58.5 grams of NaCl is one mole of NaCl.

Another way to measure concentration is moles of solute in a kilogram of solvent (mol/kg). Because one liter of pure water weighs one kilogram, this measure will be different from moles/liter if chemicals in solution add appreciably to the weight of a solution.

A third measure of concentration is *gram equivalents*/liter (usually called "equivalents" and having the symbol eq). This measure is used to represent the dissociation of molecules in a solvent. One mole of NaCl dissolved in one liter of water is "equivalent" to 23 gram equivalents of Na^+ (the atomic weight of sodium) and 35.5 gram equivalents of Cl^- (the atomic weight of chlorine) per liter. When possible, expressing concentrations as moles of dissolved particles (that is, moles of Na^+ and moles of Cl^- rather than moles of NaCl) is the same as expressing concentration in gram equivalents.

A fourth measure of concentration depends on determining the effects of solutes on osmotic pressure without knowing the chemical composition of the fluid on each side of a membrane. For example, a solvent of unknown composition can be added to one side of a membrane impermeable to solutes and a change in volume or pressure as the result of osmosis can be measured. The unit of measure for this is the osmole/liter (osmol/l).

If osmole/liter was identical to mole/liter there would be no reason to use two measures for the same thing. However, solutes have properties in solution (such as electrical charges; see below) that result in the osmolarity of a solution usually being less than its total molarity. For example, seawater has an osmolarity of 1.0 osmol/l but contains a total solute concentration of 1.122 mol/l (Table 12.1).

Table 12.1   Various measures of concentration (n/V) of solutes in seawater (from Potts and Parry, 1963)

SOLUTE COMPOSITION	mmol/l	mmol/kg	meq/l*	mosmol/l
Sodium (Na^+)	470.2	475.4	10.8	
Chloride (Cl^-)	548.3	554.4	19.4	
Magnesium (Mg^{++})	53.6	54.2	1.3	
Sulfate (SO_4^-)	28.2	28.6	2.7	
Calcium (Ca^{++})	10.2	10.3	0.4	
Potassium (K^+)	10.0	10.1	0.4	
Bicarbonate (HCO_3^-)	2.3	2.4	0.1	
Totals	1122.8	1135.4	35.1	1000

*The total number of grams of the substances listed equal their gram equivalents

Most solutes in animals and their environments are present in small amounts, and their concentrations are usually given in "milli" (1/1000) amounts, i.e., millimoles/liter (mmol/l), millimoles/kg (mmol/kg), milliequivalents/liter (meq/l), or milliosmoles/liter (mosmol/l) (Table 12.1).

Tonicity

For the case shown in Fig. 12.2 the osmotic pressure depends on the complete absence of solute movement across the membrane. In this case osmotic pressure is directly related to any concentration difference. If solute concentrations are the same on two sides of the membrane the two fluids are *isosmotic* (same osmotic pressure). If solute concentration is higher on one side, that fluid is *hyperosmotic* compared with the other, and the other fluid is *hypoosmotic* compared with the first.

Biological membranes are often selectively permeable to *some* solutes; whether a change in pressure or volume is observed depends on whether particular solutes on either side will pass through the membrane. Thus some solutions that are isosmotic across a completely impermeable membrane need not produce the same effect across biological membranes.

The term *tonicity* is used to distinguish osmotic effects on biological membranes (as contrasted with completely impermeable membranes). A solution is *isotonic* if there is no net movement of water into it or from it across a biological membrane, and it will be *hyper-* or *hypotonic* if water moves into it or from it.

SOLUTE DISTRIBUTIONS ACROSS BIOLOGICAL MEMBRANES

Figure 12.3 shows the relative distributions of some major solutes across the membrane of a cell. Potassium is usually high in concentration within cells but low in concentration in the extracellular fluid. The opposite is true for sodium. Chloride is also unequally distributed across cell membranes, as are the concentrations of various proteins that carry a net negative electrical charge in so-

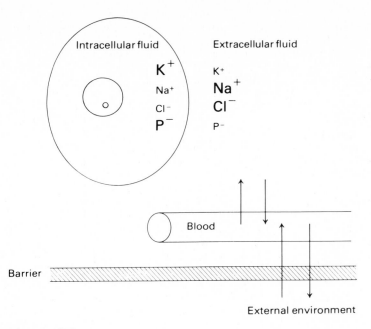

Figure 12.3
Distributions of some major solutes across the membrane of a cell. The size of the symbols indicates their relative concentrations in each compartment. Exchange of solutes between extracellular fluid and blood and between blood and the external environment influence intracellular water content as a result of changes in extracellular tonicity.

lution (Fig. 12.3; P^-). There are several processes responsible for this general pattern of solute distribution.

Gibbs-Donnan Equilibria

For a membrane through which solutes can diffuse where the solutes carry an electrical charge the ratio of solutes carrying similar charges must be equal for each side at equilibrium. Consider the example given in Fig. 12.4. For case A, when both Na^+ and Cl^- can diffuse across the membrane, at equilibrium:

$$\frac{[Na^+]_1}{[Na^+]_2} = \frac{[Cl^-]_2}{[Cl^-]_1} .$$

If a solute is then added to side 1 that dissociates into a diffusing Na^+ and a *nondiffusing* P^- (case B, Fig. 12.4), sodium ions will diffuse into side 2. This would produce an excess of Na^+ ions on side 2, so Cl^- will tend to move back to side 1 (electrical neutrality must eventually occur on each side). However, at equilibrium an unequal amount of Na^+ and Cl^- ions are distributed across the membrane. At equilibrium for the diffusing ions,

$$\frac{([Na^+]_1^{Cl} + [Na^+]_1^{P})}{[Na^+]_2} = \frac{[Cl^-]_2}{[Cl^-]_1},$$

where $[Na^+]_1^{Cl}$ is the sodium concentration on side 1 resulting from the dissociation of NaCl, and $[Na^+]_1^{P}$ is the sodium concentration on side 1 from the dissociation of NaP. Because the term in the numerator on the left is additive, more dissociation of NaP on side 1 will produce more divergence in sodium and chloride concentrations across the membrane. The Cl^- concentration will become higher on side 2, while the Na^+ concentration will become higher on side 1. Electrical neutrality will exist on each side of the membrane, but side 1 will be negatively charged compared with side 2 primarily because of the presence of P^-. These types of unequal distributions due to selective membrane permeability are called Gibbs–Donnan equilibrium effects, and they are partly responsible for unequal ion distributions across cell membranes.

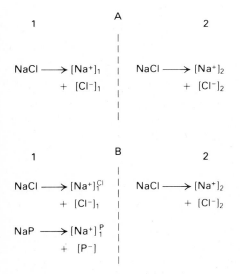

Figure 12.4
Illustration of a Gibbs-Donnan equilibrium effect on solute distributions across a membrane permeable to Na^+ and Cl^- but not P^-. At equilibrium for case A, $[Na^+]_1 \times [Cl^-]_1 = [Na^+]_2 \times [Cl^-]_2$. At equilibrium for case B, $([Na^+]_1^{Cl} + [Na^+]_1^{P}) \times [Cl^-]_1 = [Na^+]_2 \times [Cl^-]_2$. The nondiffusing P^- ion results in unequal distributions of Na^+ and Cl^- for case B.

Active Transport

The membranes of cells also contribute to unequal solute distributions by means of mechanisms requiring energy. If the metabolism of a cell is poisoned with a metabolic inhibitor that prevents ATP production, the concentration differences of Na^+ and K^+ ions across the membrane will eventually decrease. Thus some energy-using process normally contributes to high internal K^+ concentrations and high external Na^+ concentrations.

These mechanisms are conceptualized as membrane "pumps." Both Na^+ and K^+ ions are moved with energy expense toward regions of higher concentration. Several types of pumps may be involved, but all depend on coupling the movement of Na^+ and K^+ ions so that as Na^+ is moved out of a cell K^+ is moved into the cell. Hydrolysis of ATP by enzymes in cell membranes is required, and membrane "pumps" differ mainly in the relative number of Na^+ and K^+ ions exchanged at the pump sites.

Energy use for active transport across cell membranes is considerable. From 20 to 30 percent of the total rate of energy use by a cell involves expenditures to move solutes against concentration gradients. For a particular solute, the energy expended depends exponentially on the difference in concentration and whether a charged ion is moved against both an electrical and a chemical gradient. The equation that describes this condition is

$$\text{cal/mol} = RT \ln \frac{C_1}{C_2} + nFE,$$

where

R = universal gas constant,
T = absolute temperature,
C_1 = higher concentration (direction of movement),
C_2 = lower concentration,
n = electrical valence of ion (positive or negative),
F = Faraday's constant, and
E = difference in electrical potential.

Note that because this cost function depends on a logarithm of a *ratio* of concentrations, costs will increase *exponentially* with a difference in concentration $((C_1 - C_2)^{ex} \propto \log C_1/C_2$; see Appendix II).

For a case where K^+ is transported from extracellular fluid with a concentration of 20 mmol/l to intracellular fluid with a concentration of 400 mmol/l against an electrical potential of 70mV (internal negative) at 27 °C, about 3400 cal/mol must be expended. This type of energy expense and its exponential dependence on a difference in concentration has a major impact on osmoregulation because of the costs of the mechanisms to balance water exchange (see below).

The maintenance of concentration differences of Na^+ and K^+ across membranes influences other processes that contribute to differences in solute distributions. Some Na^+ "leaks" back into cells (and is pumped out), and "carrier" molecules in membranes transport amino acids and sugars into cells by means of utilization of Na^+ movement from a region of high (outside) concentration to low (inside) concentration. The energy for this transport ultimately comes from Na^+ - K^+ "pumps," which maintain concentration differences of Na^+ across the cell membrane.

ENVIRONMENTAL INFLUENCE ON OSMOREGULATION

The composition of extracellular fluid is influenced by the external environments of animals (Fig. 12.3). With the exception of some blood proteins (colloids), the solutes in blood plasma water are in equilibrium with extracellular fluid. Changes in Na^+ or K^+ concentrations in blood result in changes in their concentrations in extracellular fluid. If the tonicity of blood and extracellular fluids increase, intracellular water volume will decrease, and if the tonicity of blood and extracellular water decrease, intracellular water volume will increase.

Blood tonicity is influenced by the exchange of water and solutes across surface barriers of animals (Fig. 12.3). In many cases the surfaces of animals will be relatively impermeable to water, which will reduce the rates of gain or loss of water and solutes influencing extracellular fluids. However, some surface barriers must be permeable to water and solutes to allow the exchange of other chemicals with the external environment. Oxygen and carbon dioxide exchange across a fluid interface, meaning that respiratory exchange surfaces must be permeable to water. In addition, assimilation of chemicals from food in the digestive tract involves membrane surfaces permeable to water. When the permeability of membranes is a major variable, the equation describing the effects of concentration differences on energy expenditures (costs) for osmoregulation is

$$\text{cal/mol} = (P)\,(A)\,RT\ln\frac{C_1}{C_2} + nFE,$$

where

P = surface membrane permeability (l/cm^2 × molar difference in osmotic concentration),

A = surface area of membrane in cm^2,

and other terms are as defined previously.

Changes in the tonicity of fluids on external surfaces will influence the tonicity of blood and, in turn, the tonicity of extracellular fluid, which will produce changes in intracellular water content. For regulation of intracellular water volume within relatively narrow "optimal" limits, environmental changes that cause changes in extracellular tonicity must be counterbalanced (if costs are not prohibitive). If water is gained in excess of losses or extracellular solutes are lost in excess of gains, extracellular fluid tonicity will decrease. This inequality must be balanced by increased loss of water and/or increased gains of solutes. Conversely, if extracellular water is lost in excess of gains or extracellular solutes are gained in excess of losses, osmoregulation depends on increased gain of water and/or losses of solutes with osmotic effects.

We will consider four external environmental conditions influencing supplies and demands for gains and losses of water and solutes. We will now examine the aquatic marine environment (high environmental solute concentration), the aquatic brackish-water environment (variable decreased environmental solute concentrations), and the aquatic fresh water environment (low environmental solute concentration). In the next chapter we will examine the terrestrial environment (variable and low environmental water supplies). For each environment we will ask if regulation of osmotic concentrations of body fluids occurs, the extent to which regulation occurs depending on costs and benefits of osmoregulation, and how losses or gains of water or solutes are counterbalanced with the use of control mechanisms.

PATTERNS OF OSMOREGULATION IN AQUATIC ENVIRONMENTS

The Marine (Hypertonic) Environment

There are two general patterns of osmoregulation by animals that live in environments with high osmotic concentrations (Fig. 12.5). Total blood osmotic concentration is either *similar to* or considerably *less than* the total osmotic concentration of the environment. For both groups of animals the marine environment is stable ("unchanging") with respect to total solute concentration and composition (Table 12.1). The two general patterns of osmoregulation are illustrated for several animals in Tables 12.2 and 12.3. The different approaches to solving the problems of osmoregulation have very different consequences for the animals.

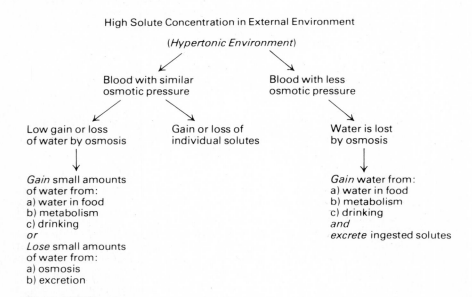

Figure 12.5
Summary of impact of different patterns of osmoregulation in marine (hypertonic) environments on gains and losses of water and solutes.

Near-isotonic Species

If total blood osmotic pressure is identical to total environmental osmotic pressure there will be no net movement of water into or out of an animal by osmosis. However, some water loss is necessary to excrete waste products. In this case the losses are balanced by small gains of water from one of three sources: (1) ingestion of food (prey contain 50-90 percent water); (2) oxidation of food substrates, which yields energy, CO_2, water, and waste products; and/or (3) drinking sea water and removing (excreting) the solutes in it to obtain water.

Table 12.2 Marine animals in which internal osmotic concentration is close to external osmotic concentration (from Potts and Parry, 1963; Bentley, 1971, and Gordon *et al.*, 1961)

ORGANISM	TOTAL SOLUTE CONCENTRATION (mmol/kg)	DISPROPORTIONATE COMPONENT* (mmol/kg)
1) Invertebrates		
Jellyfish (*Aurelia*)	1144	Chloride: 580
Sea urchin (*Echinus*)	1134	none
Mussel (*Mytilus*)	1134	none
Squid (*Loligo*)	1280	Protein: 150
Isopod (*Ligia*)	1267	Chloride: 629
Crab (*Carcinus*)	1209	Protein: 60
Crab (*Maia*)	1126	Sulphate: 14
Lobster (*Nephrops*)	1174	Protein: 33
2) Cyclostomes		
Hagfish (*Myxine*)	1185	Protein: 67
3) Elasmobrachs		
Ray (*Raja*)	1050	Urea: 444
Dogfish shark (*Squalus*)	1000	Urea: 354
4) Teleosts		
Coelacanth (*Latimeria*)	1181	Urea: 355
5) Amphibia		
Crab-eating frog (*Rana*)	830**	Urea: 350

* The component judged to be most different from its normal concentration in sea water

** For frogs maintained in an environment in which the total osmotic concentration was 800 mmol/l

Table 12.3 Marine animals in which internal osmotic concentration is less than external osmotic concentration (from Potts and Parry, 1963; Bentley, 1971, Beadle, 1943; Croghan, 1958)

ORGANISM	mmol/l in BLOOD	mmol/l in ENVIRONMENT
1) Invertebrates		
Brine shrimp (*Artemia*)	480	5150
Shrimp (*Palaemonetes*)	340	1122
2) Teleosts		
Herring (*Clupea*)	512	1122
Anglerfish (*Lophius*)	431	1122
Salmon (*Salmo*)	400	1122
Toadfish (*Opsanus*)	392	1122
Eel (*Anguilla*)	371	1122

Because the amount of water required by an isotonic species is very small, no drinking is required, and any excess water (from food and oxidation) remaining after excretion is lost by osmosis.

The total osmotic concentration of sea water is 1,135 mmol/kg (Table 12.1). For the animals listed in Table 12.2 only the sea urchin and mussel have total solute concentrations essentially identical to seawater. Other species (e.g., the crab (*Maia*), the ray, and the dogfish shark) have total blood osmotic concentrations slightly less than their environment. These animals lose some water by osmosis. The rate of water loss depends on the difference in osmotic pressure between their environment and their blood, and this difference is low. Because of slightly increased losses of water as compared with isotonic species these animals must obtain somewhat more water from food, metabolism, or drinking (and excreting solutes).

The remaining species listed in Table 12.2 have total blood osmotic concentrations somewhat greater than their environment. These species gain water by osmosis at rates depending on the difference in concentration, reducing their requirements for water from other sources (metabolism, food, drinking), and internal solute concentrations are maintained by excreting excess water with waste products.

In addition to balancing slight gains or slight losses of water, most of the species listed in Table 12.2 have different amounts of certain solutes in their blood as compared with the environment. Some of these are listed as "disproportionate components" in Table 12.2. Because of differences in concentrations, some solutes will be lost by diffusion or excretion and others will be gained by diffusion, through drinking, or in food. Losses or gains of different solutes must be balanced by reciprocal gains and losses to keep the internal osmotic environment relatively constant.

Hypotonic Species

The rate of osmotic loss of water is much greater for species that maintain low total blood concentrations in hypertonic environments (Table 12.3). For example, in brine shrimp (*Artemia*, Table 12.3), which live in saturated saline lakes, there is an enormous difference in osmotic concentration between them and their environment. Hypotonic animals must balance a relatively large osmotic loss of water with gains from food, metabolism, and drinking. Because the supply of water from food and metabolism is relatively fixed by energy demands, adjustments for large osmotic losses of water occur primarily by means of drinking, with control mechanisms to excrete the solutes in seawater that are ingested.

CONTROLS FOR OSMOREGULATION IN HYPERTONIC ENVIRONMENTS

Isotonic Species

Problems of water loss and gain are minimized for isotonic and near-isotonic species. For many species the mechanism of osmoregulation involves maintenance of relatively high blood osmotic concentrations by the addition of one or two solutes in relatively large amounts. Proteins or urea are present in high concentrations in the blood of several species (Table 12.2), and these chemicals are obtained from food or as a "waste" product of protein metabolism (urea; see Chapter 13).

The use of urea for osmoregulation is interesting because it is toxic to other species when it is present in relatively high concentrations. The specific mechanisms underlying maintenance of urea and protein concentrations for osmoregulation are poorly understood. They may involve a receptor sensitive to cellular volume changes (an *osmoreceptor*) that detects increases or decreases in cellular water content, ultimately resulting in increases or decreases in urea or protein blood concentrations to counteract the osmotic changes. In Chapter 13 we will examine details of such a negative-feedback osmotic control system in some terrestrial vertebrates for which there is much more experimental information.

Hypotonic Species

Animals with much lower blood osmotic concentrations as compared with their environment must drink seawater (Fig. 12.6). If an eel in salt water is prevented from drinking water by placing a balloon in the pharynx, it rapidly loses weight and dies from cellular dehydration in about 3 days.

The solutes in ingested seawater must be *excreted* back to the environment when an animal obtains the water required to replace osmotic losses. A great deal of interest has been focused on the tissues where salt excretion occurs and the mechanisms for transporting salts back to the environment against concentration gradients.

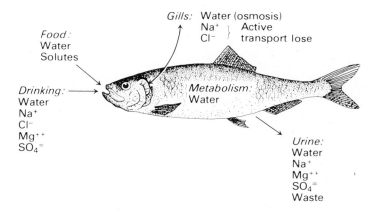

Figure 12.6
For a hypotonic fish water is gained from metabolism, from food, and by drinking. Water is lost by osmosis and excretion. Solutes are gained from food and drinking and are lost by active transport.

A variety of tissues excrete sodium in different species of hypotonic animals. Salts are excreted from the gills of teleost fishes and crustaceans and from the rectal glands of elasmobranchs. Some cells in these tissues actively transport sodium out and potassium in across surface membranes with a

Na$^+$ - K$^+$ "pump" similar to that present in other cell membranes. The movement of salts depends on the presence of the membrane enzyme Na$^+$ - K$^+$-activated ATPase, which splits ATP to ADP. This enzyme has been found in every tissue where solutes are moved against concentration gradients, including the gills of marine teleosts and elasmobranch rectal glands.

The eel (*Anguilla rostrata*) has been extensively studied to characterize the controls for hypotonic regulation. Eels migrate from fresh water to salt water and maintain remarkably constant blood osmotic concentrations despite the extreme change in the external osmotic environment. Species that have the ability to tolerate relatively large changes in their osmotic environments are called "euryhaline," and those with very limited ability to tolerate osmotic changes are termed "stenohaline." These differences are reflected in the mechanisms that have evolved to deal with osmoregulatory problems related to environmental changes.

When an eel first encounters salt water its blood osmotic concentration initially increases because of increased sodium concentration as a result of drinking seawater. This increase is followed by some cellular dehydration and an increase in the production of the hormone *cortisol* by the adrenal cortex. Cortisol produces two important effects. It results in the migration of special "chloride-secreting" cells from within the gill out to the gill surface, and it also results in the development of a large number of Na$^+$ - K$^+$ pumps in the membranes of these cells. Within a few days the development of the sodium pumping mechanism results in hypotonic regulation of blood concentration by means of the removal of ingested sodium.

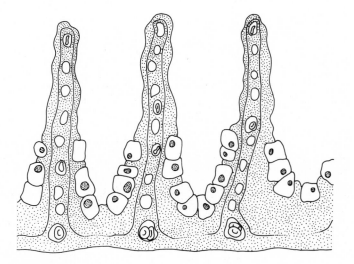

Figure 12.7
Diagram of a section through the gill epithelium of an eel from salt water. The "chloride"-secreting cells are the larger cells toward the base of the gill filaments (adapted from Keys and Willmer, 1932).

The chloride-secreting cells in gill filaments are relatively large (Fig. 12.7) and have large numbers of mitochondria. The number of Na^+ - K^+ pumps that develop depend on the difference in osmotic concentration between blood and the external environment. Thus the major mechanism for osmoregulation by hypotonic animals involves increased energy expenditure (cost) for sodium transport at an epithelial surface. The mechanism is similar to the one that maintains ion distributions across other cell membranes, but it is localized at the surface of the animal. Other solutes ingested in seawater are excreted with other waste products in urine (Fig. 12.6).

WHY TWO PATTERNS OF REGULATION?

Natural selection has led to the evolution of two very different patterns of osmoregulation in hypertonic environments: (1) maintenance of nearly isotonic blood concentrations, or (2) maintenance of hypotonic blood concentrations. Hypotheses to explain the evolution of these two patterns fall into two categories.

The Historical or Ancestral Hypotheses

These hypotheses attempt to explain osmoregulatory patterns from the ancestry of a species. We will find that many freshwater animals have blood osmotic concentrations similar to marine species with hypotonic blood concentrations. The euryhaline eel, for example, maintains a similar osmotic concentration in both freshwater and marine environments. The ancestral hypotheses argue that marine or freshwater animals have "retained" a particular internal osmotic concentration from their ancestors to some degree. For example, those marine species with relatively high osmotic concentrations in blood are presumed not to have had a recent freshwater ancestor.

These hypotheses are based entirely on ancestry and not on the relative advantages (benefits) and disadvantages (costs) of particular osmoregulatory mechanisms. They are difficult to test by experiment because they depend on independent information about an animal's ancestry. Moreover, the ancestral hypotheses do not explain why a particular osmotic concentration *should* be "retained" when animals evolve in different environments. The hypotheses are similar to "primitive" versus "advanced" hypotheses for circulatory patterns, body temperatures, and regulation of pH. They do not provide an explanantion for why a particular condition should be "advanced" or "retained," and they usually neglect possible advantages that can be associated with alternative solutions to the same regulatory problem.

The Economic Hypothesis

Both solutions to the problems of osmoregulation in hypertonic environments have benefits and costs. A benefit for both groups of animals is regulation of intracellular substrate concentrations. However, the costs for achieving this regulation are different for each group.

Hypotonic animals must expend energy to excrete ingested solutes. The cost depends on the difference in concentration and electrical potentials of excreted solutes across the membranes that actively transport ions (see equation on p. 362). This cost has been calculated for some animals in which the difference in concentration across excretory membranes changes when the osmotic concentration in the external environment is changed.

Nereis diversicolor is a marine and brackish-water polychaete that changes the composition of body fluids when the osmotic concentration of its environment changes (see Fig. 12.9). When the difference in concentration between body fluids and the environment is low, little energy is expended (Fig. 12.8). When the difference increases, the cost for maintaining a larger difference increases *exponentially* (Fig. 12.8) because energy costs depend on the logarithm of the ratio of internal and external concentrations. For *Nereis* the cost for osmoregulation by active transport increases 8 times when the difference in osmotic concentration that is maintained approximately doubles from 42 to 98 mmol/liter (Fig. 12.8).

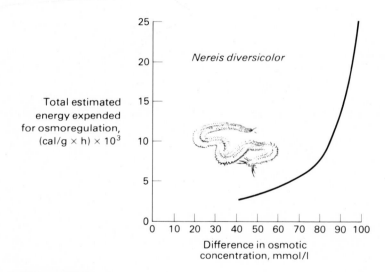

Figure 12.8
Total calculated energy expenditure resulting from osmoregulation of *Nereis* at 25 °C depends exponentially on the difference in concentration between body fluids and the environment. This dependence is a consequence of energy requirements for active transport that depend on the logarithm of the ratio of internal and external concentrations (data from Potts and Parry, 1963).

The total cost for osmoregulation by active transport depends on the total number of ions that must be moved against a particular concentration gradient. This number is lower for smaller animals, although *per gram* transport costs are higher, because rate of water loss per gram from osmosis is higher in smaller animals with membrane surface areas for exchange that are higher with respect to their volumes.

Species that maintain blood osmotic concentrations similar to their environments achieve osmoregulation without these high energy costs. However, some cost is involved from the use of "disproportionate components" to keep blood osmotic concentrations high. Because of their higher concentrations in blood, they will be lost and must be replaced at the expense of some energy (for obtaining food or oxidizing proteins), but this cost is probably low when compared with the costs for osmoregulation in hypotonic animals.

Another cost for isotonic marine species is the effect of the solute used for osmoregulation on the metabolism of the animal. For example, urea is toxic to many animals but it is not toxic to rays, sharks, the coelacanth, or the crab-eating frog (Table 12.2). Thus these animals must have had to evolve adjustments in their metabolism that can involve costs in cellular function.

Unfortunately, there is so little information on the cellular metabolism of these animals that it is not now possible to state what "trade-off" in function occurs because of high urea concentrations as compared with the costs for active transport in hypotonic animals. When the information is available, and if it can be expressed as an equivalent energetic cost, then it will be possible to decide if the two patterns of osmoregulation are economically equivalent or whether one is more effective than the other for achieving osmoregulation.

BRACKISH AND FRESHWATER (HYPOTONIC) ENVIRONMENTS

The problems of osmoregulation in brackish-water and freshwater environments are similar; they differ mainly in degree and timing. Brackish-water animals are exposed to environments that may fluctuate in osmotic concentrations. When the osmotic concentration decreases, such as when isotonic marine animals move into a tidal coastal area near a freshwater inlet, they must continue to adjust extracellular fluid concentrations as the difference in osmotic concentration between them and their environment increases.

Figure 12.9 shows the effect of decreasing environmental osmotic concentration on the concentration of extracellular fluids in several marine isotonic species. When they are exposed to a full-strength seawater environment (upper right part of Fig. 12.9) these species all have body fluid concentrations equal to or slightly higher than the environment. As the environmental solute concentration decreases (moving to the left in Fig. 12.9), body fluid concentrations decrease in these animals by different degrees.

Animals that show a change in body fluid concentrations when their environments change are called "osmoconformers," and those that show constancy of body fluid concentrations when osmotic environments change are called "osmoregulators." As you can see from Fig. 12.9, there are varying degrees of precision of osmoregulation among species. At one time these differences were attributed to "primitive" versus "advanced" characteristics relating to the taxonomy of different animals. However, a functional explanation for this diversity can be found by examining the economics of osmoregulation of species faced with variable environments.

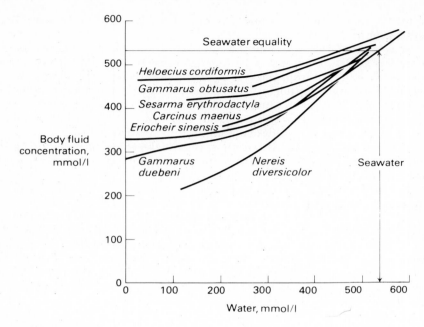

Figure 12.9
Relationship between concentration of solutes in body fluids and in the environment when environmental concentration decreases from the amount in seawater (upper right) to the amount in fresh water (left) for several marine isotonic species (modified from Beadle, 1943).

The decrease in body fluid concentration when environmental concentration decreases should reflect the costs for the mechanisms to maintain intracellular water content as compared with the benefits of regulation. When environmental solute concentration decreases, water enters by osmosis and solutes are lost from body fluids by diffusion and excretion. The water that is gained must be excreted and the solutes that are lost must be gained from the environment. When gain of solutes from the environment is by active transport *into* the animal, the cost for it can be reduced by means of a reduction in internal solute concentration as environmental concentration decreases so that the *difference* in concentration between the animal and its environment is less. However, cost reduction for osmoregulation will involve less precision of regulation of cellular water volumes.

Nereis diversicolor shows the most pronounced decrease in body fluid concentration of all the animals shown in Fig. 12.9. When external concentration is 100 mmol/l, body fluid concentration is 200 mmol/l, a difference of 100 mmol/l. With this difference the energy cost for osmoregulation in *Nereis* has increased 8 times (Fig. 12.8). If *Nereis* was to maintain an even higher difference in concentration by means of active transport, the cost would be exponentially higher.

The extent to which extracellular concentration decreases should represent a "trade-off" in costs for osmoregulation versus the benefits of maintaining intracellular water content absolutely constant as environments change. Thus "osmoconformers" should be most sensitive to osmoregulatory costs (influenced by factors such as surface areas for exchange and energy expenditures required for regulation) and "osmoregulators" should be less sensitive to osmoregulatory costs, meaning that net benefits (benefits – costs) from more precise regulation can be maintained when environments change. Note that these arguments, based on the economics of exchange, are similar to those discussed in Chapter 6 for the factors influencing the precision of temperature regulation (i.e., there are "thermoconformers" and "thermoregulators," depending on benefits relative to costs).

Freshwater Animals

Freshwater animals are faced with the most extreme problems involving osmoregulation in hypotonic environments because the osmotic concentration of the environment is very low (total of 2-3 mmol/kg or lower; see Table 12.4). Because blood osmotic concentration is greater, water continuously floods the animals (Fig. 12.10). This excess gain (along with excess gains from food and/or metabolism) must be balanced by excreting water.

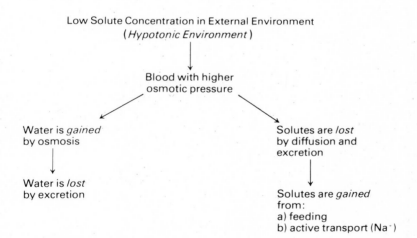

Figure 12.10
Summary of the impact of higher blood osmotic concentration in a hypotonic environment on gains and losses of water and solutes.

Freshwater animals also face the problem of loss of solutes. Some are lost in urine, and some are lost by diffusion from the animal to its environment. These losses must be balanced by gains from two sources. Food will contain some solutes, and the gills or other surface epithelial tissues actively transport

Table 12.4   Solute constituents of a freshwater environment and of the blood of some freshwater animals (from Potts and Parry, 1963)

	Na^+	K^+	Ca^{++}	Mg^{++}	Cl^-	HCO_3^-	Others	Total
A. Average composition of North American river water (mmol/kg)	0.39	0.04	0.52	0.21	0.23	1.11	0.21 (sulfate)	2.71
B. Composition of the blood of some freshwater animals								
1) Frog (*Rana*)[1]	109	2.6	2.1	1.3	78	26.6	3.5 (lactate)	223.1
2) Trout (*Salmo*)[2]	161	5.3	6.3	0.9	119	—	1.0 (phosphate)	293.5
3) Crab (*Potamon*)[2]	259	8.4	12.7	—	242	—	—	522.1
4) Insect larva (*Sialis*)[2]	109	5.0	7.5	19	31	15	152 (amino acids)	338.5
5) Lamellibranch (*Anodonta*)[1]	15.6	0.5	6	0.2	11.7	12	0.2 (amino acids)	46.2
6) Stingray (*Potamotrygon*)[2*]	150	5.9	3.6	1.8	149	—	0.5 (urea)	310.8

[1] mmol/kg water
[2] mmol/l blood
* From Thorson et al., 1967

sodium into the animal (Fig. 12.11). Sodium is quantitatively the most important extracellular solute (Table 12.4), and other solutes are obtained from food
in sufficient amounts to balance losses.

There is very little variation in the low osmotic concentration of fresh
water. The only way to reduce active-transport costs would be to lower the
osmotic concentration of extracellular fluid, and this reduction should be
limited by its effects on intracellular volume and concentration changes. Nevertheless, there are some species with comparatively low extracellular fluid concentrations. The lamellibranch *Anodonta* has a total blood osmotic concentration of only 46.2 mmol/kg as compared with values of 220-520 mmol/kg for
other freshwater species (Table 12.4). It would be very interesting to examine
cellular function in *Anodonta* to assess the impact of reduced extracellular concentrations (lower osmoregulatory costs).

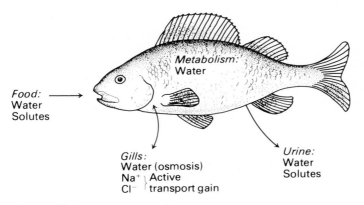

Figure 12.11
For a hypertonic fish water is gained from osmosis, from food,
and from metabolism. Water is lost from excretion. Solutes are
lost from diffusion and excretion and are gained from active
transport and food.

Controls for Active-transport Solute Gains from Fresh Water

A variety of epithelial tissues of freshwater animals have membrane Na^+ - K^+
pumps that transport sodium from the environment against a concentration
gradient into the animal. These tissues include kidney tubules or their analogues
in all freshwater animals, gills in teleost fishes and in crustaceans, anal papillae in
freshwater insect larvae, and the surface of the skin in amphibians. The mechanism depends on membrane Na^+ - K^+ ATPase in each case, and the "pump"
operates in the same manner as all other Na^+ - K^+ pumps; namely, sodium is
moved out of cells against concentration gradients while potassium is moved in.

Regardless of the direction of solute transport, all epithelial tissues that pump sodium have a concentration of membrane pumps in the same location in the cell, on the basal or "interior" (inward-facing) and lateral sides of the cell. There are no membrane pumps or only a few on the "exterior" (outward-facing) side of the cells. If this is the case, how is sodium moved *into* animals in freshwater and *out of* animals in hypertonic (marine) environments with the same mechanism?

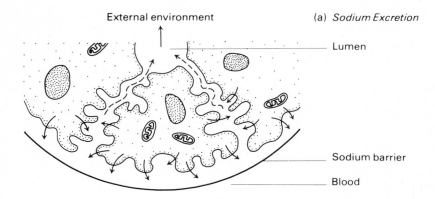

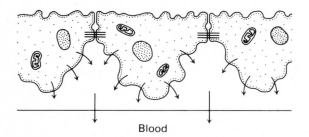

Figure 12.12
Schematic diagram of tissues that excrete sodium (a) or absorb sodium (b) by means of active transport. In both cases membrane Na^+-K^+ pumps are located on the basal and lateral parts of the cells (arrows) and transport sodium out of the cells. Excretory cells are arranged with channels between cells leading to a lumen connected with the external environment. Absorbing cells have "tight junctions" near external surfaces (adapted from DiBona and Mills, 1979).

The direction of movement of sodium depends on the morphological arrangement between cells in the transporting tissue. The differences are shown in Fig. 12.12 for tissues at which sodium movement is *out* or *in* against concentration gradients. In tissues at which sodium is excreted by active transport it is pumped into channels between cells and extracellular fluid moves to a lumen connected with the external environment (Fig. 12.12a). In addition, there is usually an effective barrier against movement of sodium into blood.

In tissues at which sodium is actively transported into an animal the cells are usually tightly bound at the interface with the external environment (Fig. 12.12b). Sodium is pumped out of cells at their bases and enters the cells from the external environment providing a net movement of sodium into the animal through the tissues. Moreover, sodium readily passes from spaces between cells to the blood.

We will examine the morphology of spacing between actively transporting cells in more detail in the next chapter because similar arrangements are involved in the excretion and resorption processes of terrestrial animals. From modifications of intercellular channels between cells and connections, the same mechanism achieves solute movement in opposite directions. In species that migrate between fresh and salt water (such as euryhaline eels and salmon) the same cells are used for transport by the same cellular mechanism but with different intercellular channels resulting in different directions of net sodium movement.

Controls for Excreting Water in Hypertonic Animals

Excess water gains must be excreted by hypertonic animals to prevent dilution of extracellular fluid. There are a number of tissues that achieve this function in different animals. In unicellular species contractile vacuoles expel fluid from the animal at intervals that depend on the rate of excess water gain. Contractile vacuoles actually represent part of a complex of structures within the cytoplasm that includes a membraneous *spongiome* surrounding the vacuole and sometimes a distinct pore from the vacuole to the external surface. Current theories of function suggest that cytoplasmic fluid is segregated by local osmotic gradients across membranes of the spongiome, solutes are resorbed to produce fluid hypotonic to cytoplasmic fluid, and this fluid is stored until it is excreted by means of the contraction of the vacuole.

Multicellular animals have tubular excretory systems. In flatworms (planarians, tapeworms, flukes), "flame" cells excrete fluid into ducts connected with the body surface. Animals with closed circulatory systems filter fluid from blood into a tubular excretory system. The excretory organs, the kidneys, are composed of many similar units called *nephrons*. A schematic diagram of a single nephron from a freshwater animal is shown in Fig. 12.13(a). The blood supply to a nephron passes through a series of convoluted capillaries called a *glomerulus*, which is enclosed in a capsule of the nephron. The capsule is given different names in different animals. In crayfish it is called a "labyrinth," in earthworms a "nephrostome," and in vertebrates a "Bowman's capsule."

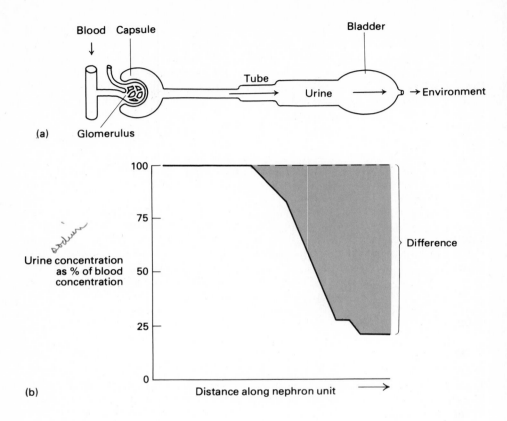

Figure 12.13
(a)   Schematic diagram of the tubular excretory nephron of a freshwater animal and
its glomerular blood supply. (b)   After filtration the nephron fluid is initially the same
concentration as that of blood. Active transport of sodium into blood reduces urine
concentration down to a limit set by costs of the increased difference in concentration
between urine and blood (shaded area) (modified from Lockwood, 1966).

The capillaries of the glomerulus are situated close to arterioles; thus
blood pressure within the glomerular capillaries exceeds blood colloid osmotic
pressure (see Fig. 4.21), and plasma fluid is filtered into the capsule and passes
down the nephron tube. Animals in hypertonic environments usually lack a
filtration apparatus because little water is lost from excretion via the kidneys.
The kidneys of many marine animals are "aglomerular" (without any glomeruli).
However, freshwater animals must filter and excrete large amounts of water
from blood to remain in osmotic balance. A 100-gram goldfish filters 2 ml of
plasma from its blood every hour. The fish has a total blood volume of 5-6 ml; it
completely filters and excretes its total blood volume every 2-3 hours.

The blood plasma filtered into the nephron capsule contains solutes, nutrients, and energy substrates. Many of these substances are recovered before the fluid is lost to the external environment. Recovery of sodium, the principal extracellular osmotic solute, takes place along the tubular portion of the nephron (Fig. 12.13a; also see Chapter 13).

Energy is expended for recovery of sodium by means of active transport in the nephron epithelium. The concentration of sodium in the glomerular filtrate is initially identical to its concentration in blood, meaning that the cost for active sodium transport is initially low. As sodium is "pumped" from urine back to the blood side, urine sodium concentration decreases and energy must be expended at an exponentially greater rate to transport it into the blood side against a larger and larger concentration difference across the nephron tubule membrane (Fig. 12.13b).

In this situation sodium is absorbed from urine to blood, and the cells that actively transport sodium have "tight junctions" between them that reduce the rate of movement of water to the blood side of the epithlium. Thus the epithelium of the nephron tubule of freshwater animals is similar to the epithelium of other tissues that "pump" sodium into the body (see Fig. 12.12b).

Despite this mechanism for regaining sodium in the nephrons, some is lost because an enormous energy expenditure would be required to reduce urine to pure water. Urinary solute losses are balanced by active-transport gains at other transporting tissues and by solute gains from food.

SUMMARY

Rate of product formation in cells depends on intracellular substrate concentrations. Rate of product formation will equal demands for products when intracellular water content is maintained within certain limits, because addition or removal of cell water changes substrate concentrations. Thus control mechanisms should counterbalance changes that tend to increase or decrease intracellular water (unless the costs of this are prohibitive).

Water moves across cell membranes by osmosis. A membrane permeable to water but not to solutes will result in hydrostatic pressure as a result of movement of water toward a region of higher net solute concentration. At equilibrium the hydrostatic pressure produces movement of water *from* a region of higher solute concentration that balances the osmotic movement of water *to* a region of higher solute concentration.

For membranes impermeable to all solutes, osmotic pressure depends on solute concentration difference across the membrane (net n/V, that is, the net difference in moles of solute particles in a volume of water). Thus osmotic pressure can be measured from concentration differences, expressed as moles/liter, moles/kilogram, gram equivalents/liter (moles of dissociated particles/liter), or osmoles/liter (when the specific concentration is unknown).

Biological membranes are usually permeable to some solutes as well as to water, and "tonicity" is used to distinguish this characteristic from membranes completely impermeable to solutes. Permeability to some solutes produces some inequality in solute distributions from Gibbs-Donnan equilibrium effects, by which permeable solutes with electrical charges distribute such that the ratio of similarly charged ions are equal across membranes. Active transport of sodium and potassium also contributes to patterns of solute distribution across biological membranes, and energy expended for "pumping" these solutes against concentration gradients is exponentially related to the difference in concentration maintained across a cell membrane.

The external environments of animals influence blood tonicity by means of solute and water exchange across surface barriers. The tonicity of the blood, in turn, influences extracellular fluid tonicity and thus intracellular water content. In aquatic environments osmoregulation depends on differences in osmotic concentration that determine whether water or solutes move into or out of an animal through its surface.

Animals in marine (hypertonic) environments show two patterns of osmoregulation. Blood osmotic concentrations are either similar to or considerably less than the environmental osmotic concentration. Nearly isotonic species experience little net movement of water by osmosis; maintenance of water exchange depends on slight gains or losses of water from food, metabolism, drinking, excretion, and/or osmosis. However, this pattern of regulation depends on maintenance of some blood solutes at relatively high concentrations as compared with the environment. This maintenance involves some costs of maintaining the high concentrations and other poorly understood possible costs of the use of certain solutes to the metabolism of cells.

Hypotonic marine species face high osmotic losses of water, and rate of loss exceeds rate of supply of water from food and metabolism, which means that seawater must be ingested and that the solutes in it must be "pumped" out to replace water losses. The exponential dependence of active-transport costs on difference in concentration results in comparatively high energetic costs for osmoregulation in these animals. Active transport occurs in epithelial tissues (gills, rectal glands) where certain cells (such as "chloride"-secreting cells) develop membrane pump sites in proportion to requirements for the number of solutes that must be transported to achieve osmoregulation.

Animals in brackish and freshwater (hypotonic) environments gain water from osmosis at a rate dependent on the difference between the osmotic concentration of their fluids and the environment. Excess gains of water are excreted. Excretion and diffusion also produce losses of solutes that must be balanced by gains from feeding and from active transport of sodium *into* the animals.

Active transport at surface epithelial tissues and within the kidney nephrons of freshwater animals involves "pumping" sodium from cells to extracellular fluid, and the location and mode of action of membrane pumps are the same for tissues that move sodium in different directions (out of animals in hypotonic species; into animals in hypertonic species). Differences in the direction of movement of sodium depend on the morphology of spaces between transporting cells. Where transport is out of an animal, sodium is pumped into fluid channels connected with a lumen to the external environment. Where transport is into an animal, "tight junctions" at the external surfaces between cells restrict water flow into blood.

Osmoregulation and Control in Terrestrial Environments

13

For the purposes of osmoregulation a terrestrial environment is one with relatively low water availability, but the amount of water in terrestrial environments varies considerably. This lack of uniformity in water distribution requires us to examine different animals in the context of their local environments. Some animals, such as earthworms or "aquatic" frogs, are in continuous contact with water; for osmoregulatory purposes they are freshwater animals. Other animals exposed to air will lose water by evaporation. Many of the problems of osmoregulation in terrestrial environments concern water supplies and demands whereby a major demand is produced by water evaporation and the major supply is determined by variable amounts of water in the external environment.

The problems of maintaining optimal cellular water volumes, or substrate concentrations are similar among terrestrial animals and aquatic animals. Table 13.1 shows the general similarity of the distribution of water in extracellular compartments and in intracellular compartments of several animals from a variety of osmotic environments. The maintenance of similar intracellular water volumes is a reflection of the importance of osmoregulation for maintaining effective cellular function.

Table 13.1 Distribution of water in the body of some vertebrates from different environments (g/100 g body weight) (from Bentley, 1971)

ENVIRONMENT	TOTAL BODY WATER	=	EXTRACELLULAR	+	INTRACELLULAR
Salt water					
Spiny dogfish	71.7		21.2		50.5
Parrotfish	73.1		16.6		56.5
Fresh water					
Freshwater shark	72.1		19.7		52.4
Paddlefish	74.0		15.6		58.4
Bowfin	74.5		18.9		55.6
Carp	74.1		15.5		55.9
Terrestrial					
Ornate lizard	73.6		25.7		47.9
Human	66.0		20.0		46.0
Camel	69.8		19.2		50.6

Although terrestrial animals have cellular osmoregulatory problems similar to those of aquatic animals, the avenues for water exchange with the environment are somewhat different. The major components of water and solute losses and gains are summarized in Fig. 13.1. Evaporation is a major avenue of water loss, and we will examine losses from respiration and across skin barriers and the mechanisms to minimize these losses when water supplies are limited. Excretion of waste products by defecation and urination also requires water (Fig. 13.1), and we will examine the ways animals have evolved that minimize these losses when water supplies are limited.

Net losses of water are balanced from two or three sources (Fig. 13.1). Food contains some water, and the oxidation of energy substrates in food will yield some additional water. Water may also be gained by drinking supplies from the external environment if water is available. However, some animals have evolved in environments with drinking-water supplies that are very low compared with demands, and we will examine regulation of water balance by animals in different environments in which the costs of achieving regulation by different mechanisms change.

Some animals exhibit thirst for water (as a consequence of demands produced by water losses) when external environmental supplies permit balancing losses by drinking. Whether drinking is the most effective way to balance losses depends on costs of control mechanisms to reduce losses of water from evaporation and excretion as compared with costs of obtaining water for drinking from the external environment.

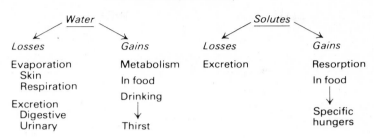

Figure 13.1
The major components of gains and losses of water and solutes in
terrestrial environments.

Solute regulation involves balancing net losses from excretion with intakes, primarily by means of food (Fig. 13.1). Water and solute regulation are
connected by the common avenue of loss from excretion. Urine contains both
water and solutes. For terrestrial animals in "arid" environments, excretion is
considered efficient when excreted fluids contain a minimum amount of water,
which means high solute concentration in the excreted fluid. We will examine
this interaction in our discussion of controls for excretion.

Regulation of solutes occurs when losses are balanced by gains from
resorption of solutes from excretory fluid and from specific hungers (Fig. 13.1).
A specific hunger for a solute occurs when excretion results in a loss of that
solute. Sodium-specific hunger can balance sodium losses; we will examine this
balance after we understand the controls governing losses of sodium from excretion.

DETERMINANTS OF EVAPORATION

Animals in terrestrial environments lose water by evaporation when the air surrounding them contains less than a maximum amount of water vapor. If the air
can contain more water vapor, water moves from the animals to their environment. The factors that determine movement of water by evaporation are those
involved in the change of state of water from a liquid to a vapor.

The pressure of water vapor in air (P_{H2O}, mmHg) depends on movement of
water molecules from a liquid into air. Molecules of water in air will also pass
back into a liquid at the interface. When the rate of escape exactly equals the
rate of condensation of vapor, the vapor is said to be *saturated*.

The pressure exerted by a vapor at equilibrium is called the *vapor
pressure*. Vapor pressure depends exponentially on temperature. This relationship is described by the Clausius-Clapeyron equation in logarithmic form:

$$\ln P = -\frac{\Delta \overline{H}_{vap}}{RT} + \text{constant},$$

where

P = vapor pressure,

R = universal gas constant,

T = absolute temperature, and

$\triangle \bar{H}_{vap}$ = heat of vaporization at temperature T (energy required to convert water from liquid to vapor).

Molecules of water leave a liquid surface at greater rates when their kinetic energy increases. An increase in temperature causes increases in molecular kinetic energy, resulting in higher equilibrium vapor pressure.

The exponential dependence of air vapor pressure on temperature (°C) is shown in Fig. 13.2. Air that is 100% saturated with water vapor contains exponentially more water when the temperature increases (upper line in Fig. 13.2). However, animals will not lose water to 100% saturated air because, at equilibrium, as many water molecules condense to liquid as are evaporated into the air. The important consideration here is that terrestrial environments seldom have air that is saturated with water, but animals can be considered to have liquid interfaces with air.

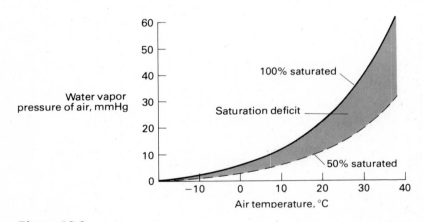

Figure 13.2
Air 100% saturated with water vapor (upper line) contains exponentially more water vapor at higher temperatures. A saturation deficit (shaded area) is shown with respect to air 50% saturated with water vapor (lower dashed line).

If vapor pressure is not the maximum possible, a *saturation deficit* will exist and more water can pass from liquid to vapor than can pass in the reverse direction. If, for example, air is 50% saturated (lower curve in Fig. 13.2), an animal will encounter a gradient of evaporation that depends on temperature. The gradient between 50% and 100% saturation is shown as the shaded area in Fig. 13.2. The larger the saturation deficit in mmHg, the larger will be the rate of water loss to an external environment by evaporation in a given animal.

Animals have evolved barriers that restrict or slow the rate of water loss to air less than 100% saturated. The skin of terrestrial animals may contain less water than other tissues and the movement of water through it may be slow. Among a number of species of lizards and snakes, lipids in the skin appear important in restricting the movement of water, because the permeability of their skin to water increases to uniformly high values after the lipids are removed.

Although some surfaces are relatively impermeable to water, most terrestrial animals must have some epithelial tissues saturated with water. This requirement comes from demands for gas exchange with the external environment. Oxygen and carbon dioxide diffuse across a respiratory surface in water, and the respiratory epithelium represents a relatively large surface area (see Chapter 2). For this reason respiration potentially involves a higher rate of water loss, and most measurements of evaporative water loss distinguish losses from respiration and losses from other surfaces of animals.

We wish to compare animals to determine what impact water losses (such as from evaporation and excretion) have on osmoregulation. Which animals should we compare? The regulation of water balance is a problem of *both* demands and supplies. If there is an excess availability of water in an animal's environment and there is little risk or cost associated with obtaining it, losses can be relatively high and can be balanced by drinking. However, if water is scarce (or costs to obtain it are high) there will be greater selective pressure for economizing losses so that restricted supplies will be sufficient to meet demands. Thus we expect more effective control over water losses when demands are high *relative to* environmental supplies.

A desert environment provides such a situation. Water is scarce, the air is dry, and temperatures can be high, resulting in large saturation deficits. One desert animal (the kangaroo rat, *Dipodomys merriami*) has been studied in great detail. Information is available for each component of loss and gain of water, meaning that we can study total water regulation. In addition, sufficient information is available about the environment and behavior of kangaroo rats to appreciate water demands when compared with supplies in a natural context. In order to decide which aspects of function represent environmental adaptations, we will compare the kangaroo rat with the laboratory rat, an animal that has its origins in more moist environments where the problems of osmoregulation should be solved in different ways (through lower-cost drinking to balance higher losses).

WATER REGULATION IN THE KANGAROO RAT AND THE LABORATORY RAT

Part of water gain for all animals comes from substrate oxidation and from water in food. To standardize oxidation water as a variable, Table 13.2 presents the components of water loss and gain for kangaroo rats and laboratory rats both fed the same amount of the same food (100 grams of barley). Different substrates yield different amounts of water when they are completely oxidized (Table 13.3). Fats contain more hydrogen to combine with oxygen, and they yield the highest amount of water for every gram oxidized. In addition, recall that fats yield the highest amount of energy for every gram oxidized (Chapter

5). Heat production as compared with water requirements will be discussed in the next chapter. For present purposes it is sufficient to note that barley is composed primarily of protein and carbohydrate; thus internal oxidation of 100 grams provides 54 milliliters of water for both the kangaroo rat and the laboratory rat (Table 13.2).

Table 13.2   Components of water balance for the kangaroo rat and laboratory rat, each one consuming 100 grams of barley as food at 25°C and in dry air (from K. Schmidt-Nielsen, 1964b)

	Kangaroo rat	Laboratory rat
A. *INTERNAL GAIN*		
Oxidation water	54ml	54ml
B. *LOSSES*		
Total evaporative loss	44ml	77ml
Loss in feces	2.5ml	13.6ml
Loss in urine	13.5ml	22.0ml*
Total losses	60.0ml	112.6ml
C. *REQUIRED ADDITIONAL GAIN FROM ENVIRONMENT*	6.0ml	58.6ml

*based on maximum urine-concentrating ability

Table 13.3   Water resulting from complete oxidation of different substrates

SUBSTRATE	g WATER/g SUBSTRATE
Carbohydrate	0.56
Fat	1.07
Protein	0.40

Evaporative Losses

To get the oxygen and release the carbon dioxide involved in the use of the substrates in 100 grams of barley, the kangaroo rat evaporates a total of 44 ml of water, but the laboratory rat evaporates a total of 77 ml (Table 13.2). What is responsible for this substantial difference in total evaporative water loss? The difference is due almost entirely to a lower rate of water loss through the skin of the desert-adapted kangaroo rat. This is demonstrated by examining the major factors contributing to similar water losses by respiration in both species.

In Chapter 7 we discussed respiratory countercurrent heat exchange as a mechanism for minimizing heat loss from the respiratory passages. It can also save water (Fig. 13.3). The illustration in Fig. 13.3 is for a kangaroo rat inspiring air at two temperatures (15°C and 30°C) where the air is initially 25% saturated with water vapor (a relative humidity of 25%). The inspired air must be warmed to body temperature (37°C) and saturated with water vapor in the lungs (arrows pointing up in Fig. 13.3), and the air that is expired is saturated with water vapor (arrows pointing down in Fig. 13.3). But by cooling expired

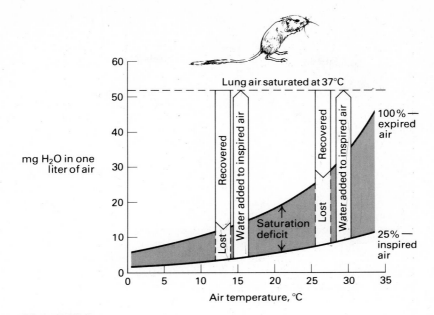

Figure 13.3
When kangaroo rats inspire air (upward-pointing arrows at 15 °C and 30 °C) that is 25% saturated with water vapor, the air must be saturated at a body temperature of 37 °C in the lungs. Some water is recovered as the result of cooling expired air (downward-pointing arrows), and respiratory water loss depends on the saturation deficit at the temperature of expired air (from Schmidt-Nielsen *et al.*, 1970).

air, some water condenses and is recovered during expiration. For the example in Fig. 13.3, air inspired at 15 °C is warmed to 37 °C and cooled to 13 °C before it leaves the body, and air inspired at 30° is expired at 27 °C.

Cooling expired air reduces the amount of water that is lost as compared with animals that expire air at higher temperatures. Also, note from Fig. 13.3 that the amount of water lost and recovered depends on the saturation deficit between inspired (25%) and expired (100%) air, and the saturation deficit depends on temperature.

The kangaroo rat has a very small nose with very narrow and convoluted passages (Fig. 13.4). The distance between the center of the air stream and the tissues of the nasal passage is very small; heat exchange is rapid and effective. During inspiration the evaporation of water from the moist nasal tissues humidifies the air. The evaporation also cools nasal tissues because evaporation of every milliliter of water requires about 580 calories of heat (the latent heat of vaporization of water, $\triangle \overline{H}_{vap}$; see the next chapter for details). During expiration the air is cooled as heat passes from the warmed air to the cooler nasal tissues. Cooler air contains less water when it is saturated, so some water condenses on the nasal tissues ("recovered" water in Fig. 13.3).

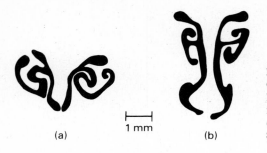

Figure 13.4
Cross sections of the nasal passages of a kangaroo rat at 3 mm (a) and 9 mm (b) from the nasal opening. Note the small dimensions and large surface area (from Schmidt-Nielsen *et al.*, 1970).

Both the laboratory rat and the kangaroo rat are effective in cooling expired air (Fig. 13.5). They both have small nasal passages, and the efficiency of water recovery in respiration is related to their small size and is independent of the environment of these animals.

Because the two species recover similar amounts of water during expiration, differences between them in *total* evaporative water loss (Table 13.2) must be due to differences in water loss through other, nonrespiratory surfaces of the body. The kangaroo rat has evolved a more effective barrier to water loss through the skin that reduces demands for water resources as compared with those of the laboratory rat. The skin of the kangaroo rat is thicker, contains less water than does the skin of the laboratory rat, and perhaps contains a more effective lipid barrier to restrict loss of water through the skin.

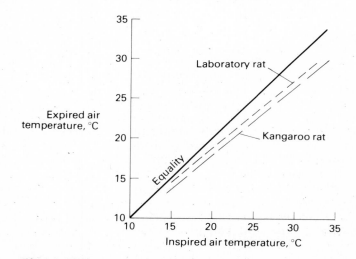

Figure 13.5
When inspired air temperature changes, both the laboratory rat and the kangaroo rat cool expired air to a similar degree. Thus they will both lose similar amounts of water during respiration (based on data from Jackson and Schmidt-Nielsen, 1964).

Excretory Losses

When wastes pass through the intestine, some water is resorbed into the blood across the epithelium of the digestive tract even when there is no osmotic difference from one side of the epithelium to the other. This isotonic water movement across cells requires energy, but the water itself is not actively transported. Water moves as a consequence of osmotic concentration differences created by active transport of sodium out of cells (Fig. 13.6). Lateral spaces between epithelial cells open and sodium is pumped into local extracellular "pockets." Water moves by osmosis from the cells into the pockets and from the lumen into the cells. When water is not moved in this manner the spaces between cells are constricted (Fig. 13.6 right).

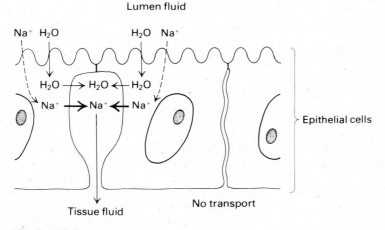

Figure 13.6
Even when there is no osmotic concentration difference across an epithelium, fluid can be transported when Na^+ is actively transported into extracellular channels. Water moves by osmosis from lumen fluid, and sodium moves by diffusion from this fluid.

This mechanism for water movement occurs in a variety of tissues including the gut, gall bladder, and kidney (see below). Note that the mechanism for isotonic water movement is similar to the mechanisms used for excreting or absorbing sodium (see Chapter 12). "Tight junctions" at external cell interfaces permit locally high concentrations of sodium in extracellular spaces; these concentrations are produced by Na^+-K^+ membrane pumps on lateral surfaces of the cells.

The kangaroo rat is more effective at removing water from the lumen fluid of the intestine (Table 13.2). The laboratory rat loses more than five times the water needed to excrete the solid wastes resulting from the ingestion of 100 grams of barley. Note, however, that the gain of water by the kangaroo rat from intestinal transport involves higher energy expenditure, a cost that is relatively less for the laboratory rat. The measurements have not been made, but it is likely that the kangaroo rat intestinal cells have more membrane pumps to produce higher sodium concentrations in extracellular pockets to absorb larger amounts of water. The cost for this absorption will increase exponentially as the local concentration across the membranes of the cells increases (see Chapter 12).

Barley contains a relatively large amount of protein, which in mammals ultimately results in the necessity for excreting urea (see below). Water losses in urine are minimized when urine concentration is maximized. We will discuss the mechanisms for urine formation and concentration in terrestrial animals shortly. When both the kangaroo rat and the laboratory rat minimize urinary water losses by producing the maximum possible urine concentration, the kangaroo rat excretes 13.5 ml of urinary water to process 100 grams of barley and the laboratory rat excretes 22 ml of urinary water (Table 13.2). For the kangaroo rat, this lower loss also involves a relatively high cost of resorbing a larger amount of water in the kidneys (see below).

Environmental Water Gains for Regulation

When both kangaroo rats and laboratory rats are exposed to dry air, water losses exceed water gain from oxidation (Table 13.2). The additional required water gain must come from other sources in the external environment. The kangaroo rat requires 6 ml to balance losses and the laboratory rat requires about ten times this amount (Table 13.2).

Food normally contains some water, but seeds contain relatively little water. The amount of moisture contained in barley seeds also depends on the moisture content (vapor pressure) of the air surrounding them (Table 13.4). More water is absorbed by seeds exposed to air with a higher moisture content. To achieve water regulation the kangaroo rat should consume seeds exposed to air with a moisture content between 10 and 33% (Table 13.4). The laboratory rat cannot balance losses with gains just from water absorbed by barley seeds and must drink water to achieve osmoregulation if it consumes this food.

Table 13.4   Amount of water that is absorbed by 100 grams of barley exposed to air having different degrees of water vapor saturation (from K. Schmidt-Nielsen, 1964b)

Saturation of air with water (25°C), %	10	33	76
ml water in 100 grams of barley	3.7	10.2	18.1

In order to interpret the importance of the external environment for water balance in kangaroo rats we need to know the temperature and water saturation characteristics of the air to which the animals and their food are normally exposed. Kangaroo rats spend the day underground in a burrow. The temperature of the air in the burrow is moderate when compared with the extremes that occur on the surface (Fig. 13.7). The kangaroo rat searches for seeds at night

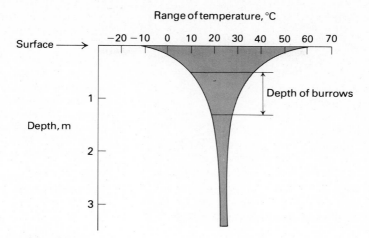

Figure 13.7
Temperature fluctuations are less extreme below the surface of the desert at depths typical of kangaroo rat burrows.
A temperature of 25 °C represents an "average" burrow
temperature (after Schmidt-Nielsen, 1964b).

when it is cooler and spends the day in its relatively cool burrow, so it is seldom
exposed to air with high temperatures or large saturation deficits.

Another important aspect of the burrow environment is the moisture content of the air. The air is essentially saturated with water vapor, which has two
important consequences. First, the seeds gathered on the relatively dry surface
will absorb water when they are stored in the burrow. Second, the water loss of
the kangaroo rat from evaporation is reduced considerably while it is in the
burrow because there is no saturation deficit.

The water regulation of the kangaroo rat can be summarized with respect
to the major environmental determinant of water gains and losses; the moisture
content of the air at a typical burrow temperature (Fig. 13.8). Oxidation gain
from 100 grams of barley is fixed at 54 ml. In addition, minimal fecal and urinary water losses resulting from the processing of 100 grams of barley are relatively fixed at about 16 ml (Table 13.2). Absorption of water by the barley increases total water intake as the air becomes more saturated with water. Total
water loss also decreases as the saturation deficit decreases because evaporative losses decrease.

Note that total water gain will equal total water loss for the kangaroo rat
when the air contains 10% water vapor or more (Fig. 13.8). Oxidation water
alone is sufficient to balance total water loss if the air contains more than 20%
water vapor. Clearly, the kangaroo rat will have a net excess of water if it remains in its burrow. The extent to which water balance is achieved under
natural conditions will depend on the magnitude of losses experienced outside
the burrow (determined by time required to obtain food) compared with the excess gains achieved in the burrow.

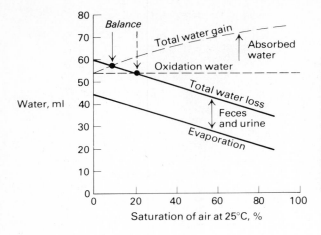

Figure 13.8
Summary of the components of total water exchange of kangaroo rats as a function of water vapor saturation of air at a typical burrow temperature (25 °C). Note that water balance occurs at relatively low values of air saturation (from Schmidt-Nielsen and Schmidt-Nielsen, 1951).

The kangaroo rat is certainly more effective in restricting water losses than is the laboratory rat. However, the laboratory rat is in no way "inferior" to the kangaroo rat in terms of osmoregulation. Differences among these (and other) animals reflect supplies as compared with demands. Because of environmental restrictions in water supplies, the kangaroo rat has evolved relatively costly mechanisms for resorbing water; these costs are less for the laboratory rat and should be due to its access to environmental water supplies. Interpretation of osmoregulation in terms of both supplies and demands provides a useful framework in which to examine some further details of water exchange.

WATER SUPPLIES AND DEMANDS WITH CHANGES IN SIZE

Water evaporates from the surfaces of animals. As size decreases, surface area increases relative to volume. We examined this fact in the context of respiration (Chapter 1) and heat exchange between animals and their environments (Chapters 5 and 7). Changes in size also influence water regulation by means of changes in the proportion of available body water lost through evaporation.

The impact of body size on evaporative water loss is illustrated for birds in Fig. 13.9; other animals face similar constraints. As size decreases, demands for water to replace evaporative losses increase within any environment with a saturation deficit. The total amount of water lost by a smaller animal is less, but the amount lost by evaporation in relation to internal supplies is larger. Thus the consequence of size for use of body water supplies are similar to those for use of energy (see Chapter 5).

Instead of the external environment generating the primary problem for osmoregulation from differences in water supply, a small size produces added problems as a result of higher demands relative to internal supplies. In addition, there are some small animals that live in arid environments. A high demand in relation to *both* internal and external supplies of water has led to the evolution of several adaptations for water economy among small terrestrial animals.

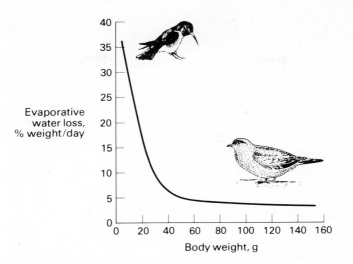

Figure 13.9
Rate of evaporative water loss as a proportion of internal
water supply increases with decreasing body size in birds.
Similar constraints apply to other small animals (from Bartholomew and Cade, 1963).

Insects and other arthropods are small, and terrestrial species have
evolved mechanisms that minimize the rate of water loss by evaporation both
from respiratory surfaces and from other surfaces of the body. Terrestrial insects
obtain oxygen and release carbon dioxide through a tubular tracheal system
(see Chapter 2). The external openings to the trachea are through the spiracle
valves. When the spiracles are open, gases and water exchange between the
animals and their environments at a maximum rate (see Fig. 2.21).

Mealworms (beetle larvae) live in a very dry environment. When they are
starved for several months and kept in dry air, metabolic rate decreases, and
gas exchange occurs with the spiracle valves opening only for very brief periods
at infrequent intervals (Fig. 13.10). When the spiracles are open the rate of

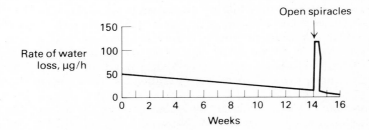

Figure 13.10
Rate of water loss is very low for mealworms in a dry environment for a period of weeks except for when the spiracles are
open (from Mellanby, 1934).

water loss increases as much as six times (Fig. 13.10), but by keeping the spiracles closed for periods of weeks the rate of water loss by evaporation is minimized.

Arthropods are covered with an exoskeleton composed of rigid chitin. It provides a barrier to water loss, and a more effective barrier occurs in those species that produce a wax coating over the chitin. Waxes are produced and secreted through the chitin in a solvent that spreads them over the outer surface. If the surface of an insect is heated, the wax will separate and become liquid (at its melting point) and the rate of evaporation will increase (Fig. 13.11).

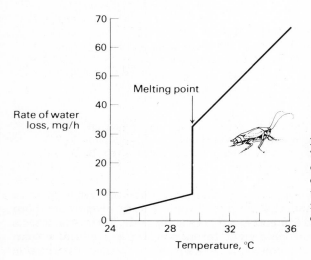

Figure 13.11
The rate of water loss from a cockroach increases dramatically when the temperature of the exoskeleton reaches the melting point of the wax coating (from Beament, 1958).

Minimizing Excretory Water Losses in Small Arthropods

The waste products excreted by arthropods from terrestrial environments are very dry. The mechanism for this process involves osmotic filtration of fluid from blood together with formation of a chemical for nitrogen excretion that is relatively insoluble and permits some resorption of water.

The excretory system of insects consists of a series of several *Malpighian tubules*, which are long blind sacs in contact with the fluid of the vascular system. The tubules drain into the digestive tract near the rectum. The fluid in the vascular system is filtered across the tubular membrane by osmosis because of active transport of potassium into the Malpighian tubular space; water and other solutes move toward the region of higher concentration in the tubular fluid.

Nitrogenous waste (primarily potassium urate) empties into the rectum where it is stored prior to elimination. Within the rectum the urate forms uric acid, which has a low solubility and precipitates from solution. After precipitation the uric acid no longer contributes to osmotic concentration; thus urine osmotic concentration decreases as compared with the osmotic concentration of the blood, and some water moves by osmosis back into the blood. Additional water is "transported" through local pockets between cells in the rectal epithelium as a consequence of the energy expense of pumping sodium (Fig. 13.6). We will find that some terrestrial vertebrates (birds, reptiles) use a similar mechanism for minimizing water losses in excretion (see below).

Water Uptake by Small Arthropods

Arthropods obtain water from oxidation, in food, and by drinking if water is available in the environment. Although a drop of water is a large supply for a small animal, the environment of a small animal is comparatively large, so drops may be hard to come by. Even when water is not available as a liquid in the environment, some arthropods can "drink" water vapor in air. Intake of water from air occurs only after dehydration or water loss. A flea (*Xenopsylla*) will rehydrate by absorption of water from air that is only 50% saturated with water vapor. The effectors for uptake of water from air have not been identified, but they appear to include glandular secretions of hygroscopic salt solutions, which may involve considerable energy expense for water uptake.

WATER REGULATION AND NITROGEN EXCRETION

Some water used to excrete waste products is lost by animals. However, the amount lost differs depending on the chemicals used to excrete some wastes. Different species have evolved different patterns of nitrogen excretion that reflect supply-demand problems of water economy. These patterns can be illustrated by examining a number of species from both aquatic and terrestrial environments.

Protein contains nitrogen atoms. Amino acids are utilized when protein is incorporated into the structure of an animal. However, when amino acids are oxidized for energy or converted in the liver to glucose or fats (Fig. 10.21), the nitrogen must be removed and transferred to other compounds. Some of these compounds are excreted and have an effect both on water balance and energetic efficiency.

A variety of chemicals are involved in nitrogen excretion including allantoin, allantoic acid, amino acids, ammonia, creatine, creatinine, guanine, hippuric acid, pyrimidines, trimethylamine oxide, urates, urea, and uric acid. Despite this diversity, most animals excrete nitrogen in the form of ammonia, urea, and/or uric acid. Many animals form some of each of these nitrogenous compounds. The major differences between animals in most cases are not what chemicals can be produced but the degree to which a particular chemical is used for excreting nitrogen.

Before examining major hypotheses to explain differences among animals in their relative use of ammonia, urea, and uric acid, it is helpful to examine some consequences of use of these chemicals based on the properties listed in Table 13.5. Ammonia is highly toxic, but it is a small molecule that diffuses rapidly and it has a high solubility in water. Because of its high toxicity, it should not accumulate in appreciable amounts but should instead be dispersed into the environment as rapidly as it is formed. Each molecule of ammonia contains only one nitrogen atom, but there is little or no energy expenditure required for its excretion by diffusion.

Urea molecules contain two atoms of nitrogen, and it is less toxic than ammonia and has a high solubility in water (Table 13.5). However, the urea molecule contains carbon, oxygen, and hydrogen atoms that are combined with nitrogen through a series of metabolic transformations (the ornithine or urea cycle) requiring energy. As a consequence, excretion of urea involves loss of chemicals that would otherwise be available for other purposes. Every gram of nitrogen excreted in urea involves loss of 64.3 kcal (Table 13.5).

Table 13.5 Some characteristics of the major chemicals used for nitrogen excretion

CHEMICAL	NUMBER OF N ATOMS	NUMBER OF C, H, AND O	kcal/g OF N	RELATIVE TOXICITY	SOLUBILITY (g/100 ml)
Ammonia	1	H_3	0	High	80
Urea	2	$C_1O_1H_4$	64.3	Low	119
Uric acid	4	$C_5O_3H_4$	116	Low	0.006

Uric acid contains four atoms of nitrogen in every molecule, and it is relatively nontoxic (Table 13.5). The solubility of uric acid is very low, which leads to water conservation when uric acid precipitates and water moves by osmosis back to blood (see above). As a result, nitrogen can be excreted with very little water loss, while urea, because of its higher solubility, requires more water for excretion. However, note (Table 13.5) that uric acid excretion involves the loss of more energy for every gram of nitrogen that is processed (116 kcal/g of N).

Recall that, as "waste" products, chemicals containing nitrogen are not necessarily excreted. Some marine species use proteins or urea as "disproportionate components" (Table 12.2) for the regulaton of blood osmotic concentration. Also, ruminants "recycle" urea into the digestive tract, where it is converted to ammonia and ultimately back to microbial protein that is assimilated (Fig. 10.15).

Despite these other uses, some nitrogen excretion must occur, and two general sets of hypotheses have been proposed to explain differences between species. As was the case with regulation of other resources (pH, circulatory patterns, body temperatures, internal osmotic concentrations), one set of hypotheses deals with past history while the other seeks an explanation based on relative efficiency of function in different environments.

The Historical Hypotheses

These hypotheses are based on a presumed correlation between general taxonomic history and type of nitrogen excretion. They are usually presented as a "phylogenetic tree" (Fig. 13.12) on which living representatives of ancestors of "higher" taxonomic forms appear to exhibit a different form of nitrogen excretion as compared with living representatives of more taxonomically "primitive" species. There are many examples of this argument, and it has been extensively applied to both vertebrates and invertebrates.

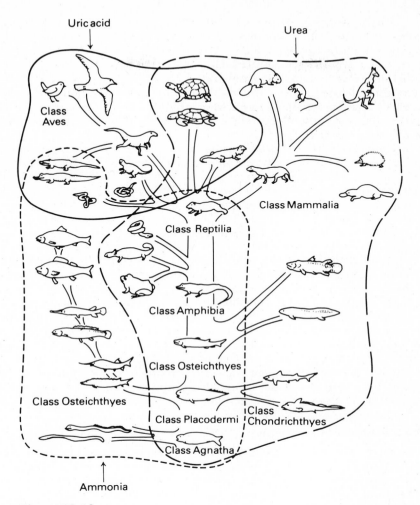

Figure 13.12
Taxonomic hypotheses for patterns of nitrogen excretion are usually illustrated with a phylogenetic "tree," with "primitive" animals or
chemicals on the bottom and more "advanced" animals or chemicals
at the top (after B. Schmidt-Nielsen, 1972).

Although there may seem to be a correlation between taxonomy and
nitrogen excretion at a gross (very general) level, the correlation breaks down
completely when individual species are examined in relation to their environments. The main reason for a correlation between taxonomy and nitrogen
excretion is due to another correlation between taxonomy and environment. At
a general level, most "primitive" species are aquatic and most "advanced"
species are terrestrial. Which is more important, taxonomy or environment?

A few examples will illustrate the problem. Mollusks are taxonomically rather "primitive" invertebrates, yet different species show two forms of nitrogen excretion. Aquatic species such as cephalopods (relatively "advanced" mollusks) excrete ammonia. Terrestrial forms such as some gastropods (somewhat more "primitive" mollusks) excrete uric acid. Within mollusks there is no clear relationship between taxonomy and type of nitrogen excretion, but there is an excellent relationship between environment (aquatic versus terrestrial) and nitrogen excretion (ammonia versus uric acid). Many other similar examples exist for a variety of other invertebrates.

The picture for vertebrates is similar. Although a gross correlation can be seen between taxonomy and nitrogen excretion (Fig. 13.12), the correlation falls apart on close inspection of relationships between different animals and their environments. Most teleost fishes excrete ammonia, but an exception is the taxonomically more "primitive" coelacanth, which produces urea for osmotic purposes. Elasmobranchs (class Chondrichthyes in Fig. 13.12) also produce urea.

Differences between teleosts and elasmobranchs have led to a confusing development of alternative historical hypotheses that illustrate problems associated with these taxonomic arguments in physiology. The confusion comes from the question concerning what should be considered "primitive," the fishes or the nitrogen molecules. Elasmobranchs are taxonomically more "primitive," yet they produce what is considered a more "advanced" excretory product (urea). The opposite is true for most bony fishes. If you wish to consider the *fishes* primitive, urea has to be hypothesized as "abandoned" in teleosts in favor of ammonia for reasons other than taxonomy. If you wish to consider *ammonia* primitive, use of urea in elasmobranchs has to be for reasons other than taxonomy. The confusion here is due to a mixing of historical hypotheses with functional hypotheses. It can be avoided simply by considering the economics involved in nitrogen excretion in different environments.

The Economic Hypothesis

This hypothesis simply states that animals have evolved to produce an excretory product most effective for a given environment. The hypothesis is completely distinct from taxonomic considerations. All the information needed to understand the economic hypothesis is in Table 13.5.

Ammonia involves no loss of energy for excretion but is usually limited to use in aquatic environments, where it diffuses rapidly from an animal and does not accumulate to toxic levels. Uric acid is most effective with respect to saving water but least effective with respect to energy (Table 13.5). Its use is predicted for animals in environments where supply-demand problems of osmoregulation are severe and outweigh the energetic disadvantages of producing uric acid. Urea represents a compromise between environmental extremes. It provides a way to excrete nitrogen in a terrestrial environment that is less expensive than uric acid in terms of energy but more costly in terms of water loss. Its use is predicted for animals with relatively less severe supply-demand problems for water or for animals in which urea serves another function besides excretion.

Note that the economic hypothesis distinguishes between excretion of urea and uric acid in terrestrial animals, based on the *degree* of severity in supply-demand problems of osmoregulation. If supplies of water are sufficient for regulation without having to produce uric acid, urea should be produced because it involves less loss of energy. Just because an animal is "terrestrial" does not necessarily mean it faces severe problems of osmoregulation. The degree of severity depends both on characteristics of terrestrial environments and on characteristics of animals that influence demands for water by different species.

This point of view provides a way to examine differences between two major groups of terrestrial animals: birds and mammals. Both are "terrestrial," yet birds excrete uric acid and mammals excrete urea. This situation is not a paradox. If mammals have evolved mechanisms to minimize water losses relative to supplies so that regulation can be achieved with urea excretion, urea should be excreted because less energy is lost. Mammals have evolved controls for water loss in kidney function that reflect the variations in severity of supply-demand problems of osmoregulation in different terrestrial environments, and the mammalian kidney is one of the more completely studied effector organs involved in osmoregulation.

CONTROLS FOR EXCRETORY LOSSES IN TERRESTRIAL VERTEBRATES

The control mechanisms for excretory losses of water and solutes depends on whether urea or uric acid is produced. We have discussed some controls involving uric acid excretion in terrestrial arthropods. Two types of mechanisms can be illustrated in vertebrates: those for mammals (urea) and those for birds and reptiles (uric acid). There is considerable experimental information for both mechanisms.

Kidney Function in Mammals

The major effector for excretion in mammals is the kidney. As was the case for aquatic animals, the kidney is composed of many structurally similar nephron units (Fig. 13.13) having a particular orientation within the kidney. Blood plasma is filtered in a glomerulus, and the filtered fluid passes from a capsule through a tube where it changes in composition before it leaves the kidney and passes to a bladder (Fig. 13.13).

The composition of the fluid in a nephron tube changes in two ways: some solutes are removed from the filtrate and returned to the body, and others are secreted and added to the filtered fluid. Both mechanisms require energy, and some solutes move in or out as a consequence of energy expended to move other solutes. Movement of a solute that is actively "pumped" is called active transport. Movement of other solutes as a consequence of active transport is called passive transport. In addition, water moves in or out of nephron tubes, depending on concentration differences and changes in the permeability of membranes to water.

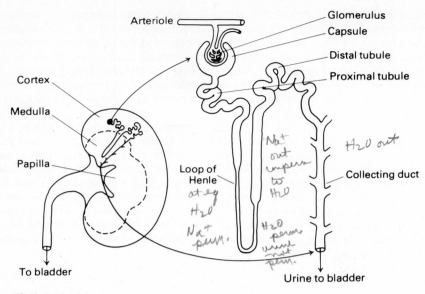

Figure 13.13
Major features of the anatomy of the mammalian kidney and a nephron unit.

Glomerular Filtration

Fluid moves from blood in a glomerulus to a nephron tubule when the hydrostatic pressure of blood exceeds the sum of the hydrostatic pressure of fluid in the nephron tubule *and* the colloid osmotic pressure of the blood plasma. The latter two tend to draw filtered fluid toward blood; thus blood pressure in the glomerulus must exceed these opposing pressures to produce filtration.

The rate of glomerular filtration can be measured by a procedure called *renal clearance*. Inulin is a molecule that is filtered from blood in the glomerulus and is neither secreted nor resorbed from the nephron tubule. Thus the amount of inulin injected into blood that appears in urine over a period of time (volume of urine/time × urine inulin concentration) must equal the rate of glomerular filtration times the concentration of inulin in plasma:

$$\frac{(\text{Urine volume})}{\text{Time}} (\text{Urine [Inulin]}) = \frac{(\text{Filtered volume})}{\text{Time}} (\text{Plasma [Inulin]}),$$

or

$$\text{Glomerular filtration rate} = \frac{(\text{Urine production rate}) (\text{Urine [Inulin]})}{(\text{Plasma [Inulin]})}.$$

Measurements taken using this procedure show that relatively large volumes of fluid are filtered from blood in mammalian kidneys. The glomerular filtration rate for humans is about 125 ml/min, and the kidneys receive a relatively large proportion of total cardiac output. The high rates of filtration are a consequence of low-vascular-resistance blood supplies to the glomeruli. Blood pressure on the arteriole side is relatively high, but pressure on the venous side is relatively low.

The final net composition of urine is a function of the rate of passage of filtered fluid through nephrons plus the rates at which chemicals are added to or removed from the fluid. Most controls over urine production involve rates of addition or removal of chemicals rather than changes in rates of filtration. However, some control of urine production is produced from factors influencing filtration rates.

A change in blood pressure at the kidney level will alter the rate of glomerular filtration. There are local control mechanisms within the kidney that detect changes associated with blood pressure and that produce adjustments; these mechanisms involve hormones that influence arteriole radius and sodium resorption (see below). Changes in plasma colloid osmotic pressure and nephron tubular hydrostatic pressure will also influence rate of filtration. Increased plasma colloid osmotic pressure is associated with decreased blood volume; thus filtration usually does not result in excessive decreases in blood volume. Increased nephron tubular hydrostatic pressure also occurs at high filtration rates in relation to blood volume and provides a local mechanism for control of filtration rates.

Tubular Resorption

An important reason for a high rate of filtration is to excrete chemicals that can only be added to urine by filtration. However, a number of chemicals are filtered that should not be lost. One example is sodium. An adult human filters about 1000 mmol of sodium in an hour, normally resorbs about 992 mmol/h, and loses only 8 mmol/h in excreted urine.

Most sodium resorption occurs within the proximal tubules, loops of Henle, and distal tubules (Fig. 13.13). Fine adjustments in the amount of sodium removed from or added to urine occur in the final passage of urine through the collecting ducts (see below).

Under normal circumstances about 75% of salt and water resorption occurs in the proximal nephron tubule. Sodium is actively transported into extracellular channels between cells in a manner depicted in Fig. 13.6. Net movement of fluid is isosmotic across the tubule epithelium because there is no measurable difference in concentration between tubular fluid and interstitial fluid bathing the upper part of the nephron (see Fig. 13.16). However, a very considerable amount of energy is expended in the proximal tubule on sodium and water movement. Resorption of a large isotonic volume requires a large number of membrane Na^+-K^+ pumps along the sides and toward the base of proximal tubule cells. The cells are extremely elongated perpendicular to fluid movement and have "brush borders," i.e., extensive membrane folds into the tubular space, along with "tight junctions" between adjacent cells (Fig. 13.6).

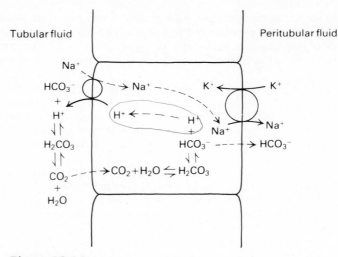

Figure 13.14
Schematic diagram of the mechanism for bicarbonate resorption. Addition of H^+ to tubular fluid drives the reaction toward
CO_2, which diffuses into cells and combines with water, yielding HCO_3^- that in turn diffuses into peritubular fluid (based
on Pitts, 1968).

Bicarbonate Resorption, Hydrogen Excretion, and Acid-base Balance

The proximal tubules also play an important role in acid-base balance. Without
a mechanism to resorb bicarbonate from tubular fluid, its concentration in
blood would decrease and shift the reaction of CO_2 with water toward carbonic
acid. This would drastically alter the blood acid-base balance and influence
respiratory ventilation, CO_2 production, and enzyme net negative charge.

The mechanism proposed for bicarbonate resorption is shown in Fig.
13.14. Cells of the proximal tubule excrete hydrogen ions (H^+) that are exchanged with sodium ions diffusing into the cells. The hydrogen ions combine
with bicarbonate ions in tubular fluid, driving the reaction toward $CO_2 + H_2O$.
The carbon dioxide in tubular fluid diffuses down a concentration gradient into
the cells where it combines with water to form carbonic acid. The carbonic acid
dissociates under the influence of carbonic anhydrase, and cellular bicarbonate
then diffuses from the cell to the peritubular fluid while cellular hydrogen ions
are available for excretion into tubular fluid (Fig. 13.14).

A number of factors influence this process. If P_{CO_2} in blood increases,
tubular P_{CO_2} will increase, resulting in more CO_2 diffusion into tubular cells, in
turn resulting in increased bicarbonate resorption. Other factors that influence
intracellular or tubular hydrogen ion concentrations will also influence bicarbonate resorption by means of the equilibrium relations shown in Fig. 13.14.

If blood and the tubular filtrate have a high acidity, maintenance of urine
acidity (excretion of H^+) will also contribute to blood acid-base balance. The
acidity of urine is determined in part by ammonium excretion. The tubular cells

produce ammonia (NH_3) from glutamine oxidation when tubular fluid pH decreases. The ammonia readily diffuses from cells to urine. Ammonia is uncharged, and it readily penetrates cell membranes. When ammonia combines with hydrogen ions to form ammonium ions (NH_4^+), it becomes "trapped" in tubular fluid. The ammonium ion is charged and does not readily penetrate membranes. Thus from a mechanism by which ammonia production depends on fluid pH, hydrogen ions are excreted in excess, which, together with bicarbonate resorption, contributes substantially to the control of blood acid-base balance.

Transport-limited Resorptions

A number of chemicals in tubular fluid are essentially completely resorbed as long as their concentration in tubular fluid does not exceed a certain value. They include glucose, phosphate, sulfate, malate, lactate, vitamin C, and certain amino acids. Glucose resorption is a particularly well-studied example.

Figure 13.15 shows the relationship between plasma glucose concentration and urine glucose concentration. No glucose is excreted (solid line) until plasma glucose concentration exceeds about 250 to 300 mg/100 ml. Up to this concentration, essentially all filtered glucose molecules (dashed line) are resorbed. When glucose is excreted, the amount parallels what is filtered with a constant difference of about 250 to 300 mg/100 ml. This is interpreted as *transport-limited* resorption. The mechanism responsible for glucose transport out of urine is saturated at a particular maximum limit.

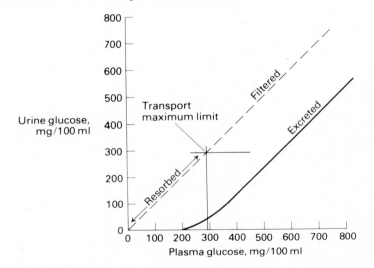

Figure 13.15
Relationship between urine glucose concentration and plasma glucose concentration. No glucose is excreted until plasma glucose reaches about 250 to 300 mg/100 ml, when the resorption transport mechanism is saturated at a maximum limit (after Pitts, 1968).

Under normal circumstances the maximum limit of transport for glucose is sufficiently high that essentially no glucose is excreted. Thus glucose in urine (glucosurea) is a symptom of illness (deficiency in insulin production or kidney function). A similar situation exists for certain amino acids resorbed from urine. However, the transport maxima for some substances, such as phosphate and sulfate, are at relatively low concentrations of filtered chemicals. This situation produces a more active role for the kidney nephrons in the regulation of the blood concentration of these chemicals at specific values. Any excesses filtered are normally excreted, contributing to maintenance of the relative constancy of their concentrations in blood.

The Concentrating Mechanism of the Kidney

By the time tubular fluid leaves the proximal tubule, its volume has been reduced considerably and certain chemicals have been resorbed or excreted, but the osmotic concentration of the urine is still essentially the same as blood. Some additional resorption and excretion occurs in the distal convoluted tubules and in the collecting ducts (see below), but concentration of urine primarily involves movement of chloride, sodium, and water.

Urine flows through the loop of Henle after leaving the proximal tube (Fig. 13.13; Fig. 13.16). A major function of the loop of Henle is to establish differences in osmotic concentration of sodium chloride in the fluid surrounding the collecting duct from the upper cortex down to the inner medulla of the kidney (Fig. 13.16). Tubular fluid that enters the first part of the loop (the descending limb) has an osmotic concentration similar to blood, and the membrane of the descending limb is permeable to water but not sodium chloride.

The membrane on the ascending limb actively transports chloride from tubular fluid to interstitial fluid bathing the nephron. Sodium passively moves from tubular fluid to interstitial fluid, but the membrane on the ascending limb is not permeable to water. As a result, sodium chloride accumulates in fluid surrounding the nephron in increasing concentration from the cortex down through the medulla. The descending limb serves to provide sodium chloride to be transported out the ascending limb. Note that water is withdrawn from tubular fluid in the descending limb.

The loop of Henle is called a *countercurrent multiplier*. Fluid flow in the ascending limb is countercurrent to flow in the descending limb and in the collecting ducts. In addition, energy is expended to produce concentration gradients along the loop and collecting ducts. Each small section of membrane on the ascending limb transports a certain amount of sodium chloride against a particular difference in concentration. The difference in concentration (and thus energy expenditure for transport) remains the same at each point along the ascending limb. The total external gradient of concentration from cortex to medulla depends on the total number of segments of ascending limb membrane where chloride is transported. This addition of small sections of membrane along the entire length of the ascending limb produces a multiplicative effect because multiplication is a series of successive additions of the same amount.

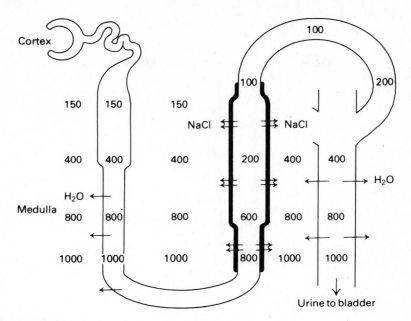

Figure 13.16
Mechanism for urine concentration from countercurrent multiplication
in the loop of Henle, illustrated with fluid Na^+ concentrations. Double
arrows on the ascending limb indicate transport of NaCl out of tubular
fluid and the establishment of a concentration gradient in interstitial
fluid from the cortex down through the medulla. Single arrows indicate
osmotic movements of water. The collecting ducts are depicted as
permeable to water.

When tubular fluid reaches the top of the ascending limb it may be slightly
less concentrated than blood plasma (Fig. 13.16), resulting in some water movement by osmosis to blood and further reduction of urine volume. As fluid passes
back down through the collecting ducts, water will move by osmosis out of
urine if the collecting-duct membranes are permeable to water. For the case in
Fig. 13.16, collecting-duct membranes are shown to be permeable to water.
Because the concentration of fluid surrounding collecting ducts increases along
their length, more and more water can be withdrawn by osmosis as fluid moves
down the collecting ducts.

The water that moves from the descending limb and the collecting ducts
must be removed from the fluid surrounding the nephron tubules; otherwise the
gradient of sodium chloride concentration would be diluted. Water removal occurs with blood flow in capillary loops called *vasa recta* (Fig. 13.17). Venous
blood leaving a glomerulus gains sodium chloride by diffusion and loses water by
osmosis on the descending limb of the vasa recta, but by looping back, the blood
loses sodium by diffusion and gains water by osmosis on the ascending limb.

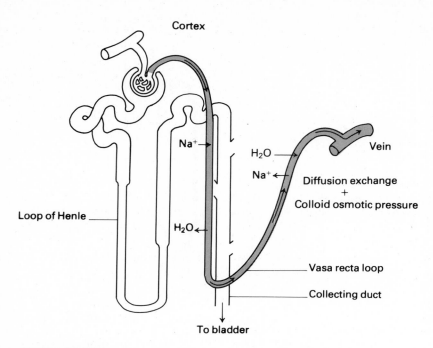

Figure 13.17
When blood leaves a glomerulus it passes through a capillary loop (vasa recta, shown as the shaded loop). Sodium is gained and water is lost on the
descending limb; water is gained and sodium is lost on the ascending limb.
There is a net gain of water by blood because of the blood colloid osmotic
pressure, particularly in the cortex where the sodium concentration of the
interstitial fluid is least (see Fig. 13.16).

Countercurrent flow through the vasa recta produces a net gain of water in
blood because the blood contains proteins that do not exchange by diffusion
and that do produce a colloid osmotic pressure (see Chapter 4). This fact is most
important in the region of the proximal and distal tubules, where the osmotic
concentration of tubular fluid is least.

Urea diffuses across all of the membranes in the nephron tubules and in
the vasa recta. Urea is transported passively and becomes more concentrated in
tubular fluid in collecting ducts as a consequence of the movement of water out
of the ducts and into blood. Although it may appear that no energy is expended
to excrete urea in the kidney per se, it is a fact that without the active transport
of chloride in the loop of Henle no concentration gradient could form for resorbing water and concentrating urine in the collecting ducts.

Control over Urine Concentration in Collecting Ducts

The final amount of water lost in urine by mammals changes, depending on the
permeability of the collecting ducts to water. With lower membrane permeability, urine concentration decreases; with higher membrane permeability,

urine concentration increases up to a maximum limit. A negative-feedback control governs osmoregulatory gains or losses of water by means of effects of a hormone on collecting-duct membranes.

If water loss from kidney excretion is not at a minimum and blood osmotic concentration increases (from loss of extracellular water or gains of extracellular osmotic solutes), osmoreceptors in the hypothalamus detect the change and signal a proportional release of the pituitary hormone *antidiuretic hormone* (ADH). Diuresis means water loss. Antidiuretic hormone increases the permeability of collecting duct membranes to water. This change increases osmotic resorption of water (up to a limit), which decreases blood osmotic concentration, removing the stimulus at the osmoreceptors.

If blood osmotic concentration decreases (from gains of extracellular water or losses of extracellular osmotic solutes), detection of the change results in reduction of ADH release from the pituitary, reducing the water permeability of collecting-duct membranes and producing increased loss of water in urine.

Adaptations for Maximum Urine Concentration in Mammals

All mammals can lose excess water by reducing ADH release and producing a dilute urine. However, not all mammals can concentrate urine to the same degree. Differences between species in maximum urine-concentrating ability reflect the severity of supply-demand problems for osmoregulation mainly because of differences in the availability of water in terrestrial environments.

Adaptations to minimize urinary water loss involve different degrees of countercurrent multiplication within kidneys. Mammals with more severe restrictions of water availability have evolved kidneys with longer loops of Henle that are capable of producing larger total concentration gradients along longer collecting ducts. Differences in the lengths of loops of Henle and collecting ducts are apparent at the level of gross kidney structure (Fig. 13.18). The papilla of the kidney is composed primarily of loops of Henle, collecting ducts, and vasa recta. Mammalian species from environments with low water supplies

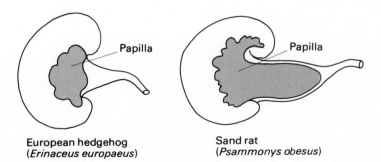

European hedgehog
(*Erinaceus europaeus*)

Sand rat
(*Psammonys obesus*)

Figure 13.18
Comparison of differences in gross kidney structure between a desert-adapted species (right) and a species from a more moist terrestrial environment (left) (after B. Schmidt-Nielsen, 1964).

in relation to losses have longer papillae. For example, the sand rat (*Psammonys obesus*) lives in the deserts of Northern Africa and consumes plants with high salt content. Both of these environmental situations place a premium on efficiency of water conservation through urine-concentrating mechanisms.

Control over Sodium Resorption in Collecting Ducts

Although a relatively large amount of sodium and water are resorbed in proximal tubules, the amount resorbed there is under little control. The extent of final sodium resorption depends on resorption in proximal tubules *and* the degree of sodium transport that occurs across the distal tubules and collecting ducts. The latter is controlled with the hormone *aldosterone*, which is produced in the adrenal cortex.

Sodium resorption increases following sodium loss from blood and extracellular fluid (Fig. 13.19). Decreased extracellular sodium concentration results in reduced blood volume because of increased osmotic movement of

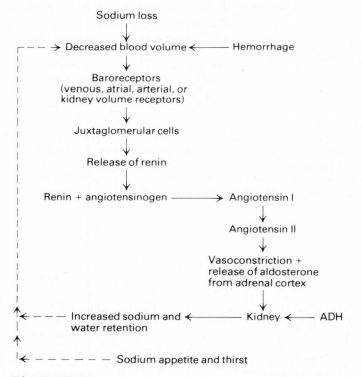

Figure 13.19
General scheme of the factors involved in negative-feedback (dashed lines) control of blood volume (extracellular sodium) by means of reduced losses of sodium, increased retention of water, and/or sodium appetite coupled with thirst.

water into cells. Reduced blood volume will decrease blood pressure. The reduced blood volume is detected by baroreceptors (isotonic hemorrhage produces a similar change; Fig. 13.19), and cells near glomeruli (juxtaglomerular cells) are stimulated to release the hormone *renin*. Renin combines with angiotensinogen in blood to produce angiotensin I, which is rapidly converted to angiotensin II (Fig. 13.19).

Angiotensin II produces two important results. It is a potent vasoconstrictor influencing blood pressure, and it stimulates release of the hormone aldosterone from the adrenal cortex. Aldosterone stimulates increased active transport of sodium in the distal tubules and collecting ducts, which increases sodium resorption. If, as a consequence of increased sodium retention, the blood osmotic concentration increases, release of ADH will lead to increased water retention (Fig. 13.19). Both of these processes can reverse a decrease in blood volume (restoring both solutes and water to the vascular system), and these kidney controls can be augmented by thirst (drinking) and sodium appetite (Fig. 13.19; also see below).

Excretory Function in Birds and Reptiles

Birds and reptiles excrete uric acid into a cloaca where it precipitates. This action results in osmotic resorption of water, which can be augmented by isotonic active transport of sodium, as in arthropods. The mechanisms of tubular filtration and resorption in nephron units are similar to those in mammals (including isotonic resorption in proximal tubules). However, less control is exerted over final salt excretion within some bird and reptile kidneys. The loops of Henle and the collecting ducts are poorly developed, and major quantities of salts that may be ingested are excreted by a separate set of *salt glands*.

The salt glands are located in the orbits of the eyes (Fig. 13.20). Ducts from each gland drain into the nasal passages. When blood osmotic concentration increases, the glands produce a secretion of almost pure sodium chloride. The fluid drains from the opening of the nasal passages in some species; in others it remains within the nasal passages (see below).

The salt gland is particularly well developed in marine birds (it is absent in many terrestrial birds in which nephron controls suffice for salt excretion). The concentration of the fluid excreted by the petrel and albatross is close to 1000 mmol Na$^+$, but the duck excretes a fluid with a concentration of only 500 mmol Na$^+$. Moreover, unlike the kidney of mammals, secretion by the salt gland always occurs at about the same concentration. The amount of salt excreted depends on whether and for how long the salt gland secretes fluid (an example of on-off control; see Introduction).

Control of salt gland secretion appears to involve osmoreceptors that respond to an increased blood osmotic concentration. This increase can result from increased loss of extracellular water or from increased concentration of extracellular osmotic solutes. The osmoreceptors cause the nervous activation of the salt glands, and secretion is maintained until the internal osmotic stimulus has been removed. A combination of salt removal through the salt glands with uric acid excretion and water resorption in the kidneys and cloaca produces osmoregulation, involving variation in salt or water net gains.

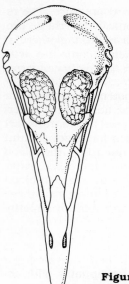

Figure 13.20
Location of the salt glands of a seagull (from Schmidt-Nielsen, 1979).

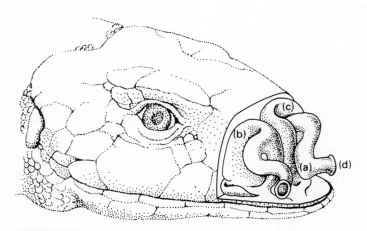

Figure 13.21
Fluid from the nasal salt glands (b, c) of the desert iguana
drains into a depression in the nasal passages (a) just inside
the openings (d) and contributes to humidification of inspired
air (after Murrish and Schmidt-Nielsen, 1970).

Some species of desert reptiles, such as the desert iguana, are able to excrete essentially pure salt. The salt gland secretes a fluid of sodium chloride that drains into the nasal passages (Fig. 13.21). The fluid accumulates in a pocket just inside the opening of the nasal passages. Water from the secreted salt solution evaporates during inspiration, which reduces demands for water required to humidify inspired air, and after the water evaporates the animals sneeze to excrete essentially pure salt. The combination of this mechanism and the precipitation of uric acid with water resorption in the cloaca minimizes water losses in environments where water supplies are limited.

ADDITIONAL CONTROLS FOR WATER AND SOLUTE GAINS: THIRST AND SODIUM APPETITE

Not all terrestrial environments are limited in the availability of water and solutes. With a sufficient environmental supply of water, balance can be achieved from the environment with less cost by drinking than by kidney or intestinal resorption. An example is the laboratory rat (or its more "natural" counterpart, the Norway rat), which has a relatively high requirement for water from drinking (see Table 13.2) that it normally obtains with relatively little cost from its relatively "moist" terrestrial environment. On the other hand, the kangaroo rat expends more energy to minimize water losses, but these costs should be less than costs required to search for and obtain water in a desert environment.

In Chapter 10 we discussed hunger for energy, and in Chapter 11 we discussed hunger for nutrients. "Hunger" for water is thirst. It is manifested through the appetitive behavior of an animal that will preferentially seek out and consume water to replenish an internal loss.

Consumption of water provides a precise proportional control mechanism to balance water losses. Drinking in most animals is quantitatively related to requirements for water needed to maintain optimal cellular hydration. When extracellular osmotic concentration increases or cellular water volume decreases, the amount consumed is usually precisely enough to restore osmotic concentrations and thus cellular water volumes (Fig. 13.22). Osmoreceptors in the hypothalamus are thought to be involved in this control; thus the degree of thirst is proportionally related to demands for osmoregulation.

If both water and sodium are lost, thirst and sodium appetite can both contribute to maintenance of blood volume (see Fig. 13.19). Both are necessary if blood volume is reduced isotonically (such as in hemorrhage) or if extracellular sodium losses result in movement of water from blood to cells. Drinking pure water would dilute blood concentration further and result in movement of more water into cells. However, ingestion of sodium chloride coupled with an appropriate amount of water would restore blood volume isotonically. This interaction occurs with remarkable precision in laboratory rats (Fig. 13.23). With reduced blood volume, only a small amount of pure water is consumed when no salt is available. Much larger amounts of a salt solution are consumed if it is the same osmotic concentration as that of blood (Fig. 13.23); the amounts consumed are sufficient to restore blood volume.

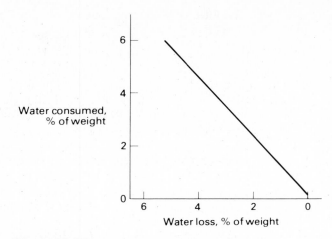

Figure 13.22
When dogs have been deprived of water to produce net losses, the amount consumed within five minutes closely matches losses (from Adolph, 1939).

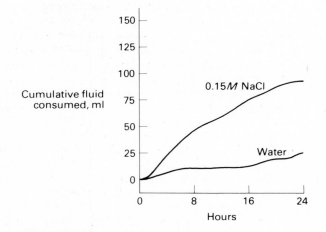

Figure 13.23
When laboratory rats undergo vascular dehydration and are given either pure water or water with sodium chloride (isotonic to blood), they consume only small amounts of water but amounts of isotonic saline solution sufficient to restore blood volume (after Stricker, 1973).

SUMMARY

Water availability in terrestrial environments is variable. Losses from evaporation and excretion must be balanced by gains from water in food, from oxidation and/or from drinking if supplies permit. Solute regulation involves balancing net excretory losses with intakes from feeding. Both the extent of osmoregulatory problems and their solutions depend on the extent to which water and solute losses must be minimized such that available environmental supplies are sufficient to meet demands.

Evaporation is a major avenue of water loss for terrestrial animals. Rate of evaporation depends on the magnitude of the saturation deficit for air, i.e., the degree to which air is not saturated with water vapor. Air can contain exponentially more water at higher temperatures at saturation, so rates of evaporation are potentially higher at high temperatures when air is not saturated. Thus animals from hot, dry environments (deserts) would be expected to show adaptations for minimizing rates of water loss.

The desert-adapted kangaroo rat loses less water through the skin than does the laboratory rat, although both species save water as a consequence of their small sizes by means of effective respiratory countercurrent cooling of expired air. The kangaroo rat also excretes less water in urine and feces than does the laboratory rat as a consequence of mechanisms dependent on energy expenditure that transport sodium against concentration gradients within epithelial tissues. The kangaroo rat can balance relatively low water losses with intakes of water in food and from oxidation. The burrow environment is important for this balance because moderate temperatures and high water vapor saturation reduce evaporative losses and increase water gains by food stored in the burrow.

The laboratory rat must drink water to balance losses. This species has evolved in terrestrial environments where water is available, and the costs of osmoregulation by means of drinking are less than the costs of resorbing water to a larger degree during excretion.

Water evaporates from surfaces, and surface area increases relative to volume as size decreases. Small arthropods reduce respiratory evaporation by timing the opening of their tracheal spiracles over long intervals. In addition, evaporation through the exoskeleton is minimal because of wax secretion on the outer surfaces. Excretory losses are minimized because of the production of uric acid for nitrogen excretion. The uric acid precipitates in the rectum, and water is resorbed by osmosis and from isotonic sodium transport. Some small arthropods can gain water by "drinking" water vapor in air when losses of water exceed gains from other sources, but the mechanism probably involves considerable energy costs.

The type of chemical used to excrete the nitrogen from ingested protein is related to both environmental supply-demand problems of osmoregulation and the energy expenditures associated with nitrogen excretion. Although there is a gross correlation between taxonomy and type of nitrogen excretion, historical hypotheses fail to account for the diversity of adaptations of different animals to their environments. An economic interpretation explains the use of different nitrogenous chemicals, based on water conservation and energy expenditures.

Ammonia is restricted to use in aquatic environments, where it can diffuse rapidly and not accumulate to toxic levels, and its use involves no loss of energy. Urea and uric acid are much less toxic but differ in solubility and energy loss from processing a gram of nitrogen. Uric acid should be produced when osmoregulatory problems due to water losses are severe and outweigh the use of a chemical requiring more loss of energy. Urea excretion involves less energy loss and should occur when the extra water required for excretion is available.

Control mechanisms governing excretory losses in terrestrial animals depend on whether urea or uric acid is the primary excretory product. In mammals, blood is filtered in a glomerulus when blood pressure exceeds nephron hydrostatic pressure plus blood colloid osmotic pressure. Rate of glomerular filtration is high because filtration is the primary mechanism for excreting most chemicals. However, many chemicals are filtered that should not be excreted. These are resorbed from active or passive transport.

Most sodium and water resorption occurs within proximal tubules, the loop of Henle, and distal tubules. Sodium is actively transported into channels between cells. Net fluid movement is isosmotic between tubular fluid and peritubular fluid, but considerable energy is expended to produce localized concentration gradients across lateral membranes of proximal tubule cells.

Bicarbonate is resorbed in proximal tubules by means of a mechanism dependent on hydrogen excretion. Increased tubular acidity increases tubular P_{CO_2}, resulting in diffusion of CO_2 into cells, where it combines with water to form carbonic acid. The acid dissociates and the bicarbonate diffuses into peritubular fluid. Urine acidity is maintained in part from production of ammonia by tubular cells. The ammonia readily diffuses into tubular fluid and associates with hydrogen ions, forming less mobile ammonium ions.

A number of chemicals are resorbed from tubular fluid by transport-concentration-limited mechanisms. Below a certain concentration essentially all of the chemicals (glucose, phosphate, sulfate, malate, lactate, vitamin C, amino acids) are resorbed, but above a "saturation" concentration they are excreted in proportion to their concentration in blood. Depending on the concentration for saturation, this selective excretion provides a mechanism for control of blood concentrations of the chemicals.

Urine concentration in the mammalian kidney occurs from countercurrent multiplication of the peritubular sodium chloride concentration along the loop of Henle. Chloride is actively transported out of the ascending limb of the loop, with each small section of the limb pumping the same amount of chloride. Thus a gradient of sodium chloride concentration is established along the collecting ducts such that water moves by osmosis from the ducts if the collecting-duct membranes are permeable to water.

The amount of water lost in urine of mammals varies because of changes in the permeability of the collecting ducts to water. Antidiuretic hormone (ADH) is released from the pituitary when blood osmotic concentration increases, increasing the permeability of the collecting ducts to water. Less release of ADH results in lower collecting-duct permeability to water and increased urinary water loss.

Maximum urine-concentrating ability is determined by the magnitude of the peritubular concentration gradient established by the loop of Henle. Mammals that have evolved in environments with limited water supplies have evolved longer countercurrent multipliers resulting from longer loops of Henle, resulting in a larger concentration gradient along the lengths of the collecting ducts for the osmotic removal of water from urine.

Final adjustments in sodium resorption also occur in the collecting ducts. When sodium loss from extracellular fluid increases, blood volume decreases, and renin is secreted by juxtaglomerular cells. Renin is transformed into angiotensin I and angiotensin II. The latter produces vasoconstriction and stimulates release of aldosterone from the adrenal cortex. Aldosterone stimulates increased active transport of sodium in collecting ducts and distal tubules.

Birds and reptiles excrete uric acid into a cloaca, where water is absorbed when uric acid precipitates. NaCl is excreted by means of a separate set of salt glands in some birds and reptiles. When blood osmotic concentration increases, the glands secrete sodium chloride solutions of constant concentrations for a period sufficient to reduce blood osmotic concentrations to normal values. The fluid drains into the nasal passages, where it either is lost or humidifies inspired air.

Mechanisms for water and solute resorption can be augmented by thirst and sodium appetite. When environmental supplies permit, drinking and salt ingestion may be more economical than resorption. Water intake by drinking is precisely related to demands for water used for the control of cellular water volume. When both sodium and water are lost (such as during hemorrhage or extracellular sodium loss), ingestion of salt, coupled with an appropriate intake of water, results in restoration of isotonic blood volume.

Water and Temperature Regulation in Hot Environments

14

A hot "environment" is any situation requiring increased heat loss (rather than minimization of heat loss) for body temperature regulation. It is usually associated with high environmental temperatures (above the zone of thermoneutrality for endotherms; see Fig. 7.4, showing effects such as those of intense solar radiation on endotherms), high internal heat production (from activity), or a combination of the two. We have delayed discussion of these situations because temperature regulation in hot environments involves important interactions between water resources and energy exchange. It is not possible to understand adaptations for temperature regulation in hot terrestrial environments without appreciating the impact that they have on osmoregulation.

By now you should appreciate the extent to which animal functions depend on *interactions* between important supply-demand relationships of resources. For example, in our discussion of the regulation of respiration we found that gas exchange and acid-base regulation of blood are closely related; this relationship is also reflected in the effects of temperature on acid-base regulation and metabolism and in the functions of the kidney. Osmoregulation, energy exchange, and temperature regulation are also intimately related in animals living in hot terrestrial environments, where survival depends on changes in priorities for the use of resources as their supplies change relative to demands.

Recall that the total heat exchange of an animal is described by the following relationship (see Chapter 7).

For constant body temperature, $\quad$ Heat gain $= H_m =$ Heat loss $= \pm\, H_c \pm\, H_r \pm\, H_w,$

where

H_c = heat exchange by conduction and convection,

H_r = heat exchange by radiation, and

H_w = heat exchange by evaporation or condensation of water.

If heat gain exceeds heat loss, body temperature increases and heat is *stored*. About 0.9 calorie is stored when the temperature of one gram of an animal increases by 1.0 °C.

Endotherms in "cold" environments exchange heat primarily through losses by conduction, convection, and radiation, and heat gain from metabolism must balance losses as a requirement for the regulation of body temperature (see Chapter 7). Endotherms and ectotherms (with no choice for heat exchange) in hot terrestrial environments must lose heat for the purpose of temperature regulation, primarily by increasing evaporation of water. For endotherms (where H_m is high), this relationship can be appreciated by examining the simple equation summarizing heat gains and losses.

For constant body temperature, $\quad$ Heat gain $= H_m =$ Heat loss $= C(T_B - T_A) \pm H_w,$

where

C = thermal conductance,

T_B = body temperature,

T_A = "ambient" or environmental temperature, and

H_w = heat exchange from water changes of state.

In this chapter we are concerned with the consequences of a high "T_A" in relation to T_B and of high heat gain from H_m in relation to total heat loss. If T_A increases such that it equals T_B, $(T_B - T_A)$ will be zero, and the only way to lose heat from metabolism is from evaporation of water. If $(T_B - T_A)$ is negative, additional heat is gained from the environment that, together with heat gains from metabolism, must be lost by evaporation if body temperature is to remain constant.

If there is some increase in body temperature when T_A is high or when H_m is high, $(T_B - T_A)$ could become positive. Thus heat storage could contribute to some increased heat loss. However, there are limits to the extent to which body temperature can increase for many animals. The temperatures of many animals are close to upper limits for effective function, so there is less of a range for an increase available for regulation than there is a range for a decrease in body temperature (torpor; see Chapter 8).

Heat storage also depends on body size. Small animals are limited by their size in the extent to which they can store heat and in the extent to which they can lose heat by evaporation of body water supplies (see below). For them, behavioral selection of cool microhabitats represents a primary mechanism for temperature regulation in hot environments (see Chapter 6). Most studies of the interactions between temperature regulation and water regulation involve animals in which water supplies can be sufficient as compared with the demands for evaporative cooling or where heat storage can contribute appreciably to temperature regulation.

Adaptations for increased heat loss when T_A or H_m is high involve evaporation of water and/or heat storage. First we will consider several mechanisms that have evolved that increase the rate of water evaporation from animals. Then we will examine the interactions between water and temperature regulation by studying responses of some animals to problems of increased heat loss when water is in short supply. Water for temperature regulation may be in short supply either from limited availability in the external environment or from limitations in the rate at which it can be used from internal supplies for evaporation when demands for heat loss are high. Under these conditions, heat storage becomes more important in dealing with problems of heat loss. Finally we will examine some special adaptations for heat loss and heat storage that have evolved in interesting relationships between predators and their prey. We will also examine some additional evidence for the possibility of temperature regulation in dinosaurs.

MECHANISMS TO INCREASE WATER EVAPORATION

Energy is required to change the state of water from a liquid to a vapor. This energy is called the *latent heat of vaporization* ($-\Delta \bar{H}_{vap}$). It is not a constant but varies with the temperature of water (from 595 cal/ml at 0°C to 539 cal/ml at 100°C). For many endotherms and some ectotherms a value of 580 cal/ml is used because it is the latent heat of vaporization at a temperature (35°C) similar to that of many body surfaces where water evaporates.

Insensible Water and Heat Loss

All terrestrial animals lose some water by evaporation when air is not saturated with water vapor, resulting in heat loss. However, this water loss is usually small, and adaptations in skin permeability and respiratory countercurrent heat exchange minimize rates of water loss when heat loss by evaporation is not important for survival. The minimum rate of water loss from an animal is called *insensible* water loss, and it varies among species and environments, depending on factors discussed in the last chapter.

The rate of insensible water loss increases when water vapor saturation deficits increase. Moreover, rate of insensible water loss through the skin depends on the pattern of blood flow to the skin. In hot environments or when internal heat production is high, blood vessels near the skin surface dilate and more internal blood vessels constrict (see Fig. 7.12). This shunting of blood flow decreases effective insulation because vascular countercurrent heat exchange is less effective and a greater proportion of warm blood is shunted to the surface. It will also increase the rate of water loss through the skin by means of increases in skin temperature and hydration.

The rate of respiratory insensible evaporation also increases when inspired air temperature increases. Expired air is saturated, and air cannot be cooled to the same extent when inspired air temperature increases. This condition leads to a greater loss of water and heat from evaporation when air temperatures increase (see Fig. 13.3).

Although these increases in rates of insensible water loss will increase evaporative heat loss, additional increases in evaporation are necessary for temperature regulation if environmental temperatures are close to body temperature and/or if internal heat production is high. Terrestrial animals have evolved two basic types of mechanisms to deal with these problems. The first involves ways to increase the rate of evaporation in respiration; the second involves ways to increase the rate of evaporation from the skin or external surfaces.

Increasing Respiratory Evaporation: Panting

More water will be lost by evaporation from the respiratory passages when air is moved into and out of the respiratory tract in excess of the amount required just for demands set by respiratory gas exchange. This increase is accomplished as a result of changing minute volume by means of panting. Panting occurs in

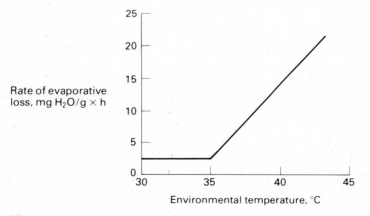

Figure 14.1
The rate of water evaporation increases in Japanese quail (*Coturnix coturnix*) above an environmental temperature of 35 °C as a consequence of panting and gular flutter (based on Weathers and Schoenbaechler, 1976).

several species of mammals including carnivores and some large ungulates. Birds and reptiles also increase the rate of evaporative heat loss by panting. In some species of birds it is accompanied by rapid fluttering of highly vascularized skin tissue in the throat region (gular fluttter). As a consequence, the rate of evaporation increases up to 7-8 times when environmental temperatures increase (Fig. 14.1).

Minute volume will increase as a result of increases in respiratory rates, tidal volumes, or both. During panting, most ventilation increases occur from increases in respiratory rates, with less adjustment in volumes respired. In addition, it is common for panting to occur at particular frequencies of breathing. For any object capable of oscillating (such as the lungs), the energy to produce a given displacement will be minimum when the object oscillates at a *resonant* frequency, i.e., a natural frequency for the object. This characteristic was studied in dogs by measuring changes in pulmonary ventilation at different rates of respiration. For the dogs studied, a maximum ventilation occurred at a respiratory rate close to 6 breaths/second (Fig. 14.2). This rate was also observed when the dogs panted.

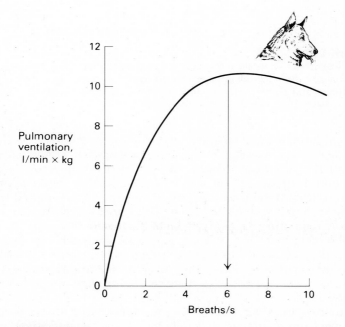

Figure 14.2
As a consequence of moving air at a resonant (natural) frequency of the respiratory system, pulmonary ventilation of dogs is maximum at a respiratory rate of about 6 breaths/ second (from Crawford, 1962).

When panting starts there is usually an abrupt increase in respiration to a high characteristic frequency (a type of on-off control; see Introduction). By respiring at a frequency at which the volume of air moved for evaporation is maximized relative to the energy (heat production) required for air movement, net loss of heat is maximized. As a result, panting can produce disproportionate increases in evaporative heat loss relative to heat production at high environmental temperatures (Fig. 14.3).

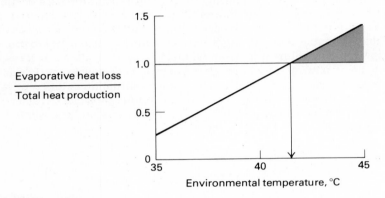

Figure 14.3
Heat loss from evaporation of water exceeds metabolic heat production (shaded area) for the Japanese quail depicted in Fig. 14.1 when environmental temperature exceeds 42 °C (when $T_B \leq T_A$) (after Weathers and Schoenbaechler, 1976).

Although panting is an effective way to increase evaporative heat loss, there are disadvantages associated with increased rates of ventilation. Increasing lung ventilation in relation to tissue CO_2 production will remove larger than normal amounts of CO_2 from blood, which will decrease the acidity of blood, i.e., shift pH upward toward more alkaline values and away from optimal values for effective enzyme function (see Chapter 3). This situation can be minimized to some extent by restricting a major portion of the increased ventilation to deadspace volume, where water is lost but where no gas exchange takes place. Nevertheless, panting may still involve a degree of respiratory alkalosis. It represents a cost to panting animals that must be balanced relative to the costs of *not* panting and, consequently, not precisely controlling temperatures in hot environments.

Under many natural circumstances the relative costs of acid-base regulation and temperature regulation may only have to be faced for short periods if increases in environmental temperatures and/or activity are short-term. Observations of alkalosis during short-term heat stress in some species suggest that temperature regulation priorities take short-term precedence over maintenance of optimal acid-base balance. However, there is little information concerning long-term interactions between temperature regulation and acid-base balance.

Another potential disadvantage associated with panting is the normal effi-ciency of respiratory countercurrent heat exchange. Recall that small dimen-sions within nasal passages produce a situation in which water conservation oc-curs from the cooling of expired air (see Chapter 13). In hot environments or during activity, priorities can change so that increased water loss rather than water conservation is a major problem.

The effectiveness of respiratory heat and water exchange changes during panting in dogs (Fig. 14.4). During panting, air is inspired primarily through the nose, where it is warmed and humidified effectively in the narrow nasal passages. Air is expired primarily through the mouth, where dimensions are larger, meaning that less cooling and condensation occurs and that more heat is

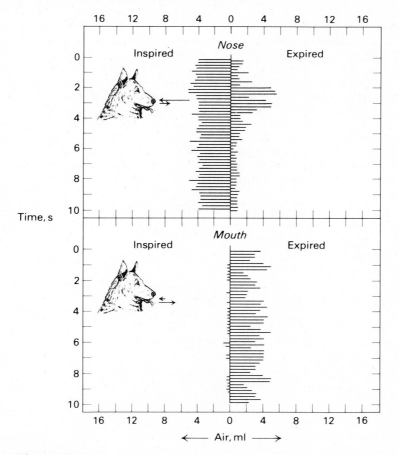

Figure 14.4
During panting in dogs most air is inspired (left) through the nose (upper), but during expiration (right) most air leaves through the mouth (lower) (from Schmidt-Nielsen *et al.*, 1970).

lost to the external environment. As a consequence of separation of channels for inspired and expired air, nasal tissues are cooled but not appreciably rewarmed during expiration. The cool nasal tissues serve to cool blood, particularly blood passing to the brain (see below).

Increasing Evaporation from Skin: Sweating

Glands in the skin of a number of species of mammals secrete fluid, composed of water and solutes, onto the skin. There are two types of sweat glands. Apocrine glands are found on the plantar areas of extremities (surface-contacting regions such as the palms of hands and the bottom of feet). Eccrine glands are distributed over the body surface in large numbers (up to $400/cm^2$) and secrete a cumulatively large amount of fluid for evaporated heat loss.

Water from eccrine sweat evaporates from the surface of the skin, cooling the skin and the blood flowing through it. Thermoregulatory sweating occurs in primates and several species of ungulates. In humans it results in a large increase in the rate of evaporation in hot environments (Fig. 14.5). Measurements of rates of water loss for men working in hot, dry environments indicate that as much as 2000 ml of water are lost by evaporation per hour, and much of this loss comes from the evaporation of sweat secreted by a large number of glands.

Sweating provides a way to cool a relatively large surface without the problems associated with respiratory alkalosis. However, fluid secreted onto skin must evaporate through any insulating layer over the skin. For this reason, sweating appears to have evolved in mammals whose fur does not represent an appreciable barrier to surface evaporation.

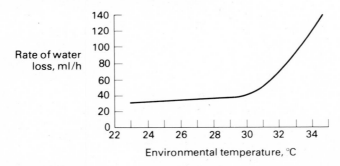

Figure 14.5
For humans, the rate of water evaporation increases at environmental temperatures above 31 °C. Much of the water loss is a result of evaporation of sweat (after Hardy and Soderstrom, 1938).

Sweat contains both water and solutes. For humans, the total concentration of solutes in sweat is about 100 mmol/l, of which 40-50 mmol/l is sodium. Because total sweat concentration is less than blood, the blood will become more concentrated as a consequence of sweating, ultimately resulting in ADH secretion, water resorption in the kidney, and thirst resulting from cellular dehydration (Fig. 14.6).

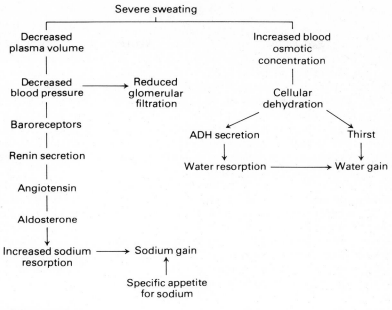

Figure 14.6
The consequences of loss of fluid by sweating involve both water loss
(right) and sodium loss (left). Net losses must be balanced by gains
from drinking and sodium-specific appetite.

Because sweat contains appreciable amounts of sodium, losses of sodium
from blood also will reduce blood volume (see Chapter 13), which will ultimately lead to either aldosterone secretion and kidney sodium resorption or
sodium appetite or both (Fig. 14.6). Therefore, even though sweating involves
some advantages when compared with panting, there are other interactions involving trade-offs for regulation, primarily with respect to osmoregulation and
cardiovascular function.

Apocrine sweat glands do not secrete for the purpose of thermoregulation.
The two types of sweat glands are controlled by different parts of the nervous
system. Thermoregulatory eccrine sweating is governed through the parasympathetic part of the autonomic (involuntary) nervous system. Apocrine sweating is controlled by the sympathetic part of the autonomic nervous system (see
Chapter 17 for details concerning the divisions of the nervous system). The
sympathetic nervous system usually deals with short-term stress (i.e., fear or
flight from immediately dangerous stimuli; see Chapter 17), and aprocrine
sweating occurs under acutely stressful situations. Many of you may be familiar
with the cold, clammy feeling due to evaporation of apocrine sweat from your
palms under stress, such as during a final exam. Experiments with a variety of
mammals suggest that apocrine sweating increases traction for running animals. When animals run on a tilted treadmill they slip backwards at a lower angle of elevation if apocrine sweating is prevented.

Increasing Evaporation from Skin: Saliva Spreading

A number of species of small mammals, such as most rodents, neither sweat nor pant but groom saliva onto their skin when body temperature increases. Grooming of saliva for evaporative cooling also occurs in many marsupials. This action is somewhat less effective than sweating for evaporative cooling because the fur must be soaked with saliva before heat can be lost from the skin. Moreover, the mechanism is only effective for relatively short time periods in small animals. Small species can sustain high rates of evaporative water loss for less time than larger animals because internal supplies of water are less in relation to rates of loss (see Fig. 13.9). However, this mechanism for increasing evaporative heat loss is important for survival when heat stress is relatively short (on an hourly or daily basis).

The localization of the effectors for increasing evaporation in discrete, relatively large salivary glands has permitted experimental studies of some of the details of control mechanisms for temperature regulation involving salivary secretion in laboratory rats. Figure 14.7 shows the extent of their evaporative water loss by means of saliva. The upper line is for laboratory rats with salivary glands; the lower line is for rats that have had their salivary glands removed (note that the y axis is logarithmic). The difference between these (the shaded area in Fig. 14.7) represents evaporation of saliva. Rats without salivary glands

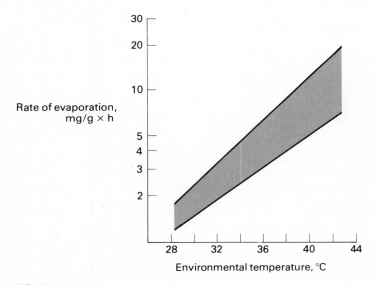

Figure 14.7
The rate of evaporation of water from saliva (shaded area), measured as the difference between rates of evaporation in rats with salivary glands (upper line) and in rats without salivary glands (lower line) (after Hainsworth and Stricker, 1970).

cannot regulate their temperatures at an environmental temperature of 40°C (at which $T_B \leq T_A$), whereas rats with salivary glands can control their temperatures by means of increased evaporative cooling for 5-6 hours at this temperature by removing saliva from their mouths and spreading it onto body surfaces.

There are two major sets of salivary glands in rats—submaxillary and parotid—and they differ as thermoregulatory effectors. When male rats are exposed to a hot environment body temperature increases above the normal 37°C value. When body temperature reaches 38.5°C, the submaxillary glands start to secrete large amounts of fluid. No increase in parotid secretion occurs if the evaporation of this submaxillary saliva is effective in preventing further increase in body temperature. However, if body temperature continues to increase and reaches 40°C, parotid secretion increases and provides an additional emergency supply of fluid for evaporative cooling. In rats, parotid saliva contains 100-140 mmol/l of sodium, and submaxillary saliva contains 5-10 mmol/l of sodium. Thus the use of submaxillary saliva for thermoregulation reduces sodium loss and the consequent problems associated with reduced blood volume and mechanisms required for sodium retention.

Salivary secretion for thermoregulation is controlled by the hypothalamus (Fig. 14.8). An increase in body temperature from the reference "set point" of 37°C to 38.5°C that is detected in the anterior hypothalamus results in activation of salivary control centers in the brain stem. Independent heating of the anterior hypothalamus of rats produces increased salivary secretion at normal body temperatures. The chorda tympani nerve connects the salivary centers with the glands. Increased secretion will decrease or stabalize an increasing body temperature if groomed saliva is effective in increasing evaporative heat loss (dashed line—negative effect—in Fig. 14.8).

Similar types of negative-feedback control systems govern the control of body temperature through sweating and panting, although many of the details of these controls are not well understood. Recall that thermostatic hypothalamic control is proportional to the difference between the hypothalamic temperature and a reference (set point) temperature for control of thermoregulatory heat production in "cold" environments (Chapter 7). Proportional control is also suggested for thermoregulatory salivary secretion because parotid saliva is added to submaxillary saliva in rats only at relatively high body temperatures (larger differences between body temperature and a reference).

Increasing Evaporation from Surfaces: Regurgitation in Honeybees

Although most mechanisms for the control of body temperature by means of heat loss have been studied in larger animals in which body size provides some advantages, there are some interesting examples of similar mechanisms among some small insects faced with problems for heat loss. Honeybees must forage for nectar from flowers, and the activity of flight can produce relatively high rates of internal heat production. When environmental temperatures are high, body temperature can be controlled effectively only from evaporation of water. At high environmental temperatures honeybees regulate the temperature of the head by means of evaporative cooling by regurgitating honeycrop contents.

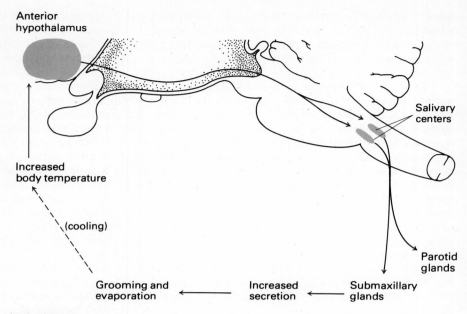

Figure 14.8
Schematic diagram of the negative-feedback relationships involved in control of
body temperature by evaporative cooling in the rat. The dashed arrow symbolizes
the negative effect in the feedback loop.

Blood pumped to the head is secondarily cooled by the resulting evaporation.
This type of mechanism is feasible for the honeybee despite its small size
because of the relatively large supply of water (obtained in its food, flower nectar) that would normally be eliminated to produce honey in the hive. We will
find that evaporative mechanisms for keeping a "cool head" in hot environments are also common in other animals (see below).

HEAT STORAGE AND INTERACTIONS BETWEEN WATER AND TEMPERATURE REGULATION

Heat storage saves water because heat that is stored does not have to be lost by
evaporation. The limits on storage are set by the adverse effects of an increase
in body temperature, that is, less precision of temperature regulation and the
availability of heat loss avenues not requiring evaporation to lose stored heat. In
addition, upper lethal extremes of body temperature set an absolute limit to
heat storage. There is considerable evidence that heat storage occurs primarily
when water is in short supply in relation to demands for temperature regulation. When supplies of water are sufficient, precision of temperature regulation
takes priority over minimization of water losses.

An example of this interaction can be seen in the heat storage patterns of camels that have water versus those that are dehydrated by prolonged restriction of the availability of drinking water (Fig. 14.9). When they are dehydrated, camels heat during the day to a higher body temperature than they do when they have access to water. The stored heat is lost at night primarily by radiation, conduction, and convection to a cooler environment, saving water. Note that body temperature not only increases more during the day for dehydrated camels but also it decreases more at night. The maximum change in body temperature for dehydrated camels is about 6 °C; for camels with water it is about half of this amount (Fig. 14.9).

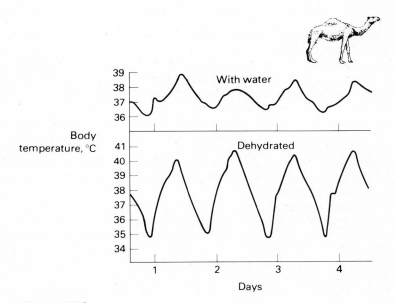

Figure 14.9
Changes in body temperature represent heat storage and loss. For camels, more heat is stored when water availability is restricted (adapted from Schmidt-Nielsen *et al.*, 1957).

The amount of water saved by storing heat and losing it later by nonevaporative means can easily be calculated. About 0.9 calorie is required to increase each gram of tissue 1.0 °C. A 500,000 gram camel that increases in temperature by 6 °C will store 2,700,000 calories. If this heat had to be lost entirely by evaporation (that is, body temperature regulation had to be absolutely precise) whereby each milliliter of water evaporated lost 580 calories, a total of 4655 milliliters would be required. Thus over 4½ liters of water are saved as a consequence of less than perfect precision in body temperature regulation.

Another way to view this heat storage is with respect to metabolic rate or heat production (H_m) because this heat must also be taken into account in hot environments. A 500-kilogram camel has a resting (standard) metabolic rate of 7400 kcal/day (see Chapter 5). The amount of heat stored when body temperature increases 6°C is 32% of this value. Increases in body temperature will increase metabolic rate above standard values, but the comparison still provides a useful way to view the value of heat storage as compared with heat production.

Body Size and Heat Storage

Body size has a pronounced impact on the utility of heat storage for water conservation and temperature regulation. This impact can be illustrated by completing a set of calculations for a 35-gram kangaroo rat similar to those done for the camel above. An increase in body temperature of 6°C stores only 189 calories and saves only 0.32 milliliters of water (189 cal/580 cal per ml = 0.32 ml) for the kangaroo rat. The resting metabolic rate of a 35-gram kangaroo rat is about 5.7 kcal/day. Storage of 189 calories represents only 3.3% of daily energy expenditure, or one-tenth that of the percentage for the larger camel for the same increase in body temperature.

Although small size sets severe limits on the advantages of heat storage with a given increase in body temperature, heat storage is still important for small desert rodents that move in and out of their burrows several times during the hot portion of a day. The burrow environment is cooler than the outside (see Fig. 13.7), so a small rodent does not have to wait until night to lose stored heat by nonevaporative means.

If a 35-gram rodent goes in and out of its burrow during the day and stores 189 calories by allowing body temperature to increase 6°C each time on ten different trips, it will store and lose the same proportion of its resting heat production (3.3% × 10 = 33%) as does the camel, which does not have the option of using a cool underground burrow. There is evidence that small desert rodents that are normally active during the day use heat storage and behavior to conserve water. Figure 14.10 shows the pattern of heat storage for an antelope ground squirrel as compared with the longer-term pattern for the camel.

Note that the solution of problems for water conservation and temperature regulation in smaller desert animals involves smaller *amounts* of heat storage and higher *frequency* of movement to cooler microhabitats. As with many other physiologically important processes (respiration, circulation, feeding), total exchange varies with size, with smaller amounts (tidal volume, stroke volume, meal size, amount of heat stored) but higher frequencies (respiratory rate, heart rate, feeding frequency, frequency of microhabitat selection) in small species.

The effective use of heat storage by small animals depends on behavior for selecting microenvironments suitable for heat exchange. Use of behavior for temperature and water regulation in these animals is similar to the use of behavior for temperature regulation by ectotherms (see Chapter 6). In both cases the animals move around in their environments depending on requirements for heat and water exchange.

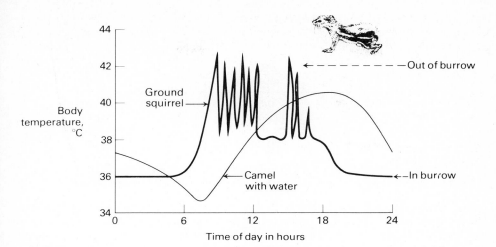

Figure 14.10
A small antelope ground squirrel stores heat when it is active out of its burrow during
the day and loses heat during frequent visits to its cooler burrow. This short-term pattern effectively saves water in a way similar to the longer-term pattern for the larger
camel (from Bartholomew, 1964).

Trade-offs between Osmoregulation and Temperature Regulation

Observations of a greater degree of heat storage in dehydrated animals (Fig.
14.9) suggests that short-term changes in priorities for osmoregulation take
precedence, meaning that temperature regulation is less precise. Thus restriction of water supply relative to demands for temperature regulation influences
the pattern and the degree of precision of temperature regulation.

Trade-offs between temperature and osmoregulation may involve changes
in substrates used or selected (consumed) for oxidation when rates of evaporative water loss are high. Part of the heat that must be lost by an animal in a hot
environment comes from internal oxidation of substrates (see Chapter 5). These
substrates differ in energy value per gram and also yield different amounts of
water when a gram is completely oxidized (see Chapter 13). Calculations showing which substrates are most effective for supplying water *relative to* the heat
produced during metabolic water formation are shown in Table 14.1.

Note that even though fat yields more water per gram, more heat is produced for every gram of fat oxidized; thus fat yields only an intermediate
amount of water relative to heat production. Protein provides the least water
relative to heat production in mammals with urea excretion but somewhat
more with uric acid production. Carbohydrates provide the most water relative
to heat production. Oxidation of carbohydrates provides 44% more water than
oxidation of protein with urea formation and 18% more water than fat oxidation
or protein oxidation with uric acid formation.

Table 14.1 Heat production, water production, and water production relative to heat production for oxidation of one gram each of different substrates (from King, 1957; Schmidt-Nielsen, 1975)

SUBSTRATE	$\dfrac{\text{g } H_2O}{\text{g substrate}}$	$\dfrac{\text{kcal}}{\text{g substrate}}$	$\dfrac{\text{g } H_2O}{\text{kcal}}$
Carbohydrate	0.56	4.2	0.13
Fat	1.07	9.4	0.11
Protein (urea excretion)	0.39	4.3	0.09
Protein (uric acid excretion)	0.50	4.4	0.11

We might expect animals with severe supply-demand problems of water and temperature regulation to use (or ingest) carbohydrate during dehydration in the heat. However, recall that other constraints due to the weight required to store glycogen with increased cellular water (see Chapter 5) could restrict the amount of energy stored internally as carbohydrate. Perhaps this situation could change for relatively inactive animals in hot environments if there is sufficient selective pressure to maximize water supplies relative to heat production.

HEAT STORAGE DURING ACTIVITY: PREDATORS AND PREY IN HOT ENVIRONMENTS

Some animals generate large amounts of heat internally when they are active. In addition, some animals must be active in relatively hot environments. Heat storage represents a way to deal with heat and water exchange, and several patterns of heat storage have evolved that depend on the rate of heat production from activity and the requirements that dictate survival by means of activity in interactions between predators (chasers) and their prey (those chased). Most of these patterns have been studied in mammals that live on the hot, dry plains of East Africa, although the patterns are not limited to those species.

The tricolored Cape hunting dog (*Lycon pictus*) is well known for its extensive running in pursuit of prey. The dogs hunt in packs and rely on endurance to run down prey over long distances in relatively hot, dry environments. A hunt may take the better part of a day and does not allow the dogs to stop and drink.

By allowing body temperatures to increase, the hunting dogs lose less water by evaporation as compared with the domestic dog (Fig. 14.11). Some heat is stored by hunting dogs that increases the difference between body and environmental temperatures, increasing the proportion of heat lost by nonevaporative means. Moreover, the hunting dog is somewhat more effective in terms of heat production required for locomotion at low running speeds. A combination of comparatively lower rates of water loss and less heat production required for locomotion can extend the hunting range of these predators over longer distances as compared with the domestic dog or other species (perhaps prey) that control temperatures more precisely by evaporation.

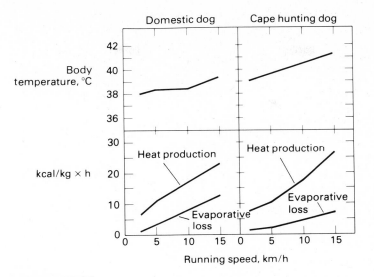

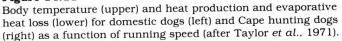

Figure 14.11
Body temperature (upper) and heat production and evaporative
heat loss (lower) for domestic dogs (left) and Cape hunting dogs
(right) as a function of running speed (after Taylor *et al.*, 1971).

Cheetahs (*Acinonyx jubatus*) rely on sprinting at high speed rather than
on low-speed, long-distance running to capture prey. They can sprint at speeds
exceeding 100 km/h (62 miles per hour) and, over short distances, are among
the fastest runners known. The amount of heat produced during a high-speed
sprint is considerable, and it greatly exceeds the capacity for heat loss by
evaporation and nonevaporative cooling within the short time of the sprint.

Although it has not been possible to study cheetahs running at 100 km/h
in the laboratory, even at relatively slow speeds on a treadmill (10 km/h) up to
80% of total heat production is stored (Fig. 14.12). This storage sets a limit to
the duration of a sprint. Cheetahs refuse to run if body temperature exceeds
40.5 °C, perhaps because of continued heat storage and the requirements for
losing stored heat after a sprint.

Cooling the Brain

If the cheetah is limited in sprint duration by heat production and storage, it is
important that it get as close as possible to its potential prey before starting a
sprint. If the prey notices the attacking cheetah soon enough and runs at maximum possible speed, it may be able to escape if the cheetah stops at a certain
body temperature, but how does the prey deal with the same problem of increased heat production required because of sprinting for survival? Must it also
stop running when body temperature increases?

This question has not been fully explored, but there is evidence that some
prey may outlast sprinting predators if they start soon enough and store heat

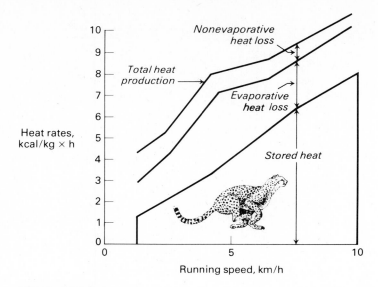

Figure 14.12
Total heat budget for a cheetah as a function of running speed
(after Taylor and Rowntree, 1973).

generated during running. Some species of antelope are able to store different amounts of heat in different parts of the body. This ability was studied by exposing them to high environmental temperatures when they were hydrated and when they were dehydrated (Fig. 14.13). When they were dehydrated (lower supply of water relative to demands for evaporative losses) the temperature of the body measured in the rectum increased to very high values, temperatures that would be lethal for most mammals. However, the brain was kept at a lower temperature than the rest of the body by means of countercurrent heat exchange involving evaporative cooling.

The internal carotid artery supplies blood to the brain. Prior to entering the brain the artery passes through the cavernous sinus, where veins from the nasal region divide into many small branches called a rete. The anatomical arrangement between arteries and veins is shown for the dog in Fig. 14.14. Water that evaporates from the nasal passages during inspiration cools venous blood passing through the cavernous sinus. The blood in the internal carotid artery is cooled by countercurrent exchange when it passes through the sinus; thus heat loss by respiratory evaporation cools the brain while the remainder of the body heats. If this type of mechanism operates in gazelles when they flee a cheetah, it could permit relatively long periods of high heat production as compared with ability to lose heat by evaporation.

Both the cheetah and their prey may possess countercurrent exchangers for cooling the brain. The trade-off involved in who catches whom depends on distance and speed (extent of heat production). Rate of heat production by the

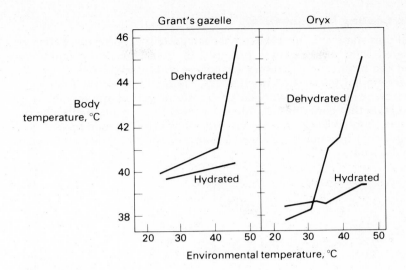

Figure 14.13
Body temperatures for two species of antelope from arid areas of
Africa at different environmental temperatures when supplies of
water for thermoregulation are changed (after Taylor, 1970).

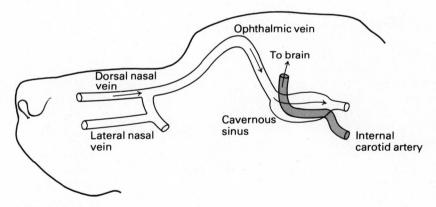

Figure 14.14
Diagram of the countercurrent relationship in dogs between cooler venous
blood from the nasal region and blood passing to the brain in the internal
carotid artery (after Magilton and Swift, 1969).

cheetah during sprinting is so high that even a vascular rete cannot contain an
increase in temperature for a long time. Prey move somewhat slower (less heat
production), which may allow heat exchange to operate longer if distances between predators and prey are initially sufficient for escape.

Brain Cooling in Dinosaurs?

Recall that some anatomical evidence suggests that some dinosaurs may have regulated their temperatures (Chapter 6). Some large dinosaurs may have cooled the brain via respiratory evaporation and vascular heat exchange in a manner similar to that described above. Figure 14.15 shows a section through the nasal crest of a hadrosaur. These dinosaurs were characterized by relatively large crests that contained channels for passage of respired air. The nasal capsule was situated close to the brain (Fig. 14.15), and it could have served a thermoregulatory function if heat loss was important for temperature regulation in these dinosaurs.

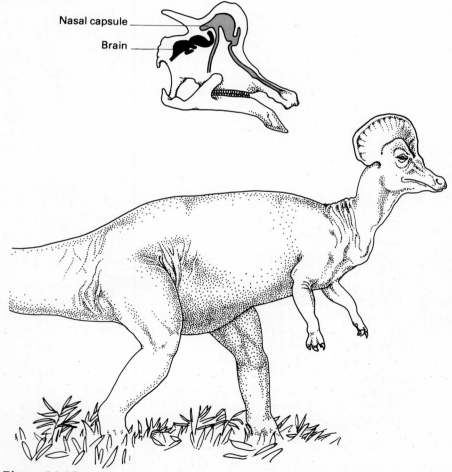

Figure 14.15
A number of relatively large hadrosaurine dinosaurs (*Corythosaurus* depicted here) had large narial crests. The insert shows the relationship between air passages within the nasal capsule and the position of the brain in the hadrosaur *Lambeosaurus* (after Wheeler, 1978).

SUMMARY

When environmental temperature is high such that $(T_B - T_A)$ is low, zero, or negative, and/or when heat production (H_m) from activity is high, the only effective way to increase heat loss for terrestrial animals is to increase the rate of water evaporation. If heat loss by evaporation does not match heat gain, body temperature will increase and heat will be stored. For every milliliter of water evaporated, 580 calories of heat will be lost (at 35 °C). All animals lose some insensible (minimum amount of) water and heat to unsaturated air, and the rate of this loss increases in hot environments (with larger saturation deficits). However, this loss is usually not sufficient for the demands of temperature regulation in hot environments.

Some animals increase the rate of water evaporation by panting, i.e., increasing respiratory ventilation above requirements for gas exchange. The increased work required to move air adds heat to be dissipated, but panting is still effective when air is moved at a resonant frequency of the respiratory system, involving minimum heat production to move a maximum volume of air. Loss of excessive CO_2 can be minimized by restricting increased ventilation to dead-space volume, but some respiratory alkalosis may occur during panting. Over short periods requirements for survival by means of temperature regulation take priority over acid-base regulation. Panting could be less effective because of respiratory countercurrent heat exchange, but inspiration through the nose and expiration through the mouth increases respiratory water loss when heat loss by evaporation is important for survival.

Evaporative heat loss from the surfaces of some animals increases as a consequence of glandular secretions (or regurgitation in bees). Some animals produce thermoregulatory eccrine sweat; small rodents secrete increased amounts of saliva, which is groomed onto body surfaces. Sweat is less concentrated than blood but contains appreciable amounts of sodium. If large amounts of sweat are produced, cellular dehydration and reduced blood volume occur, requiring resorption and consumption of water and salt.

Male laboratory rats secrete submaxillary saliva if body temperature reaches 38.5 °C. Parotid salivary secretion increases only if body temperature continues to increase to 40 °C. Thermoregulation by saliva spreading in small animals may be less effective than panting or sweating but suffices for short-term exposures to hot environmental conditions. Detection of an increase in body temperature takes place in the anterior hypothalamus, which activates control centers for salivary secretion in the brain stem. Degree of fluid secretion for thermoregulation in rats is proportional to the displacement of body temperature from a reference set point.

Heat that is stored and lost by nonevaporative means at a later time saves water. Limits for this mechanism involve both the effects on efficient function because of imprecision of temperature regulation and also upper lethal limits for heat storage. Animals that are dehydrated store more heat than do animals with access to water resources, suggesting changes in priorities between precision of osmoregulation and precision of temperature regulation. Large animals can store more heat in a given period of time (day) and lose heat by conduction and radiation at night. Smaller animals can effectively save water by heat storage over smaller time intervals when they retreat to cool microenvironments (burrows) for nonevaporative heat loss of smaller amounts of heat stored several times a day.

The interactions between predators and prey are influenced by temperature regulation and water requirements in hot, dry environments or by abilities to rapidly mobilize water for evaporation to match increased heat production during activity. Some predators that pursue prey by long-distance endurance running, such as the Cape hunting dog, store heat and have lower rates of evaporative heat loss. Sprinting predators, such as the cheetah, generate large amounts of heat internally in a short time. Most of this heat must be stored, and sprinting by the cheetah is limited by an increase in body temperature. Some prey that are pursued by sprinting predators may escape by storing relatively large amounts of heat in certain parts of the body while the brain is kept relatively cool via respiratory evaporation and countercurrent cooling of blood supplies to the brain. Certain anatomical evidence suggests that some dinosaurs may have had a similar countercurrent mechanism for cooling the brain.

ANNOTATED REFERENCES

Chapter 12: Principles of Osmoregulation and Control in Aquatic Environments

Bentley, P.J. (1971). *Endocrines and Osmoregulation. A Comparative Account of the Regulation of Water and Salt in Vertebrates.* Springer-Verlag, New York. (Comprehensive treatment of problems of osmoregulation for both aquatic and terrestrial environments.)

Croghan, P.C. (1958). The osmotic and ionic regulation of *Artemia salina* (L.) *Jour. Exp. Biol.* 35:219-233. (Account of exchange in an invertebrate with large osmotic gradients in hypertonic environments.)

DiBona, D.R., and **J.W. Mills** (1979). Distribution of Na$^+$-pump sites in transporting epithelia. *Fed. Proc.* 38:134-143. (Summary of information on location of membrane pumps and morphology of extracellular channels for sodium absorption, secretion, and isosmotic transporting tissues.)

Forrest, J.N., Jr., **W.C. MacKay**, **B. Gallagher**, and **F.H. Epstein** (1973). Plasma cortisol response to saltwater adaptation in the American eel *Anguilla rostrata. Amer. Jour. Physiol.* 224:714-717. (Mechanism of short-term changes in movement of sodium during migration of eels to hypertonic environments.)

Gordon, M.S., **K. Schmidt-Nielsen**, and **H.M. Kelley** (1961). Osmotic regulation in the crab-eating frog (*Rana cancrivora*). *Jour. Exp. Biol.* 38:659-678. (Use of urea for near-isotonic regulation in a marine amphibian.)

Hossler, F.E., **J.R. Ruby**, and **T.D. McIlwain** (1979). The gill arch of the mullet, *Mugil cephalus.* II. Modification in surface ultrastructure and Na, K-ATPase content during adaptation to various salinities. *Jour. Exp. Zool.* 208:399-406. (Functional and structural changes associated with chloride-secreting cells used for osmoregulation in different osmotic environments.)

Lockwood, A.P.M. (1963). *Animal Body Fluids and Their Regulation*. Harvard University Press, Cambridge. (Brief account of osmoregulation and some problems of osmosis.)

Patterson, D.J. (1980). Contractile vacuoles and associated structures: their organization and function. *Biol. Rev.* 55:1-46.

Potts, W.T.W. (1954). The energetics of osmotic regulation in brackish- and fresh-water animals. *Jour. Exp. Biol.* 31:618-630. (Development of a model of osmoregulatory cost that is dependent exponentially on osmotic concentration differences, and a discussion of mechanisms to reduce the costs.)

Potts, W.T.W., and **G. Parry** (1963). *Osmotic and Ionic Regulation in Animals*. Pergamon Press, New York. (Detailed description of osmoregulatory problems in a large number of invertebrates and vertebrates.)

Schmidt-Nielsen, B. (1974). Osmoregulation: effect of salinity and heavy metals. *Fed. Proc.* 33:2137-2146. (Brief summary of osmoregulatory problems and mechanisms for aquatic and terrestrial animals.)

Thorson, T.B., **C.M. Cowan**, and **D.E. Watson** (1967). *Potamotrygon* spp.: elasmobranchs with low urea content. *Science* 158:375-377. (Environmental relationship of the use of urea for osmoregulation.)

Chapter 13: Osmoregulation and Control in Terrestrial Environments

Adolph, E.F. (1939). Measurement of water drinking in dogs. *Amer. Jour. Physiol.* 125:75-86. (Early examination of the precision of the relationship between ingestion and demands for water.)

Bartholomew, G.A., and **T.J. Cade** (1963). The water economy of land birds. *Auk* 80:504-539. (Relationship between body size and rate of evaporative water loss.)

Diedrich, D.F. (1966). Glucose transport carrier in the dog kidney: Its concentration and turnover number. *Amer. Jour. Physiol.* 211:581-587. (Characteristics of the transport-limited mechanism for glucose resorption.)

Jackson, D.C., and **K. Schmidt-Nielsen** (1964). Countercurrent heat exchange in the respiratory passages. *Proc. Nat. Acad. Sci.* 51:1192-1197. (Description of temporal system for respiratory countercurrent cooling.)

Murrish, D.E., and **K. Schmidt-Nielsen** (1970). Exhaled air temperature and water conservation in lizards. *Resp. Physiol.* 10:151-158. (Importance of respiratory countercurrent exchange for water and solute balance in some reptiles.)

Pitts, R.F. (1968). *Physiology of the Kidney and Body Fluids*. Year Book Medical Pubs., Chicago. (Details of mechanisms of kidney function in terrestrial vertebrates, particularly humans.)

Roberts, J.B., and **H.B. Lillywhite** (1980). Lipid barrier to water exchange in reptile epidermis. *Science* 207:1077-1079. (Experimental removal of lipids from skin increases rates of water loss through skin in reptiles.)

Schmidt-Nielsen, B. (1972). Mechanisms of urea excretion by the vertebrate kidney. In J.W. Campbell and L. Goldstein (eds.), *Nitrogen Metabolism and the Environment*. Academic Press, New York.

Schmidt-Nielsen, B., and **L.E. Davis** (1968). Fluid transport and tubular intercellular spaces in reptilian kidneys. *Science* 159:1105-1108. (Relationship between direction and extent of salt and water transport and morphology of spaces between cells in transporting epithelia.)

Schmidt-Nielsen, B., and **R. O'Dell** (1961). Structure and concentrating mechanism in the mammalian kidney. *Amer. Jour. Physiol.* 200:1119-1124. (Countercurrent multiplication resulting from the function of the loop of Henle.)

Schmidt-Nielsen, K. (1960). The salt-secreting glands of marine birds. *Circulation* 21:955-967. (Features of osmoregulation, with salt-excretion characteristics of birds and reptiles.)

Schmidt-Nielsen, K. (1964). *Desert Animals.* Oxford University Press, New York. (Excellent summary of water balance in several terrestrial animals from arid environments.)

Stricker, E.M., **A.H. Vagnucii**, **R.H. McDonald, Jr.**, and **F.H. Leenen** (1979). Renin and aldosterone secretions during hypovolemia in rats: relation to NaCl intake. *Amer. Jour. Physiol.* 237:R45-R51. (Ingestion of NaCl for isotonic regulation of blood volume, and experimental examination of some possible controls.)

Venkatachalam, M.A., and **H.G. Rennke** (1978). The structural and molecular basis of glomerular filtration. *Circ. Res.* 43:337-347. (Examination of glomerular mechanisms for fluid filtration.)

Chapter 14: Water and Temperature Regulation in Hot Environments

Adelman, S., **C.R. Taylor**, and **N.C. Heglund** (1975). Sweating on paws and palms: what is its function? *Amer. Jour. Physiol.* 229:1400-1402. (Role of apocrine sweat in predator avoidance resulting from increased traction.)

Bartholomew, G.A. (1964). The roles of physiology and behavior in the maintenance of homeostasis in the desert environment. *Symp. Soc. Exper. Biol.* 18:7-29. (Role of heat storage and microhabitat selection by small desert animals.)

Crawford, E.C., Jr. (1962). Mechanical aspects of panting in dogs. *Jour. Appl. Physiol.* 17:249-251. (Characteristics of ventilation for maximum rate of respiratory water loss as compared with energy expenditure to move air for evaporation.)

Daniel, P.M., **J.D.K. Dawes**, and **M.M.L. Prichard** (1953). Studies on the carotid rete and its associated arteries. *Phil. Trans. Roy. Soc. London*, Ser. B, 237:173-208. (Anatomy of brain-cooling heat exchangers in various species of vertebrates.)

Hainsworth, F.R., and **E.M. Stricker** (1970). Salivary cooling by rats in the heat. In J.D. Hardy, A.P. Gagge, and J.A.J. Stolwijk (eds.), *Physiological and Behavioral Temperature Regulation.* CC Thomas, Springfield, Ill. (Review of the role of salivary glands as thermoregulatory effectors.)

Heinrich, B. (1979). Keeping a cool head: honeybee thermoregulation. *Science* 205:1269-1271. (Regurgitation mechanism for evaporative cooling via surface evaporation.)

Magilton, J.H., and **C.S. Swift** (1969). Responses of veins draining the nose to alar-fold temperature changes in the dog. *Jour. Appl. Physiol.* 27:18-20. (Some functional characteristics of the brain-cooling heat exchangers.)

Schmidt-Nielsen, K., **W.L. Bretz**, and **C.R. Taylor** (1970). Panting in dogs: unidirectional air flow over evaporative surfaces. *Science* 169:1102-1104. (Description of nasal inspiration with oral expiration during panting.)

Taylor, C.R. (1970). Dehydration and heat: effects on temperature regulation of East African ungulates. *Amer. Jour. Physiol.* 219:1136-1139. (Dissociation of brain and body temperatures in some species when internal water supplies for evaporative cooling are restricted relative to demands for temperature regulation.)

Taylor, C.R., and **V.J. Rowntree** (1973). Temperature regulation and heat balance in running cheetahs: a strategy for sprinters? *Amer. Jour. Physiol.* 224:848-851. (Mechanism for heat storage when heat production exceeds capacity for evaporative cooling.)

Taylor, C.R., **K. Schmidt-Nielsen**, **R. Dmi'el**, and **M. Fedak** (1971). Effect of hyperthermia on heat balance during running in the African dog. *Amer. Jour. Physiol.* 220:823-827. (Heat storage for increased heat loss.)

Wheeler, P.E. (1978). Elaborate CNS cooling structures in large dinosaurs. *Nature* 275:441-443. (Examination of the possible role of nasal crests for countercurrent cooling of the brain in some large dinosaurs.)

Reproduction

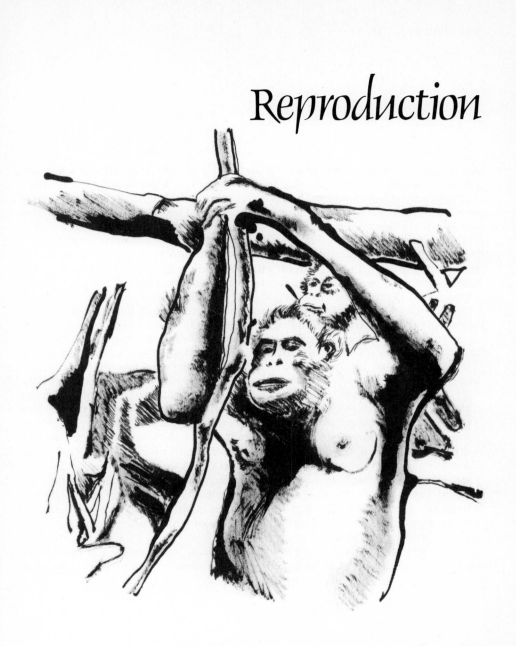

Part V

An animal is considered biologically successful if it not only survives but also reproduces and its young survive to reproduce. Our discussion of interactions between animals and environments has primarily dealt with factors influencing survival. These factors also influence reproduction because an animal that does not survive will not reproduce. Yet if an individual survives but cannot or does not reproduce it will not be biologically successful. There are some exceptions to this situation; animals can increase representation of their genetic material in the future through their relatives without producing their own offspring. We will consider an example as shown by social insects. However, most animals reproduce if they succeed in surviving.

The combination of survival and reproductive effort determines *relative biological fitness*. An individual will be relatively more fit (more biologically successful) if it survives to reproduce so that it leaves more reproductive genetic units (successful offspring) in future generations than do other individuals. Those individuals that leave fewer successful offspring are selected against because they will have less genetic impact within future populations of animals.

Note that fitness is considered to be an *individual* trait and not a separate characteristic of groups or of a species. Natural selection acts on individuals, not "species." To the extent that individuals of a species are related they will share some similar characteristics, but individual variation results in differences. Variation in survival and reproduction by individuals forms the basis for evolutionary changes by means of natural selection. A common misconception is that individuals reproduce to "ensure the survival of the species." What happens to a species as a group is a consequence of selection acting on individuals within the group. Individuals in some groups cooperate, but we will find that this apparent altruism depends on the degree to which individuals are genetically (reproductively) related to one another.

Those individuals that are more economically efficient in using resources should be more fit because their demands for resources will be lower. We have already examined a number of physiological adaptations that result in relative advantages for survival of animals in different environments. These adaptations will also ultimately contribute to fitness through reproduction. However, the ability to reproduce and the success of reproductive efforts require supplies of resources above and beyond those usually required just for survival. Extra demands are placed on animals attempting to produce other animals. Moreover, the resources required for reproduction must be obtained from environments that change over time with regard to resource availability.

The complexity of possible solutions to the problems of matching variable resource supplies with reproductive demands among animals is great. In Chapter 7 we discussed a few of the factors interacting among temperature regulation, energy demands, and the number of young influencing nesting success in altricial animals. We will examine several other cases of complex interactions in the coming chapter.

In general, the complexity of reproduction is due to the fact that every aspect of the life history of an animal can be changed, with consequences for resource supplies and demands that influence fitness. How large an animal is, how many times it can reproduce in its life, when it reproduces, how rapid development can be, how many young are produced and their size, and the extent of possible parental care are a few of many important variables that will influence fitness.

Despite this complexity, successful reproduction is determined in large part by factors that influence the *timing* and extent of reproduction relative to resource availability for animals that have evolved a particular set of life-history characteristics. Among animals (and plants), timing controls involve *hormones*, which are chemicals that influence the function and development of cells. In the following chapter we will examine the operation of hormonal control mechanisms in different animals that serve to time reproduction so that supplies of resources match demands.

Resource Regulation and Control in Reproduction

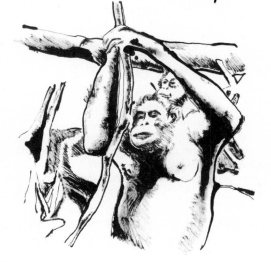

15

An individual of a given species will have a maximum reproductive capacity that cannot be exceeded and that may not even be achieved. The maximum reproductive contribution of an individual will occur in an environment that provides minimum supply-demand problems concerning the resources needed during reproduction. The maximum reproduction in the "ideal" situation is determined by the genetic traits of the animal as influenced by the natural environmental conditions with which it has evolved.

A nonideal environment will result in less than a maximum reproductive performance if there is variation in the resources necessary for the increased demands of reproduction. For example, consider the number of eggs produced by a female flour beetle (*Tribolium*; larvae are mealworms) per day when the density of adults per gram of flour changes (Fig. 15.1). The maximum reproductive output occurs at the level of slightly more than two individuals per gram of resource (flour). At lower densities males or females become limiting to each other as resources (they have difficulty getting together), and at higher densities interference between individuals for mates or food resources will lower the reproductive contribution per female.

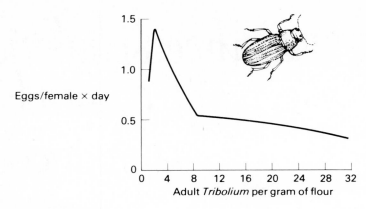

Eggs/female × day

Adult *Tribolium* per gram of flour

Figure 15.1
The reproductive success of female *Tribolium*, measured as egg
production, is maximum at a particular density of adults per
gram of food resource (flour) (after Andrewartha and Birch, 1954).

The effective reproductive fitness of an individual also depends on what
happens to the young that are produced. Some young may survive and others
may die before reproducing. The difference between the rates of births and
deaths of individuals determines the change in size of a population over time,
and biologists distinguish at least two general patterns of reproductive interactions between animals and their environments that are based on two characteristics of population growth.

r-SELECTED AND K-SELECTED REPRODUCTIVE PATTERNS

Figure 15.2 illustrates some general characteristics of a change in population
size over time. When a population is first established in an environment (time
zero, Fig. 15.2) it may subsequently grow relatively rapidly (have a high *rate* of
growth, *r*), because individuals are few relative to the resources needed for survival and reproduction. Thus more individuals are born and survive to reproduce than die. This stage is called the *r* phase of population growth.

Eventually, as the number of individuals in an environment increases
deaths will approximately equal births and the size of a population will become
constant. This stage is called the *K* (for constant) phase of population growth.
Deaths increase when environmental resources become limiting for survival
and reproduction. In the *K* phase there is more competition for resources and a
change in the general characteristics influencing how animals deal with
resources.

Animals can be divided into two groups, depending on whether they have
evolved to deal principally with environments that allow "*r*" types of population growth or environments having "*K*" types of resource limitations. These
animals are called "*r*-selected" and "*K*-selected" species, respectively. The
r-selected species reproduce primarily by increasing the numbers of individuals

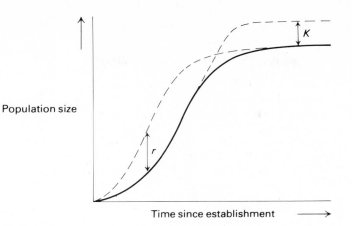

Figure 15.2
Schematic graph of a change in population size over time
since founding. The "r" phase (rate of growth) occurs soon
after founding, when population size is low relative to resource
supplies. The "K" phase (constant growth) occurs when births
and deaths are equal. The dashed lines refer to differences in r
or K that can occur because of differences in animals or their
environments (after Horn, 1978).

that are added to a population in a short time (high rate of addition of individuals). The rate of death may be high over a long time period, but during
reproduction large numbers are produced relative to deaths. An example of an
r-selected species is one that colonizes a set of resources on a seasonal or spatial
basis, such as insects that emerge from diapause in the spring having a large
resource base to exploit in a short time for growth and reproduction.

K-selected species reproduce primarily by investing limited resources in
the production of fewer, higher-quality young. An example of a K-selected
species is one that exploits a large proportion of available resources over long
periods, such as humans and many terrestrial vertebrates.

The theoretical bases for this division between animals are beyond the
scope of this book. It is possible to identify other patterns of reproduction that
depend on the nature of variation of environmental resources and the probabili-
ties of juvenile versus adult survival; additional information can be found in
current books on ecology or population biology. However, for our purposes it is
important to understand that these major patterns exist because they allow
very different predictions of the ways animals will respond to environments by
means of reproduction.

Some characteristics of r- and K-selected species that influence reproduction are listed in Table 15.1. Note that each characteristic for an r-selected
species is opposite in degree to the one for a K-selected species. Moreover, the
characteristics are interrelated, which can be appreciated by picking one important feature and briefly examining its impact on others.

Table 15.1 Some life-history characteristics for r- and K-selected organisms

r-SELECTED	K-SELECTED
Smaller adult size	Larger adult size
Rapid development	Slow development
Low competitive ability	High competitive ability
Low parental care	High parental care
Many young/effort	Few young/effort
Early breeding	Delayed breeding
Shorter adult survival	Longer adult survival
Low predator-escape ability	High predator-escape ability
Smaller young	Larger young

Usually, r-selected species are small in size. Small size results in less demand for a resource in absolute amounts (e.g., kcal/day), meaning that smaller animals will be less resource-limited (less K-type). Small size, however, will influence ability to survive with respect to competition, predators, and environmental variability (variations in temperature, food availability, pH, osmotic environments, etc.); smaller, r-selected species are usually relatively short-lived (usually on a seasonal scale or less for a complete life cycle). Shorter individual life expectancy will lead to rapid development and a relatively early investment in reproduction. In addition, small young may have to be produced by small adults, with each young having a relatively low probability of survival. This situation can be compensated for somewhat by production of more young per effort.

K-selected species are usually large in size. Large size results in higher demands for resources, meaning that larger animals are more resource-limited. Large size, however, will enhance individual ability to survive with respect to competition, predation, and environmental resource variation; larger, K-selected species are usually longer-lived. Longer life expectancy can involve slower development and a delayed investment in reproduction. Moreover, fewer but larger young may have to be produced, with each having a higher probability of survival. Survival also can be enhanced somewhat by a greater degree of parental care.

Trying to work out what influences what and why can be a tortuous task. We will simplify the process somewhat by considering the general characteristics of the underlying controls for r- and K-selected species, with a few specific examples. However, the importance of a life-history characteristic can be different for different species. Thus some caution should be used in applying general rules to specific cases.

Figure 15.3 illustrates the relationship between size (as length) and reproduction (as generation time). In general, most of the principles of control of reproduction can be considered by discussing two major size classes of animals: small invertebrates (r-selected) and relatively large terrestrial vertebrates (K-selected). We will examine the controls governing growth and reproduction

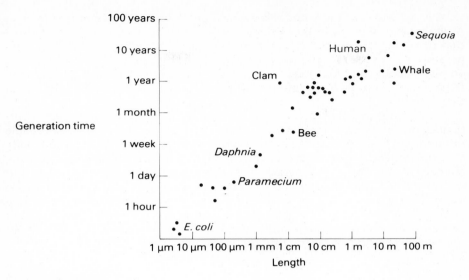

Figure 15.3
The relationship between size (measured as length) and reproduction (measured as generation time). Each point represents a separate species (after Bonner, 1965).

in insects and then we will examine the controls of some terrestrial vertebrates. However, before considering the details of each type of control system it is important to discuss some common features of both types of reproductive patterns. Both major groups of animals rely extensively on *timing* reproduction to environmental variation, and control of timing is achieved chemically through the use of *hormones*.

TIMING WITH RESPECT TO ENVIRONMENTAL VARIABLES

Reproductive timing is crucial for success in both reproductive tactics. The *r*-selected species may breed only once and should time reproduction to a certain stage of growth and with respect to a changing environment so that the number of young that succeed is "optimal." *K*-selected individuals may breed several times, but the success of each effort will depend on the availability of resources that change with time. Thus the underlying controls that result in a maximum possible reproductive effort of any animal will involve timing with respect to important environmental variables.

Timing with Photoperiods

A number of variables in the external environment have an effect on reproduction and vary with different degrees of predictability. For example, if an insect evolved so that eggs were laid with respect to environmental temperature, it could delay laying until temperatures reached a certain value. But temperatures can change abruptly and relatively unpredictably in some environments. A late frost in spring could destroy an entire reproductive effort of an animal that timed egg laying exclusively to temperature changes. Furthermore, temperature is not the only important variable influencing the success of the young. Moisture and food may be required in an appropriate combination with temperature to provide optimum conditions for reproduction and survival of the young.

Temperature, moisture, and food availability all change such that a particular time of the day or year will be "best" for reproduction. The most reliable information from the external environment on seasonal (yearly) and daily changes comes from changes in *photoperiod*. There are no sudden, drastic changes in day length, but a gradual change occurs from season to season and within a day. Long day lengths can be associated with warm temperatures, moisture, and food availability in some areas. For some species, resources may be more available on a daily basis during dark periods; for others, during light periods. Both invertebrates and vertebrates are sensitive to changes in photoperiods with respect to reproductive timing (Table 15.2)

Table 15.2 Some examples of seasonal responses of growth and reproduction in different photoperiods (from Hendricks, 1956)

INVERTEBRATES	SHORT DAYS	LONG DAYS
Aphids	laying of dormant eggs	nondormant eggs
Red spider mites	laying of dormant eggs	nondormant eggs
Mosquitoes	diapausing larvae	metamorphosis to adult
Pulmonate snails	nonreproductive growth	laying of eggs
VERTEBRATES		
Brook trout	laying of eggs	nonreproductive growth
Juncos	migration and non-reproductive growth	mating
Ptarmigan	molting and non-reproductive growth	mating
Sheep	mating	birth of young
Horses	nonreproductive growth	mating and birth of young (11-month gestation)

Photoperiod will interact with other environmental characteristics. An extreme example is the "desert" spadefoot toad, which has evolved to reproduce in localized pools of water resulting from summer rains. The toads emerge from underground soon after a rain, breed, lay eggs, and the young must then develop before the water evaporates. This timing requires information on availability of water within days as well as on the seasonal and daily photoperiod.

Timing and Predation

The biological environment of a reproducing organism will also have an impact on the timing of reproduction. The presence of predators will influence when reproduction takes place if adults and/or young are susceptible to predation. Many plants produce seeds (offspring) that are consumed by predators. One way to deal with this is by a type of "parental care" consisting of either manufacturing toxins to poison predators (see Chapter 11) or giving the endosperm a hard, protective coating.

Another more r-selected tactic is to produce large numbers of young (seeds in plants) at the same time as other individuals. This will effectively "satiate" predators because more offspring will be produced than can be consumed; thus some will "escape" predation (a type of "tribute" defense; see Chapter 11). Predator satiation requires a high degree of *synchrony* in timing between individuals such that ripe seeds are released from a large number of individuals at the same time, because one individual may not have sufficient resources to produce enough young to satiate all predators.

The impact of predation can also lead to very long-term timing patterns. For example, "century" plants bloom and produce seeds over long time intervals. This has been called a "big bang" reproductive pattern because resources are sequestered over long periods and then shifted into a reproductive effort over a brief period. Timing reproduction over long intervals will reduce predation if most predators have shorter life spans; it is more difficult for the predators to evolve with respect to a resource present only in certain generations.

An interesting example of "big bang" long-term timing among animals due to predation is shown by several species of periodical cicadas (*Magicicada* spp.). They have the longest known life cycle for insects, and all individuals in a population are essentially the same age. Adults emerge from the ground every 13 years in the South and every 17 years in the North. They mate, lay eggs, and die within the same few weeks, and separate species emerge synchronously. The eggs develop into nymphs that live underground by sucking sap from the roots of plants for 13 or 17 years.

The impact of this reproductive timing on cicadas and their predators is illustrated in Fig. 15.4. Underground predators (moles) can use nymphs for food when the nymphs reach a minimum size, but when the cicadas emerge the moles are without this food for about two months, followed by the availability of only very small nymphs of the next generation. The aboveground predators only have access to adult cicadas for a short time every 13 or 17 years, and the short-term synchrony of emergence tends to satiate aboveground predators (increasing predator success in the short term); thus many cicadas escape predation.

It may also be important that the periodicity of emergence occurs over intervals of time that are prime numbers (numbers divisible only by themselves and one). This timing requires a predator that evolves to use cicadas as a primary food resource at a particular stage of its life cycle to have the same life span or longer. A 12-year cicada could be used by every other generation of a 6-year predator at the same stage of its life cycle; a 13- or 17-year cicada cannot. Because life span is related to size (Fig. 15.3), a relatively large size for the

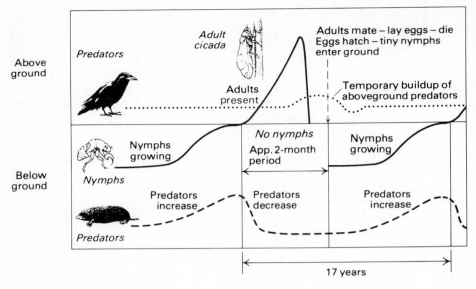

Figure 15.4
Summary of the interactions between predators and cicadas as a consequence of long-term timing of development and reproduction (from Lloyd and Dybas, 1966).

predator would be required, which would make the relatively small cicadas less suitable as a food resource.

Timing and Sex

Timing with respect to reproduction can also involve which organism is male or female at different times during development. For many animals sex determination takes place early in development, and an individual is male or female for the remainder of its life. Sex determination in most vertebrates depends on the presence of sex chromosomes; a pair of X chromosomes results in a female in mammals (a male in birds), and an XY combination results in a male in mammals (female in birds). The probability of fertilization of an egg containing an X chromosome by sperm with either an X or Y chromosome is about 50%, meaning that equal numbers of males and females will be generated if there is no mortality differential between sexes.

In some animals, fitness changes with age or size. Small or younger individuals are less likely to survive or may not have access to important resources required for reproduction. One way to deal with this problem is to delay reproductive maturity until a certain age or size. In some species this delay has led to the evolution of considerable flexibility in when an individual is male or female within local groups, depending on changes in its social environment.

A number of species of tropical fishes from three families show age-dependent sex reversal. The coral reef fish *Labroides dimidiatus* is particularly well known. The fish is called a "cleaner fish" because it removes ectoparasites from the skin of other fishes. Within any population there are many more females than males, and males are always the largest individuals. The males defend territories in which are included several females of a harem, and there is a linear hierarchy of aggressive interactions depending on size, with the socially most dominant male being the largest.

If the dominant male dies, the largest female becomes a male. The change takes place rapidly, and within a few hours the largest female begins to act like a male. Within one or two days the new male can fertilize the eggs of the remaining females.

Each individual reef fish contains both ovarian and spermatogenic tissue. Ovaries are functional only in the smaller, subordinate "females," and removal of the large, socially dominant male results in rapid development of sperm production in the largest "female." Males are most likely to mate with larger, older females, so this flexible pattern of sex change results in the timing of reproduction by individuals at an age when their fitness is likely to be maximum.

Sex change could occur if one sex can gain an advantage in fertility much more rapidly with age than can the other sex. Thus it "pays" individual reef fish to be female while small and while increasing in size and to delay being male until the individual is older and larger. Many larger animals, such as deer, sheep, and goats, do *not* change sex even though older males of these species monopolize a large proportion of available matings (with a harem system of mating). Why do these species not time a sex change so that individuals can produce more offspring by being female when they are younger? The cost of changing the reproductive anatomy of animals that employ internal fertilization and development would be quite high. In fishes, which have external fertilization and development, the cost for a sex change as compared with its advantages may be sufficiently low to explain the timing of sex changes in appropriate circumstances.

These examples of different patterns of timing reproduction should convince us that a great variety of characteristics of animals can change over both short periods and relatively long periods to influence individual fitness. Timing over short (daily) and long periods (monthly, seasonally, or longer) usually involves chemicals. Chemicals are produced in one part of an animal that have an effect on the rate of development or function of another part of the animal.

Chemicals that function internally in this way are called *hormones*. When chemicals are released from an animal to the external environment and have an effect on another animal of the same species they are called *pheromones*. There are many examples of hormones and pheromones involved with the control of reproduction in which functions must be timed with respect to environmental variation.

CONTROLS FOR DAILY PHOTOPERIOD TIMING

Before considering the controls for longer-term reproductive timing in different major groups of animals, it is important to examine some controls for timing over daily intervals. Animals and plants must process information on seasonal changes from photoperiod variations that occur from one day to the next. Most organisms display some form of *circadian* periodicity for a variety of functions. For example, recall that we discussed daily patterns of energy intake and expenditure in Chapter 10. We noted that a daily pattern of net energy gain could contribute to a seasonal (*circannual*) pattern of timing energy use for events such as seasonal hibernation (Chapter 8).

Timing on a daily basis provides functions similar to timing on a seasonal basis. Food availability may change daily or predator activity may vary daily. Daily survival and longer-term reproductive success can depend on timing activities with respect to these changes, and there are large numbers of examples of circadian rhythms for virtually every type of organism and virtually every type of physiological process. Some species of *Drosophila* emerge near dawn in a circadian rhythm that may enhance ability to obtain resources. When mating occurs in a number of species of fishes and amphibians it is timed to a particular part of the day that may be optimal for short-term survival of either adults or their young. At a general level many species can be classified as diurnal, nocturnal, or crepuscular, depending on the times of day during which they are normally most active (during the day, at night, or at dusk and dawn). These daily activity patterns may reflect changes in food resources, competitors, mates, or predator activities. At a more specific level, circadian changes in digestive secretions, cellular metabolism, and susceptibility to poisons have been described that may represent "built-in" mechanisms to synchronize internal functions with daily changes in the external environment.

What could be responsible for the control of such a diversity of daily timing patterns? There is some evidence from studies of activity rhythms in sparrows (*Passer domesticus*) and some other vertebrates that daily timing is hormonally (chemically) mediated by means of an interaction between two internal oscillators ("biological clocks") and information from the external environment. Although the specific mechanism described below for sparrows may only apply to vertebrates in some respects, the general picture of chemical control may ultimately be extended to all organisms. The experimental procedures also illustrate the general approach used for studies of circadian rhythms.

Because so many functions exhibit daily periodicity, one can be selected that is easy to monitor in sparrows. Locomotor activity can be monitored in isolated individuals by providing perches that are connected through switches to recorders that monitor the activity. Usually, external environmental variables, such as photoperiods, temperatures, food availability, and the presence of other individuals, are carefully controlled so that changes that are observed can be attributed to some change occurring *within* individuals.

Figure 15.5(a) illustrates some typical results for a sparrow monitored for activity on a 12-light-12-dark photoperiod. Each line represents the activity pattern for 24 hours; note for this diurnal species that most activity occurs during the light phase and that the activity change is synchronized to the photoperiod.

If the sparrows are placed in constant darkness they show the pattern of daily activity in the upper part of Fig. 15.5(b). There is still a daily periodicity to activity, but it does not occur at exact 24-hour intervals. In this case the activity pattern shows a period slightly in excess of 24 hours, and the activity period shifts from left to right. In other species the period may be less than 24 hours

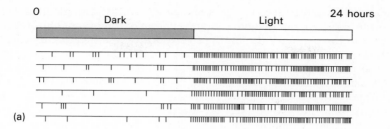

Figure 15.5
(a)   Pattern of locomotor activity typical of house sparrows exposed to a
12-hour light and 12-hour dark photoperiod . Each line represents 24 hours;
each movement of a perch produces the smaller lines extending downward.
(b)   Pattern of locomotor activity observed for house sparrows under constant darkness when a "free-running" rhythm occurs that is slightly longer
in period than 24 hours. The arrows denote the day when the pineal gland
was removed. This was followed by completely aperiodic behavior (based
on Menaker and Zimmerman, 1976).

under constant photoperiod conditions and would shift from right to left; it is
called a *free-running rhythm*. Because it occurs under constant environmental
conditions, the free-running rhythm has been thought to represent an expression of the operation of an endogenous (internal) self-sustained oscillator
("clock") that drives the locomotor activity of the animal. When an environmental cue such as light is present, the internal clock is thought to be synchronized ("set") to daily time. An external cue that operates in this manner is
called a *Zeitgeber* ("time-giver").

Further evidence for endogenous control and its chemical nature comes from some experiments involving manipulation of the *pineal gland*. The pineal is located in the center of the brain of vertebrates and has the appearance of a pine seed. It produces the hormone *melatonin*. The arrows in Fig. 15.5(b) show when the pineal gland of the sparrow was removed. Following removal, the sparrow's activity becomes *completely aperiodic* in constant darkness. This is a very important result, indicating that the sparrows receive no information on time from either internal or external sources (under constant laboratory conditions) without the pineal. If the pineal gland from another bird is cut slightly and placed in the anterior chamber of the eye, a free-running rhythm can be restored within a day in some birds, suggesting that the control is chemical in nature. It is perhaps normally mediated from a circadian periodicity of melatonin production by the gland.

If a bird that shows aperiodic activity patterns following removal of the pineal is then exposed to a photoperiod, it will show a normal, synchronized rhythm similar to that shown by normal birds (Fig. 15.5a). Thus there appear to be two "clocks." The oscillator associated with the pineal gland is *self-sustained* and can influence another oscillator that is not self-sustained but that can be activated by a Zeitgeber such as light.

The general features of an endogenous, self-sustained, chemically mediated oscillator have been studied in other species, although the experimental manipulation of the control is not as complete. The control of daily patterns of timing in some insects appears to involve specific lateral areas of the brain that normally receive information on changes in light intensity. Within plants and unicellular organisms attempts have been made to identify daily changes in cellular functions, such as protein synthesis, that might provide a clue to the nature of a cellular biological clock.

In the remainder of this chapter we will be concerned primarily with longer-term hormonal control mechanisms. In many cases effective long-term timing with respect to resources involves controls by which information on daily changes is integrated with other, more seasonal factors that influence reproductive success.

CONTROLS FOR TIMING INSECT GROWTH AND REPRODUCTION

Insects are divided into two groups, depending on the pattern of development from larva to adult (Fig. 15.6). *Hemimetabolous* forms progress through several molts in the larval stage and then develop reproductive ability after the last molt, in which the only major differences between adult and juvenile forms are a somewhat larger size and developed reproductive abilities. *Holometabolous* forms progress through several molts in the larval stage and then arrest development in a pupal stage (Fig. 15.6), which is a mechanism to deal with adverse environmental conditions. For example, a number of insects overwinter as pupae. The pupae have very low metabolic demands, and they are usually encased in rigid chitin. Under appropriate environmental circumstances, metamorphosis occurs and a reproductive adult with very different morphological features emerges (see Fig. 15.7).

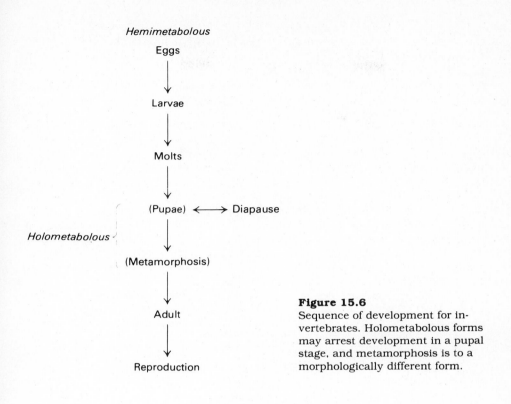

Figure 15.6
Sequence of development for invertebrates. Holometabolous forms may arrest development in a pupal stage, and metamorphosis is to a morphologically different form.

The transformation from larval caterpillar to adult moth or butterfly has fascinated biologists for centuries. Usually the larval and adult forms of holometabolous insects can be considered ecologically and physiologically separate animals. The larvae are primarily feeding machines that are adapted to consume and process particular resources; the adults are primarily reproducing machines that are adapted to a different environment. The transformations within larvae and from larvae to adults depend on two sets of hormonal controls that act with information integrated between the state of the internal and external environments.

The transformations in hemimetabolous insects are similar but less dramatic to the naked eye. Differences between these two types depend partly on factors influencing reproductive success. Hemimetabolous forms may lay their eggs prior to stressful changes in the environment. In holometabolous forms pupation may be used to deal with nonoptimal environments, and the adults are normally more mobile than are the larvae. Mobility permits mate selection, and it also permits dispersal of young by means of the selection by females of hosts for the next generation of feeding machines.

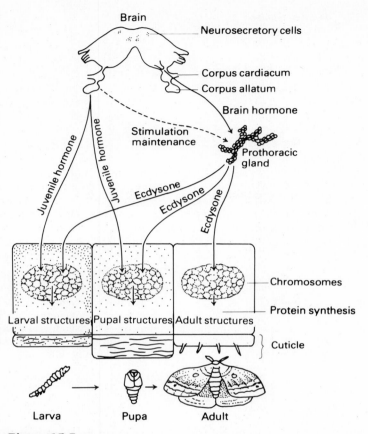

Figure 15.7
Summary of the major hormonal controls involved in development and reproduction for holometabolous insects. Juvenile hormone and ecdysone influence the molts in larval stages; inhibition of juvenile hormone produces metamorphosis to adults (based on Gilbert, 1964).

Hormonal Controls for Insect Growth

The external features of insects change in spurts as a consequence of their rigid exoskeleton. Internal growth is continuous, but an expanded covering must be produced by molting at intervals.

Molting is under the control of a hormone called *ecdysone*, also known as prothoracic gland hormone (PGH), produced in the prothoracic gland (Fig. 15.7). The secretion of hormone into the blood increases just prior to a molt. If the prothoracic gland is removed, molting will not occur, and if extra ecdysone is introduced earlier in development by transplanting the prothoracic gland of one individual into another, molting will occur earlier than usual.

Ecdysone has been shown to induce puffs in salivary-gland polytene chromosomes of the midge *Chironomus*. In *Drosophila*, protein-synthesis patterns change following such chromosomal puffing. These observations led to the suggestion that ecdysone produced its effects by acting directly on the nucleus to influence gene expression. Ecdysone is a steroid hormone (see below), and this mechanism is thought to be involved in steroid-hormone action (see below and Fig. 18.1). Proteins that bind ecdysone with high specificity have recently been localized on insect cell membranes, and the proteins are thought to represent receptors specific for ecdysone. The hormone-receptor complex has been shown to enter the cell nucleus, where it could influence gene expression (Fig. 15.7).

Another hormone interacts with ecdysone during molting to determine reproductive maturity during growth; it is called *juvenile hormone* (JH, Fig. 15.7). Juvenile hormone is secreted from the corpus allatum, a collection of nerve cells connected with the brain (Fig. 15.7). When juvenile hormone is present in sufficient amounts it inhibits or prevents development of adult reproductive ability. So long as juvenile hormone is produced, the larval (or juvenile) form is retained and molting only involves increases in size. Without juvenile hormone, ecdysone results in metamorphosis to a reproductive adult (Fig. 15.7). The relationships are depicted for a hemimetabolous insect in Fig. 15.8.

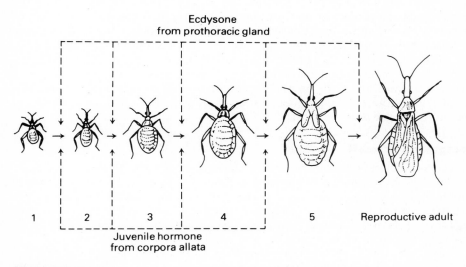

Ecdysone
from prothoracic gland

1   2   3   4   5   Reproductive adult

Juvenile hormone
from corpora allata

Figure 15.8
Summary of the consequences of interaction between juvenile hormone and ecdysone for the hemimetabolous *Rhodnius* (from Schmidt-Nielsen, 1979).

These hormonal interactions have been studied experimentally by manipulating the corpora allata. Removal of these tissues from a young larva leads to precocious (early) development of adult reproductive characteristics. For example, if the corpora allata are removed from *Rhodnius* (Fig. 15.8) after the second molt, a small reproductive adult results after the third molt rather than after the fifth molt. If the corpora allata from a second-molt individual are transplanted into a fifth-molt individual the next molt will result in a larger nonreproductive juvenile, and the adult that results from subsequent molts (after juvenile hormone production decreases) will be much larger than normal.

Brain Hormone from Integration

Clearly, smaller or larger individuals would be produced as reproductive adults, depending on the timing of ecdysone and juvenile hormone production. For maximum reproductive potential it is important that these hormones be either produced or suppressed at appropriate times during development and in appropriate amounts with respect to changes in stimuli from external and internal environments. Information from various sources influencing growth and reproduction is *integrated* (put together) to determine responses. As with some other control systems we have discussed, integration takes place in the brain. Soon we will examine some environmental stimuli that are integrated.

As a result of integration of environmental information, some cells in the brain produce *brain hormone*, which is stored in an area close to the brain called the corpus cardiacum (Fig. 15.7). Release of brain hormone from this area is necessary to produce the release of ecdysone because larvae without the corpus cardiacum will not molt. A hormone that influences the production or release of another hormone is called a *trophic* hormone. We will examine several other trophic hormones when we consider vertebrate reproduction.

Brain integration also influences juvenile hormone production in the corpora allata. The cells that manufacture juvenile hormone are connected with neurons in the brain. If the neurons are cut in the last larval stage, juvenile hormone will continue to be produced rather than being inhibited, and a molt will result in a larger juvenile form.

Integrated Stimuli

Investigations concerning the nature of the stimuli that are integrated primarily involve factors ultimately influencing ecdysone or juvenile hormone. There is much information concerning juvenile hormone because of interest in the dramatic events during metamorphosis to adults. However, the processes of integration of stimuli can all be interpreted with respect to the timing of hormone changes to optimal environmental conditions.

Rhodnius (Fig. 15.8) is a bloodsucking insect that molts only after it has ingested a blood meal. Information from digestive processing is passed to the brain, resulting in the release of brain hormone and then in the release of ecdysone (Fig. 15.9). If a blood meal occurs in the fifth larval stage, digestive information also results in brain inhibition of juvenile hormone, allowing metamorphosis to the reproductive adult to occur.

The body wall stretches with a blood meal and provides the specific stimuli to the brain through neurons (Fig. 15.9). Cutting the neural connection to the brain prevents molting, even after a large meal. The timing of molting

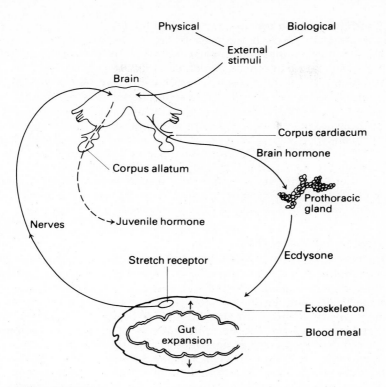

Figure 15.9
Summary of information integration influencing the major hormonal controls for development in *Rhodnius*. Integration of information from physical and biological external environments occurs in
the brain along with information from the internal environment.

and metamorphosis to the internal availability of energy and nutrient substrates ensures that development occurs when resources are optimal for growth
and reproduction.

The internal availability of energy and nutrient substrates is also integrated with a variety of stimuli from the external environment that influence
reproductive success (Fig. 15.9, "external stimuli"). Deviations from optimal
conditions of temperature and humidity can result in arrested development
despite food intake. Long-day photoperiods and/or warm temperatures stimulate growth if food is available. Short-day photoperiods and/or cold temperatures produce arrested development.

Information from an animal's biological environment is also integrated in
the brain and influences development (Fig. 15.9, "biological external stimuli").
The presence or absence of members of the same species can influence rate of
development. At low densities molting activity of cockroaches is stimulated
whereas at crowded densities molting is suppressed. This control will adjust

timing, within limits, to produce numbers of individuals optimal for resource use in reproduction (such as the peak in Fig. 15.1). The mechanism may involve certain pheromones (chemicals) produced by individuals that inhibit brain hormone release in other individuals.

Predators (part of the biological external environment, Fig. 15.9) also influence the timing of development and reproduction. If a predator amputates part of an immature individual early in an intermolt interval, the next molt is inhibited (delayed) long enough to permit regeneration of the lost part. There is also a feedback between other organs and the brain; developing organs influence the rate of molting. Amputations inhibit brain hormone release, and repeated amputations result in larger-than-normal individuals in the next molt, indicating that adjustments are made with respect to the state of development of other tissues. This hormonal control effectively regulates development by maintaining appropriate timing despite injury caused by predators.

Hormonal Control of Death

For many r-selected species, death of the adults is associated with reproduction. The most familiar example of this association is represented by some species of salmon that expend considerable effort to swim up freshwater streams, lay and fertilize their eggs, and die shortly after reproducing. Why do the adults die?

There are two interpretations for the association of reproduction and death: (1) death may be a direct consequence of extreme expenditures of resources required to maximize reproductive output, or (2) death of the parents may make the probability of survival and reproductive success of their young more likely than if the parents continued to live. The second hypothesis is suggested by some results of an experiment on the hormonal control of reproduction and death in the octopus, *Octopus hummelincki.*

Glands near the eyes (the optic glands) of the octopus are the only hormonal tissues known for this animal. The glands produce hormones that are required for egg or sperm production at a certain stage of development. Females and males copulate, eggs are laid, and the female decreases her feeding and constantly protects and cleans the developing eggs. Brooding of the young continues until the eggs hatch in about one month, and the female dies about 10 days later.

This pattern is changed if the optic glands are removed after eggs are laid. The results are abandonment of the eggs, increased feeding and growth of the female, and some resumption of copulation but without any egg laying. The life span of an adult female octopus can be extended from 51 days to as long as 9 months after egg laying if the optic glands are removed just prior to when death would normally occur. Clearly, continued survival of the female is possible despite exertions associated with reproduction; thus the alternative explanation, the evolution of hormonal control of death in the octopus to permit increased reproductive success, is favored. It is not yet clear how the death of the adult promotes the survivorship or reproductive success of the young. Perhaps the young have access to resources that would be unavailable if the adults lived, or perhaps predators are less likely to notice young octopus when the adults are not present.

CONTROLS FOR REPRODUCTION IN SOCIAL BEES: SEX AND PHEROMONES FOR FITNESS

The high degree of cooperative behavior ("altruism") in some social insects is both fascinating and puzzling. The caste system of worker bees that tend a reproductive queen and raise her developing young seems counter to the principle that fitness through natural selection is an individual reproductive trait and should not be a characteristic of groups of animals. Why should the daughters of the queen bee tend their sisters and brothers instead of producing their own offspring?

Recently a hypothesis has been developed that makes individual fitness potentially explainable in cooperative social groups. The hypothesis depends on the extent to which individuals in a group are genetically related to each other, and it provides an interpretation of some controls that have evolved governing growth and reproduction in social bees through the use of pheromones.

If natural selection operates on individuals, an individual should not evolve to benefit another individual if, as a consequence, it experiences a decrease in reproductive success. However, because the guiding principle of fitness is representation of *genetic* material in future generations, and relatives share some genetic material, relatives could benefit each other. This kin-dependent cooperation should occur if cooperation results in a greater degree of future genetic representation than would occur without cooperation.

According to this hypothesis, the extent of altruistic behavior should depend on the *degree of genetic relatedness* (r_o), which can easily be determined for sexual organisms from the average fraction of genetic material shared by relatives. For example, diploid offspring are related to each parent by ½ because each parent contributes half of the genetic material to the offspring. Similarly, sisters and brothers are related by ½, half-sisters by ¼, first cousins by ⅛, and so on. For diploid sexual animals (including humans), altruism is expected to be highest among family (kin) groups. This principle forms one of the bases for a recent, controversial development in biology called *sociobiology*, which basically considers humans as animals subject to genetic effects in relation to reproductive fitness.

Social bees have a somewhat different sexual system; they are haplodiploid. Unfertilized eggs laid by the queen develop into males (the drones). Fertilized eggs develop into females (the workers or another queen; see below). Thus males are haploid and females are diploid. Because of this condition, the degree of relatedness among sister workers is ¾ instead of ½ because the sisters all have common genes with their father; that is, each worker receives a complete set of the haploid drone's genes. The other half of the worker bee's genes comes from half of the genes of the diploid queen; thus

$$r_o \text{ (among sisters)} = (1 \times \tfrac{1}{2}) + (\tfrac{1}{2} \times \tfrac{1}{2}) = \tfrac{3}{4}.$$
$$\phantom{r_o \text{ (among sisters)} = (1} \underset{\text{♂}}{} \phantom{\times \tfrac{1}{2}) + (} \underset{\text{♀}}{}$$

Sister workers can increase their individual fitness more by caring for their younger sisters than giving an equal amount of care to their own offspring, to whom they would be related by only ½.

Drones and workers (male - female siblings) are related only by ¼, meaning that cooperation of workers to raise drones would not normally be predicted. However, the drones are the only vehicle for reproduction available to the sister workers except for a new queen, for the workers do not produce their own offspring.

Because sister bees are more related to each other than to their queen mother (¾ versus ½), sisters could also increase their fitness by caring for a queen that was a sister. This circumstance does not occur. The queen controls the sexual activity of the workers with a pheromone called *queen substance* (9-oxodec-2-eonic acid). This chemical is produced in the mandibular glands of the queen and is ingested by workers when they transfer honey among themselves in the hive. The chemicals used by the queen to manufacture the pheromone are found in "royal jelly," a protein-rich food made from pollen that is fed by the workers to the queen for egg production. Larvae destined to become workers receive only a small amount of royal jelly for development. A larva destined to become a new queen receives large amounts.

When queen substance is present, female larvae that undergo metamorphosis to adults develop normally except that the ovaries do not mature. The pheromone inhibits the effects of ecdysone on the ovaries. In addition, control of the reproductive state of a hive is achieved with this pheromone when environmental conditions change. A new queen is produced when the hive reaches a certain size. The female larva that is fed large amounts of royal jelly produces queen substance, which is attractive to male drones. Copulations take place after a nuptial flight that carries the new queen and drones away from the old hive. A new queen stores large quantities of sperm in a storage sac near the oviduct for fertlization of the eggs destined to become workers in the new hive.

CONTROLS FOR TIMING TERRESTRIAL VERTEBRATE GROWTH AND REPRODUCTION

A number of aquatic vertebrates are relatively small and show r-selected characteristics of reproduction. For example, many fishes and amphibians lay large numbers of small eggs, with each egg having a low probability of survival. There are interesting exceptions, e.g., viviparous fishes, mouthbreeding fishes, and brooding frogs that show degrees of parental care. However, for simplicity we will examine the extreme case of control of K-selected patterns by studying reproductive control in relatively large terrestrial vertebrates.

Hormonal control of reproduction in invertebrates and vertebrates is remarkably similar in general features. The similarity is the result of the common requirement of effective timing for reproductive success. We have all passed through puberty and can attest to the "metamorphosis" that occurs at that time, during which we pass through the sometimes difficult transition to adult reproductive ability.

Controls for K-selected vertebrates are somewhat more complex and more difficult to piece together because of the longer time scales involved in K-selected reproduction and the greater degree of control required for parental care (a major feature for K-selected species). However, the mechanisms still involve the features of brain integration involving internal and external environmental information, "brain hormones," trophic hormones that influence other developmental and reproductive hormones, and pheromones.

The Pituitary Gland and Hormonal Controls for Growth

The pituitary is the site of storage and release of many trophic hormones. It is divided into two anatomical parts: the neurohypophysis, situated close to the hypothalamus, and a larger anterior lobe called the adenohypophysis. The hormones secreted by each part are listed in Table 15.3. We have discussed control of osmoregulation involving ADH (see Chapter 13). With the exception of adrenocorticotrophic hormone (ACTH), all hormones of the pituitary are involved in some aspect of control of growth and reproduction. ACTH is primarily involved in the control of secretions of hormones such as aldosterone and cortisol from the adrenal cortex.

Table 15.3   Hormones released from the pituitary

NEUROHYPOPHYSIS
ADH
Oxytocin

ADENOHYPOPHYSIS
Gonadotrophic Hormones
Follicle-stimulating hormone (FSH)
Luteinizing hormone (LH)
Growth hormone
Thyroid-stimulating hormone (TSH)
Melanocyte-stimulating hormone (MSH)
Adrenocorticotrophic hormone (ACTH)
Prolactin

The importance of the pituitary for effective development and reproduction is apparent from the consequences observed when it is removed or from clinical symptoms associated with reduced pituitary function (hypopituitarism). Removal or reduced function result in a dramatic slowing of growth in young, atrophy of the adrenal cortex, atrophy of the thyroid glands, and absolute reproductive failure in both males and females.

The brain influences production and release of most pituitary hormones. Some cells that produce hormones are located in the hypothalamus. In addition, certain areas of the hypothalamus and brain influence release of pituitary hormones. Some of these areas, indicated in Fig. 15.10, influence the pituitary through nerves or through other trophic hormones called "releasing factors."

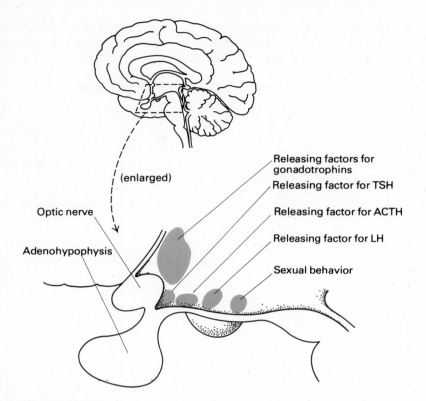

Figure 15.10
Schematic diagram of the location of some areas within the hypothalamus
of vertebrates that influence gonadotrophins (based on Tepperman, 1962).

The releasing hormones were originally thought to be exclusively localized in the hypothalamic region of the brain close to the pituitary. However, recent highly sensitive methods (utilizing radioimmunoassays) have shown that
they are widely distributed throughout the nervous system of mammals and
some other vertebrates. One "releasing" hormone (called *somatostatin*) that is
responsible for inhibiting the release of growth hormone is localized in the
gastrointestinal tract of vertebrates, where it may function in the control of
secretion of gastrointestinal hormones. In addition, thyrotropin-releasing hormone (TRH) has been found in relatively large amounts in the skin of frogs. This
evidence strongly suggests that these releasing factors may have more general
functions than just control over pituitary release of hormones. One important
role that has been suggested is that of chemical transmission of information between neurons (see Chapter 17 for details of this process).

The neurohypophysis is connected with the hypothalamus by means of
neurons, whereas the major connection with the adenohypophysis is through a

portal circulation (the hypothalamic-hypophyseal portal system). The hypothalamic connections provide a means for release or inhibition of hormones as a consequence of integration of information in the brain. For example, the timing of reproduction in a number of vertebrates involves photoperiods (Table 15.2). Information on photoperiod from photoreceptors influences reproductive condition by a long-term effect on releasing factors (see below).

The cellular mechanisms involved in release of some pituitary hormones have recently been studied in detail. We will consider the several ways secretory cells are controlled in Chapter 18 (see Fig. 18.1). The hormones involved in release of mammalian follicle-stimulating hormone, luteinizing hormone, and thyroid-stimulating hormone stimulate the production of cytoplasmic cyclic adenosine 3′5′-monophosphate (cAMP), which in turn influences the function of cytoplasmic protein kinases and, ultimately, the enzyme activity for protein synthesis and release of synthesized products. Note that this mechanism of action differs from that postulated for ecdysone and for vertebrate steroid hormones (see below and Fig. 18.1).

The Pineal Gland and Melatonin

Recall that the pineal gland has been implicated as the site of an endogenous oscillator involved in the control of short-term daily rhythms (see Fig. 15.5). Melatonin has been shown to be produced by the pineal with a circadian periodicity, and the hormone appears to accumulate in the region of the hypothalamus where it may have an effect on the longer-term releasing factors infuencing growth and reproductive development in vertebrates. This type of an influence could provide a mechanism whereby daily timing mechanisms (circadian rhythms) lead to seasonal timing (circannual rhythms) in vertebrates. However, there is little information on the long-term effects of the pineal or its chemical products on seasonal rhythms. The possible feedback relationships between the hypothalamus, pituitary, and pineal represent exciting possibilities for research on the controls for timing and the controls for linking short-term timing to long-term timing.

The characteristics of K-selected reproduction involve delayed development of reproductive ability, parental investment in the high quality of relatively few young, and several potential cycles of reproductive effort (Table 15.1). We will consider controls for these traits by discussing hormones involved in the timing of both growth and reproduction in males and females.

Growth

Growth hormone influences the rate of incorporation of net energy and nutrient gains into new body substance. Juveniles produce more growth hormone than do adults, and abnormalities of pituitary function can lead to abnormally low or high growth rates in humans (Fig. 15.11). Some growth occurs without growth hormone, but the hormone is necessary for coordinated development over long time periods.

The effects of growth hormone are often compared with the effects of insulin and glucagon, two hormones involved in very short-term control of food

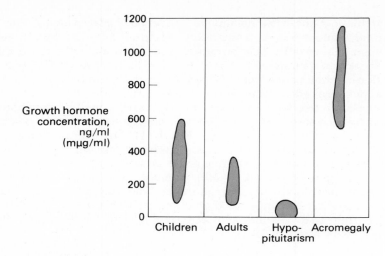

Figure 15.11
The production of growth hormone is somewhat higher in chil-
dren than in adults and is abnormally low in cases of hypopituitarism and abnormally high in cases of acromegaly (based on
Tepperman, 1962).

processing (see Chapter 10). Injections of growth hormone produce effects
similar to glucagon and opposite to insulin, such as release of fatty acids and
production of glucose from intermediary metabolic conversions in the liver.
However, growth hormone does not normally act over short periods but is produced to influence rate of development over months or years. The long-term
characteristics of the hormone have made it very difficult to study except in extreme disease states.

Growth hormone also has the short-term effect of increasing amino acid
uptake by cells. This increase is important for growth in the young, and lower
growth-hormone production in adults is associated with higher use of proteins
for energy requirements or egg and sperm production. Moreover, growth hormone interacts with thyroid-stimulating hormone (TSH) and thyroxine. Thyroxine influences the rate of use of assimilated food substrates and the general
level of metabolism, both of which have an effect on growth rate. TSH and thyroxine are also involved in the control of metamorphosis of amphibians from
aquatic to semiterrestrial forms.

Growth in terrestrial vertebrates is not associated with periodic molts;
otherwise, growth hormone is quite similar to ecdysone. Because this similarity
exists, we might expect long-term interactions between growth hormone and
hormones involved in sexual development similar to the interactions between
ecdysone and juvenile hormone in invertebrates.

For *K*-selected species, reproductive maturity is delayed because of the
consequences of size-related factors on survival and reproduction. Although
there is little experimental information, growth hormone may be involved in

the control of this long-term reproductive timing by means of an inhibitory effect on sexual tissues. The stimuli that are integrated for this control and the possible mechanisms of action represent exciting possibilities for future research efforts. For example, the effect of the environment on growth and reproductive development in humans is suggested from long-term changes in average size and age of puberty. Human size has increased over the years, and the average age of puberty has decreased by several years, particularly in females. Because these factors can have effects on reproductive success, it would be very interesting to understand how they are controlled and what aspects of environmental change represent effective stimuli for the changes.

Gonadotrophins and Male Hormones

At puberty the male metamorphosis is associated with the increased production of sexual steroid hormones known as androgens. The principal androgen is *testosterone*, which is produced by cells in the testes that are called the interstitial cells of Leydig. Production of testosterone is under the control of the gonadotrophic (trophic hormone influencing production of gonadal hormones) adenohypophyseal *luteinizing hormone* (LH), which gets its name from its function in females, where it was first discovered (see below). Puberty is associated with increased adenohypophyseal sensitivity to luteinizing-hormone releasing hormone (LHRH).

The androgen produced in the testes has three major effects. It is necessary for the production of sperm by the seminiferous tubules of the testes; it has a negative-feedback effect on the release of the gonadotrophic LH from the adenohypophysis; and it has an effect on the development of secondary sexual features associated with "maleness" (structures and behaviors influencing competition for mate selection). These relationships are shown diagramatically in Fig. 15.12.

The male androgens (and female estrogens discussed below) are *steroid* hormones, derived by synthesis from cholesterol. These hormones (also including progesterone and steroids secreted by the adrenal cortex as well as ecdysone in invertebrates) influence tissues primarily by affecting nuclear genomic activation (see Chapter 18, Fig. 18.1) rather than by changes in cytoplasmic cAMP. The steroid hormones form a complex with a membrane receptor that ultimately influences messenger-RNA formation in the nucleus and subsequent gene expression.

The feedback between a hormone and a gonadotrophic hormone maintains appropriate concentrations of the hormone. A drop in testosterone concentration in the blood is detected in the hypothalamus and/or adenohypophysis (Fig. 15.12). It results in increased release of the trophic LH, which stimulates increased testosterone production in the testes. Increased secretion of testosterone results in less LH release, with the net effect that testosterone remains at a level appropriate for effective reproduction. What is effective (the set point for control) depends on environmental changes integrated in the brain that influence the level of trophic hormone release (Fig. 15.12; see below).

Sperm production is under the control of the gonadotrophic *follicle-stimulating hormone* (FSH). Although feedback may occur between the

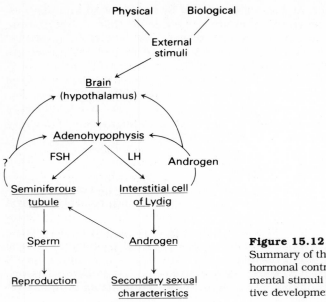

Figure 15.12
Summary of the integration of the major hormonal controls with the environmental stimuli that influence reproductive development in male vertebrates.

seminiferous tubules and the brain or pituitary (the hormone involved is not clearly understood), the timing of sperm production is more periodic than the relatively constant "maleness" features resulting from androgen production. In animals that mate several times, it is important that the rate of sperm production be increased soon after an ejaculation. Increased sperm production should occur until an amount optimal for reproductive success has been produced, and the rate of production should then decrease. Thus FSH should be most important for stimulating sperm production just after copulation. This stimulation is particularly important for some large vertebrates of which the males mate with several females (polygyny) of a harem (pheasants, horses, cattle, buffalo, sheep, elk, etc.); the probability of male reproductive success depends on multiple matings within a short time.

The association of androgens with secondary sexual features and with primary sperm production is also important for reproductive success. Females exercise considerable choice over mates, and the quality of a potential mate must be assessed in a relatively short period before reproduction to ensure the highest possible quality of the few young produced. A male that signals a high degree of "maleness" by means of aggressive behavior and other secondary sexual features (antlers, horns, size, etc.) produced from the effects of the same hormone-determining sperm production will thereby signal some information to a female on numbers and quality of sperm.

Integrated Stimuli

As was the case with invertebrates, timing reproductive activities to environmental stimuli involves integration of information in the brain, and the information comes from an animal's physical and biological environment (Fig. 15.12). Sexual activity must be coordinated between males and females. Timing is usually more dependent on the reproductive condition of the female (see below) because of the imbalance in reproductive investment between males and females for many K-selected species. Even though eggs are more expensive to produce, sperm production still represents a cost to a male that can be minimized by timing production to optimal environmental conditions for mating.

Use of seasonal photoperiod changes is common for timing reproduction in male vertebrates (Table 15.2). Males of many K-selected species are reproductively "refractory" (no sperm production) during short-day photoperiods, when gestation periods dictate birth of the young in a short time (smaller species) or in seasonal multiples (larger species), so that development of the young will occur under optimal conditions with respect to food and temperatures. Species with intermediate seasonal periods for gestation time sperm production and mating to short day lengths (Table 15.2), and young are also born at times optimal for their development.

In some songbirds there is evidence that "measurement" of specific daily periods influences male reproductive activity. Male white-crowned sparrows kept on a 6-hour light—18-hour dark photoperiod showed maximum testicular development when the dark period was interrupted by a brief light period 11 hours after the onset of darkness (Fig. 15.13). In this case the daily (circadian) rhythms become integrated with seasonal (circannual) reproductive rhythms under particular daily conditions.

A direct environmental effect of temperature can also influence male reproduction in some mammals. The testes of a number of animals are located in a scrotum, which provides a temperature environment 1-6°C cooler than the core of the body. In some species the testes descend and ascend at certain times, and changes in testicular temperature will influence sperm production and viability.

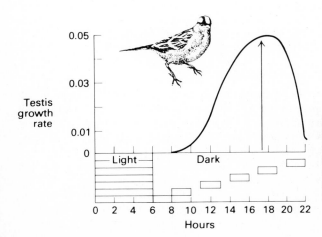

Figure 15.13
For the white-crowned sparrow (*Zonotrichia leucophrys*), maximum testicular development occurs when the dark period is interrupted after about 11 hours (after Farner, 1955).

A vascular countercurrent heat-exchange system in the scrotum can maintain testicular temperatures somewhat below body temperature (Fig. 15.14). Heat exchange takes place between venous and arterial blood. Temporary male sterility has been produced from exposure of laboratory animals to high environmental temperatures, and domestic rams (*Ovis aries*) typically become sterile in the summer. This type of external environmental influence will interact with internal hormone controls to produce seasonal timing of reproduction in males.

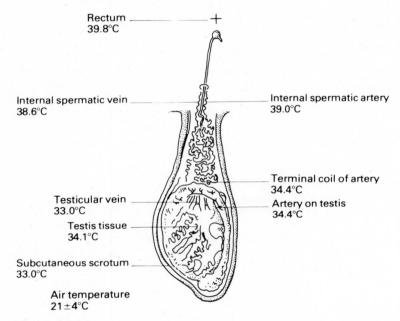

Figure 15.14
Measurements of temperatures at various points within a testis enclosed in a scrotum. Vascular countercurrent heat exchange contributes to cooler temperatures, which makes sperm viability possible (based on Nalbandov, 1958).

Gonadotrophins and Female Hormones

Reproductive timing is much more important in female vertebrates because of a relatively high degree of reproductive investment by them. Eggs are more expensive to produce than are sperm, and females of *K*-selected species invest much more time and energy in young during gestation and following birth, during the period of parental care. Males assist with the care, depending on the impact of resources on the probability of success of the female effort, which has led to the evolution of a variety of mating systems, including promiscuity, monogamy, polygyny, and polyandry. However, the main burden for reproduction is unequally distributed between the sexes, requiring more elaborate controls for vertebrate females.

Like the testes, the ovaries produce both gametes and hormones. Figure 15.15 indicates the degree of complexity (consider the number of arrows) of the control system governing female reproduction in humans, partly a consequence of differential reproductive investment.

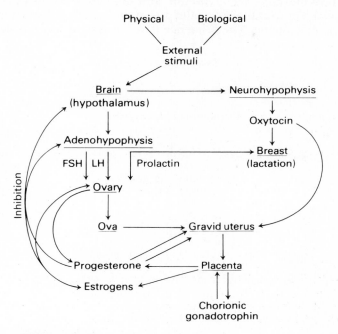

Figure 15.15
Summary of the integration of the major hormonal controls with the environmental stimuli that influence reproduction in female vertebrates. Note the higher degree of complexity as compared with male reproduction (see Fig. 15.12).

Puberty is initiated in females with an increase in the output of gonadotrophins (FSH and LH) from the adenohypophysis that is produced by increased sensitivity to hypothalamic releasing factors (LHRH). In humans, increased LH secretion occurs in "pulses" during early puberty, particularly during sleep. In sheep, puberty is associated with periodic release of LH at intervals of less than one hour. Increased FSH amounts stimulate ovaries to produce *estrogen*. Estrogens are female steroid analogues of androgens and produce several results by means of effects on gene expression. Initially, estrogens result in rapid growth of the ovaries and the development of eggs (ova). Estrogens also stimulate development of secondary sexual characteristics associated with "femaleness" (breast development, pelvic changes, etc.).

The estrogens produced by the ovaries also influence the hypothalamus and adenohypophysis (Fig. 15.15). Low levels of estrogen initially stimulate more FSH release. When estrogen levels increase, FSH output is reduced and the output of LH increases. When LH concentration reaches a sufficient level (from the frequency of pulse releases) it stimulates ovulation (release) of the egg(s) from the ovaries. The tissue that had surrounded the egg in the ovary then develops into the *corpus luteum*, and this tissue produces estrogen and *progesterone*. The combination of estrogen and progesterone inhibit LH and FSH release in humans. The sequence in an ovulation cycle is shown in Fig. 15.16.

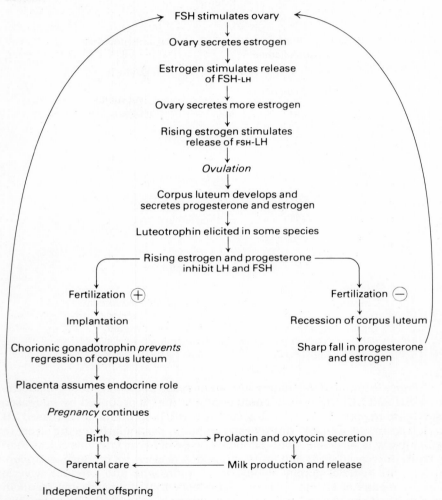

Figure 15.16
Summary of the sequence of events involved in hormonal control of ovulation, pregnancy, and postnatal female parental care (based on Tepperman, 1962).

Further changes in the sequence depend on whether fertilization takes place (Fig. 15.16). If no fertilization of the egg occurs, the cycle is repeated at some future time after the corpus luteum degenerates and releases inhibition of subsequent production of FSH and LH. If fertilization does occur, the egg implants in either the oviduct or the uterus. Meanwhile, these tissues have changed under the influence of estrogens and progesterone (Fig. 15.17). If implantation occurs, the uterus produces *chorionic gonadotrophin*, which inhibits breakdown of the corpus luteum such that FSH and LH are inhibited through continued estrogen and progesterone production. In mammals the placenta maintains this inhibition in later stages of pregnancy. The inhibition ensures that a limited number of eggs (potential offspring) are dealt with at any given time; that is, the number of eggs is low enough to avoid extreme internal supply-demand problems for reproduction.

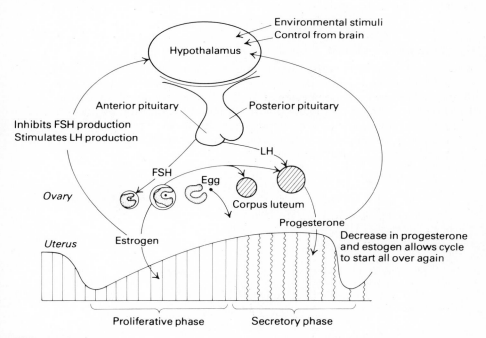

Figure 15.17
Schematic diagram of the sequence of events and interactions between tissues of an ovulatory cycle in a mammalian female.

This complicated cycle is subject to considerable selection for reproductive success, and species differ in the timing of female fertility. The average length of a cycle for egg production is 4-6 days for rats, 14 days for guinea pigs, 20 days for cows, 28 days for humans, 36 days for chimpanzees, and once a year for species such as marmots or bats that may hibernate for long periods.

The minimum period for a cycle represents limits on the ability to produce an optimum number of eggs from available resources. The maximum period for a cycle reflects environmental restrictions on the probability of success of reproduction. Within these broad limits, females are most fertile within a short time of ovulation. The rest of the cycle after ovulation represents physiological commitments to young that must be made if fertilization occurs.

If environmental conditions are optimal for the subsequent development of young, it is important that copulation occur close to ovulation. This timing is reflected in the sexual behavior of females of many species that only become "receptive" to males and exhibit mate selection close to the time of ovulation, in part governed by hormones and by integration of information in the brain, which influences hormones (see below). In some species, such as baboons, the females signal the stage of their reproductive cycle by changes in the color of external "sexual skin," influencing sperm production and mating behavior of the males.

Integration of Stimuli

A variety of signals from the internal and external environment are integrated in the brain and influence the ovulatory cycle. Photoperiod influences whether ovulation occurs in some species as well as the frequency of ovulation (Table 15.2) by means of its relationship with variations in optimal physical environmental conditions. Females of some species (rabbits, ferrets) can be manipulated to cycle continuously if they are continuously exposed to long-day photoperiods.

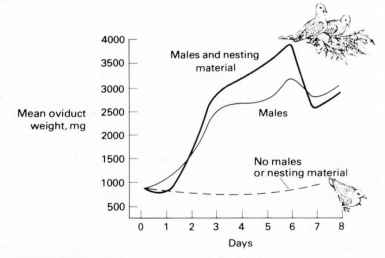

Figure 15.18
When female ringdoves can see and hear males it stimulates ovulation. Control females, which receive no exposure to males, do not ovulate. The presence of nesting material can also influence female timing of reproduction (based on Lehrman, Brody, and Wortis, 1961).

Visual and tactile stimuli can also influence ovulation timing in some species. Rabbits and cats, for example, time ovulation to the occurrence of copulation by detection of the mechanical stimulation of the vagina. Ringdoves (*Streptopelia risoria*) start ovulatory cycles in response to the sight of nesting material and the sight or sound of reproductively active males (Fig. 15.18).

Pheromones can also influence timing. Odors can signal the stage of a reproductive cycle between sexes. College females in dormitories have been reported to cycle in synchrony, which has been interpreted as a mechanism for competition for mates, possibly involving pheromones. In general, integration of information on diverse environmental conditions will lead to effective allocation of resources for reproduction, particularly in the sex for which reproductive investment is high.

Gonadotrophins in Other Vertebrates

Similar hormones appear to be involved in the control of reproductive timing in fishes, amphibians, and reptiles, although there is much less information concerning their function for these animals. One problem of assessing the function of hormones in these animals comes from a method of experimenting by injecting hormones (sometimes not specific to the species) rather than assessing changes in hormones during natural reproductive cycles. In amphibians, LH influences ovulation, testicular androgen secretion, and ovarian progesterone secretion, but these functons appear to be under the control of FSH in a number of reptiles. There appears to be a high degree of similarity in the chemical structure of hormones from different vertebrate groups, meaning that differences in function may reflect differences in receptors at the membranes of tissues influenced by the hormones. Reptiles, amphibians, and fishes show a wide diversity of reproductive patterns, and more information on the mechanisms of hormonal controls in them will provide interesting information.

Hormones and Development of Sexual Behavior in Vertebrates

The male and female sex steroid hormones discussed above (androgens and estrogens) are produced during development and influence a variety of characteristics, including sexual behaviors. From experiments on some birds and mammals there is evidence that development of copulatory behaviors is influenced by the action of sex hormones on the brain during an early "critical period" of development; the effects are different for birds and mammals.

When chicken and Japanese quail embryos are treated with either androgens or estrogens, the genetic males are subsequently feminized, i.e., they show adult copulatory behaviors indistinguishable from normal genetic females. There is evidence suggesting that androgens are converted to estrogens within the brains of these animals, indicating that the brain is initially "male," and it normally develops female function when exposed to estrogens produced by the developing animal. The opposite situation occurs in mammals. Treatment of embryos of mammals with either sex hormone masculinizes adult behavior. Thus the female is the initial ("neutral") sex in mammals, and male behavior normally results from fetal androgen production.

Why are birds and mammals different in their response to development of sexual behaviors? In both classes the "neutral" sex is the homogametic one (the sex with two similar chromosomes, denoted as XX). Thus in mammals the homogametic (and the neutral) sex is female; in birds it is male. In addition, it has been suggested that internal development in mammals exposes the embryo to female hormones. If the male sex in mammals was "neutral," embryos could be feminized by their mothers.

Recent techniques have shown that certain cells of the brain in a variety of vertebrates "concentrate" steroid sex hormones. Cells that selectively bind sex steroids have been found in the hypothalamus and may be involved in neuroendocrine control of gonadtropins or in sexual behavior (Fig. 15.10). The hormones may interact with specific areas of the brain to produce specific neural "circuits" during a critical period of development, and the circuits may control such functions as copulatory behaviors. The hormones involved may differ among species, depending on which sex is "neutral" and on the stage of development of the organism.

Pregnancy and Other Parental Care

During pregnancy in humans the corpus luteum and/or the placenta maintain relatively high levels of estrogen and progesterone secretions. This high-level maintenance prevents further ovulation and involves the principle of some birth control pills, which mimic pregnancy by keeping estrogen and progesterone levels high.

The high levels of estrogen and progesterone during pregnancy promote development of tissues associated with the growth and maintenance of the young (oviductal, uterine, and/or placental tissues). These hormones also influence the initial development of structures used for the care and feeding of young following birth. The mammary glands of mammals begin increased development during pregnancy, and the incubation patch of birds develops toward the end of pregnancy.

Progesterone and estrogen levels decrease just prior to birth. The decrease in progesterone secretion results in production of *prolactin* by the adenohypophysis (Fig. 15.15). Prolactin production results in active secretion of milk in mammary glands and increased growth of epithelial cells in the crop of doves. The epithelial cells break off and form a "crop milk," which is regurgitated to the young after birth. The relationship between changes in the oviduct and crop tissues during a reproductive cycle for a ringdove is shown in Fig. 15.19. The sequential timing produced by hormone changes permits shifts in internal allocation of resources as the demands of the developing young change.

In mammals, *oxytocin* is released from the neurohypophysis during labor. It stimulates contractions of uterine muscles to force the young through the birth canal (Fig. 15.15). Oxytocin also plays a role in parental care following birth. When young mammals suckle a nipple the tactile stimuli influence the preoptic hypothalamus, where a releasing factor results in release of oxytocin. The oxytocin, in turn, stimulates the milk "let-down" reflex, producing a flow of milk from the mammary glands to the young.

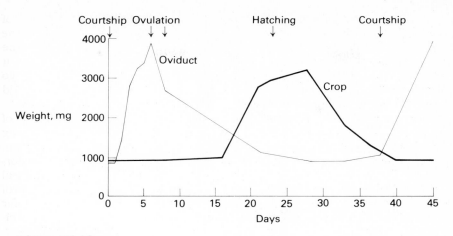

Figure 15.19
Relationship in ringdoves between the cycles for ovulation (potential investment in offspring) and crop milk production (committed investment in care of offspring).
From "The reproductive behavior of ring doves," D.S. Lehrman. Copyright © 1964 by Scientific American, Inc. All rights reserved.

Oxytocin also plays an important role in timing additional reproduction during the period of parental care following birth. An interesting example of this reproductive timing that is dependent on environmental factors has been studied in red kangaroos (*Megaleia rufa*). Irregular and unpredictable periods of drought place severe pressure on a female to provide a supply of milk for developing young, and a red kangaroo can have up to three young, "staggered" in various stages of development at one time.

The first estrus and mating of a breeding season produces (after fertilization) a young "joey" that migrates to the marsupial pouch after a 33-day gestation. The female can breed again two days later. Fertilization occurs, but the suckling of the first joey results in arrested development (diapause) of the second uterine young caused by the effects of oxytocin-inhibiting progesterone secretion from the arrested development of the corpus luteum (Fig. 15.20). While in the pouch, the young feeds on high-protein milk. After the first young leaves the pouch it will return to suckle from mammary glands that provide low-protein, high-fat milk. Oxytocin production is not involved in release of this milk; thus the uterine young resumes development and is born in about a month.

There are two ways to interpret this type of control: from the point of view of the parent, and from the point of view of the offspring. The two interpretations are not identical. The pattern of reproductive timing in the kangaroo will maximize the rate of production of young and will result in rapid replacement of young that die if the supply-demand problems of parental care cannot be met because of environmental variations. When all young survive, the pattern of staggered development also distributes resources between the young in a manner optimal for the reproductive success of the female.

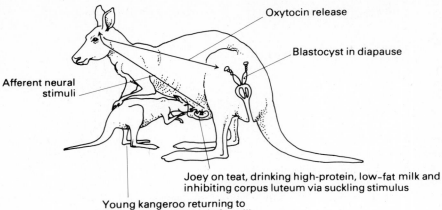

Oxytocin release

Blastocyst in diapause

Afferent neural
stimuli

Joey on teat, drinking high-protein, low–fat milk and
inhibiting corpus luteum via suckling stimulus

Young kangeroo returning to
drink low-protein, high–fat milk

Figure 15.20
Diagram of the controls over the timing of offspring development in red kangaroos
such that young are staggered at three stages of development (from Short, 1972;
after Gunderson, 1976).

The other interpretation of this control is that it ensures survival with
respect to the young; that is, the young that obtains the high-protein milk can
be thought of as competing for resources with its mother and future siblings to
ensure its own survival. The two interpretations are similar but differ in emphasis, depending on degrees of relatedness (r_o) between parent and offspring
($r_o = \frac{1}{2}$), between offspring and offspring ($r_o = \frac{1}{2}$), and between an offspring
and itself ($r_o = 1$). This point of view provides an explanation hypothesized for
events that accompany the termination of parental care in K-selected species.

The period of parental care must ultimately come to an end. This usually
occurs at a time when the young have developed the ability to maintain
themselves; the time varies with size and development period. The termination
of parental care represents a conflict situation between parent and child. The
biological interests of a K-selected parent are to terminate care to one offspring
so that others can be produced, thus increasing its reproductive fitness, but to
delay doing so until the survival of the young already invested in is relatively
ensured. The biological interests of the young are to continue to use a source of
resources supplied by another individual (a form of parasitism); this use can
enhance its survival and reproductive success. However, the young should also
have some interest in the reproduction of the parent, because the young are
related to their parents and to possible future siblings.

Parents and offspring are, on average, related by $\frac{1}{2}$ in diploid animals, but
each individual (parent and offspring) is entirely related to itself ($r_o = 1$).
Because an offspring is completely related to itself, it may demand parental care
until the cost to the mother (where $r_o = \frac{1}{2}$) is two times the benefit to itself. If
cumulative parental care decreases with time (parents control access to
resources) the period of conflict will start when the ratio of benefits to costs

"perceived" by the parent is 1 and will continue until the ratio of benefits to costs "perceived" by the young are ½ (twice the costs "perceived" by the parent).

This relationship is shown graphically in Fig. 15.21, showing that two different "perceived"-cost functions (one for parent and one for child) with the same cumulative-benefit function from the parent result in different maximum differences for each, producing a conflict "zone." This type of interpretation also applies to other aspects of reproducive timing, such as the "competition" between siblings for developmental resources supplied by the parents. In this case also, siblings are related to themselves by 1 and to each other by ½; thus there should be a degree of sibling "rivalry." The argument can be extended to a number of events that occur following fertilization when the degree of relatedness is established. For example, it would be interesting to see if there are controls that may differ governing the competition between twins for maternal resources in the uterus, depending on whether the twins are identical ($r_o = 1$) or not ($r_o = ½$).

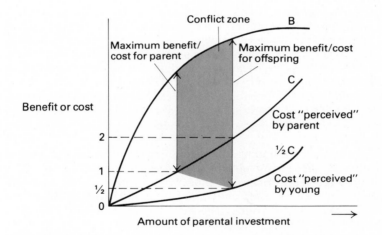

Figure 15.21
The points for maximizing benefits (B) that are cumulatively decreasing relative to costs (C) for changing parental investment differ for parents and offspring, depending on "perceptions" for what costs are, based on degrees of relatedness. This difference results in a "zone" of conflict (shaded area) between parent and child (modified from Trivers, 1974).

You should recognize the arguments based on r_o. They are similar to other hypotheses we have considered in which regulation and control are interpreted according to "economic" principles (benefits versus costs) that influence efficiency. Regulation and control in reproduction are a bit more complex because efficiency involves not only short-term survival but also ultimate reproductive success. However, the biological principles underlying the evolution of patterns of long-term reproduction and development are usually similar to those underlying respiration, circulation, acid-base balance, feeding, temperature regulation, and osmoregulation.

SUMMARY

Fitness, i.e., ultimate biological efficiency or success, involves survival and reproduction, which allows an individual's genetic material to have a better-than-average representation in future generations. Increasing or maintaining fitness involves both adaptations for survival and the additional demands of obtaining extra resources for reproduction. Virtually every aspect of an animal's function and behavior can change, with consequences for relative fitness, and biologists recognize at least two general patterns of reproductive tactics based on supplies of environmental resources as compared with the biological demands of reproduction: r-selected and K-selected reproduction. These patterns are based on two aspects of population growth.

Species that are r-selected have characteristics normally suitable for exploiting relatively large environmental resource supplies. The characteristics include small size, rapid development, early breeding during growth, lower adult survival, and production of many small young with low levels of parental care. K-selected species have features normally suitable for exploiting relatively limited environmental resource supplies. These features include large size, delayed development and reproduction, higher adult survival, and production of few young with higher levels of parental care. Some r-selected species are represented by invertebrates; some K-selected species are represented by terrestrial vertebrates.

Timing of reproduction with respect to environmental variation is crucial for reproductive success for both r- and K-selected species. Species that are r-selected may breed only once and should have evolved mechanisms to ensure that reproduction occurs when daily and seasonally available resources are optimal for survival of the young. K-selected species should have evolved to minimize the extra supply problems associated with reproduction.

Seasonal and daily timing occurs by means of responses to photoperiod. Day length represents a reliable indicator of the availability of other, more variable, physical environmental resources (temperature, moisture, food availability). The biological environment can also influence timing. Predators can result in synchronous breeding in some species to produce "predator satiation." Predators can also cause long-term timing of reproduction by which the "prey" delay reproducing to beyond the life cycle of some predators (e.g., 17- or 13-year periodical breeding in cicadas). The controls for the timing of reproduction involve the use of chemical hormones and/or pheromones that influence development.

Controls for daily (circadian) timing in vertebrates (sparrows) suggest that hormones may be involved in producing some endogenous rhythms. The daily rhythms of a variety of functions of organisms may represent interactions between a self-sustained oscillator, another internal oscillator (not self-sustained), and environmental Zeitgebers. House sparrows become completely aperiodic in locomotor behavior during constant darkness following removal of the pineal gland, but a Zeitgeber (such as photoperiod) can still entrain a rhythm. An endogenous, circadian production of melatonin by the pineal of vetebrates may be involved in the control of internal oscillations on a daily basis that, in turn, can lead to timing of seasonal (circannual) rhythms.

Insects that are r-selected grow in spurts of molting when the exoskeleton is replaced at intervals. The last of a series of molts involves the development of adult reproductive abilities. In hemimetabolous forms, the adult is similar in morphology and function to the juvenile forms. In holometabolous forms, the adult is morphologically and functionally a "different" animal. The pupal stage in the latter usually involves arrested development (diapause) in response to adverse environmental conditions for reproduction. In both forms, development through molting and metamorphosis to adults must be timed to changes in the external and internal environments.

Ecdysone produced in the prothoracic gland together with juvenile hormone produced in the corpora allata determine development. When both are present, a molt occurs when ecdysone levels increase, and the result is a larger juvenile form. When juvenile hormone is inhibited, ecdysone production results in metamorphosis to a reproductive adult. Integration in the brain results in release of brain hormone from the corpus cardiacum, which stimulates ecdysone release. Integration can also result in inhibition of juvenile hormone release at the last molt. The stimuli that are integrated involve changes in the environment that signal optimal conditions for development and/or reproduction. Brain hormone release and/or inhibition of juvenile hormone will not occur in some species unless feeding has occurred at suitable photoperiods and temperatures. Predators can also influence brain hormone release and inhibition of juvenile hormone; amputation of a limb leads to delayed development and regeneration of lost parts.

Reproduction in social bees involves sister worker bees caring for their younger sisters and brothers instead of producing their own offspring. This "altruism" can be explained through the concept of degree of genetic relatedness (r_o). Care of the young sisters increases genetic representation in future generations for the workers because they are more related to their sisters ($r_o = $ ¾) than they would be to their own offspring ($r_o = $ ½) (males are haploid and females are diploid). The queen mother bee prevents sister worker bees from producing a queen who is a sister (who would be more related) by producing the pheromone "queen substance" that inhibits development of the ovaries of the workers.

The pituitary of K-selected terrestrial vertebrates produces most of the trophic hormones involved in the control of timing development and reproduction. Growth hormone influences the rate of incorporation of net gains of energy and nutrients into new tissue; it is analogous to ecdysone. As such, it may also interact with other hormones to generate the long-term timing of development of adult reproductive ability.

Male metamorphosis in vertebrates occurs at puberty when increased frequency of luteinizing hormone (LH) production in the adenohypophysis stimulates androgen production by the interstitial cells of Leydig in the testes. Androgens and follicle-stimulating hormone (FSH) from the adenohypophysis are necessary for sperm production by the seminiferous tubules of the testes. Androgens also influence the hypothalamus and adenohypophysis as well as the development of secondary sexual features associated with "maleness" (e.g., antlers, horns, size) that have been selected for as indirect indicators of sperm quality and quantity by females in making a choice of mates. Sperm production can be influenced by brain integration of information and its effects on FSH and LH release. Releasing factors in the brain can be stimulated or inhibited as a consequence of environmental stimuli such as photoperiod or the presence of reproductively receptive females. Environmental temperatures can also directly influence sperm production and viability in species with a scrotum if those temperatures exceed values low enough for effective maintenance of testicular temperatures cooler than the body (as a result of vascular countercurrent heat exchange).

Females invest more in gamete production than do males and also invest in initial care of developing young during pregnancy; thus timing reproduction is important to ensure sufficient resource supplies for the demands of reproduction. Puberty occurs when the adenohypophysis produces FSH and LH as a result of increased sensitivity to releasing factors. FSH stimulates estrogen production by the ovaries, which, in turn, influences FSH production and, ultimately, LH production, leading to ovulation and development of the corpus luteum, which produces estrogen and progesterone. The combination of estrogen and progesterone inhibits FSH and LH release, meaning that only a limited number of potential offspring are dealt with for a given reproductive attempt.

If no fertilization occurs, the corpus luteum degenerates, and this degeneration will lead to another cycle, depending on resource effects on egg production and/or optimal conditions for development of the young. If fertilization occurs, commitments to the developing young continue by suppression of additional ovulation as a result of maintenance of the corpus luteum with chorionic gonadotrophin.

Timing reproductive activities to the ovulatory cycle involves female receptivity for mating at a time close to ovulation. This timing can be achieved by integrating information on availability and quality of resources from stimuli such as photoperiods, availability of nesting material (e.g., in ringdoves), and the presence of reproductively active males as detected from sight, sound, or as a consequence of copulation.

The development of sexual behavior in vertebrates depends on secretion of sex steroid hormones during an early "critical" period of development. The homogametic sex (XX) is the "neutral" sex (females in mammals; males in birds). Production of sex hormones along with their concentration and conversion in certain areas of the brain normally leads to the development of sexual behaviors for the opposite sex.

The pregnancy phase of parental care is terminated at birth when progesterone and estrogen production decrease, resulting in production of prolactin by the adenohypophysis and oxytocin production by the neurohypophysis.

Prolactin stimulates production of food for the newborn in the form of mammary or crop milk. Oxytocin stimulates uterine contraction during birth and the release of milk from the mammary glands during nursing in mammals.

The period of parental care following birth continues with parental investment in young to increase the probability of their survival and exploitation of a food resource by young for their biological fitness. Because young are completely related to themselves but only related to their parents and siblings by ½, conflicts arise over termination of care. Young should have evolved to demand care for a longer period than parents have evolved to provide it. This type of interpretation is based on efficiency (different "perceived" costs and benefits from reproduction for parents and offspring). It provides a way to view some controls that have evolved for the timing of offspring production by which parents and offspring (siblings) may compete for limited resources influencing their ultimate efficiency as living organisms.

ANNOTATED REFERENCES

Chapter 15: Resource Regulation and Control in Reproduction

Adkins, E.K. (1978). Sex steroids and the differentiation of avian reproductive behavior. *Amer. Zool.* 18:501-509. (This paper is one in an extensive symposium on brain-hormone interactions in the same volume.)

Austin, C.R., and **R.V. Short** (eds.) (1972). *Reproduction in Mammals.* Vol. III, *Hormones in Reproduction.* Cambridge University Press, New York.

Barrington, E.J.W. (1975). *An Introduction to General and Comparative Endocrinology.* Oxford University Press, New York.

Bonner, J.T. (1965). *Size and Cycle: An Essay on the Structure of Biology.* Princeton University Press, Princeton. (Patterns of timing dependent on animal and environmental characteristics.)

Dawkins, R. (1976). *The Selfish Gene.* Oxford University Press, London. (Delightful book concerning the basis of evolutionary fitness as a representation of future genetic material.)

Gilbert, L.I., **W.E. Bollenbacher**, and **N.A. Granger** (1980). Insect endocrinology: regulation of endocrine glands, hormone titer, and hormone metabolism. *Ann. Rev. Physiol.* 42:493-510. (Review of diverse information on hormonal controls for insect development.)

Hamilton, W.D. (1972). Altruism and related phenomena, mainly in social insects. *Ann. Rev. Ecol. Syst.* 3:193-232. (Application of principles of degrees of relatedness to apparent biological altruisms among insects.)

Horn, H.S. (1978). Optimal tactics of reproduction and life history. In J.R. Krebs and N.B. Davies (eds.), *Behavioural Ecology: An Evolutionary Approach.* Sinauer Associates, Sunderland, Mass. (Review of ideas concerning *r*- and *K*-selected reproductive tactics.)

Lehrman, D.S. (1964). The reproductive behavior of ring doves. *Sci. Amer.* 211:48-54. (Hormonal and environmental factors influencing female investment in offspring.)

Licht, P. (1979). Reproductive endocrinology of reptiles and amphibians: gonadotrophins. *Ann. Rev. Physiol.* 41:337-351. (Review of information for nonmammalian vertebrates.)

Lloyd, M., and **H.S. Dybas** (1966). The periodical cicada problem. I. Population ecology. *Evolution* 20:133-149. (Long-term timing effects on interactions between predators and prey.)

Menaker, M., and **N. Zimmerman** (1976). Role of the pineal in the circadian system of birds. *Amer. Zool.* 16:45-55. (Review of effects of removal of the pineal on daily activity patterns. This is one paper in a series on the endocrine functions of the pineal in the same volume.)

Trivers, R.L. (1974). Parent-offspring conflict. *Amer. Zool.* 14:249-264. (Application of theories of degrees of relatedness to the period of parental care.)

Warner, R.R., D.R. Robertson, and **E.G. Leigh, Jr.** (1975). Sex change and sexual selection. *Science* 190:633-638. (Relationship between sex change and dependence of reproductive success on age and/or sex.)

Wilson, E.O. (1975). *Sociobiology: The New Synthesis.* Harvard University Press, Cambridge. (The "bible" for application of theoretical genetic bases of behavioral phenomena.)

Wodinsky, J. (1977). Hormonal inhibition of feeding and death in *Octopus*: control by optic gland secretion. *Science* 198:948-951. (Experimental basis for hypothesis of parental death contributing to reproductive fitness.)

Information Processing for Regulation

Part VI

Survival often depends on speed. When the environment of an animal changes, its ability to respond rapidly to the change could make the difference between life or death. Some environmental changes have obvious, immediate consequences for survival, and others are more long-term. For example, when a predator detects prey items, the ability of the prey to escape or the predator to capture them may depend on speed and evasive action for which fractions of a second are important.

When an environmental change involves a change in supply of or demand for a resource, the importance of response speed to the change depends on the importance of the rate of change in the net quantity of the resource to the animal. For example, the predator should be able to respond rapidly to detection of prey if it is "hungry" because speed will ensure sufficient net gains of energy. Whether a rapid response is observed depends on the levels of its energy reserves as compared with demands. However, if reserves can be low or demands high at *some* times, rapid and controlled responses should be possible.

Most animals are believed to be limited by the availability of food at some times in their lives; thus there should be strong selective pressures for both predators and prey to detect each other and respond rapidly in ways that favor survival. Moreover, rapid responses are not limited to food resources. Changes in oxygen or other resources involving immediate consequences for survival or reproductive success should have resulted in selection for mechanisms that can produce rapid and controlled responses to environmental changes.

In the last part we discussed a variety of timing mechanisms important for relatively *long-term* (daily, seasonal) responses to environmental changes. Most of these controls involve hormones or pheromones. The release of a chemical into and transport via the circulatory system produces effective control over relatively long periods. However, this type of mechanism alone is not suitable for rapid controls because of the time required for a chemical to be transported from its site of production to the tissues it influences.

More rapid timing requires a system with faster response capabilities than can be achieved from circulatory transport. The nervous systems of animals provide this ability. Certain cells can respond to rapid or slow changes in internal and external environments, and these receptors filter information from environmental changes that may be either long-term or short-term. The filtered information is transmitted along specific pathways (neurons) so that only certain cells are influenced; they can be influenced very quickly and with a high degree of control. The transmitted information may be combined with other information from other parts of the animal, and the results of integration can be transmitted along other neurons to effectors, where coordinated action is initiated as a consequence of environmental changes. From detection to response, the entire process is variable in duration, but where short-term survival is involved it can occur within seconds or even fractions of a second.

The complexity of requirements for rapid timing responses to environmental changes depends on the complexity of both animals and their environments. As animals have evolved in complexity (e.g., size, abilities to move), their information-processing mechanisms have followed suit. However, mechanisms for relatively rapid information processing in animals involves an elaboration on a basic theme or pattern. Nervous-information processing is similar among different animals; most neurons function similarly in a way that reflects the relative importance of speed and specificity of control.

In Chapter 16 we will examine the principles of information transmission by neurons. We will also consider the characteristics of receptor function that lead to information input in different animals. Contrasting with many similarities among animals in neural function are many differences in types of receptors and their accessory structures that reflect different types of information filtering for environmental changes influencing survival. In the last two chapters we will follow the information-processing pathway from reception and transmission to integration of information for effective responses (behavior), and finally we will consider adaptations in effectors that produce responses.

Information Transmission and Resource Detection

16

Electrical events can occur rapidly, and they form the basis for nerve-cell function. Cells are bathed in fluids, and their membranes create a separation of different concentrations of ionized (electrically charged) solutes across the cell membrane (see Chapter 12). As a consequence of unequal distributions and movements of positively and negatively charged ions, an electrical-potential difference exists across the membrane of a cell.

Changes in this electrical-potential difference resulting from the movements of ions form the basis for nervous-information transmission. Before examining the changes that occur in nerve cells, we will establish the basis for observed electrical-potential differences for any cell that is not transmitting information. A nerve cell that is not transmitting information is said to be "resting"; we will be concerned here with the basis for the *resting membrane potential* of any cell.

EQUILIBRIUM MEMBRANE POTENTIALS

All cells have unequal distributions of ions across their membranes. Figure 16.1 shows a possible distribution of various solutes across a cell membrane, with arrows denoting the directions of concentration gradients at equilibrium.

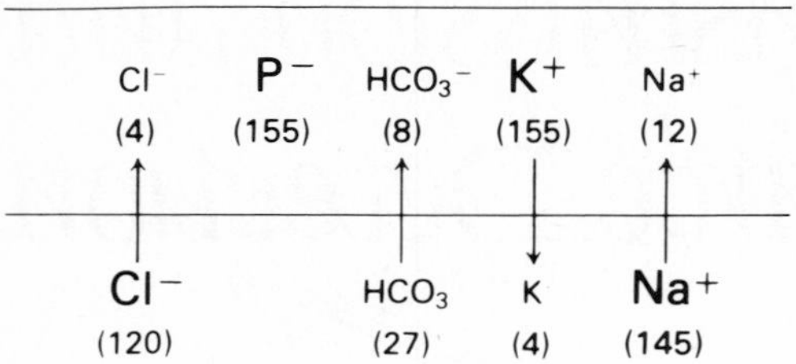

Figure 16.1
A distribution of some major ions across a cell membrane. Numbers in parentheses give concentrations in millimoles/liter. P^- refers to solutes such as proteins that carry a net negative charge and that do not normally diffuse across cell membranes (based on Roeder, 1967).

The importance of unequal distributions of ions to electrical potentials can be illustrated from a simple experiment (Figure. 16.2). If a membrane is selected that is permeable to K^+ but not to Cl^- and a voltmeter is connected across the membrane, the electrical potential will be zero when there is an equal concentration of KCl on each side (Fig. 16.2a). However, if the concentration of KCl is increased on the left side, K^+ will diffuse to the right. Because Cl^- cannot diffuse across this membrane, a Gibbs-Donnan equilibrium will occur (see Chapter 12), and the electrical charge on the left side will become negative relative to that on the right side.

Dissimilar electrical charges attract and similar charges repel; eventually (at Gibbs-Donnan equilibrium, Fig. 16.2b) the attractive electrical force on the left and the repulsive electrical force on the right (the net difference in potential from left to right) will *balance* the force for K^+ movement down a concentration gradient. At equilibrium, the movement of K^+ from left to right by diffusion is equaled by electrical forces acting to move K^+ in the opposite direction.

If a membrane is permeable to a particular cation but not to anions (to K^+ but not to Cl^-), the magnitude of the equilibrium potential can be calculated from the *Nernst* equation

$$E_{(ion)} = \frac{RT}{nF} \ln \frac{[ion]_{high}}{[ion]_{low}},$$

where

$$E_{(ion)} = \text{electrical potential at equilibrium for a particular ion (assumed to be equal to effects of concentration differences; see Chapter 12);}$$

R = universal gas constant;

T = absolute temperature;

F = Faraday's constant;

n = valence of ion;

$[ion]_{high}$ = higher concentration of ion; and

$[ion]_{low}$ = lower concentration of ion.

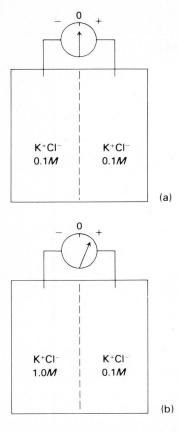

(a)

(b)

Figure 16.2
With compartments separated by a membrane permeable to K^+ but not Cl^-, the electrical potential from one side to the other is zero when the concentration of KCl is the same on both sides (a). When the concentration of KCl is increased on the left (b), K^+ diffuses to the right until an equilibrium is reached (when electrical forces balance diffusion forces). Because potassium carries a positive charge, the potential at equilibrium is negative on the left relative to the right (based on Eckert and Randall, 1978).

Note that this equation involves the logarithm of a ratio of concentrations; thus $E_{(ion)}$ depends exponentially on a difference in concentration across a membrane (see Chapter 12).

Biological membranes are normally permeable to more than one ion species, and the permeabilities for various ions are different. In this case both the permeabilities of ions and their concentration differences must be accounted for to calculate the equilibrium membrane potential. Both of these variables are included in the *Goldman* equation. For K^+ and Na^+ ions, the equation is

$$\begin{array}{c}\text{Potential}\\\text{difference}\end{array} = \frac{RT}{F} \ln \frac{P_K[K^+]_o + P_{Na}[Na^+]_o}{P_K[K^+]_i + P_{Na}[Na^+]_i},$$

where

R = universal gas constant;

T = absolute temperature;

F = Faraday's constant;

P_K = permeability constant for potassium;

P_{Na} = permeability constant for sodium;

$[\ ]_o$ = concentration of ion outside membrane; and

$[\ ]_i$ = concentration of ion inside membrane.

If measurements are made at a constant temperature of 18°C so that the first term in the Goldman equation (RT/F) becomes a constant, and natural logarithms are converted to logarithms to the base 10, the equation for the K^+ and Na^+ concentrations shown in Fig. 16.1 becomes

$$\text{Potential difference} = 0.058\ \text{Log}\ \frac{P_K[K^+]_o + P_{Na}[Na^+]_o}{P_K[K^+]_i + P_{Na}[Na^+]_i}$$

$$= 0.058\ \text{Log}\ \frac{P_K(4) + P_{Na}(145)}{P_K(155) + P_{Na}(12)}\ .$$

The permeability of a resting membrane to sodium (P_{Na}) is about 100 times less than the permeability to potassium (P_K), meaning that the potential difference due to sodium is smaller than the difference contributed by potassium, and

$$\text{Potential difference} = 0.058\ \text{Log}\ \frac{(4) + 0.01\ (145)}{(155) + 0.01\ (12)}$$

$$= 0.058\ \text{Log}\ \left(\frac{5.45}{155.12}\right)$$

$$= 0.058\ \text{Log}\ 0.04$$

$$= -0.084\ \text{volts} = -84\ \text{millivolts (mV)}.$$

The electrical potential due to the equilibrium potentials of K^+ and Na^+ is -84mV on the inside of a cell membrane relative to the outside.

The equilibrium-potential difference calculated from K^+ alone is -92 mV for this example; inward movement of Na^+ under resting conditions contributes $+8$ mV to the inside of the cell. Most of the resting membrane potential is determined by the distribution of potassium and the higher permeability of cell membranes to its movement. However, we will find that changes in membrane permeability to sodium (changes in P_{Na}) as well as potassium (changes in P_K) have very important consequences for the events leading to transmission of neural information (see below).

Measuring Membrane Potentials

If the resting potential across a cell membrane is essentially the calculated equilibrium potential for potassium, we should be able to measure a potential similar to it that depends on potassium concentrations. Membrane potentials can be measured with the equipment shown in Fig. 16.3. Glass tubing heated and pulled to a very fine point (0.1 – 0.5 μm in diameter) can be filled with a conducting salt solution (usually KCl) and connected to an amplifier. The amplifier magnifies electrical signals and displays them on an oscilloscope.

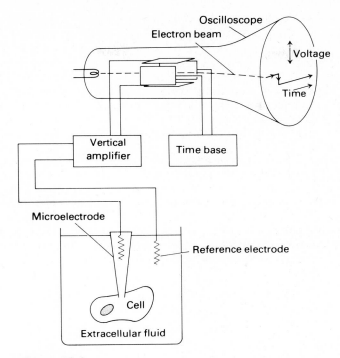

Figure 16.3
The potential difference across a cell membrane is
measured by placing a microelectrode in the cell and
amplifying the difference in potential between it and an external reference electrode. The amplified potential is
displayed on an oscilloscope by applying voltage to vertical
plates, deflecting a negatively charged electron beam that
is forced to move across the face by a timed application of a
voltage to horizontal plates.

Because all electrical potentials involve a potential difference between one
location and another, a reference electrode is placed in the fluid bathing the outside of a cell. When a very small electrode is inserted into a cell, it can be used to
measure the potential between it (inside) and the reference electrode (outside).
The oscilloscope permits the measuring of changes in potential over very small
time periods. The electron beam is sensitive to electrical charge on two sets of
plates surrounding the beam (Fig. 16.3). An alternating voltage is applied to the
plates on each side of the beam over specified time periods to sweep the beam
rapidly back and forth across the face of the oscilloscope. The amplified potential from the measuring electrodes is applied to the plates above and below the
beam so that a displacement of the beam in the vertical direction represents a
change in electrical potential measured across the membrane.

When a microelectrode is inserted in a "resting" cell, a negative potential
is measured on the inside of the cell relative to the outside. The change in poten-

tial occurs immediately when the electrode penetrates the membrane. The potential does not change if the electrode is moved within the cell, indicating that the resting membrane potential is due to characteristics of the surface membrane that influence movements of ions across the membrane. These characteristics are thought to involve "channels" or "pores" that permit passage of only certain ions out of or into cells across the membrane. The extent to which these channels are open determines permeability for a particular ion. Opening and closing of membrane channels may involve conformational changes in proteins lining the channels that influence the rate of passage of particular ions through the pores.

The potential that is measured is usually not quite as negative as predicted by equations from the equilibrium potential for potassium alone. For example, the measured potential may be a few millivolts more positive because of the contribution of other ions to the total resting potential (such as from sodium; see above).

The resting potential depends primarily on the distribution of potassium, which can be studied experimentally by changing the concentration of potas-

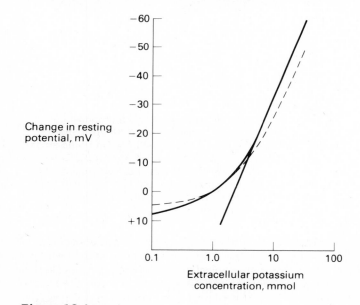

Figure 16.4
The resting membrane potential depends primarily on the difference in K^+ concentration across a cell membrane; thus when extracellular K^+ concentration is changed by orders of magnitude, the measured resting potential changes (dashed line). Measured values differ from predictions based on equilibrium potentials for K^+ (straight solid line) and K^+ plus other ions (solid curved line) because P_K changes when the resting potential changes (modified from Shapiro, 1977).

sium in the fluid bathing a cell and measuring the change in the resting membrane potential, as shown in Fig. 16.4. The straight line is the predicted (calculated) change in membrane potential from potassium alone. The solid curved line predicts changes when other ions (Na^+ and Cl^-) are taken into account. The actual measured line (dashed curved line) differs from both predictions somewhat because the permeability of the cell membrane to potassium changes; thus it is not a constant. P_K increases when membrane channels for K^+ change so that permeability to K^+ increases. This change occurs when the resting potential becomes more positive on the inside.

Maintenance of the Resting Membrane Potential

If a microelectrode is left in a cell, the resting membrane potential remains unchanged as long as the cell is healthy, but if the metabolism of the cell is poisoned, the resting membrane potential will deteriorate; i.e., ions will become more equally distributed across the membrane. Normally potassium ions leave and sodium ions enter a "resting" cell, and for the concentration of these ions to remain essentially constant, energy is expended to move them in directions opposite to their directions of movement by diffusion.

The Na^+-K^+ "pumps" in the membranes of cells maintain differences in the concentrations of these ions (see Chapter 12). There are a number of pumps, all of which require energy (from hydrolysis of ATP with Na^+-K^+-dependent ATPase) to move chemicals against diffusion gradients. If Na^+ and K^+ are exchanged across a membrane in a ratio of one for one, there is no net change in electrical charge, and this type of pump is called an "electrically neutral pump." However, a number of Na^+-K^+ pumps exchange three Na^+ for two K^+. The exchange results in a net loss of positive charge from a cell, and this type of pump is called an "electrogenic pump." As a consequence of energy expense for active transport together with opposite diffusion movements, the net concentrations of K^+ and Na^+ remain relatively constant over long time periods for "resting" cells.

ACTION POTENTIALS

For some cells the resting membrane potential is maintained for the life of the cell; in others the membrane potential changes when the characteristics of the membrane pores that influence permeabilities to different ions change.

Two types of electrical changes can occur: the inside of the cell may become either more positive or more negative relative to the outside. For example, an increase in membrane permeability to Na^+ leads to increased movement of these positively charged ions into the cell by diffusion such that the membrane potential becomes more positive. This type of change is called a *depolarization*. A change toward more negative charge inside a cell, such as from increased permeability to outward diffusion of positively charged ions, is called *hyperpolarization*. Both are defined with respect to the resting membrane potential, and cells that show the changes are called *excitable* because the changes have an influence on whether information transmission occurs.

If an electrode is inserted into an excitable cell, it is possible to experimentally inject positive charge into the cell by passing an electrical current through the electrode. Another electrode can be used to measure the changes in membrane potential (Fig. 16.5). When a membrane is depolarized in this manner, one of two events will occur. If the depolarization is slight, the membrane potential will return to a resting value when the depolarizing stimulus is stopped. However, if the depolarizing stimulus current exceeds a critical value a dramatic change in membrane potential occurs. The potential continues to increase in the positive direction, reaches a maximum value, and then decreases back to the resting value in a few milliseconds. This rapid, disproportionate change in membrane potential is called an *action potential*. The extent to which the resting membrane potential must be displaced to produce an action potential is called the *threshold potential* (Fig. 16.5).

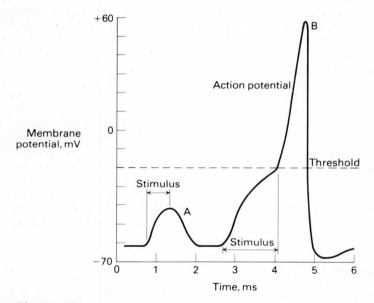

Figure 16.5
When the membrane potential of an excitable cell is depolarized below the threshold, the membrane potential returns to resting values when the stimulus is removed (A). However, if threshold depolarization occurs the membrane potential will undergo action-potential formation (B).

Once the threshold membrane potential is reached, a number of events occur that lead to the same type of action potential each time one is generated. The initial, rapid increase in positive charge above threshold depolarization inside the membrane results from a rapid increase in the membrane permeability to Na^+. The evidence for this increase comes from several sources. With special electronic equipment it is possible to keep the voltage across a membrane constant at different values (called "voltage clamping") and examine the effects of replacing ions that bathe the cell and thus changing the difference in their concentration across the membrane. When sodium concentration is reduced outside the membrane (the difference in concentration across the membrane is reduced), the change in membrane potential for a constant amount of depolarization decreases. Moreover, the maximum positive potential developed at the peak of a normal action potential is similar to the equilibrium membrane potential for Na^+ if P_{Na} was high, i.e., the membrane potential that would occur if the membrane was completely permeable to Na^+ alone.

Once a threshold depolarization is reached, the "gates" within Na^+ channels in the membrane are thought to open because of the effects of electrical charge on the structure of the proteins lining the channels. This opening of pores produces a rapid influx of Na^+. The poison tetrodotoxin (from the Japanese globefish) has been found to selectively block these sodium channels; one molecule of poison blocks one sodium channel and hinders action-potential formation at threshold depolarization.

Once the threshold is reached, the Na^+ channels change and increase Na^+ permeability in a *positive-feedback* manner (see Introduction). The more Na^+ that enters and thus the more depolarization that occurs above threshold, the more the permeability of the membrane to Na^+ increases, i.e., the more the "gates" within Na^+ channels open. This increased permeability occurs up to an upper limit set by the equilibrium membrane potential for sodium with high permeability (at about $+50$ to $+60$ mV). At this point the sodium channels close and the permeability to sodium rapidly falls to values typical for a "resting" cell membrane. Figure 16.6 shows the changes in sodium permeability that contribute to part of an action potential.

The membrane permeability to K^+ also changes during an action potential, but it is delayed with respect to the changes in Na^+ permeability, and the changes are not as large (Fig. 16.6). The use of radioactive K^+ indicates that K^+ starts to leave the cell at a higher rate after Na^+ has entered. The increase in positive charge within the cell as sodium moves in increases the tendency for K^+ to move out (increased P_K) as a result of effects on the channels selective to K^+ outflow through the membrane. After these events the permeability of the membrane to K^+ returns to "resting" values.

An action potential is a characteristic event. Once it is started it goes to completion in the same manner without major variation in the characteristics of the potential, exhibiting what has been described as the "all-or-none" principle of action potentials. It is a consequence of positive-feedback features that change membrane permeabilities to limits set by equilibrium potentials. If a threshold is not reached, no action potential is produced; if it is reached, a complete action potential is produced.

There is a special exception to this general rule. Only one action potential can be produced at the same location on a membrane at one time, and for a brief period following the production of an action potential a membrane will not produce another at that location. This period is called the *absolute refractory period*. Despite continued "threshold" depolarization, no action potential re-

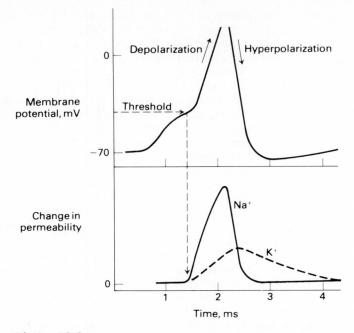

Figure 16.6
When threshold depolarization occurs (upper), an action-potential depolarization occurs when Na$^+$ permeability increases rapidly. Hyperpolarization of the action potential and repolarization of the resting membrane involves rapid decreases in Na$^+$ permeability and somewhat longer-term changes in K$^+$ permeability.

sults until the membrane has been "reset," which is shown in Fig. 16.7. Following the absolute refractory period it is possible to generate action potentials of decreased magnitude for some small amount of time. This period is called the *relative refractory period* (Fig. 16.7). Following *both* of these periods, the membrane produces the same action potential every time a threshold depolarization occurs. The lengths of the absolute and relative refractory periods vary from one excitable tissue to another, but they are usually a few milliseconds or less in duration.

Propagation of Action Potentials

For the rapid changes in membrane potential to carry information they must move from one region of a cell to another. When an action potential occurs, the membrane depolarizes at that site by an amount that far exceeds the threshold (see Fig. 16.5), that is, up to +50 to +60 mV. This strong depolarization produces current flow and depolarization of the immediately adjacent segments of membrane so that they exceed the threshold, and the action potential moves (is propagated) from one segment of membrane to another.

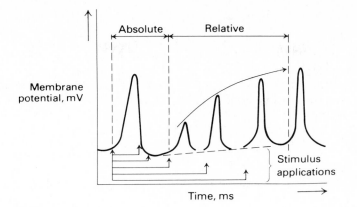

Figure 16.7
One stimulus followed at different intervals by a second
stimulus of the same magnitude (lower arrows) illustrates
the period of absolute refractoriness followed by the period
of relative refractoriness.

The electrical events for a tubular cell (similar to parts of many nerve cells;
see Chapter 17) are depicted in Fig. 16.8. At the site of an action potential a current of Na^+ flows in. To complete the electrical circuit, current must flow out at
some distance from the inward current, displacing the membrane potential
toward the threshold at other locations by an amount that depends on distance.
For a membrane stimulated in the center (Fig. 16.8) an action potential spreads
from that point in each direction, and the membrane at the initial site is absolutely refractory for a period such that current flow at the original site is not
sufficient to exceed the threshold at that location.

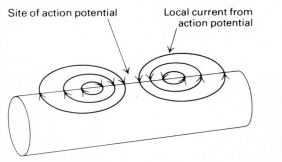

Figure 16.8
Electrical currents spread from the
site of an action potential in a
tubular nerve and result in displacement of adjacent membrane sites
toward threshold depolarization.

The speed of action-potential propagation depends on the rate at which
other segments of the membrane reach threshold depolarization. This rate is
determined by the resistance of the membrane to current flow because, from
Ohm's law, current = $\triangle$ voltage/resistance. There are two important characteristics of neurons influencing the speed of information transmission that depend

on resistance to current flow: the size of neurons and, for some, their insulating sheaths.

The internal resistance to current flow depends on the size of a tubular neuron. Resistance is inversely proportional to the square of the tube radius; thus current flow with a given voltage change increases in proportion to the square root of the diameter of the tube. Therefore, an adjacent membrane site depolarizes to the threshold more rapidly as tube diameter increases, shown as the curved line describing conduction velocity (speed of action-potential propagation) as a function of nerve fiber diameter in Fig. 16.9 (the curve for "nonmyelinated" nerves).

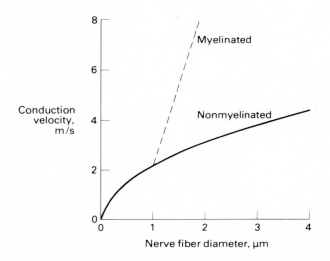

Figure 16.9
The velocity of conduction of an action potential depends on the square root of nerve-fiber diameter for nonmyelinated nerves but increases linearly with fiber diameter for myelinated nerves (from Rushton, 1951).

With this type of relationship we would predict that nerves involved in rapid responses should be larger than other nerves because information is transmitted faster in them. In many invertebrates, predator-escape responses involve large ("giant") nerve fibers. For example, the giant nerve fibers of the squid ultimately produce jet-propulsive (escape) movements. The nerves are so large (about 1 mm in diameter) that they were originally thought to be blood vessels, and much of the information we have discussed concerning action potentials is based on the ease with which these large nerves have been studied.

In vertebrates, nerves transmit action potentials over much greater distances, and resistance to current flow is influenced by insulation on the outside of some nerve-cell membranes. Some vertebrate neurons are surrounded

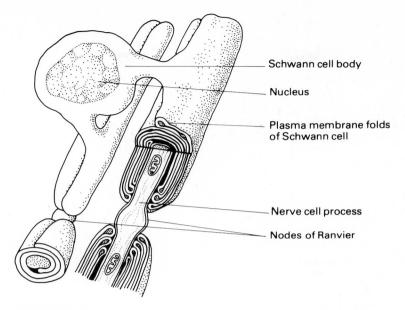

Schwann cell body

Nucleus

Plasma membrane folds
of Schwann cell

Nerve cell process

Nodes of Ranvier

Figure 16.10
The myelin insulation around neurons results from extensive folds of
the cell membrane of Schwann cells (after Bunge *et al.*, 1961).

with a sheath of *myelin*. Separate cells, called Schwann cells, develop membrane folds around tubular neurons at intervals (Fig. 16.10). The myelin in-
sulates the membrane from current flow, but the myelin is periodically broken
at areas called the nodes of Ranvier (Fig. 16.10). These regions provide relatively
low resistance pathways for curent flow. Thus local currents in the neuron
tubes are concentrated at the nodes of Ranvier, resulting in a "jumping" of action potentials from one node to another (Fig. 16.11). This jumping has been
called *saltatory conduction*.

When the diameter of myelinated neurons increases the distance between
nodes of Ranvier increases, and the speed of action-potential propagation increases with diameter (action potentials "jump" over larger distances). The
relationship between conduction velocity and nerve-fiber diameter for myelinated neurons is shown in Fig. 16.9. Note that the conduction velocity for myelinated nerves increases linearly, starting at a point where the increase in conduction velocity for nonmyelinated nerves begins to slow as diameter increases.

There is an additional advantage of the use of myelin insulation. This construction is more economical for the requirements of short-term changes
leading to action potentials. The concentration of action potentials at the nodes
of Ranvier requires fewer Na^+ and K^+ channels along the entire length of a
nerve and fewer Na^+-K^+ pumps to restore imbalances in ion concentrations as-

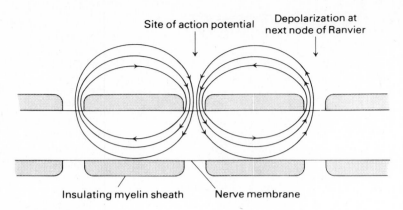

Site of action potential

Depolarization at next node of Ranvier

Insulating myelin sheath   Nerve membrane

Figure 16.11
As a consequence of external insulation, electrical currents are concentrated at nodes of Ranvier; action potentials are propagated as they "jump" from one node to another.

sociated with action potentials. The reduction of channels reduces the metabolic requirements of myelinated nerves, although it is not yet clear whether this benefit is equivalent to the cost of producing and maintaining the myelin insulation.

INFORMATION RECEPTION AND INITIATION OF ACTION POTENTIALS

Our discussion of the mechanisms for information transmission has been based on experimental procedures that alter neuron (nerve cell) membrane potentials with electrodes to influence current flow. Whether an action potential occurs in a neuron under normal (natural) circumstances depends on one of two factors. A nerve cell may be associated with either a receptor that influences its membrane or other nerve cells. In each case there will be, at times, some stimulus input influencing the extent to which current flow across a neuron membrane produces threshold depolarization.

The behavior of an animal is influenced by stimuli that affect receptors for both the external and internal environments. Receptors for the external environment are called *exteroceptors*; those for the internal environment are called *interoceptors*. Information on stimuli is coded in neurons that pass *into* central nervous systems, and collections of these neurons are called *afferent* nerves. Integration in a central nervous system may ultimately produce responses of effectors by means of *efferent* nerves that pass from the central nervous system to effectors. We will follow this normal sequence by first considering interactions between receptors and afferent neurons. The mechanisms of integration to produce effective responses will be discussed in the next chapter.

General Receptor Characteristics

Receptors convert, or *transduce*, information on energy in an environment to electrical events that influence action-potential formation in the afferent neurons with which the receptors are associated. It is a general characteristic that a given receptor is selectively sensitive to only *one* form of energy. Thus it is possible to classify receptors according to the *type* of energy they transduce. There are specific receptors for different forms of chemical energy (pH, olfaction, taste, etc.), mechanical energy (sound, touch, inertia), electromagnetic energy (light, magnetic fields, electrical fields), and thermal energy (heat, cold).

If a receptor is stimulated with another form of energy it may result in production of action potentials in afferent neurons, but the amount of energy required is much greater. For example, the photoreceptors in an eye transduce energy from just a few photons of light. You can also produce a sensation of "light" ("seeing stars") by pushing strongly on your eye (mechanical energy), but the energy required is much greater. Alternatively, the touch receptors in your skin are very sensitive to mechanical energy changes but essentially insensitive to light.

Receptor sensitivity comes from *amplification* of a particular type of energy into a relatively large amount of electrical energy by means of ion movements in neuron action potentials. For example, the energy in one action potential may involve 100,000 times more energy in the light required to *form* an action potential from transduction through a photoreceptor. The amplification involves changes in membrane characteristics in individual receptor cells (such as from positive-feedback changes in membrane pores that influence permeabilities to ions) that are specialized for the detection of certain types of energy.

Differences among Receptors

There is a great diversity of receptor types among animals and of some characteristics that influence their transduction of energy into action potentials in afferent neurons. Our immediate perception of our external environment occurs through the sensory modalities of sight, hearing, touch, smell, and taste, yet other animals sense infrared heat, magnetism, electricity, ultrasound, and other forms of energy in ways that far exceed our capabilities. To understand this diversity as well as some other characteristics of information processing, receptor functions can be interpreted with respect to characteristics important to the survival of different animals.

One important difference between receptors that transduce and amplify the same general form of energy is their range of sensitivity to energy changes in the environment. For example, one type of chemoreceptor may respond only to a narrow range of changes in chemical concentration (such as for pH or sugar concentration), whereas those in other animals may respond over a broader range of concentration. Bees see light in the ultraviolet, which we humans cannot. Bats detect sounds at much higher frequencies than do most other animals.

Figure 16.12
Flower (*Rudbeckia hirta* var. *bicolor*) photographed by simultaneous illumination
with ultraviolet light and "visible" light to mimic the visual sensitivity of a bee. The
lower photograph illustrates the "extra" ring of color detected by bees by means of
localized UV reflection (at the outer edge) and absorption (toward the center). Photo
courtesy of Kenneth D. McCrea and Morris Levy, Purdue University.

These differences reflect the evolution of detection capabilities important for the survival of particular animals. Bees use information from ultraviolet reception to detect food sources because many flowers have evolved to reflect ultraviolet light (Fig. 16.12) as a means to attract bees as pollinators (a coevolved system of communication). Bats fly at night using navigation by detecting high-frequency sound echos (sonar) that provide a mechanism for orientation and food capture. In addition, the ability of an animal to detect a particular range of chemical concentrations should reflect the range of concentrations in its environment that normally have an impact on its rate of net energy gain.

It is perhaps obvious that animals that have evolved in different environments will have evolved receptors for detecting different changes in resource supplies and/or demands. These differences form some of the best examples of adaptations in function. We will examine some in detail later in this chapter when we consider characteristics of specific types of receptors. However, first it is important to consider some common characteristics of receptor function that influence abilities of animals to respond effectively to changes in their environments.

Receptor Potentials and Sensory Coding

In order to produce an action potential in an afferent nerve cell, a receptor must operate to influence the membrane potential of the neuron, which occurs in one of two ways. The energy impinging on a receptor cell opens or closes ion channels in the membrane that influence its membrane potential by means of changes in receptor-membrane permeability to specific ions. This polarization may either (1) electrically influence the membrane of an afferent nerve cell directly or (2) cause the release of a chemical from the receptor cell that results in depolarization of the membrane of an associated afferent neuron by influencing its membrane permeability to Na^+ and K^+.

Figure 16.13 shows a potential produced across the membrane of a receptor by the movement of ions when stimuli of different intensities are applied to the receptor. The change in membrane potential is called a *receptor potential* or a *generator potential* (because it usually leads to generation of action potentials in afferent nerves). Note that this potential is not propagated. It occurs locally in the receptor cell, and its intensity (magnitude) depends on the strength of the stimulus (it is a graded potential and not an "all-or-none" potential). Moreover, the receptor potential changes with time.

What are the relationships between changes in the receptor potential, changes in stimulus strength, and the information generated in afferent neurons? Figure 16.14 shows the change in receptor potential for a receptor sensitive to mechanical stretch (mm of displacement). In this case the membrane potential of the receptor cell was measured at the peak of the receptor potential (Fig. 16.13) for different amounts of stretch. Note that the increase in receptor potential as the stimulus intensity increases is *nonlinear*. At low values of stretch the increase in receptor potential is much greater than at high values of stretch.

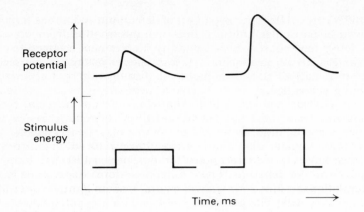

Figure 16.13
Measurements of membrane potentials for receptor cells
show that the receptor potential is graded; i.e., it depends
on the magnitude of stimulus energy. Note also that the
receptor potential changes with time while stimulus
energy remains constant.

This relationship has very important consequences for the *coding* of sensory information. The potential generated in the receptor cell produces action potentials in an afferent neuron at a *frequency* that depends directly on the magnitude of the receptor potential. The larger the receptor potential, the longer it exceeds the threshold of the nerve-cell membrane and the greater the area of nerve-cell membrane influenced.

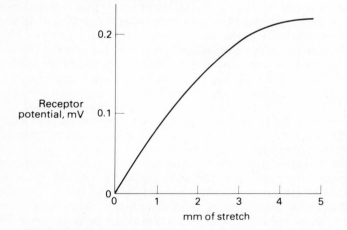

Figure 16.14
Measurement of the receptor potential of a stretch receptor
shows that the magnitude of the potential is a logarithmic
(nonlinear) function of stimulus intensity (modified from
Katz, 1950).

This type of function becomes linear if the magnitude of the receptor potential is plotted as a function of the logarithm of stimulus intensity. A logarithmic X axis involves units that increase by orders of magnitude (1, 10, 100, etc.); thus as stimulus energy increases, more and more energy must impinge on a receptor cell for the same amount of increase in receptor membrane potential.

Recall that each action potential is the same as others, meaning that the only information coded in an individual action potential involves the fact that a particular *type* of stimulus is present (and does not give its magnitude). Information on both stimulus strength and changes in stimulus strength is coded in the *number* of action potentials produced in a time period and changes in their frequencies. Because the number of action potentials/time depends directly on receptor-potential magnitude, and receptor potential magnitude depends on the logarithm of stimulus strength, the frequency of action potentials in the afferent neuron will depend on the logarithm of stimulus strength.

This type of relationship is observed in virtually every type of receptor. Why is it important? One way to answer this question is to ask what would occur if stimulus strength was coded in some other ways. Some possibilities are shown in Fig. 16.15. If the relationship was linear with a high slope, the ability of an afferent neuron to carry information would become saturated at relatively low stimulus intensities, such as when the neuron became refractory at an upper limit of action-potential frequency. The advantage to this type of response is very high sensitivity to stimulus changes at low levels of stimulus energy.

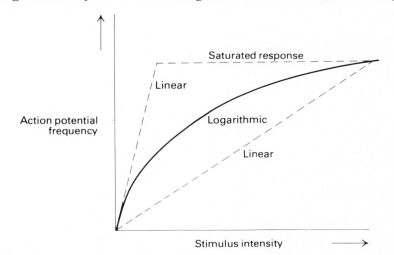

Figure 16.15
Models of the consequences of coding sensory information in different ways. Linear coding with a high slope would provide high sensitivity with no information at high intensities (saturation). Linear coding with a lower slope would provide information at high intensities but less sensitivity to stimulus changes. A logarithmic function represents a compromise between these extremes.

If the relationship was linear with a low slope, the neuron would not become saturated—it would show an advantage to respond over a wide *range* of stimulus-energy values. However, the ability to detect *changes* in stimulus intensities would be much less at low stimulus intensities as compared with a higher slope because fewer additional action potentials would be generated as stimulus intensity increased. Thus the logarithmic relationship represents a "compromise," or a "trade-off," between not processing enough information at stimulus intensities of high levels (when saturation occurs, determining the range of response) and processing a relatively large amount of information (high sensitivity to stimulus changes), when the level of environmental stimulus intensity is low.

Why is it so important for animals to be able to detect changes in stimulus intensity with higher sensitivity (more precision) if the stimulus intensity is low (high curvature of the logarithmic line in Fig. 16.15)? For almost all types of energy impinging on animals from their external environments, survival depends more on detecting *changes* in stimuli when intensities are low than when they are high. For example, in the chapter on energy regulation by feeding we discussed the importance of detecting small changes in energy values of foods (e.g., differences in chemical concentrations) when the food an animal had been able to obtain was "poor," i.e., low in energy value (low stimulus intensities). Whether it survived or starved would depend on abilities to detect a *slightly* better food if foods were "poor" in value. However, if it had achieved a "rich" supply of food (high stimulus intensities), the same small difference in stimulus intensity would have less impact on whether it survived.

The survival value of information from the environment depends to a large extent on the level of intensity of the information. As another example, consider interactions between predators and prey. Whether a prey detects an approaching predator with visual or auditory receptors with sufficient time to escape depends on its ability to detect very small changes in stimulus intensities when the level of stimulus intensity is low. The farther away a predator is, the less intense are the sounds and changes in light intensity reaching the prey because emitted energy dissipates rapidly over distance. To gain maximum advantage for survival, the prey's receptors should provide the earliest possible warning and respond with high sensitivity at low stimulus intensities (within the constraint of still providing some information at higher levels of stimulus intensity). Alternatively, the predators should minimize these signals and have similar receptor characteristics in order to detect prey over distances.

Sensory and Receptor Adaptation

Note in Fig. 16.13 that the receptor potential decreases with time even when a stimulus is applied at constant strength. This characteristic is called *receptor adaptation*, and any change in information in afferent neurons with time while a constant stimulus is applied to their receptors is called *sensory adaptation*. Sensory adaptation may involve receptor adaptation and/or changes in function between events occurring in a receptor and its afferent neuron. We are all familiar with sensory adaptation. For example, even though the light intensity in a room remains constant, it is perceived as brighter when we initially enter a light room from a dark room, and the perceived brightness changes with time.

These types of changes have caused humans considerable problems because the changes mean that our sensory mechanisms do not always precisely reflect our environments. For this reason we must use more reliable instruments to deal with problems requiring precision of information about absolute levels of energy. However, sensory adaptations serve important functions for some problems of survival, which are basic and do not necessarily always involve obtaining precise information.

In general, the extent to which sensory adaptation occurs also depends on the importance to animals of obtaining information on *changes* in environmental stimuli. Figure 16.16 shows differences in the degree of adaptation in the sensory nerves of some different mechanical receptors. The stretch and pressure receptor systems adapt very slowly and are called *tonic* receptor systems. Touch and hair receptor systems adapt more rapidly and are called *phasic* receptor systems. Tonic receptor systems provide more reliable information on the persistence of a stimulus and its absolute magnitude. For example, the stretch receptor in Fig. 16.16 is located within muscles, where continuous information on the degree of stretch determines the extent of muscle contraction (see Chapter 17).

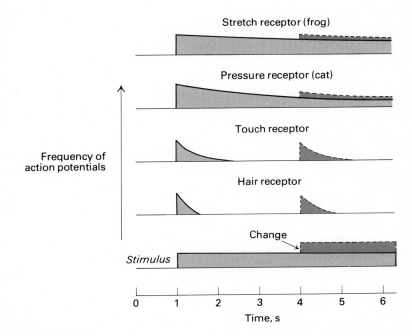

Figure 16.16
Mechanoreceptive systems may be tonic (stretch, pressure) or phasic (touch, hair), depending on the speed of adaptation to a constantly applied stimulus. Phasic receptor systems respond to a change in a stimulus (shown at 4 seconds) with more information input because of prior adaptation (modified from Adrian, 1928).

Phasic receptor systems increase information transmission more after adaptation has occurred, but tonic receptor systems are limited in the ability to provide information on stimulus changes. For example, with a slight change in the stimulus in Fig. 16.16 occurring at 4 seconds, the touch and hair receptors respond with a much larger increase in action-potential frequency (dashed lines in Fig. 16.16). In general, phasic receptor systems are common where detection of environmental changes are most important. Many of these systems exist to obtain information from the external environment (exteroceptors). In this way both the extent of sensory adaptation and the logarithmic-intensity coding mechanism serve similar functions.

What mechanisms are responsible for sensory adaptation? Several possibilities are usually recognized. First, the interaction of stimulus energy with the receptor membrane may result in a saturation of membrane receptors responsible for permeability changes to ions. Second, the changes in ionic concentrations occurring during a receptor potential may reach certain limits. For example, the local amount of positively charged ions available for depolarizing currents may be limited. Third, the "perception" of sensory information could change as information is processed in a central nervous system because of interactions between neurons (see Chapter 17). Fourth, the accessory structures of receptors could modify the nature of receptor potentials by means of effects on the energy influencing membrane potentials. This last mechanism deserves further comment because it suggests important functions for the structures associated with receptor cells.

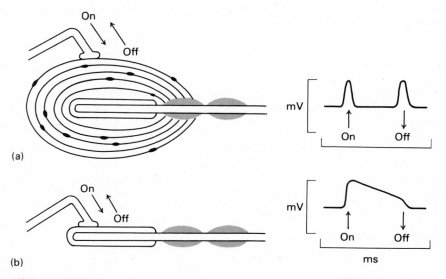

Figure 16.17
Information monitored in the afferent neuron from a Pacinian corpuscle without accessory membranes shows less dependence on rate of change (b) than when accessory membrane structures filter information on pressure changes (a) (based on Loewenstein, 1960).

The *Pacinian corpuscle* is a receptor organ that detects the rate of change of pressure. The corpuscles are composed of a neuron terminal surrounded with an "onion-like" layer of membranes (Fig. 16.17a), and the receptor organs are distributed in the skin of many mammals. The neuron produces action potentials only as the membranes surrounding it are being displaced in either direction. However, if the membranes are peeled away, the neuron potential changes more with respect to the magnitude of applied force than with respect to its rate of change (Fig. 16.17b). Thus the accessory membranes associated with the afferent neuron produce a rapidly adapting response with respect to applied force.

ADAPTATIONS IN ORGAN SYSTEMS FOR RECEPTION

Sensory adaptation is not the only possible role for accessory receptor structures (structures other than the receptor membrane itself). In a number of cases accessory structures are involved in stimulus filtering and amplification and occasionally in the generation of the energy received at a receptor membrane. We can examine a variety of adaptations in receptor-organ structures and functions by considering some well-studied examples.

Chemoreceptors

Chemicals influence all cells, and the ability of certain cells to change membrane potentials in response to specific chemicals usually involves relatively little association with elaborate accessory structures. Chemical effects usually occur on membrane surfaces (see Chapter 18); therefore, the most obvious site for reception of chemicals involves receptor-membrane surface areas.

The olfactory receptors of some insects provide an example. The male silkworm moth (*Bombyx mori*) detects and orients to females by means of extremely sensitive chemoreception of the pheromone *bombycol* that is produced and released by reproductively active females. The receptors for the pheromone are located on the featherlike olfactory antennae of the males. Each antennal filament contains a large number of small "pegs" with pores, and each antenna contains up to 20,000 separate receptors spread over a relatively large membrane surface area. As few as 40 receptor cells receiving one molecule of bombycol per second suffices to produce movements by the male.

The mechanism for the reception of chemicals at receptor-cell membranes is believed to involve a type of "lock-and-key" relationship between a particular molecule (key) and a specific site on a receptor membrane (lock). If a particular molecule "fits" into a specific membrane site, it will result in a receptor potential because of alterations in membrane permeability (see Fig. 18.1 for similar mechanisms involved in general hormone effects).

These types of interactions occur on a molecular scale, but they are extremely common. For example, the communication of one neuron with another involves release and "reception" of chemicals across small spaces between neurons (synapses); understanding the details of neuron-integration mechanisms partly depends on specifying the nature of these molecular receptions (see Chapter 17). In addition, the specificity of certain hormones for certain tissues involves similar interactions (see Fig. 18.1).

In some cases chemicals in the external environment are more localized than are olfactory stimuli. For example, foods usually occur in patches, and quality (concentration) is assessed just prior to ingestion by sampling a small amount to determine its chemical composition and concentration. The chemo- receptors in the hairs of blowflies are a particularly well-studied example (see Chapter 10). The hairs extend through the integument and serve as small capil- laries to sample whatever fluids are contacted; the hairs are located on parts of the fly most likely to contact food.

There are five sensory neurons at the base of each hair (Fig. 16.18). One is a mechanoreceptor that detects movement of the hair. The other four have receptor processes that extend up the tube of the hair. One detects water, one detects sugars, and the other two detect salts of different concentrations. The localization of separate receptor cells in the same "sampling" location also oc- curs in the taste buds of vertebrates, where different cells transduce informa- tion normally related to different aspects of food quality concerning energy (sugars), some nutrients (salts), and poisons (bitter; sour).

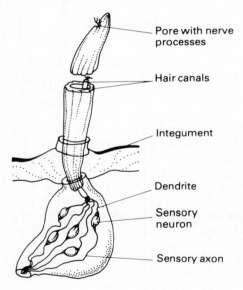

Pore with nerve processes

Hair canals

Integument

Dendrite

Sensory neuron

Sensory axon

Figure 16.18
Chemoreceptor hairs from the mouth of a blowfly contain both the processes from four receptor cells to detect the quality of ingested fluids and a mechanoreceptor cell that detects movement of the hair (from Wohlbarsht and Hanson, 1965).

Thermoreceptors

With a few exceptions, thermoreceptors are also usually not associated with elaborate accessory structures. The functions of all cells are coupled to their thermal environments because of Q_{10} effects on rate processes (see Chapter 6). The rate at which action potentials are formed in neurons depends on temperature as do other rate processes (e.g., temperature is a variable in the Goldman equation for equilibrium membrane potentials).

In the anterior hypothalamus of vertebrates, where body temperature is detected, certain neurons respond very strongly when the temperature is changed. The Q_{10} for the rate of action-potential formation in these cells is greater than in other nerve cells, which produces information on body-temperature changes with a high degree of sensitivity.

Recall that information on body temperature is combined with information on skin temperatures in endotherms exposed to cold environments to determine the rate of heat production (see Fig. 7.18). Figure 16.19 shows the frequency of action potentials measured from afferent nerves of "cold" and "warm" temperature receptors in the skin. The general separation of temperature receptors into these two categories reflects differences in the mechanisms (controls) underlying responses for the regulation of temperature via either heat loss or heat gain (Chapter 7 and Chapter 14).

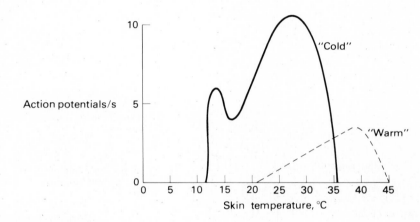

Figure 16.19
Different receptors in the skin provide information on low ("cold") and high ("warm") temperatures to mechanisms that control heat loss or heat gain (modified from Zotterman, 1953).

An exception with regard to thermoreceptors associated with an accessory structure occurs in the prey-detection mechanism of pit vipers (Crotalinae) such as rattlesnakes, copperheads, and cottonmouths. These snakes have a pair of thermoreceptive sense organs, called *facial pits*, located below and in front of the eyes (Fig. 16.20). Neurons in the membrane of the pit are extremely sensitive to infrared heat at wavelengths of 1.5-15 μm. Temperature changes as small as 0.002°C are detected. The information is used to detect and localize prey animals that emit more radiant heat energy than their surroundings.

The prey are localized as a result of the position of the facial pits at the anterior end of the snake and from some overlap in the area from which thermal energy is received in the right and left receptors. This arrangement filters information from the environment before it reaches the temperature-sensitive mem-

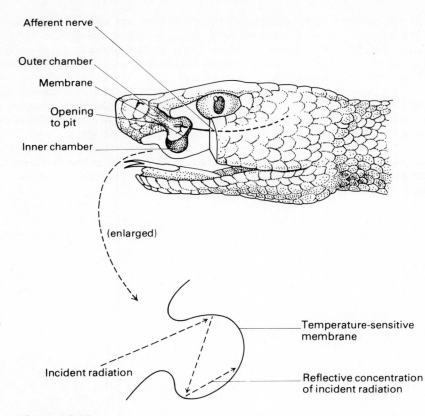

Afferent nerve

Outer chamber

Membrane

Opening
to pit

Inner chamber

(enlarged)

Temperature-sensitive
membrane

Incident radiation

Reflective concentration
of incident radiation

Figure 16.20
The facial pit organs of pit vipers contain membranes with neurons that
provide information on infrared heat. The position of the pits provides directional information, and the internal curvature of the pit amplifies information from the environment (based on Bullock and Diecke, 1956).

branes. Moreover, the curved inner surface of the pits function to concentrate
radiant heat entering the pits, providing some amplification of environmental
information.

Mechanoreceptors

Within any animal there are normally a large number of receptor types sensitive
to some form of mechanical energy. The Pacinian corpuscles are sensitive to
touch because of effects of accessory structures that produce sensitivity to the
rate of pressure change (Fig. 16.17). Receptors within muscles detect the degree
of applied force (stretch receptors, Fig. 16.16; see Chapter 17). Baroreceptors in
the circulatory system detect changes associated with blood pressure (Chapter 4).
Volume receptors in the kidney (juxtaglomerular cells, Chapter 13) or the hypoth-

alamus (Chapter 13) detect information related to the extent of hydration (volume) of the extracellular or intracellular fluid compartments. However, the best-studied examples of the role of accessory sensory structures used for information processing involve the reception of information related to waves of alternating pressure (e.g., sound waves).

A number of receptor-organ systems in different animals detect alternating waves of pressure. Lateral-line systems in fishes and amphibians are sensitive to pressure waves in water. Terrestrial animals have evolved sensory-organ systems to detect relative motion in three dimensions (equilibrium detectors such as semicircular canals) and sound waves in air (ears). In all of these systems there are a variety of accessory structures that filter and amplify mechanical energy.

At the level of receptor cells, hairlike projections function to filter and amplify information related to pressure changes in particular directions. Figure 16.21(b) shows a diagram of a hair-cell mechanoreceptor. At one end there may be a relatively large filament called a *kinocilium*. To one side of this filament are a series of *stereocilia* of different lengths. The kinocilium may be attached to the tallest stereocilium with small fiber bridges.

If fluid moves across the top of a hair cell and displaces the kinocilium away from the stereocilia, membrane depolarization occurs (see Fig. 16.21a) and results in release of chemicals from the hair cell that depolarize the afferent neuron and increase its rate of action-potential formation (see Chapter 17). If the fluid moves the hairs in the opposite direction the receptor membrane hyperpolarizes, less chemical is released, and the action-potential frequency in the afferent neuron decreases. Note that even with no applied force, the afferent neuron still carries some information. This situation is called a "spontaneous" rate of discharge and provides a point of comparison, or set point reference, for increases or decreases in the rate of action-potential formation.

The hairs on a mechanoreceptive cell filter information on certain types of pressure changes, i.e., only hair cells with kinocilia on certain sides of the cell produce increases in action-potential frequency from displacement in a particular direction. The hairs also amplify mechanical energy because they are small levers hinged at the cell membrane (a lever is a mechanical amplifier of force). The hairs may also be associated with other accessory structures that filter or amplify certain aspects of pressure changes. Perhaps the most elaborate set of accessory structures is found in the mammalian ear.

Structure and Function of the Mammalian Ear

The gross structure of the human ear is shown in Fig. 16.22(a). The *pinna* of the outer ear serves to funnel sound waves from specific directions. Thus it functions as a megaphone in reverse and gives a directional filtering component to sound-information processing. The eardrum at the base of the external auditory canal vibrates when sound waves hit it. It is mechanically coupled to another membrane (the *oval window*) by three small bones (the *auditory ossicles*) called the malleus, incus, and stapes (Fig. 16.22a). These bones function (as mechanical levers) to amplify sound waves at the eardrum because the area of the oval window is smaller than the area of the eardrum.

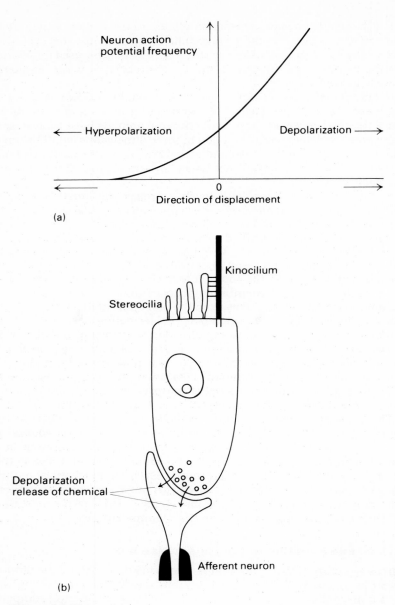

Figure 16.21
Hair-cell mechanoreceptive cells (b) depolarize afferent neurons by means of the release of chemicals when hairs bend in a certain direction (toward the kinocilium), causing an increase in action-potential frequency (a). Deformation in the other direction decreases action-potential frequency (based on Flock, 1971).

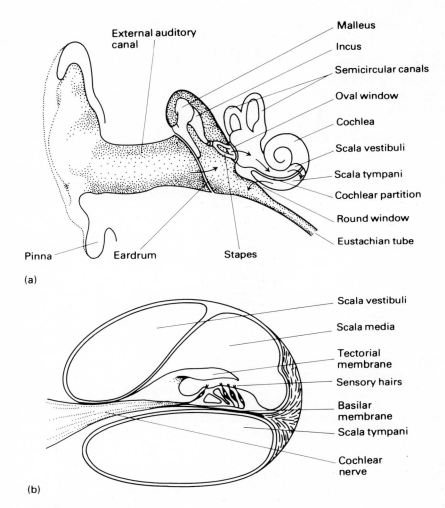

Figure 16.22
(a) Sound waves are directionally filtered at the pinna and coupled to the
oval-window membrane by the oscillation of the eardrum and auditory ossicles. Fluid waves in the cochlea result in vibration of the cochlear partition.
(b) The cochlea in cross section, showing the hair cells located in the organ
of Corti (after Shapiro, 1977).

The oval window is the site of sound-energy input to the inner ear. The inner ear is filled with fluid and contains the semicircular canals, with a canal in
each of three planes. Inertial changes in the fluid in these canals moves the
hairs on hair cells and provides information on relative placement in three
dimensions (equilibrium).

The other major structure in the inner ear is the *cochlea* (Fig. 16.22a), which is a coiled tube. If the coil were unwound it would have the cross-sectional structure shown in Fig. 16.22(b). The oval window communicates with the *scala vestibuli*, which is continuous with the *scala tympani* by means of an opening between them at the top of the cochlear spiral. The scala tympani communicates with the *round window*, completing the circuit of oscillation of sound energy within the inner ear.

The scala vestibuli and scala tympani are separated by a third chamber along most of their length, known as the *scala media* (Fig. 16.22a). The base of the scala media is composed of the *basilar membrane*, on which sits the *organ of Corti*. The organ of Corti is composed of a set of hair cells connected with afferent neurons of the eighth cranial nerve (auditory nerve). The hairs (stereocilia) are attached to the upper *tectorial membrane* of the organ of Corti. This elaborate set of accessory structures provides a degree of information processing for differences in sound-wave frequency, i.e., the *pitch* of sound. A sound wave produces a traveling wave of displacement along the length of the basilar membrane because of oscillations in pressure in the fluid in the scala vestibuli and scala tympani. However, the basilar membrane moves by a *maximum* amount at a *specific* location that depends on the frequency of sound waves, shown for a human ear in Fig. 16.23. Sound waves of low frequency displace the basilar membrane most toward the tip of the cochlea; sound waves of high frequency displace the membrane most toward the base of the cochlea. Thus some information on pitch is filtered, depending on which hair cells are deformed. Information on sound intensity at a given frequency depends in part on the amount of the oscillation at a particular location, which results in movement of the hairs fixed in the tectorial membrane.

Further "tuning" of pitch discrimination occurs from interactions between neurons in the central nervous system. However, the structures of the ear do serve some amplification and filtration functions, and there have been some elaborate modifications in some animals with specialized auditory systems.

A variety of nocturnal animals orient by using a sonar system for navigation. The animals studied most are bats. They emit very-high-frequency sounds at intervals when they fly. Because the frequency is too high for humans to hear, it is called *ultrasound*. Production of ultrasonic sound for echolocation requires very intense sound; high-frequency sounds dissipate more readily than do low-frequency sounds. However, the directional information content of a returning echo can be more precise because lower-frequency sounds spread more from their source of origin and reflect off more objects, giving less detail.

By projecting intense, high-frequency sounds ahead of itself, a bat can use information from returning echos to "calculate" its position with respect to objects in its environment (including prey items; see Chapter 17). In many respects the principles are analogous to those developed for sonar use by humans.

The accessory auditory structures of bats reflect their more complicated use of sounds for information processing from their environments. The pinna of the ear is very large and can be rotated or positioned in some species for maximum effective reception of returning echos. The cochlea of bats are mechani-

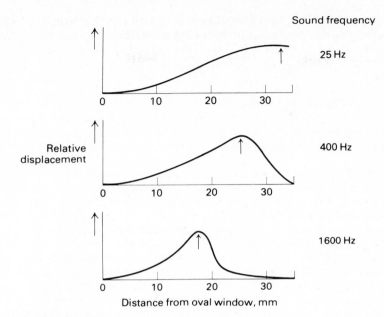

Figure 16.23
As sound frequency (pitch) increases, maximum displacement of
the basilar membrane in humans occurs at different locations on
the cochlea (after Vander *et al.*, 1975).

cally constructed to oscillate at the frequencies of sounds produced for navigation. In addition, those parts of the central nervous system involved in processing
information from the auditory nerve show a high degree of development (number of nerve cells) in animals that rely on sound to give them a large proportion
of the total information they receive from their environments.

Photoreceptors

Specialized photoreceptive cells detect light energy. "Light" usually refers to a
portion of the electromagnetic spectrum between wavelengths of 300 and 1000
nanometers (nm). Whether a particular receptor is selectively sensitive to electromagnetic energy of a particular wavelength depends on the presence of
visual pigments within the cells. There are four kinds of visual pigments in
humans: *rhodopsin*, which is very sensitive to low levels of light energy, and
the three primary-color visual pigments, *erythrolabe* (red), *chlorolabe* (green),
and *cyanolabe* (blue).

The function of visual pigments can be illustrated by rhodopsin. It is composed of two principal components: a carotenoid pigment called *retinaldehyde*
(the aldehyde of vitamin A_1) that is combined with a lipoprotein fragment called
opsin. Changes in sensitivity to light of different wavelengths (colors) is due to
changes in the nature of the lipoprotein component of the visual pigments.

All photopigments have the unique property of absorbing light energy, which for rhodopsin results in the following reversible reaction:

$$\text{rhodopsin} \underset{\text{dark}}{\overset{\text{light}}{\rightleftharpoons}} \text{retinaldehyde} + \text{opsin}.$$

Light energy changes the configuration of retinaldehyde from the *cis* to the *trans* mode of isomerization, which leads to the dissociation of the components. Dissociation is followed by a series of reactions leading to regeneration of rhodopsin from its components in the absence of light (the "dark reversal" of the reaction).

For rhodopsin in humans the maximum reaction occurs at a particular wavelength, slightly above 500 nm. As the wavelength increases or decreases from this value, the transformation of rhodopsin is less effective. This curve is shown in Fig. 16.24. It is important to note that this curve also precisely describes the sensitivity of rhodopsin containing receptive cells (rods in the human eye), which provides evidence that the information-transmission functions of these visual receptors are directly linked to the reception of light energy by photopigments. Moreover, the visual pigments for colors have distinct maximum sensitivities at 455, 530, and 625 mm, meaning that separate receptor cells containing these "primary" pigments (cones) can directly produce color vision, depending on the degree to which information from each is added after the information is received.

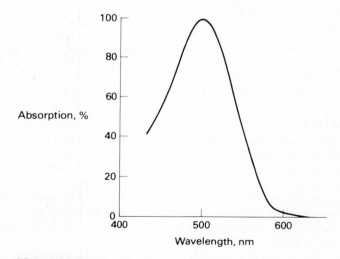

Figure 16.24
The absorbance curve for rhodopsin as a function of light wavelength. This curve also precisely describes the response of the visual receptor cells of rods to changes in light wavelengths (based on Wald and Brown, 1958).

Photosensitive pigments are located on membranes within visual receptor cells. The transformation of the chemicals by light is thought to influence membrane channels, with subsequent modification of the membrane potential of the receptor cell. This modification would, in turn, influence the membrane potential of the afferent neurons associated with the receptor. The type of change observed in the receptor potential is different in invertebrates and vertebrates. In invertebrates, changes in membrane permeability following rhodopsin activation lead to receptor and afferent neuron depolarization. In vertebrates the receptor hyperpolarizes, which is converted to a depolarization at a subsequent neuron in the afferent pathway (see Chapter 17).

The Invertebrate Compound Eye

The eyes of arthropods are composed of many similar receptor and accessory structures, collectively called *ommatidia*. The structure of an ommatidium from the compound eye of *Limulus polyphemus* (horseshoe crab) is shown in Fig. 16.25. Light is focused through a crystalline cone lens. Below the lens are 15 cells (*retinular cells*) surrounding a central extension of the afferent neuron. In the vicinity of the central neural extension, each retinular cell membrane is extensively folded in a region called the *rhabdomere*, the area where visual pigment is most concentrated. Thus each segment of the compound eye is composed of about 15 receptor cells that converge on one afferent neuron.

Measurements of membrane potentials in the afferent nerve cell show that light results in electrical depolarization of the nerve-cell membrane by the retinular receptor cells (Fig. 16.25). Considerable amplification occurs since one photon of light ultimately results in the movement of millions of Na^+ ions. Moreover, the response to light shows relatively rapid sensory adaptation.

One aspect of information processing in invertebrate compound eyes has a clear morphological representation in membrane accessory structures. A number of species of arthropods detect the plane of polarized light. Light entering an eye vibrates primarily in one plane, depending on the position of the sun in the sky. Thus information processing to detect the plane of light polarization is used to determine the sun's position (sun-compass orientation) even when the sun is not directly visible. The membranes in the rhabdomere are arranged in specific patterns in the ommatidia of arthropods that detect the plane of light polarization. Membrane folds are arranged as small microvilli (tubes) all in parallel and at right angles to each other. Nerve cell depolarization in these eyes is most effective when the plane of light polarization is parallel to the orientation of the tubular membranes in the rhabdomere. Thus the tubular membranes filter information about this particular component of incident light energy.

The Vertebrate Eye

The visual receptor cells of vertebrates are located in the *retina* of the eye (Fig. 16.26a). Considerable processing of visual information takes place by means of the functions of accessory structures in the eye, and considerable processing of visual information takes place within the retina after photoreception. The neural components of the retina are considered a part of the central nervous

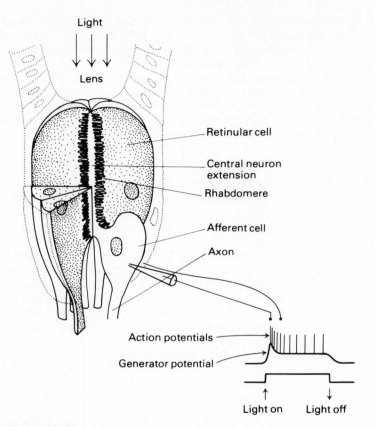

Figure 16.25
Cutaway view of the structure of an ommatidium. Rhodopsin is
concentrated in membrane folds of the rhabdomere region of
the retinular cells that surround an extension of the afferent
neuron. Measurements of changes in the afferent neuron reveal
action potentials shown on the receptor (generator) potential
(after Eckert and Randall, 1978; based on Miller *et al.*, 1961,
and Fuortes, 1959).

system (from which the retina originates during development). A considerable
amount of nerve-cell—nerve-cell interaction occurs in the retina, which we will
examine in detail in the next chapter.

Many accessory structures of the vertebrate eye operate on principles
similar to those of a camera. The amount of light reaching the retinal photoreceptors depends on its passage through the center of the *iris diaphragm*, i.e.,
the *pupil*. The size of the pupil changes in a reflex manner (automatically under
neural control), expanding in dim illumination and contracting in bright illumination; it changes the amount of light reaching photoreceptors and accounts for
some sensory adaptation.

There is feedback between the level of light intensity at receptors and the size of the pupil, meaning that photoreceptors exert some control over the amount of energy they receive. In general, the feedback operates to maximize the information input that is related to changes in the external environment. For example, pupil constriction minimizes the likelihood that *all* rhodopsin is transformed, leaving none available to detect environmental changes.

The lens focuses light entering through the pupil, and some focusing occurs at the outer corneal surface because of its high curvature. Focal adjustments for objects at different distances from the eye are accomplished in most vertebrates by changes in lense dimensions (called *accommodation*) rather than by changes in the distance between the lens and retina (as occurs in cameras and some species of fishes). Close objects are focused on the retina when the *ciliary muscles* on each side of the lens contract, causing the *zonule fibers* to move toward the lens and remove tension on it so that the lens becomes thicker. Relaxation of the ciliary muscles focuses distant objects by causing the lens to become thinner. These changes serve to amplify visual information because contrast detail is enhanced with fine focusing of images on the retinal surface.

Visual-information processing also depends on where an image falls on the retina. For example, in humans there are two types of visual receptor cells distributed differently in the retina (Fig. 16.26b). The *rods* contain rhodopsin and are very sensitive to light intensity. The *cones* contain visual pigments for colors and are less sensitive to light intensity but are more capable of distinguishing one point in space from another; that is, they exhibit higher visual *acuity*. In humans, rods are found in highest concentrations away from the visual axis of the eye in the periphery of the retina, where light intensity is lowest. Cones are found in greatest concentration at the *fovea*, where light intensity is highest and where information on image detail is best resolved.

Note that the fovea, where cones are closely spaced, has a curvature opposite to the curvature of the lens (Fig. 16.26a). It functions as a "negative lens." When two lenses of opposite curvature are placed together, considerable image magnification (amplification) occurs, which is the principle that governs the design of telephoto lens systems for cameras. It is interesting to note that the foveal depression is relatively large in falconiform birds, which are known to have very high visual acuity and can spot small moving prey from high elevations or over long distances.

Electroreceptors

A number of species of fishes from at least seven families emit electrical energy into their environments. The most commonly known examples are the electric *Torpedo*, the electric eel (*Electrophorus*), and the electric catfish of the Nile. These species discharge electrical energy of sufficient magnitude (50-60 volts at high current levels) to "stun" or electrocute some prey. The earliest human experience with electricity in animals came from these fishes, and they were used in ancient civilizations for a type of electroshock therapy.

With the advent of sensitive instruments to measure electricity it was found that these fishes and a variety of other species also produce a lower-

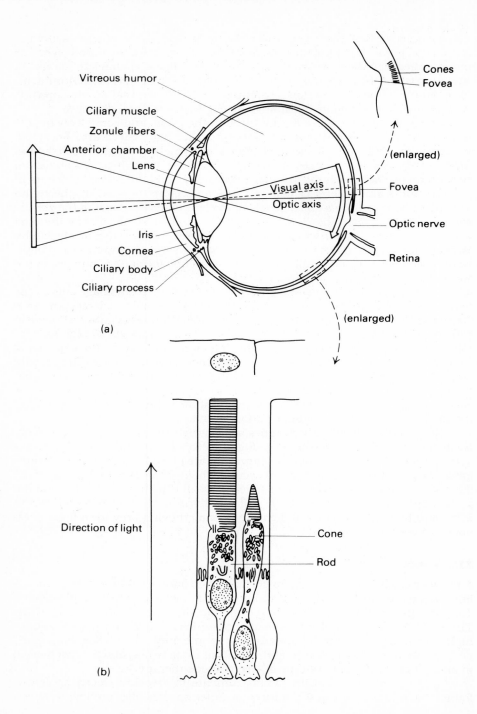

(a)

(b)

Figure 16.26
The gross structure of the mammalian eye (a), with enlargements illustrating the reverse curvature of the fovea and the structure of the visual-receptor rods and cones (b is based on Young, 1970).

voltage discharge of electricity. The electricity is generated by modified muscle cells in an electric organ in the tail of the fish. These organs consist of cells stacked in series with one another. Each cell is a small battery capable of producing a discharge of about 50 millivolts. Because they are arranged in series the voltages add; depolarization of 1000 cells with 50 mV each produces an electrical potential of 50 volts in larger fishes.

What is the function of the weaker discharge? There appears to be two possibilities: navigation and communication. Some weakly electric fishes live in particularly muddy water where light intensities are low. They have poorly developed eyes and navigate electrically. The cells of the lateral line are sensitive to very slight changes in the electrical field established from periodic discharge of the electric organ. The cells are less resistant to current flow than is the surrounding skin; thus the discharge of the electric organ provides the energy for depolarization of the electroreceptors.

How is electricity used for navigation? The electroreceptors are distributed along the length of the fish with a large concentration at the head. Experiments where fish were trained to receive food rewards by making choices showed that they detect objects in their environment that differ in electrical conductivity. Objects of high conductivity distort an electrical field established by the "battery" in the tail such that more current is concentrated on the part of the fish near the conductor (Fig. 16.27). Objects of low conductivity distort the field such that receptors near the object are not stimulated while those on each side are (Fig. 16.27).

Some electric fishes produce continuous discharges of their electric organs at frequencies that do not vary except with changes in temperature, meaning that the information content of the discharges is limited to the type of species producing the discharge. Other species vary the rate of discharge (frequency), which allows the information content to be much higher.

Among species that vary the frequency of discharge, some appear to change frequency in response to the discharges of electric organs of other individuals. The effect of one individual on another suggests communication of information by electricity; the electric organ discharge may represent a type of action potential transmission between individuals.

Sensitivity to Magnetism

Within recent years a variety of experiments on the navigation and orientation abilities of some animals have suggested that they can detect extremely slight variations in the earth's magnetic field and utilize this information for orientation. Two animals have served as models for this research: homing pigeons and honeybees.

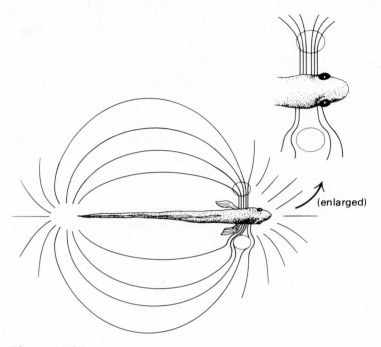

(enlarged)

Figure 16.27
The electrical field established from discharge of the electric organ
in the tail is distorted by objects of higher (upper) or lower (lower)
conductivity than water, resulting in differences in information input by
means of electroreceptors in the lateral line (after Lissmann, 1963).

Some homing pigeons can use the sun as a compass. If they are "clock-shifted" by exposure to an artificial photoperiod different from "real" sun time, they will fly in a shifted direction on sunny days. However, some pigeons can home under totally overcast skies and at night. These animals have been hypothesized to have a "backup" compass, and evidence suggesting that the compass is magnetic comes from experiments in which the birds were equipped with magnetic coils. The orientation of the birds in flight was influenced by the experimentally manipulated direction of the local magnetic field around their heads.

Suggestive evidence for a sensitivity to magnetism in bees comes from several sources. Variation in the "waggle dance" performed in the hive is highly correlated with slight variations in the magnetic field of the earth. When bees dance on a horizontal honeycomb in the dark (with no gravity or sun information), the dances indicate the directions of the eight points of a magnetic compass. In addition, there is some evidence that bees set their circadian rhythms (see Chapter 15) with daily variations in the strength of the earth's magnetic field. Bees deprived of other obvious cues (such as from photoperiod and gravity) leave the hive to forage with a precise 24-hour periodicity, with no evidence for the free-running rhythms typical of other animals deprived of environmental cues (see Fig. 15.5). However, when the magnetic field surrounding the hive is experimentally increased, the circadian periodicity of foraging is disrupted.

This and other evidence has led some investigators to hypothesize that these animals have a sensory mechanism to detect a magnetic field or slight changes in it. A major difficulty in accepting this hypothesis has been the attempt to identify where the receptors are located and how they might transduce information into neural action potentials. However, some recent exciting developments offer the first clues to this "missing link." By using supercooled detectors (by which electric current can be detected with minimum resistance to current flow), it has been found that bees contain relatively large numbers of small magnetic crystals. The crystals appear to be *magnetite* (FeO • Fe_2O_3, or lodestone). The magnetism detected in bees is equivalent to 10^6 small crystals of magnetite per bee, and the crystals are located in the anterior part of the abdomen. In pigeons, magnetite magnetism equivalent to 10^8 crystals has been detected from tissue located between the brain and the skull toward the front of the head.

The discovery of these magnets within animals and their approximate locations now permits analysis of structure and function for this elusive sensory system. Once the cells containing magnetite crystals have been localized, it will be possible to examine the relationships between receptor potentials and the information coded in afferent neurons when magnetic stimuli are applied to the receptors. This information can then be applied to natural situations to learn more about the information-processing abilities of animals used for complex behaviors such as orientation and navigation.

SUMMARY

If a resource changes rapidly in net quantity, the ability to respond rapidly can be important for survival and reproductive success. Rapid timing in animals involves controls mediated primarily by nervous systems. Certain information on environmental changes is filtered and amplified through receptor function; the information can be rapidly transmitted along nerves, integrated in a central nervous system, and effectors can be activated in short time periods. The speed of response is governed by changes in electrical events resulting from changes in the distribution of electrically charged ions across membranes of excitable cells.

For any resting cell, the electrical potential across the cell membrane (equilibrium potential) depends both on the concentration difference of particular ions across the membrane and the permeability of the membrane to movements of ions because of concentration gradients and electrical charge separations (described by the Goldman equation). For most cells the observed (measured) resting membrane potential is due primarily to K^+ because membranes are quite permeable to this ion and its concentration across the membrane is maintained at values that contribute to a potential difference. Permeability to Na^+ is about one-hundredth that for K^+; thus the inward movement of this ion resulting from normal concentration differences contributes only a small positive charge to the negative (inside) resting membrane potential. However, changes in membrane permeability to Na^+ and K^+ in excitable cells form the basis for information processing in animals.

Depolarization of excitable cells occurs when the membrane potential becomes more positive inside the cell, whereas hyperpolarization occurs when the membrane potential becomes more negative inside. If an excitable cell membrane is depolarized by a critical (threshold) amount, a large increase in membrane permeability to Na^+ results in initiation of an action potential. At threshold depolarization, membrane channels for Na^+ open such that more Na^+ enters by diffusion, increasing the depolarization. This increase, in turn, increases Na^+ permeability (positive feedback) until the membrane potential rapidly reaches the equilibrium potential for Na^+ with high permeability ($+50$ to $+60$ mV). At this point the Na^+ channels close, reducing Na^+ permeability to low values. Depolarization also results in changes in K^+ permeability.

For a short time the site of action-potential production is absolutely refractory to the formation of another; that is, the membrane must be "reset" before another action potential can be produced. Following the absolute and relative refractory periods, each action potential produced is the same in magnitude and duration. The magnitude of depolarization during an action potential depolarizes adjacent membrane areas in excitable cells to the threshold, and action potentials are propagated at a speed that depends on resistance to current flow at adjacent membrane sites, which determines the time taken to reach threshold depolarization.

In nonmyelinated nerves, resistance to current flow is inversely proportional to the square of the (tubular) nerve radius, and conduction velocity increases in proportion to the square root of the tube diameter. In myelinated (insulated) neurons, conduction velocity increases linearly with nerve-fiber diameter because action potentials are propagated between nodes of Ranvier, which are spaced farther apart in larger neurons. Responses of animals that depend on speed for survival (e.g., predator-escape mechanisms) involve relatively large nerves that conduct information rapidly.

Under normal circumstances, afferent neurons are depolarized to the threshold via receptor potentials, i.e., selective amplification of energy forms impinging on receptor cells. Different receptors have evolved to transduce chemical, mechanical, electromagnetic, and thermal energy forms to action potentials in afferent neurons, and there is great diversity among animals in the type and range of sensitivity of receptors to environmental energy forms. These differences reflect different requirements for detecting environmental changes important for survival, and the coding of sensory information in afferent neurons also reflects environmental characteristics important for survival.

For most receptors the change in receptor potential is logarithmically related to energy magnitude at the receptor; as stimulus strength increases, more and more energy must impinge on a receptor for the same change in receptor depolarization. The frequency of action potentials produced in afferent neurons resulting from receptor potentials depends linearly on the magnitude of the receptor potential; thus information on stimulus strength is coded in the action-potential frequency related to the logarithm of stimulus strength. A reason for this relationship is the importance of detecting *changes* in stimulus strength at low levels of stimulus magnitude together with a requirement for obtaining some information on stimulus magnitudes at high levels of stimulus strength. For most forms of energy from the external environment, survival depends to a great degree on detecting small changes when the levels of stimulus energy are low.

The receptor potential is not propagated, and it changes in magnitude with time even when an applied stimulus is maintained at a constant level. This condition is called receptor adaptation. The extent of receptor adaptation also depends on the importance of detecting *changes* in environmental stimuli versus detecting their absolute magnitudes. Tonic receptor systems adapt slowly and generate information for responses that are related to absolute levels of stimulus strength. Phasic receptor systems adapt rapidly and generate information on environmental changes as a consequence of decreases in response to an applied stimulus.

The receptor membrane is the site of receptor-potential formation, but considerable information processing of important environmental energy characteristics within receptors influences receptor potentials and thus the information generated in afferent neurons. The *accessory structures* associated with receptor membranes filter and amplify certain characteristics of environmental energy, and a great deal of interest in the field of sensory physiology concerns the functions of structures in receptor systems.

Chemoreceptors may be associated with accessory structures to increase receptive membrane surface areas (amplification) in very sensitive olfactory receptor systems, such as for detection of some pheromones, or chemoreceptors may be localized to filter certain information, such as the location and quality (chemical composition) of ingested food. Thermoreceptors may also be localized to filter information on temperature at specific sites, such as the skin and anterior hypothalamus in vertebrates. In pit vipers infrared thermoreceptors are associated with more elaborate accessory structures (the pit organs) that amplify thermal information from prey and filter information on location of prey because of the directional sensitivity of the organs.

There are a variety of receptors within most animals to detect different forms of mechanical energy. Those involved in detecting alternating waves of pressure have the most elaborate sets of accessory structures. At the level of receptor cells, a kinocilium and stereocilia (hairlike projections from cell membranes) filter information on the direction of applied force as a result of a dependence of depolarization or hyperpolarization on the direction of hair displacement. The "lever coupling" of hair projections to cell membranes also amplifies mechanical energy.

The mammalian ear contains a number of accessory structures used for filtering and amplifying mechanical stimuli associated with sound waves. Hair cells in the organ of Corti on the basilar membrane of the cochlea are mechanically stimulated to a maximum degree by sound waves of certain frequencies. Sound waves establish traveling waves of the largest amplitude toward the base of the cochlea at high frequencies and near the tip of the cochlea at low frequencies, providing some filtering of information on pitch.

Further filtering and amplification occur in the ear because of the directional nature of sound funneling in the pinna and the coupling of pressure changes with the inner ear by means of the auditory ossicles.

Receptor potentials in photoreceptors are generated when membrane permeabilities change as a consequence of the reception of light energy by photochemical pigments (rhodopsin, erythrolabe, chlorolabe, and cyanolabe). Rhodopsin is involved in the transduction of information on light intensity; other pigments transduce information on color (hue). Both the sensitivity of receptors to light and the dependence of chemical transformations on light wavelengths are identical, indicating that photochemical reactions directly underly receptor function. Within receptor cells, photochemicals are associated with elaborate accessory membrane structures, probably resulting in the considerable amplification of light energy. In addition, the structural arrangement of membrane folds within the rhabdomeres of some invertebrate ommatidial photoreceptors filters information on the plane of polarization of incident light.

Photoreceptors in vertebrate eyes are located in the retina, and considerable information processing occurs as a result of the function of accessory structures before light energy reaches the retina. The cornea and lens focus images on the retina. The lens accommodates (changes thickness) to change focal information about objects at different distances from the eye. The amount of light reaching the retina depends on pupil diameter. Information filtering also depends on where light hits the retina. Centrally placed cones provide information of high acuity but are less sensitive to light intensity than are more peripherally placed rods. The concentration of cones at the central fovea also amplifies (magnifies) images because of the negative curvature of the foveal depression.

Electroreceptors have evolved among fishes that live in waters where light intensity is low. Discharge of an electric organ composed of modified muscle cells arranged in series in the tail of an electric fish establishes an electric field around the fish. Electroreceptors in the lateral line are depolarized by current flow from the electric field, and distortions of the field by objects of different electrical conductivity changes information input through the electroreceptors; the information can be used for navigation. In some species of fishes that have control over the rate of electric organ discharge, there is some evidence that communication between individuals may occur, based on coding information in the frequencies of electrical discharges.

Suggestive evidence for a magnetic sense in some animals has been increasing, based on experiments with homing pigeons and bees. The recent identification of magnetic (lodestone) crystals within these animals offers the promise of identification of the receptors and study of the mechanisms for the transfer of information on magnetic energy to neural action potentials.

Information Integration and Behavior

17

Once information from an environment is filtered and amplified in a receptor, the coded information is conducted along afferent neurons to a central nervous system where integration (*summation*) of different information takes place. We will follow this sequence. To start, we will consider the interactions between neurons that influence summation of information in time and space in our discussion of *synaptic transmission*—transmission across the interfaces between neurons. This process forms the general basis of how most nerve cells interact, but the interactions occur in different types of nervous systems. We will consider the types of neuron-neuron interactions involved in central nervous systems, the major somatic (voluntary) nervous systems and autonomic (automatic) nervous systems. With this background information we will examine the processes of neural integration in terms of production of effective coordinated responses (behavior) as a consequence of environmental changes.

Most behaviors studied from the point of view of mechanisms of neural integration involve rapid information processing, which has an immediate impact on survival (see Chapter 16). We will examine several spinal reflex (automatic controlled) responses as examples of rapid information integration for movements in which the complexity of integration is minimum. We will then examine information integration within the visual and auditory systems of a few animals for which there is sufficient information to assess neural integration for producing responses needed to survive. This knowledge permits analysis of the behavioral interactions between some predators and prey in terms of the underlying mechanisms of neural controls for rapid responses.

NEURON-NEURON INTERACTIONS THROUGH SYNAPSES

Once an action potential is produced in an afferent neuron associated with a receptor, it is propagated without change along the length of the neuron until it reaches the end of that nerve cell axon, which is closely associated with another neuron. The general morphology of this sequence is shown in Fig. 17.1, along with a summary of the changes that accompany interactions between neurons.

The graded receptor (generator) potential influences a part of the afferent neuron particularly sensitive to a change in membrane potential. This region is usually located on a portion of the nerve cell near the cell body where an axon (nerve-cell process) originates, which is called the axon hillock. Action potentials are propagated along the axon into the animal until the axon ending is reached (Fig. 17.1). At the end of the axon is the *synapse* (gap or space) between a neuron and either another nerve cell or an effector.

Whether a response (effector activation) occurs as the result of a stimulus ultimately depends on whether action potentials are produced in the efferent neurons leading to effectors (see Chapter 18). In general, sensory reception usually results in some information *input* into a central nervous system. Whether there is information *output* to effectors depends on what information is transmitted across synapses at a given moment.

An action potential arriving at a synapse may result either in depolarization or hyperpolarization of the next nerve-cell membrane, and whether threshold depolarization occurs in the next nerve cell depends on the *net effect* of events occurring in a large number of synapses of several neurons on a particular cell. Both positive (excitatory, depolarizing) and negative (inhibitory, hyperpolarizing) changes occur. The addition (summation) of these positive and negative effects in time and space is neural integration. We have had several occasions to discuss integration in previous chapters in terms of the addition of information that determines responses in control mechanisms for regulation. The mechanisms of neural integration in large part involve the functioning of synapses.

Transmission Across Chemical Synapses

One neuron may influence the level of depolarization of another neuron either electrically or chemically. Chemical synapses and electrical synapses differ morphologically (Fig. 17.2). The *presynaptic* axon terminals are usually expanded (they appear "swollen"), and in chemical synapses presynaptic terminals contain a relatively large concentration of mitochondria and presynaptic

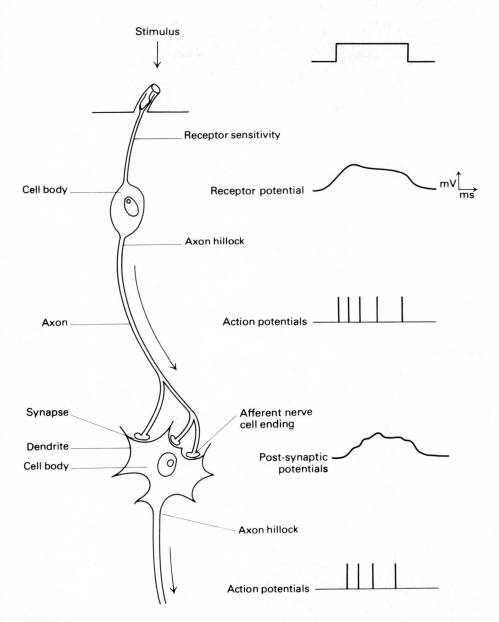

Figure 17.1
Schematic summary of the major events leading to information transmission from an afferent neuron across synapses to influence post synaptic potentials and action-potential formation in a second nerve cell (based on Eckert and Randall, 1978).

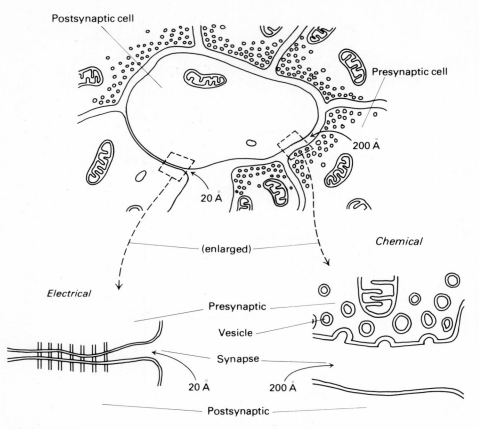

Figure 17.2
Upper Drawings of two types of synaptic associations. The neuron terminals with vesicles are presynaptic axons of chemical synapses. The neuron terminal with few vesicles forms an electrical synapse of smaller width. *Lower* The enlarged drawings show proposed intercellular channels for electrical synapses and the fusion of vesicles with external presynaptic membranes for chemical synapses.

vesicles. The postsynaptic region of a cell usually does not contain these vesicles, and presynaptic electrical synapses also contain few vesicles (Fig. 17.2; see below).

There is considerable evidence that presynaptic vesicles contain chemicals that are released into the synaptic space following the arrival of an action potential, and these chemicals influence the degree to which the postsynaptic membrane is depolarized. The morphological evidence comes from high-resolution electron micrographs that show vesicles fused with the nerve-cell membrane at chemical synapses (Fig. 17.2, bottom). The mechanism is called *exocytosis* (the opposite of pinocytosis). Chemicals sequestered in a vesicle are released into a synapse when an action potential arrives as a consequence of membrane changes (see below). The effects of the chemical on the postsynaptic membrane provide the transmission link either from one nerve cell to another or from an efferent neuron to an effector cell.

The functional evidence for the role of chemical transmission comes primarily from detailed studies of synapses between effector neurons and muscle cells. In vertebrates, motor neurons to skeletal muscles form a relatively long synapse called a *motor end plate.* When a motor neuron reaches a muscle fiber, it extends along a fiber for a distance that may exceed several millimeters. Synaptic vesicles are concentrated at intervals along these synapses and appear to release chemicals into folded regions of the postsynaptic membrane known as *junctional folds* (Fig. 17.3). Because these synapses are relatively large, it has been possible to study the electrical events accompanying synaptic transmission with comparative ease. The function of other synapses is thought to be similar in general features, although the nature of synapses can greatly differ (see below).

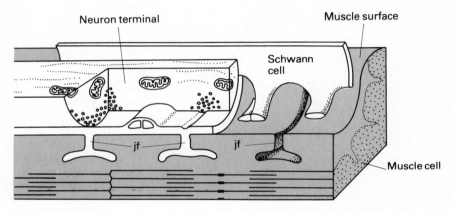

Figure 17.3
Schematic illustration of the synapse between a vertebrate motor neuron and a skeletal muscle cell. Junctional folds (jf) in the postsynaptic muscle membrane are opposite the areas of concentration of presynaptic vesicles in the neuron terminal (based on Peper *et al.*, 1974).

The chemical transmitter for vertebrate neuromuscular synapses in skeletal muscle has been identified as acetylcholine (ACh). If an electrode is placed inside the postsynaptic membrane of the excitable muscle cell and small amounts of ACh are injected into a synapse through a small microelectrode, the postsynaptic membrane will depolarize in a graded manner by an amount that is related to the quantity of ACh. If depolarization is sufficient to reach the threshold for the postsynaptic membrane, an action potential results.

Under normal circumstances, each action potential in a presynaptic motor neuron results in the release of amounts of ACh sufficient to cause threshold depolarization of the postsynaptic muscle membrane. However, it is possible to experimentally alter the characteristics of this synapse to illustrate some of the features of chemical transmission. One way is by bathing the synaptic region with fluid low in calcium concentration (see below). When this is done, each action potential that arrives results in the release of less ACh, meaning that several action potentials arriving in a short time may be required for threshold depolarization.

The effects on the postsynaptic membrane potential are illustrated in Fig. 17.4. With one action potential (arrows in Fig. 17.4), the graded postsynaptic membrane depolarization may not reach threshold and will rapidly decay. The small depolarization is called a "miniature excitatory postsynaptic potential" (mEPSP). If several occur in a short enough period to add up to a sufficient amount of depolarization, threshold is reached and an action potential results (Fig. 17.4). The "quantum" nature of mEPSPs and their decay over time have been observed in normal neuromuscular synapses, and each mEPSP is thought to represent the presynaptic release of the contents of one presynaptic vesicle.

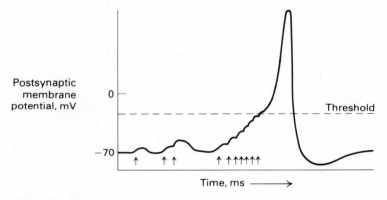

Figure 17.4
Illustration of the short-term additive effects of miniature excitatory postsynaptic potentials (mEPSPs) on postsynaptic membrane depolarization. Each mEPSP represents release of the contents of a presynaptic vesicle.

The Mechanism of Presynaptic Chemical Release

The experiments of Otto Loewi on control of heart rate (see Chapter 4) demonstrated neuron—effector chemical transmission, but the understanding of the mechanisms that result in release of chemicals from presynaptic vesicles required relatively recent technological advances to experimentally manipulate the membrane potentials of presynaptic axon terminals. One biological system particularly well suited for the study of presynaptic events is the squid stellate ganglion. The neurons in this structure are sufficiently large to allow reliable monitoring of pre- and postsynaptic membranes to study the events leading to release of chemical transmitters.

When a presynaptic membrane in the stellate ganglion is electrically depolarized in a controlled manner and the potential of the postsynaptic membrane is monitored, it can be seen that postsynaptic membrane depolarization depends on the extent of depolarization of the presynaptic membrane (Fig. 17.5). There is a very sharp zone of effect between about 50 and 150 mV of presynaptic depolarization. Thus as the presynaptic membrane becomes depolarized during arrival of an action potential, little change occurs at the postsynaptic membrane until presynaptic membrane depolarization has reached a certain value (50 mV in Fig. 17.5). Very rapid changes occur up to a saturation of the postsynaptic potential over a relatively small range in presynaptic potential.

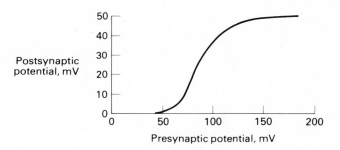

Figure 17.5
The effect of the presynaptic potential on the postsynaptic potential in the squid stellate ganglion is most pronounced over a relatively narrow range of change in presynaptic potential (based on Katz and Miledi, 1967).

The extent to which the presynaptic membrane reaches and exceeds the 50-mV potential required for the chemical-transmitter release that affects the postsynaptic potential involves the movement of calcium ions into the terminal region of the presynaptic axon when an action potential arrives. If the usual Na^+ and K^+ movements are blocked in this region by using chemicals (such as tetrodotoxin) that selectively block their ionic channels, an action potential that depends on the concentration difference of calcium across the neuron membrane can still be produced at the axon terminal. If calcium is removed and replaced with Mg^{++}, the potential is reduced and transmitter release is impaired (see above). Thus the "calcium potential" (entry of positively charged calcium ions) is thought to be necessary for release of the membrane-bound chemical transmitter into the synapse.

The events leading to transmitter release take time. In the neuromuscular junction about 0.6 millisecond is required between pre- and postsynaptic depolarization. Although this seems to be a very short time, it is relatively long compared with the speed of action potential propagation. In addition, if a response to a stimulus involves transmission through several synapses, the delays add.

In general, responses for very rapid timing involve either a minimum of synapses (see the discussion of reflexes below) or transmission through electrical synapses with shorter delay (see below).

The Mechanism of Postsynaptic Chemical Depolarization

Once a chemical is released from a presynaptic terminal and has diffused across the 200-angstrom ($\mathring{A}$)($= 2 \times 10^{-8}$m) space of a synapse it must produce a specific change in the ionic permeability of the postsynaptic membrane. This change is produced in the vertebrate neuromuscular junction by the brief combination of ACh with specific receptor sites on the postsynaptic membrane, which results in a brief opening of ionic channels for both Na^+ and K^+.

If the postsynaptic membrane is voltage-clamped and the extracellular concentrations of K^+ and Na^+ are changed, both ions can be demonstrated to participate in the change in postsynaptic membrane potential. Because K^+ moves out while Na^+ moves in, the extent of postsynaptic depolarization is intermediate between the equilibrium potentials for the two ions. In the neuromuscular junction the permeability values for K^+ outflow and Na^+ inflow result in a maximum depolarization from a resting potential of about -80 mV up to a membrane potential of about -15 mV, which is sufficient to exceed threshold depolarization for muscle membranes.

The interaction of ACh with specific receptor sites on the exterior postsynaptic membrane is similar to the interaction between an enzyme and a substrate. The receptor is the enzyme to which the substrate ACh molecules bind to result in ionic channel openings for brief periods. The poison curare produces its toxic effects by also binding to the receptor sites but without an effect on ionic channels. In addition, ACh is released from the membrane receptor soon after the ionic channels are opened, and it diffuses from the receptor. When diffusion occurs the ACh is usually hydrolyzed into acetate and choline by the enzyme *acetylcholine esterase*. This enzyme contributes to the brief, graded, and localized influence of ACh on the postsynaptic membrane.

Without ACh esterase the ACh would continue to recombine with receptor sites, and a single action potential will produce disproportionate effects on postsynaptic membranes, disrupting the short-term nature of synaptic transmission. Moreover, without the enzyme the ACh could move from one synapse into the circulatory system and be transported as a hormone to influence other nerve cells in other locations that may not be part of a specific control sytem.

A large number of neuron-neuron synapses involve ACh as a neurotransmitter (see below). On the very small scale of individual synapses the chemical is similar to a hormone, but its effective action depends on *localized* specificity of action. Only certain neurons should be influenced by means of synapses for certain control responses. Thus it is quite important that ACh effects be limited to specific synapses. If some does diffuse into the circulatory system it is destroyed by additional acetylcholine esterase located on the exterior membrane surfaces of red blood cells. Some hormones that do have a neurotransmitter function are released into the circulation, but these hormones involve a portion of the nervous system (the sympathetic autonomic nervous system) in which a large number of synapses are normally activated throughout the body at the same time (see below).

The sequence of events involved in synaptic transmission of depolarizing synapse is summarized in Fig. 17.6. With this understanding of the events leading to excitation (depolarization), it is important now to consider synapses where opposite effects occur on postsynaptic membranes.

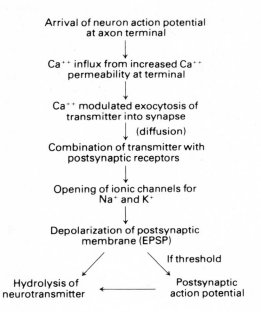

Arrival of neuron action potential
at axon terminal

Ca^{++} influx from increased Ca^{++}
permeability at terminal

Ca^{++} modulated exocytosis of
transmitter into synapse
(diffusion)

Combination of transmitter with
postsynaptic receptors

Opening of ionic channels for
Na$^+$ and K$^+$

Depolarization of postsynaptic
membrane (EPSP)

If threshold

Hydrolysis of
neurotransmitter

Postsynaptic
action potential

Figure 17.6
Summary of the sequence of events involved in synaptic transmission across excitatory chemical synapses.

Inhibitory Postsynaptic Potentials (IPSPs)

Any interaction of transmitter with postsynaptic membranes that has the effect of preventing (inhibiting) the postsynaptic membrane from reaching threshold depolarization will be inhibitory to production of action potentials in postsynaptic membranes. This inhibition can be accomplished by having a transmitter-membrane receptor interaction selectively change ion permeabilities so that a membrane either hyperpolarizes (moves away from threshold) and/or depolarizes (where the depolarization is maintained below threshold). For example, an increase in K$^+$ permeability without a change in Na$^+$ permeability changes the membrane potential toward the -92-mV equilibrium potential for potassium (see Chapter 16).

This fact brings us to the essence of neural integration. Integration is the addition of EPSPs and IPSPs in *time* over a particular *space*, the occurrence of which determines whether a postsynaptic threshold depolarization occurs. Summation in time is important because postsynaptic potentials are graded and change rapidly with time. Several miniature EPSPs may be required at a particular membrane site within a short time to achieve threshold depolarization (Fig. 17.4). IPSPs occurring at the same time decrease the impact of EPSPs on threshold depolarization (Fig. 17.7b).

Summation over space is important because postsynaptic potentials decay in magnitude over distance from their sites of origin. Action potentials are usually generated at the axon-hillock region (Fig. 17.7a), where threshold is lower than at other parts of the nerve-cell-body membrane. The same EPSP or IPSP at different distances from this site have different effects on the extent to which threshold depolarization is achieved.

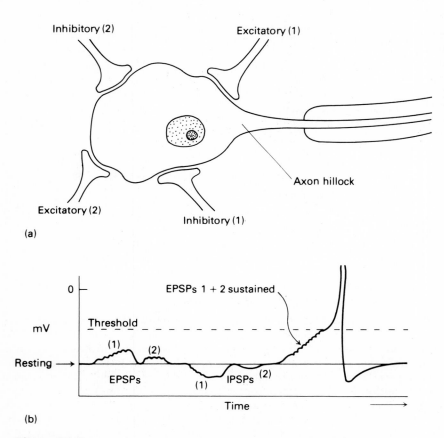

(a)

(b)

Figure 17.7
Illustration of the integration of excitatory (EPSP) and inhibitory (IPSP) events in a nerve cell. Whether action potentials are generated at the axon hillock depends on addition of transient excitatory (depolarizing) and inhibitory (hyperpolarizing) ionic currents that are graded and depend on distance from the axon-hillock region. They are reflected in (b) by distance (EPSPs or IPSPs from synapses 1 or 2) and time effects on membrane potential at the axon-hillock region.

From these effects it should be clear that an action-potential output in a postsynaptic cell depends on both the morphological positioning of synapses and their extent of activity at any moment. Synapses farther from an axon hillock must exhibit greater transmitter release than must synapses closer to the hillock to achieve the same effects. In addition, subthreshold effects at any given moment will influence the extent to which threshold is reached. When subthreshold effects are excitatory, they are called *facilitation* because they enhance the effects at other excitatory synapses involved with threshold depolarization.

Presynaptic Inhibition

Invertebrate voluntary muscles are innervated by both excitatory and inhibitory neurons. Vertebrate skeletal muscles are innervated only by excitatory neurons, and inhibition occurs within the central nervous system that determines whether an excitatory motor neuron will carry action potentials at a given time. Control over contraction occurs more at the level of the muscles in invertebrates, but there is also evidence that inhibitory neurons in some invertebrates have inhibitory synapses on axon terminals of excitatory neurons connected to muscle, as illustrated in Fig. 17.8.

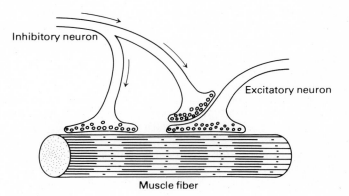

Inhibitory neuron

Excitatory neuron

Muscle fiber

Figure 17.8
Schematic illustration of synaptic and presynaptic inhibition in invertebrate neuromuscular synapses. IPSPs at the inhibitory synapse (left) prevent action-potential formation in the postsynaptic membrane, and presynaptic inhibition (right) decreases the amount of neurotransmitter released by lowering the depolarization of presynaptic membranes if an action potential arrives (after Lang and Atwood, 1973).

The arrival of an action potential at both inhibitory synapses, shown in Fig. 17.8, will produce IPSPs at both postsynaptic membranes. The inhibitory synapse on the presynaptic excitatory-neuron axon terminal will effectively prevent contraction of muscle at the excitatory location by decreasing the magnitude of any neuron action-potential depolarization that reaches the terminal while the IPSP is in effect. This decrease blocks postsynaptic excitation by decreasing the amount of transmitter released into the excitatory synapse.

Electrical Synapses

In chemical synapses, the distance across a synapse (about 200 Å) results in a considerable decrease in postsynaptic electrical currents coming from the presynaptic action potential itself. Current is shunted away from the postsynaptic membrane in the lower-resistance path of the synapse, and chemical transmission provides the electrical-transmission link.

Direct electrical transmission from one cell to another can occur if the low-resistance synaptic gap and the high resistance of the postsynaptic membrane can be bypassed. Synaptic gaps are much smaller in electrical synapses than in chemical synapses, usually about 20 Å wide instead of 200 Å. In addition, the high resistance of membranes is bypassed by means of conduction of ionic currents through small intercelluar channels built into the synaptic junction. Thus the action-potential current in the presynaptic cell has a low-resistance path for more direct electrical depolarization of the postsynaptic cell.

An advantage of electrical synapses is their speed of transmission; the 0.6-millisecond synaptic delay for exocytosis of neurotransmitters is eliminated. Many electrical synapses involve very rapid response mechanisms to stimuli. Two well-studied examples are the electrical synapses in the Mauthner cell of goldfish that produce the rapid "startle" tailflip escape response to sudden environmental changes and the rapid propulsive tailflip response of crayfish.

In addition to speed, electrical synapses produce the advantage of a high degree of certainty in the continuity of information transmission along a specific pathway. However, this advantage produces a disadvantage for other functions. Integration is difficult to achieve with electrical synapses, meaning that functions depending on information comparison (summation of positive and negative effects) operate through chemical synapses at the expense of synaptic delays.

ACh is only one neurotransmitter. It was the first one discovered, and it is the transmitter involved in a number of major divisions (parts) of the vertebrate nervous system. Note also that ACh can be either excitatory or inhibitory, depending on the nature of the postsynaptic receptors with which it interacts.

Most information on functional organization within nervous systems comes from studies of relatively complex invertebrates and vertebrates, and the nervous systems are divided into several anatomical and functional components (see below). The components differ in the nature of the information they subserve, involving whether information processing is voluntary or involuntary (automatic) and the type of environmental information processed. These differences are also partly reflected by the nature of chemical transmitters involved in integration.

CENTRAL NERVOUS SYSTEMS

Afferent neurons enter the central nervous system. In vertebrates the central nervous system is composed of a huge collection of nerve-cell bodies, with their axon fibers located within the spinal cord and brain. The complexity of neuron-neuron interactions is enormous, and the sheer number of possible interactions has proved to be the major obstacle to understanding brain function. Nevertheless, it has been possible to identify major "centers" of integration within the brain, such as we have discussed in several previous chapters (e.g., respiratory and cardiac centers in the medulla and temperature regulatory centers in the hypothalamus). In addition, some parts of the central nervous system are more accessible to studies of mechanisms of neural integration than are others, and we will consider the details of integration in spinal-cord reflexes and in retinal visual-information processing shortly.

There is little direct information on the nature of chemical transmitters within the vertebrate brain. A variety of subtances have been found that appear to function as neurotransmitters in other species or in other areas of the nervous system. These substances include ACh, norepinephrine (see below), γ-aminobutyric acid (an inhibitory neurotransmitter in crustaceans), serotonin (5-hydroxytryptophan), and dopamine.

Despite the complexity of the problem, there is considerable interest in attempting to understand which neurotransmitters act where within central nervous systems because diseases such as certain types of mental illness and drug addiction are thought to be related to imbalances in neural chemistry that may involve neurotransmitters. The analogy between the short-term chemical effects of depressant and stimulant drugs and the longer-term states of depression and euphoria without drugs has added impetus to the study, as have observations that drugs such as LSD can apparently permanently alter brain chemistry. Some cells of the brain produce natural morphine-like analgesic agents (endorphins) that may exert their effect by means of an impact on brain chemistry.

In both invertebrates and vertebrates the process of *cephalization* within central nervous systems is evident, in which concentration of sensory information input at an anterior end is associated with enlargement of the central nervous sytem. Nerve-cell bodies are clustered in areas called *ganglia*. Within invertebrates the axons and dendrites of ganglionic cells mingle extensively in a central area within a ganglion called the *neuropile*, where all synaptic interactions occur.

The Insect Central Nervous System

The functions of various parts of the insect central nervous system have been studied with experiments that involve removal of different ganglia. Experiments with the praying mantis are particularly interesting. The brain consists of the supraesophageal ganglia, which are connected with the lower subesophageal ganglia, which in turn are connected through the ventral nerve cord with the thoracic ganglia. Although the praying mantis is normally very sedentary, removal of the supraesophageal ganglia produces continuous locomotion. However, if the subesophageal ganglia are then removed, the animal becomes completely immobile (except for mature males; see below) despite the fact that the thoracic ganglia involved in producing coordinated movements are still intact and can be artificially stimulated to produce movement. Thus the thoracic neural coordinating system is normally inactive unless neural commands arrive from the subesophageal ganglia. In a normal insect the subesophageal ganglia are usually *inhibited* by the supraesophageal ganglia; removal of this inhibition transforms the sedentary mantis into a continuous walker.

If half of the supraesophageal ganglia is removed, the insect turns continuously in a tight circle toward the intact side. This action involves forward movement by the legs on the "brainless" side along with backward movement of the legs on the intact side. If the brain is cut into two halves (a "split-brain") the mantis will not move, but continuous reaching movements are made with the legs. These additional results suggest that the inhibition from the supraesophageal ganglia involves a right- and left-turning component imposed from each half of the brain on a forward command originating from the subesophageal ganglia. More refined techniques have localized certain dorsal areas of the insect brain, called the "mushroom bodies," as sites of inhibition of locomotion and as sites of learning in ants and bees.

Although somewhat crude in approach, these results illustrate the basic features of neural integration in the insect central nervous system—namely, the interplay between excitation and inhibition in determining the nature of an effector response. Moreover, recall that effector responses for feeding in the blowfly involve an interaction in the brain between excitatory neural information from external environmental receptors and inhibitory neural information from the digestive sytem (see Fig. 10.16).

There is an interesting addition to the pattern for the praying mantis that involves the reproductive behavior of the insect. The female must capture enough prey to provide protein and nutrients for her developing eggs; occasionally the male provides this resource. Usually a male will "stalk" a female from behind and jump on her back. If the female has fed recently, the two will copulate and the male will depart. However, if the female is "hungry" she will attempt to grab the male with her large forelegs, usually grasping his head, which she will start to consume. Even after the male's brain has been removed he is still capable of performing the complicated abdominal movements required to inseminate the female, and the movements will continue for hours in mature decapitated males. Thus release of inhibition from the brain helps to ensure the fitness of the male despite the fact that he has "lost his head" over the female.

The Vertebrate Central Nervous System

The vertebrate central nervous system is more complex and more difficult to fathom in terms of integrative function because of its size and complexity. Much information comes from studies in which parts of the central nervous system are removed or are stimulated in order to observe changes in behavior.

The central nervous system is normally divided into four anatomical units: the spinal cord (which we will discuss below), the hindbrain, the midbrain, and the forebrain. The hindbrain consists of the *medulla oblongata* and the *cerebellum*. The medulla contains a variety of sensory and motor paths as well as a variety of "centers" for regulatory controls. The cerebellum is involved in complex motor coordination and is most developed among birds and mammals.

The midbrain includes the *tectum*, which is a principal sensory-integration area in reptiles and amphibians. The midbrain also includes part of the *reticular system*, a diffuse collection of nerve cells receiving information from various sensory sources that is important for maintaining states of wakefulness and sleep by means of effects on the cerebral cortex.

The most complex areas of the vertebrate brain involve the forebrain. It has two major parts, the *diencephalon* (thalamus, hypothalamus, and posterior pituitary) and the *telencephalon*. The latter includes the olfactory bulbs and the complex *cerebral hemispheres*. The cerebrum consists of a surface layer of nerve-cell bodies (gray matter) with underlying layers of nerve fibers (white matter). It represents the largest part of the human brain and contains an estimated 10 billion neurons, each with several hundred synaptic inputs. The cerebral cortex subserves sensory, motor, and complex integrative functions. Although certain areas of the cerebrum have been assigned specific sensory or motor functions corresponding to specific areas of the body from the findings of ablation and stimulation experiments, large areas of the cerebrum remain unspecified in function. The posterior occipital lobes are involved in integration of visual information, and the lateral parietal lobes process auditory information. Removal of the *hippocampus* below the lateral part of the cerebrum can influence short-term and long-term memory. Despite some localizations of function, however, the cerebrum is more noted for its lack of specificity, especially given its size and apparent importance in humans.

THREE VERTEBRATE EFFERENT NERVOUS SYSTEMS

There are two major divisions of the vertebrate efferent nervous system, and one of them is divided into two parts. The major divisions consist of the *somatic* and the *autonomic* nervous systems. The autonomic nervous system is divided into the *sympathetic* and *parasympathetic* nervous systems.

The Somatic Nervous System

The somatic efferent nervous system is composed of all the nerves going from the central nervous system to skeletal-muscle-cell effectors. The cell bodies of the neurons are all located within the brain or spinal cord, and all axons of somatic efferent neurons synapse directly on skeletal muscle. Thus they form the "final common pathway" for the activation of the effectors for body movements. The transmitter released at muscles is ACh, and the neurons are always excitatory in vertebrates. Somatic efferent nerves are also called *motor* nerves. They are involved in the production of coordinated movements; we will discuss the mechanisms for these movements shortly (see "reflexes" below).

The Autonomic Nervous Systems

Nerves of the autonomic nervous system innervate smooth muscle, cardiac muscle, and gland effectors. In addition, the neurons that leave the central nervous system synapse on a second nerve cell that synapses on an effector. A third distinction is that each effector usually receives two types of autonomic neurons, one that is excitatory and another that is inhibitory. This *dual innervation* forms the basis for the division of this nervous system into two components.

The sympathetic nervous system   The sympathetic division of the autonomic nervous system subserves automatic, involuntary functions most closely related to survival in short time periods. Cardiac muscle, smooth muscle, and glands are influenced to maximize effective internal responses to immediate changes in the external environment.

The external environmental changes are usually described as "stressful." For example, when a predator imminently threatens to capture a vertebrate prey, the heart rate of the prey increases, blood flow to muscles involved in escape movements increases by means of arteriole smooth-muscle dilation, pupil diameter decreases, energy substrates are mobilized from storage deposits, and sweat glands on plantar surfaces secrete fluid (Chapter 14). At the same time, blood flow to the digestive tract decreases, stomach motility decreases, secretion of digestive enzymes decreases, and whatever is in the digestive tract may be rapidly voided. All of these changes represent the mobilization of internal control mechanisms for "flight or fight" functions.

The neurons of the sympathetic nervous system leave the spinal cord in its central (thoracic and lumbar) regions (Fig. 17.9) and synapse in ganglionic cell clusters. The nerve fibers passing to the ganglia are called preganglionic; those passing from the ganglia to effectors are called postganglionic. The ganglia of the sympathetic nervous system are located close to the spinal cord in a chain called the sympathetic trunk. The neurotransmitter for the preganglionic neurons of the autonomic nervous system is ACh (Fig. 17.10). The postganglionic neurons of the sympathetic nervous system produce norepinephrine, the neurotransmitter that influences the effectors (Fig. 17.10). Whether norepinephrine is excitatory or inhibitory depends on the receptor sites at different effectors (see Chapter 18). For example, norepinephrine increases depolarization of some arteriole smooth muscle, but it inhibits contraction of some smooth muscle in the digestive tract (see above and Chapter 18).

One ganglion of the sympathetic nervous system has no postganglionic neurons but does secrete hormones into the circulatory system (the *adrenal medulla*; Fig. 17.10). The secretion is a combination of norepinephrine and epinephrine. Epinephrine is also known as adrenaline and has an effect similar to norepinephrine. As a consequence of a stressful situation, the hormone is transported to all tissues of the body, particularly those to which blood flow has been increased as the result of local sympathetic activity. This hormone functions to maintain continued, strong sympathetic activity as long as the "stress" stimuli persist by saturating sympathetic receptor sites at the effectors.

The evolution of a hormone analogous in function to the postganglionic sympathetic neurotransmitter is related to the survival values (benefits) of *syn-*

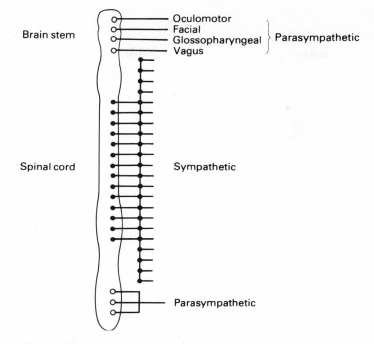

Figure 17.9
Diagram of the structure of parasympathetic and sympathetic
divisions of the autonomic efferent nervous systems. Parasympathetic preganglionic neurons originate in the brain stem or
base of the spinal cord, whereas preganglionic sympathetic neurons originate in the central spinal cord and synapse in the sympathetic trunk close to the spinal cord (from Vander *et al.*, 1975).

chronous and *sustained* activation (or inhibition) of effectors that increase effective internal response to immediate stress from the external environment. In
contrast, the other major neurotransmitter of the peripheral nervous system,
ACh, does not have a circulatory hormonal analogue because the neurons that
produce ACh at effectors subserve functions in which effective responses depend on *sequential* timing and *selective* activation of effectors. An example is
the function of the parasympathetic nervous system.

The parasympathetic nervous system   This division of the autonomic nervous system subserves internal functions usually opposite to those of the sympathetic nervous system. Increased activity in parasympathetic neurons (e.g.,
in the vagus nerve) decreases heart rate, relaxes smooth arteriole muscles in
blood vessels going to the gut, increases digestive motility, and increases secretion of digestive glands such as the salivary glands and the pancreas. In general,
the system operates to maximize the effectiveness of digestive processing,
which takes time and usually occurs under "nonstressful" conditions.

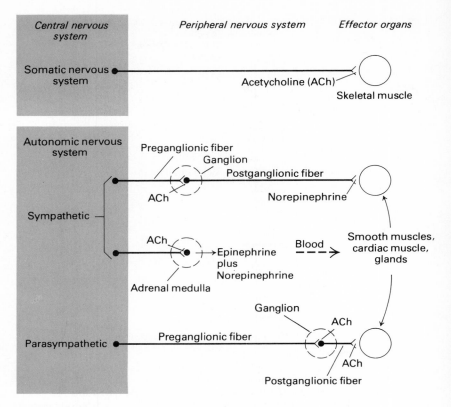

Figure 17.10
Summary of the organization of the somatic and autonomic nervous systems
of vertebrates (from Vander *et al.*, 1975).

Digestive functions are sequential and selective and do not involve rapid
or synchronous activation of a large number of different effectors. Salivary glands
secrete just prior to and during consumption of a meal, followed by stomach
secretions and motility, which are in turn followed by intestinal secretions, absorption, and processing of assimilated substrates (see Chapter 10).

The preganglionic neurons of the parasympathetic nervous system
originate in the brain stem and the base of the spinal cord (Fig. 17.9). They have
long neural fibers that synapse in ganglia located at the various autonomic effectors, where localized release of ACh influences effectors in sequential patterns
by means of the operation of local control mechanisms at each stage of digestive
function.

INTEGRATION FOR EFFECTIVE BEHAVIOR: SPINAL REFLEXES

Integration occurs as a consequence of synaptic events that sum in space and time so that action potentials are either produced or are not produced at a postsynaptic site. Integration is also an important general component of a negative-feedback control system (see Introduction) in which summation of information from various sources determines whether or not a response is produced that has an influence on (or is related with) a stimulus. This relationship is summarized in Fig. 17.11 for neural control mechanisms, and the simplest neural control mechanisms involve integration at only one synapse, which occurs in certain rapid, automatic spinal-reflex movements.

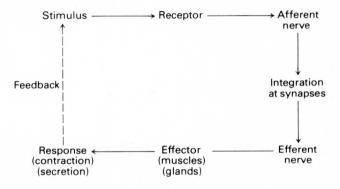

Figure 17.11
Components of a negative-feedback control system involving nerves and neural integration.

The Stretch Reflex

The most intensively studied neural control mechanisms in vertebrates are those involved with the integration of information in the central nervous system at the level of the spinal cord. A number of automatic, coordinated movements are still possible even when the brain is completely cut off from the spinal cord. These movements are called reflexes, and a simple reflex involving integration at only one synapse is diagrammed in Fig. 17.12. The name of this reflex—the stretch reflex—comes from the stimulus that leads to the muscle-contraction response.

The receptors for the detection of muscle stretch are located within *muscle spindle organs* (Fig. 17.12). A spindle organ consists of a centrally located receptor region with afferent nerve fiber endings. The central receptive region is connected on each side to muscle fibers called *intrafusal* (within the spindle) muscle fibers. The entire spindle organ is positioned within skeletal muscles such that it is arranged in parallel with the major mass of muscle fibers responsible for contraction, the *extrafusal* muscle fibers (Fig. 17.12).

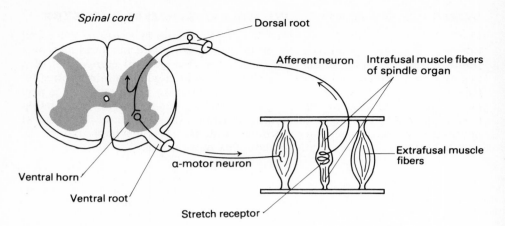

Figure 17.12
Schematic illustration of the structural arrangement of afferent and efferent neural circuits for production of a stretch reflex. This reflex involves only one synapse between afferent neurons from the stretch receptor and the alpha motor neurons going to extrafusal muscle fibers. The spinal cord is drawn in cross section.

The afferent neuron from the spindle organ passes into the spinal cord through the dorsal root (Fig. 17.12), which carries sensory information into the central nervous system from certain segments of the body. The afferent neuron synapses on a motor neuron in the ventral horn of the spinal cord (Fig. 17.12), which is a collection of nerve-cell bodies of effector neurons. The motor neuron on which the spindle afferent neuron synapses is called an *alpha* (α) motor neuron if it innervates extrafusal muscle fibers as shown in Fig. 17.12; motor neurons leave the spinal cord through the ventral root.

If an external stretch stimulus is applied to a muscle, the action-potential frequency increases in stretch-receptor afferent neurons, resulting in depolarization of and increased action-potential frequency in the α motor neurons to that muscle in a proportional manner. The degree of contraction by the extrafusal muscle fibers depends on the frequency of action potentials reaching them through α motor neurons; thus the response (contraction) is related to the extent of the applied stretch. The contraction removes the externally applied stretch stimulus on the receptor because the spindle organs are arranged in parallel with exrafusal muscle fibers.

This arrangement is a very simple neural feedback loop, but with no other components it will produce very jerky and uncoordinated movements. For example, with just these components a muscle will alternately stretch, contract, relax, and stretch again if an external stretch stimulus is continuously applied to a muscle. This type of movement is the kind produced by the "knee-jerk" reflex that results when the patellar tendon and extensor muscle of the leg are briefly stretched by a blow with an object. However, many movements are smoothly coordinated, which involves control over the degree of *internal* stretch on muscle stretch receptors by means of contraction of the muscle fibers on each side of the stretch receptor (the intrafusal muscle fibers; see Fig. 17.12).

The Gamma-efferent Control System

A separate set of smaller motor neurons originate in the ventral horn of the spinal cord (Fig. 17.13). These are *gamma* (γ) motor neurons, or gamma efferents. They innervate the intrafusal muscle fibers on each side of the spindle stretch receptor, meaning that action potentials in gamma efferents result in internal stretch on the central stretch-receptive region of the spindle organ.

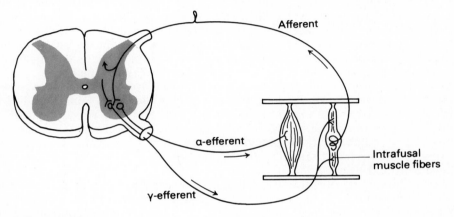

Afferent

α-efferent

Intrafusal
muscle fibers

γ-efferent

Figure 17.13
Schematic illustration of the gamma-efferent system for producing contraction. Activity in gamma-efferent motor neurons results in contraction of intrafusal muscle fibers, which places stretch within the spindle organ on stretch receptors.

The degree of activity in the gamma efferents determines the degree of internally applied stretch on the receptors. Thus graded and smoothly coordinated contraction of the extrafusal muscle fibers occurs by means of the gamma-efferent system when excitatory neurons that synapse on gamma-efferent neurons in the ventral horn of the spinal cord result in different amounts of stretch by contraction of intrafusal muscle fibers and thus different amounts of afferent activity and resultant contraction of extrafusal muscle fibers.

In essence, activity in the gamma-efferent system determines the "set-point" for the degree of an internal stretch stimulus. The neurons that synapse on gamma-efferent neurons are located in other areas of the central nervous system, such as certain areas of the spinal cord or the motor cortex of the cerebrum. Information from these areas determines the timing and degree of contraction of particular muscles by means of the internal activation of the stretch-receptor reflex system. Note also that this method for producing contraction is entirely dependent on local sensory input to the spinal cord through the dorsal root (Fig. 17.13). For this reason, injury to the dorsal root can cause paralysis of extrafusal muscles despite lack of damage to the motor nerves in the ventral root.

Reciprocal Inhibition

Although the gamma-efferent system produces smooth and graded muscle contraction, normal movements of the limbs of animals depend on the addition of an inhibitory component to the central integration process in the spinal cord. The reason for this requirement can be appreciated after close examination of the muscles on your limbs. Movement only occurs when muscles contract or shorten in length, and movements of limbs in different directions involve alternate contraction and relaxation of sets of *antagonistic muscles*, muscles which result in movement of a limb in opposite directions. The antagonistic pairs of muscles for a limb are called *flexors* and *extensors*, depending on whether contraction produces movement of a limb toward or away from the body.

The requirement for inhibition is due to the stretching of the opposite muscle whenever one of the antagonistic muscles contracts. Contraction of a flexor always stretches the antagonistic extensor and vice versa. Without a mechanism to inhibit the stretch-receptor system in an antagonistic muscle, no movements of limbs would be possible because whenever one muscle started to contract and stretched another, the second one would contract and prevent movement of the first.

Inhibition occurs within the spinal cord by means of inhibitory interneurons that synapse on alpha efferents in the ventral horn (Fig. 17.14). The inhibition is called *reciprocal* because the opposite member of an antagonistic pair is forced to remain relaxed as a result of inhibition whenever the other member of the pair is contracting. The following sequence is typical (see Fig. 17.14): (1) initiation of contraction occurs through a gamma efferent; (2) gamma-efferent control of intrafusal muscle-fiber contraction places stretch on a stretch receptor in, for example, an extensor; (3) action-potential frequencies increase in afferent neurons coming from the extensor stretch receptors; (4) within the spinal cord the afferent neurons branch into two segments—one branch depolarizes alpha efferents to the extensor and the other depolarizes inhibitory interneurons that synapse on alpha efferents going to the antagonistic muscle; (5) the extensor contracts while the contraction of the stretched flexor is inhibited.

Note that excitatory input to a stretched antagonistic muscle's alpha motor neurons still occurs. Normally this results in no output of action potentials because summation with the inhibitory events at the postsynaptic membrane on alpha-efferent neurons prevents threshold depolarization. However, if stretch of the antagonistic muscle is extreme the greater excitatory input to the alpha efferents will override the inhibition and some contraction will occur when threshold is reached despite inhibition. This usually prevents damage to antagonistic muscles when extreme stretch forces are applied to them.

Another inhibitory input also prevents damage to muscles; it comes from the *Golgi tendon reflex*. Sensory endings in the tendons of muscles are arranged in series with the muscle, and the afferent neurons from these receptors synapse on inhibitory interneurons, which in turn synapse on alpha-efferent

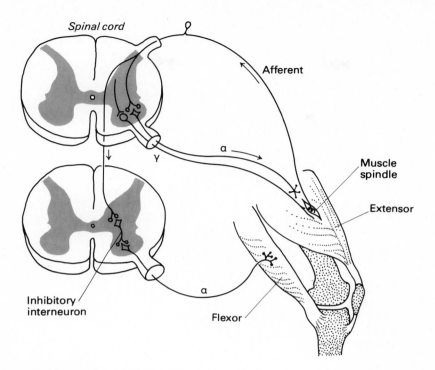

Figure 17.14
An inhibitory interneuron has been added to the spinal-reflex integration system to produce movement of a limb involving stretch of an antagonistic muscle (in this case the flexor is inhibited).

neurons to the *same* muscle. Thus a potentially damaging force at a muscle tendon results in inhibition of contraction of that muscle, reducing the force on the tendon.

Adding Complexity with Additional Neurons

With the basic elements of addition of excitation (depolarization) and inhibition (hyperpolarization) for integration, it is possible to construct neural "circuits" that provide the basis for a variety of movements. A few more examples will illustrate this fact.

In some cases a limb will either flex or extend depending on the type of stimulus applied to it. For example, the *withdrawal reflex* involves rapid flexion of a limb in response to a noxious stimulus on the skin such as heat or pain or a stimulus usually associated with damage. This reflex involves a circuit in which afferent neurons from specific receptors on the skin synapse directly on flexor alpha efferents. With only one synapse, the delay between stimulus and response is minimized.

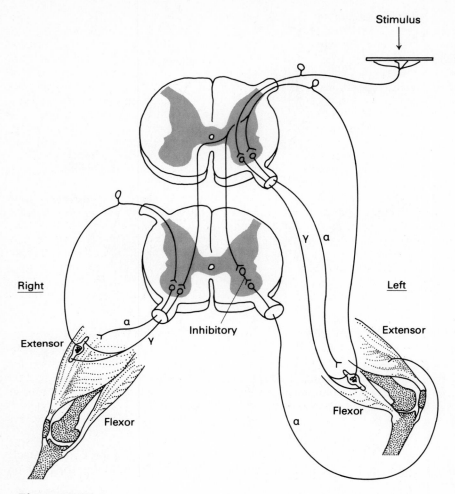

Figure 17.15
Some neural circuits involved in the crossed extension reflex with the reciprocal inhibition of antagonistic muscles. By tracing the circuits from the stimulus you should be able to visualize flexion movement of the *left* limb with simultaneous extension of the *right* limb.

If the stimulus happens to be pressure applied to the surface of a palm or foot, the response is extension of the limb to increase contact pressure on the touching surface. This is the *magnet reflex*. It involves a circuit in which touch-receptor afferent neurons synapse directly on extensor alpha efferents. The normal function of the magnet reflex is to "brace" a limb as it contacts the ground so that it will bear the weight of the animal.

A withdrawal reflex in one limb usually results in extension of the opposite limb. For example, if a foot is pricked or burned such that it is withdrawn, the limb on the other side will be extended at the same time. This is the *crossed extension reflex*, and it functions to provide support for the body while one limb is no longer touching the ground. The circuit involves excitatory connections between the flexor afferent neurons on one side of the spinal cord and the extensor efferent neurons on the opposite side. Inhibition of antagonistic muscles within a limb is also involved. The circuit is partially diagrammed schematically in Fig. 17.15. Note that the complexity of circuit connections has increased considerably, but they have been constructed with a minimum of excitatory and inhibitory synapses. You should be able to trace the circuits in Fig. 17.15 and visualize the characteristic movement of the reflex without any extra required components.

The crossed extension reflex is also partly involved in normal terrestrial locomotion (walking and running). The additional complexity required is alternate *crossed activation* of flexors when extensors on the opposite side are activated. Thus one limb is flexed and extended forward while the opposite limb is being extended downward and then flexed upward. If you attempt to diagram the simplest possible circuit for this you should be impressed with the extent of the circuitry, and you can appreciate that even with an economy of connections the vertebrate central nervous system involves a maze of neural interactions.

INTEGRATION FOR PROCESSING VISUAL INFORMATION

In the last chapter we discussed a variety of functions of visual accessory structures that result in some filtering and amplification of light energy *before* the energy reaches visual receptors. Information processing also occurs *after* light receptors have been stimulated. Integration occurs at various levels of interactions between nerve cells such that specific characteristics of a visual image on the receptive surface of an eye are passed farther into the central nervous system for processing and production of responses by effectors.

As we discussed with other receptor characteristics (Chapter 16), one very important stimulus characteristic is environmental *change*, and certain *types* of environmental change are filtered from visual sensory information via neural integration mechanisms.

Lateral Inhibition

One type of change in light energy that carries important information for animals is the *contrast* between light and dark areas falling on receptive surfaces. This contrast provides information on the presence or absence of light, and it also carries information about the *geometry* (size and shape) of objects that produce dark-light contrast. The interactions between visual neural structures filter and amplify information used for contrast detection; the mechanism has been extensively studied in the compound eye of *Limulus* (the horseshoe crab). It involves a process called lateral inhibition.

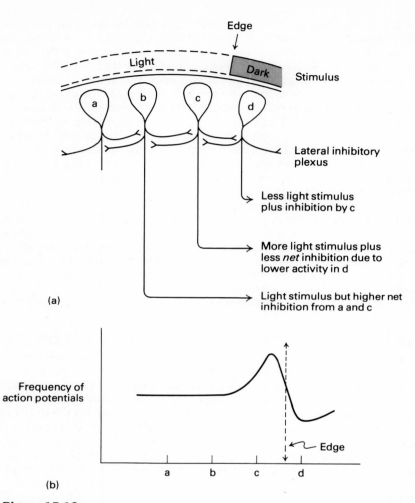

Figure 17.16
(a) Schematic diagram of the relationship between an edge stimulus and the consequences of lateral inhibition for relative amounts of afferent information from different ommatidia in a compound eye. (b) As a consequence of less inhibition at "c" and more inhibition at "d," changes of action-potential frequency coming from receptors at an edge "amplify" information on visual contrast.

The mechanism for lateral inhibition is illustrated schematically in Fig. 17.16. Collateral axons (branches) from afferent neurons associated with receptor cells extend laterally in a "lateral inhibitory plexus" (Fig. 17.16a). When a given ommatidium is stimulated, adjacent ommatidial afferent neurons are inhibited, and the magnitude of the inhibition depends on the magnitude of the stimulus and the distance from the site of stimulation. The result is enhanced (amplified) information on contrast (edge detection) because afferent neurons on each side of the edge carry more or fewer numbers of action potentials than do their respective neighboring afferent neurons (Fig. 17.16b). The enhanced activity at the edge of the light area is largley due to less *net inhibition* of afferent information from that area than from other light-stimulated areas.

Retinal Integration of Visual Information

The nerve cells in the retina of vertebrates are part of the central nervous system, and there are four types of nerve cells located in layers within the retina just above the visual receptor cells (Fig. 17.17). They are arranged into two cell types that run laterally in the retina (the horizontal and amacrine cells) and two cell types that run vertically (the bipolar and ganglion cells). The axons of the ganglion cells form the optic nerve, which transmits information to other parts of the central nervous system (the optic tectum and/or occipital cerebral cortex).

Because the axons of ganglion cells are easily accessible for the purposes of manipulation and monitoring of electrical transmission, most studies of visual neural integration concern the nature of stimuli on the retina that give rise to activity (action potentials) in axons of ganglion cells. The ganglion cells are also the first cells in the visual pathway along which action potentials are generated. Recall (Chapter 16) that visual receptors in vertebrates produce hyperpolarizing currents when stimulated (they produce depolarization when a light stimulus is *not* present). These graded potentials influence graded potentials of two types in bipolar cells. One type maintains the hyperpolarization to ganglion cells and is inhibitory, and the other type changes hyperpolarizing currents to depolarizing currents in ganglion cells and is excitatory.

There is considerable *convergence* of structure and information processing in the retina. For example, in the human eye there are a total of about 100 million visual receptors, but there are only 1 million ganglion cells. Thus when an electrode is placed on the axon of a ganglion cell, the activity represents the result of considerable interaction among many receptors and their intervening neural connections in the retina. Moreover, the extent of the interaction varies at different locations on the retina, ranging from regions such as the fovea, where numbers of receptors and ganglion cells are about equal, to the periphery, where there are many more receptors than there are ganglion cells.

When an electrode is placed in a single ganglion cell axon and light stimuli are projected onto different parts of the retina, the region of the retina that produces a change in the action-potential frequency of that ganglion cell is called the *receptive field* of that cell. Note that receptive fields are defined functionally. Each has a morphological structure on the retina, but the morphologies vary considerably for different receptive fields.

There are two general types of receptive fields that have been found in all retinas studied, and there is one type of receptive field found in the retina of the frog that provides a mechanism for explaining some species differences in responses to certain specific visual patterns (see below).

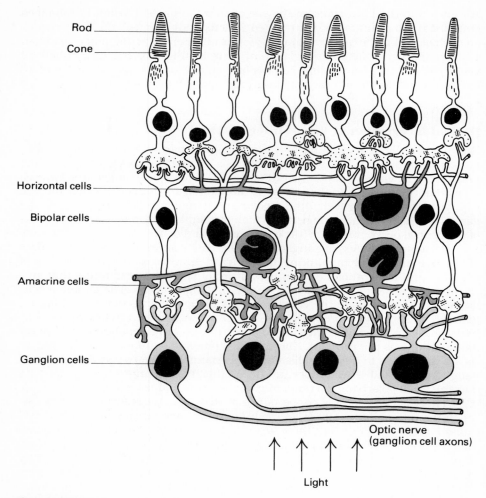

Figure 17.17
Illustration of the neural structure of the vertebrate retina. Information in the form of action potentials in the optic nerve is the result of integration between several layers of nerve cells. Note that light penetrates these layers before it stimulates photoreceptors (based on Kuffler and Nichols, 1976).

On-center and Off-center Receptive Fields

The two general types of ganglion receptive fields are illustrated in Fig. 17.18. An "on-center" receptive field responds maximally to a light-spot stimulus at the center of a circular field. Stimulation of the surrounding region with light results in no response but does result in an increased response when the light is turned off. The "off-center" receptive field functions oppositely, with a decrease in action-potential frequency when the center is illuminated and an increase in frequency when the surrounding region is illuminated (Fig. 17.18).

The interaction of the center and the surrounding region of these receptive fields is due to lateral inhibition via horizontal cells. As a consequence, these receptive fields provide selective information about (1) whether a light stimulus is present (on-center stimulation), (2) whether a light stimulus is absent (off-center stimulation), and (3) whether there is *movement* in an animal's environment. Movement information is produced by an "on-off" and/or "off-on" response as a contrast edge moves across a given receptive field. Any *change* from off to on or on to off by a center and its surrounding region only occurs for the movement of an edge across the field.

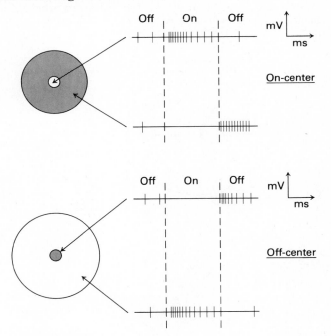

Figure 17.18
Summary of the mode of operation of "on-center" and "off-center" ganglion-cell receptive fields. Action potentials in single ganglion cell axons are represented by "spikes." A light stimulus in the center of an "on-center" field results in increased action-potential frequency, whereas light in the surrounding region inhibits action potentials but produces an "off" response. The opposite occurs for an "off-center" receptive field.

"Moving-bug-detection" Receptive Fields

Retinal integration for detection of any movement is important, but detection of the movements of certain types of objects provides more information. This shape detection involves the third type of receptive field found in the frog's retina. Certain ganglion axons in the optic nerve of the frog respond only when a *convex edge* is *moved* across the receptive fields of those ganglion cells (Fig. 17.19). For the shape in the upper part of Fig. 17.19, a response is produced when it is moved from left to right across the receptive field, but no response is produced when it is moved from right to left (which amounts to the movement of a *concave* edge).

Small moving convex edges represent food (flying insects) to hungry frogs (bottom part of Fig. 17.19), and the integration mechanisms within the retina provide a filtering mechanism specific for this crucial piece of environmental information.

What is important for a frog is not necessarily important for other animals. Is there, then, any information on visual integration in other species that can provide a basis for differences between animals in responses to different types of visual stimuli?

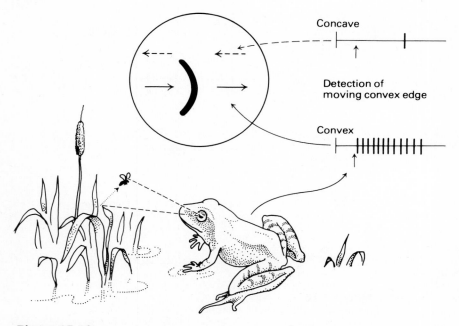

Figure 17.19
Summary of the response of the frog's "moving-bug-detector" receptive fields. Some ganglion cells respond only when a moving convex edge passes across the receptive field. Thus the movement of the object shown in the upper part of the figure is detected only when the object moves from left to right (based on Lettvin *et al.*, 1961).

Students of animal behavior have described types of visual stimuli that are called "releasing" stimuli. A certain pattern, or characteristic, of environmental visual stimulation produces a fixed or stereotyped response that is typical for all individuals of a species. An example is the "flight" response of baby chicks to the pattern shown in Fig. 17.20. If the model shown is pulled over a barnyard such that it casts a shadow, the chicks will run for cover if it is pulled from left to right, but they will show no response if it is pulled from right to left. Running from the first stimulus is adaptive because it provides visual information on the close proximity of a flying predator (e.g., a hawk); the alternative represents visual information on the proximity of a bird that is no threat to survival (a long-necked bird such as a goose).

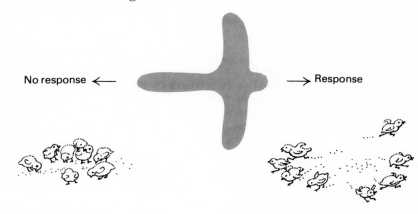

Figure 17.20
A generalized releasing stimulus that produces a fixed pattern of escape behavior in chicks when it is moved from left to right, but not when it is moved from right to left (after Tinbergen, 1948).

The analogy between this type of stimulus and reponse and the "moving-bug-detection" receptive field of a frog is striking. Behavior (response) is released only if a visual image moves and the moving image is of a certain geometry.

There are many examples of these types of species-specific released behaviors related to certain types of stimuli. For example, toads respond by orienting to stimuli that are wormlike (elongated) and that display a jerky (on-off) movement pattern, young herring gull chicks beg for food by pecking at a red spot on their parent's bill, and certain postures between aggressive individuals can produce fighting or submissive responses for a variety of animals. Although specific information on the nature of retinal neural integration is lacking, the retinal processing of a species-specific visual pattern in the frog strongly suggests similar types of stimulus filtering for other types of environmental stimuli in other species.

The integration of visual information for specific patterns of stimulation is not restricted to the retina. The optic nerve passes to the optic tectum in the brain of frogs, and in birds and mammals information is passed ultimately to the visual cortex of the cerebrum. The nerve cells in these areas occur in morphological and functional layers much like those in the retina, and the interactions between cell types have been summarized as a hierarchy of integration resulting in either "simple," "complex," or "hypercomplex" retinal receptive fields for neurons in the visual cortex.

A "simple" receptive field for a cortical neuron could respond to a bar of light of a certain width oriented at a certain angle. "Complex" receptive fields appear to respond best to vertical edges. "Hypercomplex" receptive fields for cortical neurons provide information on the *specific geometry* of contrast on the retina, and they are usually sensitive to movements in specific directions. The latter are particularly well suited to filter visual information for certain stereotyped responses at the level of the cerebral cortex.

AUDITORY INTEGRATION AND BEHAVIOR FOR SURVIVAL

The interactions between predators and their prey can occur rapidly, and survival for both may depend on rapid information processing by means of neural integration. One predator-prey interaction illustrating the features of stimulus filtering and rapid control mechanisms has been extensively studied—the interaction between bats and certain insects that have evolved ears to detect and avoid bats.

Recall that bats navigate at night by echolocating objects in their environment. Flying insects are detected and consumed by bats in midair, and bats can locate and orient to small flying objects within about 20 feet, depending on the size of the objects.

Several species of night-flying insects have evolved ears that are selectively sensitive to the ultrasonic sounds produced by bats. These include hawkmoths, lacewings, and noctuid moths. The ears of noctuid moths are particularly well studied. They are located on the thorax below the wings and consist of a membrane and two receptor nerve cells associated with it (Fig. 17.21).

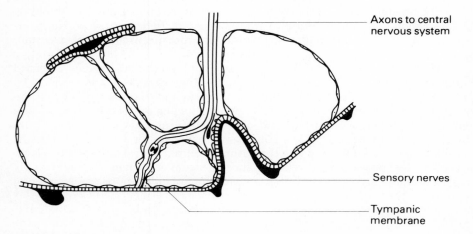

Axons to central
nervous system

Sensory nerves

Tympanic
membrane

Figure 17.21
Diagram of the structure of a noctuid moth's ear. The tympanic membrane oscillates in response to ultrasound, and two axons of sensory neurons transmit information into the central nervous system of the moth (based on Roeder, 1967).

Because of the simplicity of this ear, it has been possible to monitor the electrical activity produced in afferent neurons caused by the presence of hunting bats and to relate the information to the evasive responses of the moths.

Two Types of Evasive Behavior

Bats are rapid fliers as compared with moths, and it is difficult to closely observe the interaction between the two in the dark. However, the responses of moths to bats can be studied by constructing a stationary "artificial bat" composed of a pole with a loudspeaker through which bat cries can be broadcast at will. If this "bat" is placed near a light used to attract moths, their reactions to bat sounds can be observed easily.

Two types of evasive behavior are observed (Fig. 17.22). If the "bat" is close (i.e., moths are close to the speaker) when it is turned on, the moths abruptly cease flying and rapidly flutter to the ground in a "power dive." This reaction makes the moths difficult for bats to detect and pursue in the vegetation on the ground (if they make it). However, if the moths are farther away from the "bat" when it is turned on, they turn and tend to move in a path that increases the distance between them and their predators (Fig. 17.22).

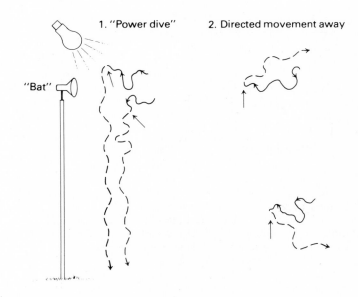

Figure 17.22
Illustration of the two types of evasive behaviors shown by moths with ears to bat sounds broadcast from a speaker on a pole. If moths are close when the "bat" is turned on they "power dive" to the ground. If they are distant they turn and fly away from the source of the bat sounds. The arrows indicate when the "bat" was switched on (based on Roeder, 1967).

The "war" between bats and moths is one of distance and speed. Bats fly more rapidly than moths and can easily catch up with them if they can be detected, but the sound navigation system of bats has a limited range for detecting moths. For many bats the maximum range is about 20 feet. If a moth detects an approaching bat outside this range, it will increase its chances for survival by moving away from a bat that has not yet detected it.

Distance results in lower sound intensity at the ears of a moth. To determine the range of sensitivity of a moth ear, action potentials were monitored in moth ears while the moths were positioned on a table directly in the flight path of bats known to roost in a barn. As a bat left the barn and flew toward a stationary moth, the frequency of action potentials in the ear increased. At first only one nerve cell was active (distant, Fig. 17.23). As the bat approached within a few feet (close, Fig. 17.23) the other nerve cell generated action potentials, and the information pathway into the moth's central nervous system became saturated with action potentials (see Fig. 16.15).

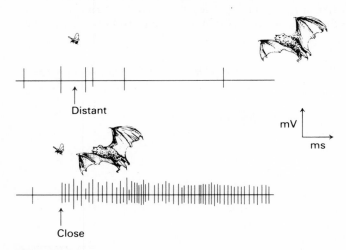

Figure 17.23
Action-potential frequency from the two nerve cells in a moth's ear is higher for bats that are close (high-intensity sounds, lower graph) than it is for bats that are distant (lower-intensity sounds, upper graph) (based on Roeder, 1967).

The evasive behavior of a moth can be partially interpreted with respect to this pattern of information transmission. A "power dive" response should occur when action-potential frequency in both neurons (from both ears) reaches a certain upper value, which is equivalent to a bat closing in for the kill. The integration of this information can lead to inhibition of flight muscles and a directed flight behavior downward away from a bat.

The "power dive" behavior suggests an advantage to moths that have evolved ears for the detection of bats at close range. The survival value of the integration-and-response mechanism was assessed with some simple observations. Floodlights were set up to attract moths, and hungry bats were observed as they passed through attempting to catch these insects. A total of 402 "encounters" were observed between bats and moths at close range. Whether the moths had ears or not was judged by whether they took evasive action. For every 100 moths that reacted to bats evasively and survived, only 60 nonevasive moths survived. Thus the close-encounter mechanism of moths of bat detection and avoidance provided a selective advantage of 40 percent.

Information for Distance Orientation

The maximum distance for detection of bats by moths was about 200 feet, measured from changes in action-potential frequency in the "barn" experiment. It is obvious that the moth has evolved an "early-warning" detection system that is effective for their relatively slow movements. However, considerably more information than is involved in using gravity to reach the ground must be integrated for a moth to determine the position of a bat and then move in the opposite direction.

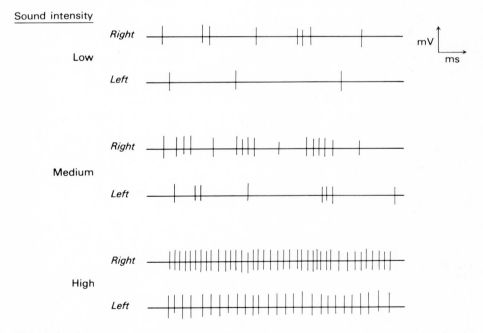

Figure 17.24
If a bat is approaching a stationary moth from its right, the right ear receives sounds of higher intensity than does the left ear. Differences in action-potential frequencies from the two ears is greatest when sound intensity is low (when bats are distant) (based on Roeder, 1967).

To localize a bat in three dimensions, a moth should receive information about whether the bat is to its left or right and whether the bat is above or below its position. Left-right position information is generated by means of the comparison (integration) of information input from the left and right ears. The body of the moth acoustically shields the ear away from the bat so that sound intensity is less at that ear. Thus action-potential frequency will differ in the afferent neurons coming from the two ears.

The difference in information input from the two ears is greatest when sound energy beamed at a moth is low in intensity (i.e., when a bat is distant). This point is illustrated in Fig. 17.24 for a bat approaching a moth from the moth's right. After the integration of simultaneous input from the left and right ears, wing muscles can be differentially activated to steer the moth in a direction away from the ear with the greatest action-potential frequency.

Up-down localization of bats requires more information integration. The position of the wings must be accounted for in processing the simultaneous input from two ears. The ears are located below the wings; as the wings beat they differentially shield the ears to sounds coming from different directions. To determine the importance of wing position for ear sensitivity, the wings of stationary moths were fixed in one of three positions, and bat sounds were beamed at the moth from different positions around the moth while the action potentials generated in the right ear of the moth were monitored.

Figure 17.25 summarizes some results of this laborious work. Note that the areas for maximum and minimum sensitivity in the moth's right ear shift during a beat of the wings. With information on action-potential frequencies in the right and left ears and simultaneous wing position, a bat can be localized in three dimensions. For example, more action potentials in afferent neurons from the right ear relative to the left when the wings are up and a decrease in action-potential frequency when the wings are down indicate a bat to the moth's right and above. An effective evasive maneuver based on this information is activation of effectors to move the moth down and to its left.

Mites That Live in Moth Ears

Moths are pursued not only by bats but also by other predators as well, and one of them provides an interesting natural experiment to investigate some of the mechanisms of integration for distance orientation.

Moths with ears may feed on nectar secreted by flowers. As they suck the nectar, small parasitic mites may jump on their backs and eventually work their way to one ear where they lay eggs. When the eggs hatch the larvae consume the ear as food, thus reducing the information input to the moth. However, it can still behave in some ways that may be effective as compared with having no ears at all. A "power dive" is still possible if a remaining ear becomes saturated with sound energy, but distance localization is impaired. The moth with only one ear always behaves as if bats were located toward the side with a functional ear. This behavior produces *circling* movements because the moth always turns away from the only ear providing sensory information. It can be more effective than no directional movement, but it may be less effective than the directed responses of moths with two intact ears.

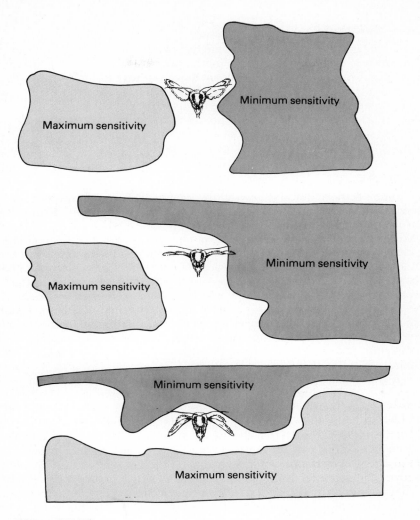

Figure 17.25
Results of some measurements of the sensitivity of a moth's *right*
ear indicate that the areas of maximum and minimum sensitivity in
the environment change as the wings move (based on Roeder,
1967).

It is important to note that the parasitic mites have evolved to infest only
one of the moth's ears. The first mite to land on the back of a moth apparently
makes a random "choice" of which ear to infect, and it appears to lay a
chemical trail (pheromone) to that ear; other mites follow the trail. Their young
develop on a host in which the risk of mites being eaten by bats is minimized
within the constraint of the requirements of their young for food.

Audio Mimicry?

Finally, the interactions between some moths and bats suggest audio advertisement by moths. Some species of moths produce ultrasonic clicks when bat sounds are beamed at them. This behavior could serve to confuse a bat by "jamming" its sonar, i.e., producing "echos" different from those normally received. Alternatively, the sounds could serve to advertise distastefulness to the sound-oriented predators. The latter interpretation has been favored as a result of experiments suggesting that, perhaps as a consequence of learned aversions (Chapter 11), bats avoid consuming some moths that produce ultrasound.

SUMMARY

Integration is the addition of positive and negative effects in time and space that determine whether an output (response; behavior) will occur. Neural integration depends on the transmission of information across synaptic spaces between neurons, and synapses may be either chemical or electrical. In chemical synapses, presynaptic axon terminals contain relatively large concentrations of mitochondria and presynaptic vesicles. The arrival of an action potential at a presynaptic terminal produces a "calcium potential," an influx of Ca^{++} ions across the presynaptic membrane, which leads to exocytosis of the contents of presynaptic vesicles into the synaptic space. The neurotransmitter diffuses across the synapse and influences information transmission by means of its effect on the postsynaptic membrane potential. The entire sequence, from arrival of an action potential to postsynaptic effects, takes about 0.6 millisecond (the synaptic delay).

A neurotransmitter (substrate) combines with specific receptor sites (enzymes) on the external surface of the postsynaptic membrane. In neuromuscular synapses in vertebrates the transmitter is acetylcholine (ACh). Combination of ACh with membrane receptors results in the opening of ionic channels for both K^+ and Na^+ such that the muscle membrane depolarizes. If depolarization reaches threshold, an action potential is generated in the postsynaptic membrane. Depolarization of a postsynaptic membrane is called an excitatory postsynaptic potential (EPSP). The contents of single presynaptic vesicles lead to quantal miniature EPSPs that add in time at the postsynaptic membrane.

In the excitatory neuromuscular synapse, ACh detaches from the membrane receptor soon after ionic channels are opened, and it diffuses from the receptor. At this time it is usually destroyed by the enzyme acetylcholine esterase, which localizes the activity of the neurotransmitter to a specific synapse and prevents long-term or more general synaptic transmission effects.

An inhibitory postsynaptic potential (IPSP) occurs whenever a neurotransmitter-receptor combination results in prevention of threshold depolarization of the postsynaptic membrane. This condition can involve membrane permeability changes that result in hyperpolarization or depolarization so long as the net change in membrane potential is less than threshold.

Nervous integration is the summation of EPSPs and IPSPs in time on the postsynaptic membrane; the summation determines whether threshold depolarization occurs. Both EPSPs and IPSPs are graded and decay in magnitude over short periods and over the distances from the localized origins on postsynaptic membranes. The threshold for action-potential formation is usually lowest at the initial segment of a nerve-cell axon (the axon hillock region), and if the summation of EPSPs and IPSPs at this location is sufficient to exceed threshold an action potential will result. If not, no output occurs despite any synaptic input to a neuron.

The nervous systems of vertebrates are organized into functional and anatomical components that reflect information integration for responses to different general types of environmental information. All afferent neurons pass to the central nervous system, where integration occurs. A variety of neurotransmitter chemicals may be involved (ACh, norepinephrine, γ-aminobutyric acid, serotonin, dopamine), although specific information on sites of action for integration is usually limited to descriptions of functional "centers" (respiratory, cardiovascular, thermoregulatory) and reflex functions of the spinal cord.

The efferent nervous systems of vertebrates are composed of the somatic and autonomic divisions. The somatic division innervates all skeletal (voluntary) mucles; ACh is the neurotransmitter for these neuron-effector synapses. Inhibition of somatic efferents occurs within the central nervous system, whereas efferents of the autonomic nervous system exist in pairs of excitatory and inhibitory fibers of the sympathetic or parasympathetic divisions.

The sympathetic system involves excitation of cardiac muscle, some smooth muscle, and glands for the purpose of responding to external environmental information indicating an imminent threat to survival. This system also inhibits effectors involved in digestive and nonstressful maintenance functions. The parasympathetic efferent system acts to maximize internal coordination of effectors involved in digestive and maintenance functions, for which sequential timing is important.

The neurotransmitter of the sympathetic system is norepinephrine in postganglionic neuron-effector synapses, and the adrenal gland secretes hormones (norepinephrine and epinephrine) under conditions of stress so that sympathetic effects are synchronous and sustained. The neurotransmitter of the parasympathetic system is ACh at neuron-neuron and neuron-effector synapses, which produce localized specificity of effector function for sequential timing of maintenance functions.

The simplest possible neural integration involves one synapse between afferent input and efferent output. This arrangement occurs in the spinal cord of the central nervous system for some types of reflex movements. For the stretch reflex, information on a stretch stimulus is detected at a receptor in the muscle spindle organ. The afferent neuron enters the spinal cord through the dorsal root and synapses in the ventral horn of the spinal cord on alpha-efferent motor neurons going to extrafusal muscle fibers. The synapse is excitatory, meaning that stretch leads to contraction, which removes the stretch stimulus because the spindle organs are arranged in parallel with extrafusal muscle fibers.

Coordinated movement depends on additional controls involving synaptic integration. The degree (set point) of extrafusal muscle contraction depends on internal stretch applied to stretch receptors resulting from contraction of intrafusal muscle fibers caused by activation of gamma-efferent motor neurons.

In addition, inhibitory synapses from interneurons within the spinal cord prevent contraction of specific muscles, such as antagonistic muscles (flexors, extensors) involved in limb movements in different directions. This reciprocal inhibition forms the basis for simultaneous contraction and inhibition of muscles involved in a variety of reflex movements of limbs (withdrawal reflex, magnet reflex, crossed extension reflex).

Information processing by neural integration has also been studied in the vertebrate retina. Nerve cells within the retina are found in layers composed of horizontal components (horizontal and amacrine cells) and vertical components (bipolar and ganglion cells). The ganglion cell axons form the optic nerve, which can be monitored for the types of visual stimuli on the retina that lead to changes in afferent information. There are normally fewer ganglion cells than there are receptors; considerable information processing occurs within the retina between stimulus reception and a ganglion-cell response.

By monitoring single ganglion cell axons, their receptive fields can be defined in terms of the functional properties of stimuli that result in changes in action potentials. Lateral interactions between neurons in the retina form the basis for two general types of ganglion-cell receptive fields. The center of an "on-center" field responds to a "light-on" stimulus, whereas the surrounding region responds to a "light-off" stimulus. "Off-center" fields respond oppositely. These fields provide information on environmental changes in the presence or absence of light and of the movement of contrast edges across the receptive fields.

Studies of the retina of the frog revealed the existence of a receptive field specific for the movement of convex edges. For frogs this field provides information about movements of prey items in their environment. This type of species specificity for retinal integration also suggests that similar neural integration can form the basis for a variety of species-specific "released" behaviors. These behaviors involve stereotyped (fixed-pattern) responses to visual stimuli of particular geometrical and movement characteristics.

Another neural integration system that has been extensively studied involves the predator-detection and escape mechanisms of moths that have evolved ears sensitive to the ultrasonic echolocation sounds produced by insect-eating bats. If bat sounds are broadcast to flying moths, the moths perform a "power dive" if sound intensity is high (indicating a bat close to the moth), or they turn and fly away from the source of sound if intensity is low (a bat distant from the moth).

The ear of a moth contains two nerve cells that can be monitored for information transmission. When bats are close (high sound intensity), both neurons in an ear produce action potentials at high frequency. When sound intensity is low, only one nerve cell responds with a slightly higher action-potential frequency. A "power dive" could involve inhibition of flight muscles when action-potential frequency in both neurons is high.

A directional response by the moth to move in the direction opposite to approaching bats requires integration of information about the left-right and up-down position of the bat with respect to the moth. Because the moth can detect a bat earlier than the bat can detect the moth (but moves more slowly), effective localization in three dimensions can produce a survival advantage by means of movements away from bats. Left-right information comes from the simultaneous comparison of input in the left and right ears because the ear away from a bat is shielded from sound energy by the body. Up-down localization involves the comparison of this information with information on the position of the wings, which shield the ears from sounds coming from certain directions.

Effectors and Responses

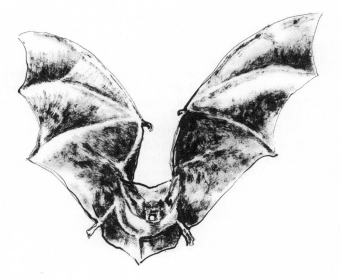

18

We have followed the components involved in neural control systems from reception and information transmission to integration of information for producing effective (adaptive) responses. Responses by animals to changes in their environments depend on the action of effectors, the muscles and glands that act on an environment to produce the feedback for control (see Introduction). The division of effectors into two general categories of cells specialized for *secretion* and *movement* represent the two major mechanisms for responses to environmental changes: chemical and mechanical actions.

The actions produced by cells at the end of an efferent pathway should be related to the characteristics of an environmental stimulus if responses are to be effective in adjusting resource supplies to demands. For both glands and muscles the relationships between environmental stimuli and responses depend mainly on which effectors are activated rather than on the mechanisms responsible for producing secretion or movement in cells. For glands, the chemicals released from cells depend on the *type* of gland that is activated, but the mechanisms of cellular secretion are similar among chemical effectors. For muscles, there is also a striking similarity in cellular mechanisms for producing movement, and differences in responses to stimuli depend on which *types* of muscles are activated and how the muscles are arranged in an animal such that movements are effective.

In this chapter we will review the major types of effectors that produce chemical and mechanical responses. We will examine details of mechanisms for producing secretions and movement at the level of cellular function. Following a discussion of the basic mechanisms for producing movement, we will examine differences in muscle types in terms of the importance of different responses for controls through movements.

GLAND EFFECTORS FOR SECRETION

A gland is usually defined as any tissue that exhibits selective uptake of certain chemicals and modification of those chemicals in cells, with subsequent release of chemical products in secretions as a consequence of specific environmental changes. We have had occasion to consider many types of glands and secretions in previous chapters in discussions of controls for regulation, particularly in Chapters 10 and 15.

Glands are divided into two general categories, depending on where secretion takes place. *Endocrine* glands secrete hormones within the body into the blood; *exocrine* glands secrete chemicals across epithelial tissues into the digestive tract or into the external environment. Table 18.1 briefly summarizes a large number of endocrine glands and their secretions in vertebrates; we have discussed many of these secretions. Note that each gland represents a mechanism for responding to different types of environmental stimuli, either short-term or long-term. In large part, the specificity of responses produced from secretion of endocrine hormones is due to chemical stimuli-target tissue interactions in which specific chemicals produce effects only in certain types of cells (see below).

Table 18.2 summarizes some exocrine glands and their secretions. Again, differences in function are reflected in differences in the nature of the chemicals produced by different tissues. For example, many of the exocrine glands serve a digestive function. Stimulation of the gland effectors may be under the control of the autonomic nervous system in which sequential parasympathetic activation results in effective digestion as a consequence of step-by-step actions of different secretions on food passing through the digestive tract.

The Mechanisms of Target-Tissue Activation

A chemical stimulus must interact with cells of particular tissues for the production of a specific response. Thus prolactin specifically influences mammary glands, while ADH influences specific parts of the kidney, etc. (Table 18.1). The specificity of effector action is thought to be due to the presence of unique *receptor sites* on the external surfaces of cell membranes. There are currently considered to be at least three types of membrane receptor sites for different types of chemical stimuli.

Table 18.1   Endocrine glands, their secretions, and some effects

GLANDS	SECRETION(S)	EFFECT(S)
Brain nerve cells	Releasing factors	Stimulation or inhibition of hormone release from pituitary
Adenohypophysis	Trophic hormones	Stimulation of other endocrine tissues
	Growth hormone	Tissue growth
	Prolactin	Mammary gland stimulation
Neurohypophysis	ADH	Kidney water retention
	Oxytocin	Uteral contraction; milk ejection
Pancreas		
Beta (β) cells	Insulin	Cellular uptake of glucose
Alpha (α) cells	Glucagon	Substrate mobilization
Thyroid	Thyroxine	Cellular metabolism
	Calcitonin	Calcium mobilization
Parathyroid	Parathyroid hormone	Calcium and phosphorus mobilization
Adrenal medulla	Norepinephrine; epinephrine	Sympathetic stimulation
Adrenal cortex	Steroid hormones (aldosterone, cortisol)	Cellular metabolism; Na^+ retention
Pineal gland	Melatonin	Pigment aggregation; circadian timing
Kidneys	Renin	Vasoconstriction; adrenal cortex stimulation
Gut	Gastrin	Smooth muscle motility
	Secretin	Pancreatic stimulation
	Cholecystokinin	Gastric motility
Testes	Androgens	Development of maleness
Ovaries	Estrogens	Femaleness
Corpus luteum	Progesterone	Uterine maintenance

   The hypothesized mode of action of the three receptor types is summarized in Fig. 18.1. The *β-type* receptors interact with norepinephrine (and epinephrine to a lesser degree) in tissues such as the salivary glands (producing amylase secretion), cardiac muscle (producing increased contractility and heart rate), smooth muscle (producing relaxation), the pancreas (causing insulin secretion), adipose tissue (producing lipolysis), and the liver (causing glucose production from glycogen). In some tissues, other hormones interact with their receptors in a similar way. A number of protein hormones (such as hypothalamic releasing factors, angiotensin, insulin, and glucagon) have a similar mechanism of action, and in the liver glucagon produces a *β*-type receptor action (Fig. 18.1).

Table 18.2 Exocrine glands, their secretions, and some effects

GLANDS	SECRETION(S)	EFFECT(S)
Salivary	Saliva	Food processing; lubrication; heat loss
Stomach	HCl: digestive enzymes	Food digestion
Intestines	Digestive enzymes	Food digestion
Pancreas	Bicarbonate	Acid neutralization
Gall bladder	Bile	Fat emulsification
Lachrymal	Tears	Corneal lubrication; expression of emotion
Sweat	Sweat	Palmar traction; heat loss
Prostate	Seminal fluid	Sperm transport
Mammary	Milk	Offspring maintenance and growth

The β-type receptors result in release of *adenylate cyclase* from the inner membrane surface following chemical stimulation. This results in the production of *cyclic AMP* (cAMP: adenosine 3′,5′-monophosphate) from ATP within the cellular cytoplasm. The cAMP influences enzyme formation (such as phosphorylase in Fig. 18.1), which in turn influences product formation within the cytoplasm.

The *α-type* receptors (Fig. 18.1) were originally hypothesized to occur because of some opposing effects of epinephrine on certain tissues. Thus epinephrine in combination with an α-type receptor produces contraction of most smooth muscle, inhibits insulin secretion from the pancreas, and inhibits lipolysis in adipose tissue. The opposite functions occur when epinephrine combines with β-receptors (see above). Following α-receptor stimulation, an unknown "alpha signal" is thought to be released into the cytoplasm from the inner surface of the cell membrane (Fig. 18.1). This release appears to produce an effect by means of a change in intracellular Ca^{++} concentration, either from a change in cell-membrane permeability to Ca^{++} or from release of internal Ca^{++} supplies (perhaps from mitochondria) or from a combination of these. A number of cellular enzymes are influenced by Ca^{++} concentrations, including those producing phosphorylase in liver cells (Fig. 18.1).

The third type of membrane receptor does not result in a cytoplasmic mechanism for modifying function but influences function by means of effects on gene expression within the nucleus (Fig. 18.1). The steroid hormones (ecdysone, testosterone, progesterone, estrogen, aldosterone, cortisone) and thyroxine produce their effects in this manner (see Chapter 15). The extent to which the three types of membrane receptors are present and the extent to which enzyme functions can be induced or modulated via these mechanisms will determine the extent of response to specific chemical stimuli.

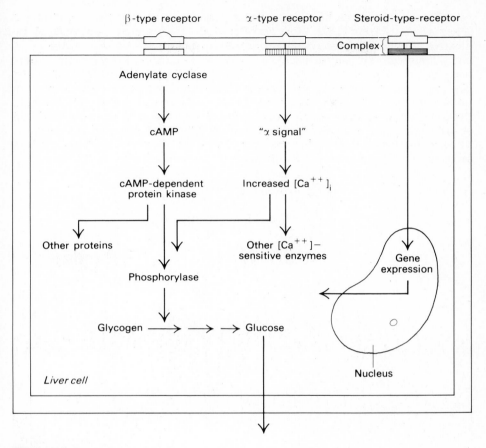

Figure 18.1

Schematic summary of the proposed mechanisms of action of three types of cell-membrane chemical receptors. Beta-type receptors influence proteins (enzymes) by production of cAMP by means of release of membrane adenylate cyclase. The example is for glucose production from glycogen via the effects of phosphorylase. Alpha-type receptors result in changes in cytoplasmic enzyme function via the effects of cytoplasmic calcium concentrations. Steroid-type receptors form a membrane "complex" that ultimately influences function by means of gene expression (based partly on Exton, 1980).

The Mechanisms of Secretion

Secretion from cells within a gland following activation occurs as a consequence of efferent hormonal or neural transmitter activation of the cells after reception and integration of information on some environmental change. Although tissues differ in the nature of the chemicals produced, the general mechanisms of cellular production and release of secretions are similar in most glandular tissues.

The mechanisms of chemical production and release are summarized schematically in Fig. 18.2. Most secretory cells have a structural polarity indicative of chemical production at one end and release (secretion) at the other end. Chemicals are synthesized in the rough endoplasmic reticulum, where proteins are elaborated on messenger-RNA templates. From this region polypeptides pass into the smooth endoplasmic reticulum, where the reticulum buds off to form small reticulum vesicles. The reticulum vesicles move through the Golgi apparatus in which further chemical synthesis and vesicle elaboration occur. Condensing vacuoles are formed on the other side of the Golgi apparatus. These vacuoles produce an osmotic concentration of vacuole contents to form secretory vesicles containing the products to be released (Fig. 18.2).

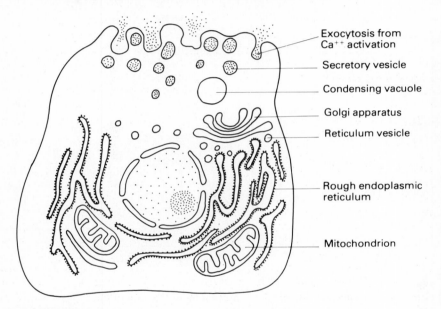

Figure 18.2
Schematic diagram of the general morphology of a secretory cell, showing the polarity between sites for chemical-product production, packaging, and release (after Eckert and Randall, 1978).

Whether release of secretory-vesicle contents occurs and the *amount* that is released in response to a stimulus depends on the information reaching a secretory effector from efferent neurons or hormones. For example, the amount of ADH released from the neurohypophysis as a consequence of an osmotic stimulus depends on the magnitude of a detected stimulus change. Osmoreceptors in the hypothalamus respond to increased cellular water loss by increasing the frequency of action potentials. The action potentials result in release of neurotransmitters at neuron-effector synapses in an amount that depends on the frequency of action potentials. The release of the hormone depends in turn on the extent and duration of postsynaptic membrane depolarization; thus secretion is ultimately proportional to a stimulus.

Efferent neurons or hormones result in depolarization of the membrane of a secretory cell, leading to *exocytosis* of secretory-vesicle contents through the cell membrane. The mechanism is remarkably similar to the sequence of events leading to exocytosis of presynaptic vesicles in neurons (see Chapter 17). The depolarization of the secretory-cell membrane depends on inward movement of Ca^{++} ions. Procedures that modify the normal concentrations of Ca^{++} inside or outside cell membranes modify the secretory response. Modifications that increase Ca^{++} influx by means of neurotransmitter effects on Ca^{++} channels in secretory-cell membranes result in the release of secretions in proportion to the increased Ca^{++} influx. Modifications that reduce Ca^{++} influx (such as reduced extracellular calcium concentrations) reduce secretory release.

The triggering and control of the magnitude of an effector response by the amount of change of calcium concentration inside a cell is not limited to secretory effectors. We will find that this change is also an integral part of the mechanisms underlying effector controls for movements (see below).

EFFECTORS FOR MOVEMENTS

Types of movement depend on the types of effector cells that produce movement and their arrangement within animals. In multicellular animals movements occur as a result of the contraction of striated, cardiac, or smooth muscles. In addition, some cells may move, or move parts of their environments, by ciliary or flagellar contractions. Cilia occur in some protozoans and in some cells within animals where bulk transport is important. For example, the fallopian tubes are lined with ciliated cells that transport the eggs. Flagellar movements occur in many types of sperm and some protozoans.

Analysis of the cellular mechanisms responsible for generating the forces for movements depends to a high degree on cellular structures associated with those movements. Most theories of mechanisms for movement are based on forces generated between two types of filaments within cells. The filaments provide the structural basis for coupling chemical energy to mechanical energy, usually involving hydrolysis of ATP. Cellular mechanisms of movement are most completely understood for effector tissues with the highest degree of organized cellular structure (striated muscles, cardiac muscle) and for cellular organelles with particular filament structures (cilia and flagella). The cellular mechanisms of movement are least understood for cytoplasmic streaming and amoeboid movements, where underlying structures are not quite as obvious.

Cyclosis and Amoeboid Movements

Protoplasmic streaming (cyclosis) can be observed in a variety of relatively large cells and is important for bulk transport of substances within cells. The "streaming" of cytoplasm represents an extreme case of more general movements of a variety of internal cellular structures. Although it may not be as obvious, metabolic products in vesicles, mitochondria, and other organelles move within cells, although perhaps not rapidly. For example, in cells whose functions are localized, such as secretory cells (Fig. 18.2), mitochondria are found in regions of high metabolic activity, and products are "packaged" and transported to other areas of a cell.

All animal and plant cells contain two filamentous proteins called *actin* and *myosin*. They form the basis for the mechanisms of muscle contraction when one filament slides with respect to the other (see below). Thus the movements on a larger scale are the consequence of mechanical and chemical interactions between small filaments on a microscopic scale.

In some cells there are obvious associations of cell-organelle movements with intracellular filaments. For example, in some microvillar extensions of actively transporting epithelial-tissue cells, mitochondria from other areas in the cells that become localized near transporting membranes are associated with sets of intracellular filaments.

The movements of amoebae have been used as a model of movements within cytoplasm. Similar types of movement occur in a variety of cells including leukocytes and slime molds, and the movement of amoeba-like cells appears to represent locomotion by means of directed protoplasmic streaming. Amoeboid cell movement has been described as an extension of cytoplasm in a particular direction resulting from changes in the degree of fluidity of the cytoplasm. Some regions of the cytoplasm are more fluid (*sol*) than others (*gel*), and sol-gel-sol transformations are thought to be necessary for directional streaming of the pseudopodial extensions of amoeboid cells. When a pseudopodium is formed, the liquid-like internal protoplasm (endoplasm), the sol, streams into the extension. This event is followed by the transformation of this protoplasm to a less liquid gel (ectoplasm), preventing further extension in that direction.

Filaments of actin and myosin have been found in the gel-like ectoplasm of amoeboid cells. In addition, the addition of ATP with magnesium and calcium causes contraction of ectoplasm in these cells, perhaps from the increased mechanical interactions between actin and myosin filaments. This type of a mechanism helps to explain the sol-to-gel transformation, but the mechanisms for the reverse process and the formation and control of pseudopodial extensions are not as clearly understood.

CILIARY AND FLAGELLAR MOVEMENTS

The movements of cilia and flagella are more completely understood than those of cyclosis in large part because of detailed studies of the fine structure of these organelles and changes that occur in structure during movement. The fine structure of cilia and flagella are the same; the organelles differ only in size (flagella are usually longer) and numbers on a cell (cilia are usually more numerous).

Fine Structure

Each cilium or flagellum consists of an outer membrane surrounding nine sets of peripherally located microtubular pairs and a centrally located set of two (occasionally one or three) microtubules. This 9 + 2 arrangement occurs along the length of the organelle and is shown in cross section in Fig. 18.3.

The two central and nine peripheral sets of microtubules and their structures are called an *axoneme*. The microtubules are composed of a protein called *tubulin*. In the peripheral pairs of microtubules the larger tubule is *tubule B*; the smaller, attached tubule is *tubule A*. Tubule A has two *arms* (inner and outer) composed of a protein called *dynein*. This protein acts as an enzyme that hydrolyzes ATP when magnesium is present; the arms are thought to represent the site of chemical production of mechanical activity between adjacent sets of microtubular pairs that results in movement.

The central pair of microtubules are attached in a *central sheath* from which radial arms extend to the tubule A components of the peripheral microtubules. Extending from the central sheath outward, the arms are composed of components called *transitional links, linkage heads*, and *radial links* (Fig. 18.3).

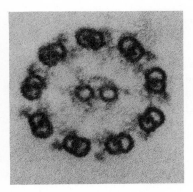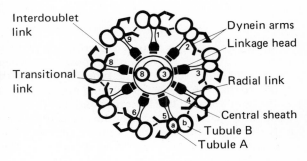

Figure 18.3
Cross-sectional structure of an axoneme. The outer membrane in the electron micrograph (left) has been removed. The drawing illustrates the relationships between major structural units of an axoneme (modified from Warner, 1972). Photo courtesy of F. Warner, Syracuse University.

Movement Patterns

All of the different patterns of movement of cilia and flagella can be characterized as bending or propagation of bending along the length or part of the length of the axoneme organelle. Some types of movement are shown in Fig. 18.4 for flagella. For flagellar movement, a cell moves in a direction opposite to the direction of wave propagation (left part of Fig. 18.4). In some cells (such as *Peranemia trichophorum*), a flagellum bends only at its tip in a *power stroke* and a following *recovery stroke* (right part of Fig. 18.4). This pattern is also typical of ciliary beating, and the movements of large numbers of cilia on the

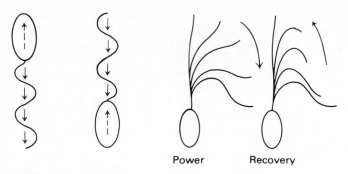

Figure 18.4
Some patterns of flagellar bending in protozoans. The two organisms to the left move (dashed arrows) in opposite directions to the directions of bend propagation along the lengths of their flagella (solid arrows). The organism to the right moves only during a power stroke at the tip of its flagellum, which is followed by a recovery stroke. This pattern is also typical of ciliary bending (based on Jahn *et al.*, 1964; Jahn and Bovee, 1967).

surface of an animal such as *Paramecium* may exhibit a degree of synchrony (coordination) of movements in power and recovery strokes.

Detailed studies of the types of bending that occur in cilia and flagella indicate that the forces for bending arise locally *within* the axoneme and are not the result of whiplike action initiated at the base. Cilia and flagella show normal bending patterns after they have been detached from a cell. In addition, the bending may normally occur from the tip toward the base in some species (Fig. 18.4), and even when bending occurs from the base the amplitude of bending does not decrease toward the tip. Therefore, these patterns suggest that explanations for bending should include interactions between the microtubular units within the axoneme.

The Bending Mechanism

Studies of the structure of flagella bent in different directions indicate that the total length of the microtubules remains the same but that microtubules on different sides of a bending flagellum slide with respect to each other. This fact is shown in Fig. 18.5, which diagrams the *tip* of a flagellum when it is not bent (center) or when it is bent in the same plane in either direction. Two of the nine sets of microtubules are shown, one on each side. Note that the pair on the side toward a bend moves *upward* in the flagellar shaft, whereas the pair on the opposite side moves *downward* in the shaft (the dotted line provides a point of reference).

The sliding of microtubular pairs with respect to each other has also been demonstrated in special preparations in which the outer membrane of a flagellum was selectively removed (as in the electron micrograph in Fig. 18.3).

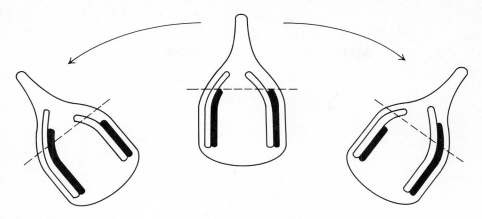

Figure 18.5
Schematic illustration of the movements of microtubular pairs at the tip of a flagellum during bending. The dotted line in each position is drawn at the same location. Note that microtubular pairs move up or down in the shaft, depending on the direction of bending (adapted from Satir, 1968).

When magnesium and ATP are added to the exposed microtubules they slide with respect to each other. This type of preparation permits partial analysis of the three events believed to be necessary for microtubular sliding: (1) bridging of the dynein arms to neighboring B subfibers, (2) generation of the sliding force accompanied by ATP hydrolysis (chemical-mechanical coupling), and (3) detachment of the dynein arms.

When magnesium alone is added to isolated preparations of microtubules the attachment of arms without evidence of sliding (step 1 above) can be seen in electron micrographs. A similar bridging is observed when Ca^{++} is used. If ATP is then added to a "bridged" preparation, sliding occurs. Thus divalent cations (Mg^{++}, Ca^{++}) appear necessary to form a bridge attachment (step 1). This action then permits activation of dynein ATPase to generate the chemical energy for sliding (step 2) when ATP is present, leading to detachment of the arms (step 3). The divalent cations may be involved in bridge formation by direct linkage, by induction of a conformational change in the dynein arms or B subtubule that permits linkage, or from enhancement of an interaction between these structures.

How does sliding of microtubular pairs result in the characteristic bending that is observed in all flagellar and ciliary movements? A possibility is diagrammed in Fig. 18.6. If the dynein arms of adjacent microtubular pairs slide and attach in a *straight* region of a flagellum of cilium, bending will occur if the dynein arms remain *fixed* in the bent region. Bending can propagate along the length of the organelle if the areas of sliding and fixed attachments change along the length of the axoneme.

You can illustrate this mechanism for yourself by holding two strings in parallel between your fingers perpendicular to the ground. When you twist or

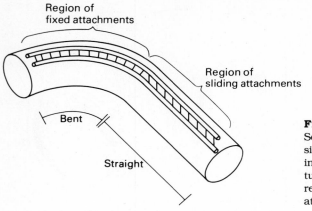

Figure 18.6
Schematic illustration of a possible mechanism for bending involving the sliding of microtubular pairs in a "straight" region, with fixed dynein-arm attachment in a "bent" region.

bend the strings over your fingers by rotating your wrist, the ends toward the ground will move with respect to each other as in Fig. 18.5. The two strings slide with respect to each other in the straight region but are fixed in position where they are bent at your fingers.

What leads to sliding of attached dynein arms to produce bending, and how is it propagated along an axoneme? Chemical energy from ATP is required to produce sliding once linkage has occurred, and a mechanical force is sufficient to cause sliding because cilia can be made to execute a propagated bending wave if they are bent with an externally applied force. Thus mechanical force in one segment of an axoneme could trigger dynein ATPase activity in an immediately adjacent region, leading to propagation of a bending wave. This sequence is summarized in Fig. 18.7 along with other components involved in the control of microtubular movements.

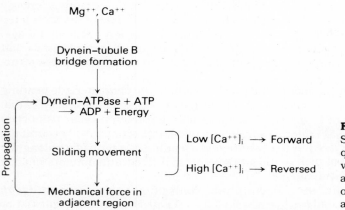

Figure 18.7
Summary of the sequence of events involved in sliding interactions between pairs of microtubules of an axoneme.

Although the sequence in Fig. 18.7 can produce propagated bending, it does not account for all aspects of ciliary and flagellar movements. For example, what is responsible for the initiation of bending? Furthermore, studies of ciliary motion indicate that there can be a high degree of synchrony in movements of cilia on a cell, both in terms of the patterns of bending and the directions of beating, the latter of which are a response to environmental stimuli.

Control of Movements

There is little information on the origin of initiation of bending forces. However, within multicellular animals, where ciliated cells are effectors for movements of fluids, the frequency of ciliary beating may be under the control of chemicals that change the permeability of the cell membranes to ions.

In some ciliated protozoans the cilia reverse direction of beating under some circumstances. For example, if a cell hits an object or enters an area of unfavorable temperature or osmolarity, the cilia reverse their beating movements and the animal moves backward. Reversal of beating occurs when the environmental change results in a membrane depolarization caused by an influx of calcium. However, the mechanism by which this influx influences the sliding microtubular dynein arms has not been described. A ciliary beat returns to a "normal" direction when internal calcium concentration returns to a low value (Fig. 18.7).

The synchrony of beating of thousands of individual cilia on the surface of a protozoan results in a net force in a particular direction that is greater than would occur if beating were not synchronized. Individual cilia on the surface beat in waves (called *metachronism*) such that forces required for movement in a particular direction are *minimized*, which results in minimum transport costs for swimming locomotion by the cell. (The analogy with the movements of fish in a school is striking—see Chapter 9). How the beating of many cilia is synchronized in waves remains a fascinating mystery, but the net effect of generating movement in which rates of energy expenditure (from ATP-dynein interactions) are minimized reemphasizes the importance of efficiency in locomotion even at this level of microstructure.

STRIATED MUSCLE

Striated muscle includes skeletal (voluntary) and cardiac muscle, and these tissues have become model systems for the study of the mechanisms for contraction because of the high degree to which structure can be related to function, even at the fine level of interactions between individual molecules.

Gross and Fine Structure

The gross and fine structure of striated muscle is summarized in Fig. 18.8. Each muscle is composed of many individual *muscle fibers* that have a striated appearance. Individual muscle fibers are arranged in parallel within a muscle and are formed from the fusion of several cells. Each striated muscle fiber is in turn composed of many smaller *myofibrils*, which also have a striated appearance.

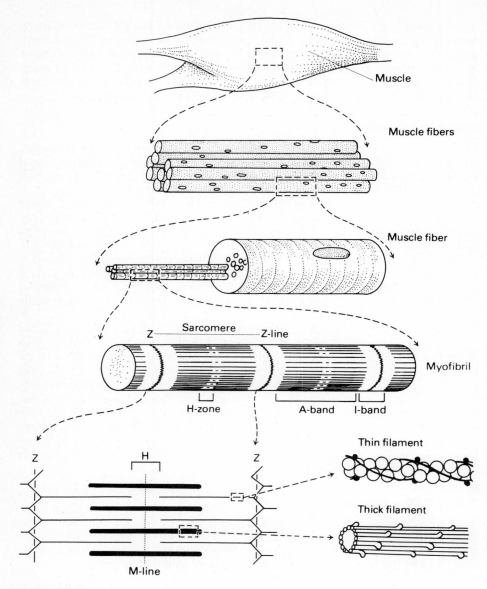

Figure 18.8
Gross and fine structural organization of vertebrate skeletal muscle (based on Bloom and Fawcett, 1975).

The striation of myofibrils and muscle fibers is due to the repetition, along the length of each myofibril, of the fundamental units for muscle contraction—the *sarcomeres* (Fig. 18.8). Each sarcomere is bounded at each end by a *Z-line*, or a thin line in the center of a band, the *I-band*, that appears relatively light. Between I-bands in the center of a sarcomere is the *A-band*, which appears relatively dark and is wider than the I-bands. In the center of the A-band is a region that appears lighter, the *H-zone*. The H-zone is in the center of a sarcomere.

All of these components of a sarcomere can be seen with the light microscope. If higher-resolution images are made with an electron microscope, the sarcomere is seen to be composed of two sets of filaments (Fig. 18.8). The A-band is composed of *thick filaments*. The I-band is composed of a central Z-line that interconnects longitudinally arranged *thin filaments*. The lighter region in the center of the A-band (the H-zone) represents that part of the A-band where thick and thin filaments do not overlap. In the center of the H-zone is another thin line, the *M-line*, that is thought to provide interconnections between thick filaments within a sarcomere.

Cross sections through various aspects of a sarcomere confirm the two-filament organization (Fig. 18.9). A cross section through the I-zone (left part of Fig. 18.9) shows thin filaments arranged at the points of hexagons. A cross section through the H-zone (right part of Fig. 18.9) shows thick filaments arranged at the points of triangles. A cross section through the A-band in which thick and thin filaments overlap (center part of Fig. 18.9) shows that each thick filament is in the center of the hexagon formed by thin filaments such that each thin filament is in the center of the triangle formed by three thick filaments.

At high magnification the thick filaments appear "bumpy" in longitudinal section (Fig. 18.8), and projections can be seen from thick filaments in cross sections on each side of the central H-zone (Fig. 18.9). The projections extend from the thick filaments toward the thin filaments and are known as the *cross bridges*.

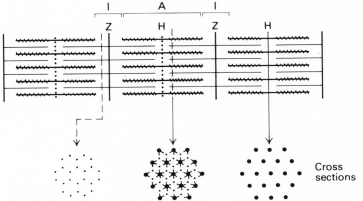

Figure 18.9
Organization of sarcomere thick and thin filaments seen in longitudinal (upper) and cross sections (lower) through various regions of a sarcomere. The central cross section shows the relationships between thick and thin filaments where they overlap and where cross bridges extend from thick filaments toward thin filaments.

The structure of thick and thin filaments has been studied in detail using chemical techniques. The thick filaments are composed of two types of protein wound together that can be separated within certain enzymes. When the proteins are joined chemically, they have the appearance of a globule at one end with a long "tail" at the other (Fig. 18.10, upper). The "head" region is composed of heavy meromyosin (HMM), which is joined to light meromyosin (LMM) "tails." When preparations of these myosin polymers are mixed, the LMM-HMM units combine "tail-to-tail" in such a way that the HMM "heads" project from the formed thick filaments at regular intervals (Fig. 18.10). These aggrega-

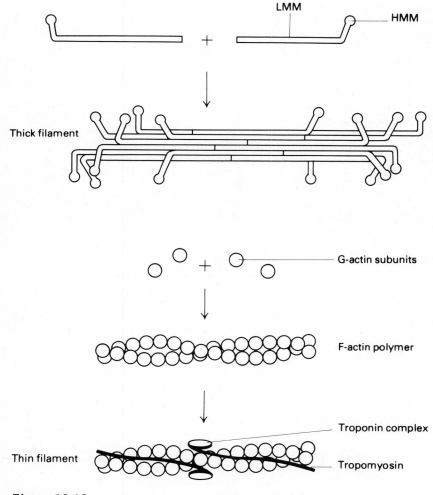

Figure 18.10
Summary of the chemical organization of thick filaments and thin filaments.

tions of myosin molecules are structurally and functionally equivalent to the thick filaments of sarcomeres, and the HMM "head" projections are the cross-bridge extensions of thick filaments.

The thin filaments are composed of at least three protein subunits (bottom part of Fig. 18.10). The bulk of a thin filament is composed of a helical polymer of G(globular)-actin subunits. The helical polymer is called F(filamentous)-actin. Fitted between the G-actin units on the F-actin helix is a protein called *tropomyosin*. At intervals along the helix the tropomyosin has an attached complex of proteins called the *troponin complex* (Fig. 18.10). Interaction between the thick-filament cross bridges and the thin filaments occurs in the region of the troponin complex (see below).

The Sliding-filament Theory of Contraction

The current theory for muscular contraction comes directly from observations of changes in the structure of sarcomeres during contraction. The changes are summarized in Fig. 18.11. In a contracted muscle the A-bands remain the same

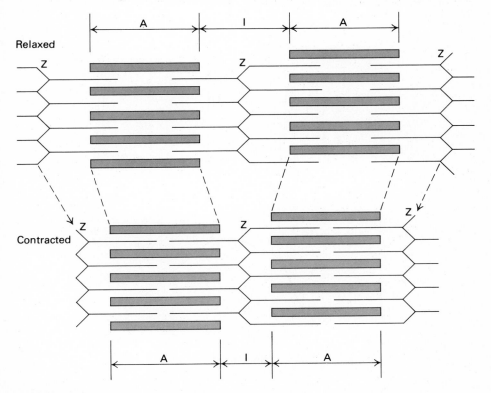

Figure 18.11
Schematic illustration of the changes in morphology of sarcomeres from a relaxed (upper) to a contracted (lower) state.

length but move closer together. The I-bands decrease in width, and the H-zone decreases in width and may disappear. In addition, the Z-lines move closer together (Fig. 18.11).

The simplest possible event that can explain all of these changes is the movement of the thin filaments with respect to the thick filaments. According to this sliding-filament theory, thin filaments slide toward the center of the thick filaments during contraction while both sets of filaments maintain the same length. The sliding involves interactions between the thick and thin filaments via cross bridges, and this interaction provides the basis for an experimental test of the theory.

The Length-tension Relationship

If contraction involves filament interactions, and if the tension (force per centimeter) of a contraction depends on the number of cross bridges that can be formed, then experimentally altering the structure of a sarcomere to influence the number of cross bridges that can be formed should change the force produced during contraction.

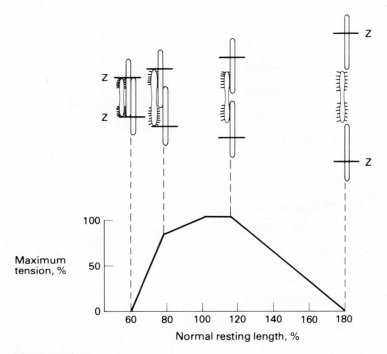

Figure 18.12
Relationship between the extent of thick-thin filament overlap for cross-bridge linkage and the tension developed during contraction. The diagrams of filament structure within a sarcomere correspond to the different degrees of resting length on the *x* axis of the graph (based on Gordon *et al.*, 1966).

The results of such an experiment are shown in Fig. 18.12. In this case several muscle sarcomeres were forced to assume different lengths relative to the normal resting length of 100% on the x axis. This change was accomplished by either pulling on muscle fibers or pushing on them. They were then contracted against a large load so that they did not change length during contraction, and the force developed was related to the extent to which cross bridges could form between thick and thin filaments.

When sarcomeres were stretched such that no overlap between thick and thin filaments was possible (extreme right part of Fig. 18.12), no tension could be measured. Proceeding to the left in Fig. 18.12, as overlap was permitted more and more tension could be developed up to a maximum that corresponded to the attachment of a maximum number of cross bridges to each thin filament near the normal resting length of the sarcomere. When the sarcomere was pushed together from this point, the thin filaments at one end began to overlap with the thin filaments from the other side such that fewer cross bridges could be attached, and the tension decreased. With extreme pushing the thin filaments overlapped completely, the thick filaments began to fold up against the Z-lines, and tension decreased again rapidly toward zero (Fig. 18.12).

Note that maximum tension development from cross-bridge formation occurs at the normal resting length for a striated muscle (Fig. 18.12). Thus the interaction between thick and thin filaments usually produces a maximum possible force along the length of a muscle filament.

The Mechanism of Cross-bridge Action

As with ciliary and flagellar sliding, three events must be accounted for in the sliding of thin filaments with respect to thick filaments: (1) cross-bridge attachment, (2) movement from the cross-bridge—thin-filament interaction (chemical-mechanical coupling), and (3) cross-bridge detachment. As with cilia and flagella, the first step in muscle activation involves calcium and the last step involves ATP hydrolysis.

Muscles will not contract without small amounts of calcium in the region of thick and thin filaments; at least 10^{-6} molar calcium is required for the production of a maximum force during contraction. Calcium ions selectively bind to parts of the troponin ("tr") complex on the thin filaments (Fig. 18.13), caus-

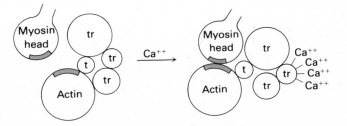

Figure 18.13
Schematic illustration of changes in the tropomyosin-troponin complex (resulting from calcium binding) that permit myosin cross-bridge linkage to actin filaments. The structures are shown in cross section (based on Cohen, 1975).

ing a shift in tropomyosin ("t") and troponin molecules near an active site on the F-actin that permits attachment of the HMM cross bridge.

Once a cross bridge is attached, sliding (step 2) is thought to involve rotation of the HMM head toward the center of a thick filament (Fig. 18.14). Cross-bridge rotation appears to occur without the mediation of other chemicals following attachment; that is, sliding occurs whenever a cross bridge has attached regardless of the chemical environment in the region of thick and thin filaments. Rotation and sliding appear only to involve interactions between HMM and actin.

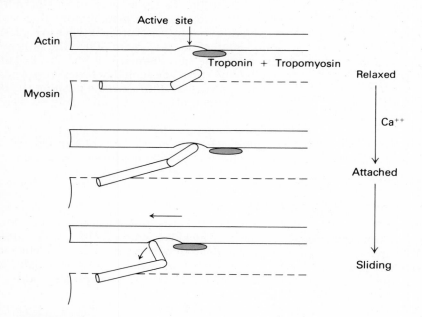

Figure 18.14
Schematic diagram of myosin cross-bridge attachment and rotation that leads to sliding of thin actin filaments with respect to thick myosin filaments.

Release of a cross bridge from actin (step 3) involves hydrolysis of ATP in the presence of magnesium. Without ATP, muscle becomes rigid because of this requirement for chemical energy for release of cross bridges (*rigor mortis* occurs following death when calcium diffuses into sarcomeres after ATP supplies within muscle have been exhausted). The interaction between HMM and actin filaments enhances this detachment when sufficient ATP is present. Isolated HMM shows ATPase activity, but its enzymatic activity increases when it is associated with actin. Following rotation of the HMM head, detachment will normally occur if ATP supplies are sufficient, followed by repetition of the cycle at new active sites along the actin filament (as a result of sliding) if sufficient calcium ions are present to initiate further cross-bridge attachments.

Excitation-contraction Coupling

An action potential that arrives at a neuromuscular synapse usually results in formation of an action potential on the excitable postsynaptic muscle membrane in vertebrates. It is quickly followed by a brief contraction-relaxation cycle (a *twitch*) in a muscle fiber. Therefore, there must be a connection (a coupling mechanism) between the electrical events in muscle membranes and the three-stage sequence of filament interaction within sarcomeres.

Most sarcomeres are found deep within vertebrate muscle fibers and sufficiently far from the surface membrane that depolarization just at the surface would not influence the sarcomeres. In many vertebrate skeletal muscles the surface membrane is folded *into* muscle fibers in narrow tubes that pass inward to individual sarcomeres in myofibrils and are called the *transverse tubules* (Fig. 18.15). An action potential is conducted down many transverse tubules to produce localized effects at the level of individual sarcomeres.

The transverse tubules are closely associated with a separate set of membranes that surrounds each sarcomere (Fig. 18.15), the *sarcoplasmic reticulum*. It extends longitudinally along and around the thick and thin filaments of a sarcomere, and arrival of an action potential via the transverse tubules results in release of calcium stored within the reticulum by means of depolarization of the sarcoplasmic reticulum membrane.

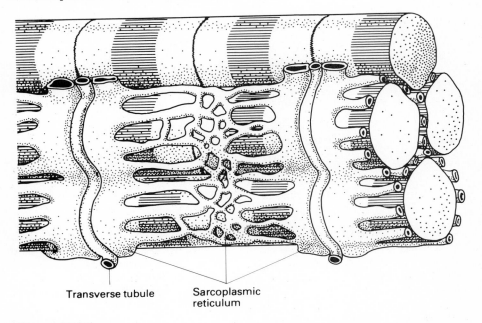

Transverse tubule   Sarcoplasmic
                           reticulum

Figure 18.15

Diagram of the structure of membranes associated with individual sarcomeres of striated muscle. Action potentials are propagated down transverse tubule invaginations of muscle-fiber surface membranes, resulting in depolarization and release of calcium from the sarcoplasmic reticulum (after Peachy, 1965).

The amount of calcium released from the sarcoplasmic reticulum depends on the magnitude and duration of an action potential, but it is usually fixed because of the all-or-none property of action potentials. Thus each action potential ultimately results in the release of a "pulse" of calcium ions, which causes cross-bridge linkage (step one in contraction).

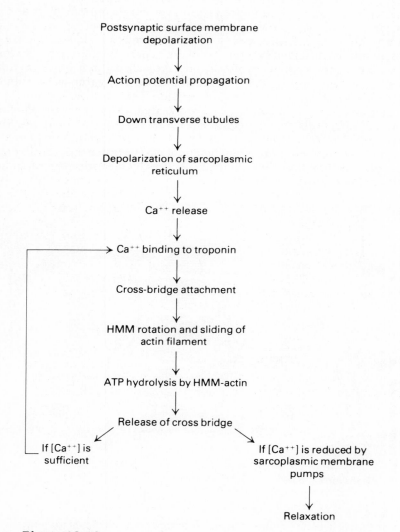

Figure 18.16
Summary of an event sequence leading to twitch contraction $\longrightarrow$ relaxation, starting with postsynaptic membrane depolarization.

The membrane of the sarcoplasmic reticulum also pumps calcium from the region of the filaments back into the sarcoplasmic reticulum. Energy from ATP hydrolysis is required to move calcium against a concentration gradient. As a consequence of this active transport, the calcium concentration in the region of the filaments decreases and results in relaxation by means of inhibition of cross-bridge attachment by the tropomyosin-troponin on actin filaments.

The temporal sequence of events involved in production of a twitch (contraction → relaxation) within a myofibril set of sarcomeres is summarized in Fig. 18.16. This mechanism forms the basis for all skeletal muscle contraction, and differences in effector responses due to contraction involve elaborations on this pattern. Some muscle fibers are "fast" and some are "slow." Other muscle fibers are capable of different degrees of sustained contraction. Interpretations of these differences involves understanding structure and function for entire sets of different muscle-fiber types.

THE MECHANICS AND CHEMISTRY OF MUSCLE-FIBER CONTRACTION

Interpretation of differences in function of muscle fibers depends partly on their morphologies and partly on the chemistries that provide the energy and other (chemical) changes that occur in contraction. The importance of these differences can be outlined by selecting a particular type of muscle fiber (e.g., striated muscle) and examining variations in the patterns of its contraction in response to depolarizing stimuli.

There are two variables that reflect whole-muscle function: the *velocity* of shortening during contraction, and the *tension* developed during contraction. Velocity depends primarily on changes in length, whereas tension depends primarily on cross-sectional area or thickness. Both the length and thickness of a muscle are determined by two components: the number of sarcomeres in series (length) and in parallel (thickness), and the noncontractile components of muscle, such as tendons and connective tissue, either in series (along length) or in parallel (across width).

Velocity and tension are usually studied using two types of experimental preparations. If a muscle is stimulated to produce a twitch when no external load must be moved other than the muscle mass, it will shorten with maximum velocity (Fig. 18.17). If different loads of increasing mass are attached to the muscle, the velocity of contraction decreases to a point where the muscle will no longer shorten. However, contraction still results in development of tension even when an external load is not moved. When a muscle shortens against a constant load it is called *isotonic* contraction because the tension developed for movement of a constant load is the same. When a muscle contracts against a load that cannot be moved it is called *isometric* contraction because the length of the muscle remains constant.

The velocity of isotonic contractions among muscles depends primarily on the number of sarcomeres arranged in *series*, that is, the number of sarcomeres along the length of muscles. Therefore, longer muscles contract more rapidly against movable loads, and most movements of animals that depend on speed involve relatively long muscles. For examle, muscles associated with limb movements are usually long.

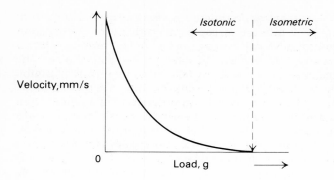

Figure 18.17
Relationship between the velocity of contraction and load, or force, during contraction. When a muscle cannot shorten, contraction becomes isometric instead of isotonic (based on Wilkie, 1968).

The tension of isometric contractions among muscles depends primarily on variations in cross-sectional areas, i.e., the number of sarcomeres across the width of muscles. Most muscles that generate maximum forces during contraction are relatively thick. Moreover, velocity and tension need not be represented separately; variations in length and thickness occur within similar muscle types.

Twitch Characteristics

Figure 18.18 shows the characteristics of single-twitch isotonic and isometric contractions in response to single depolarizing stimuli. For both types of twitch there is a *latency* time between application of a stimulus and initiation of contraction. For isotonic contractions this latent period is longer for heavier loads, and a muscle shortens less for heavier loads (Fig. 18.18).

An action potential in a muscle lasts for 1-2 milliseconds, but the contraction-relaxation of a twitch takes longer (Fig. 18.18). Thus several action potentials occurring within a few milliseconds of each other will lead to successive twitches. When this succession occurs the tension developed in isometric contraction builds, shown in Fig. 18.19. For a given muscle type, there is a maximum tension that cannot be exceeded by an increased frequency of stimulation (lower part of Fig. 18.19). It is called maximum *tetanic contraction*.

The explanation for the summation of tension during tetanic contractions involves noncontractile components in a muscle. Z-lines, connective tissues, and tendons act as springs that must be stretched before tension develops in a muscle. The collective effect of these elements is called the *series elastic component* of muscle. In a single twitch, cross-bridge interactions are not maintained long enough to completely stretch the series elastic components to a maximum extent. With several successive depolarizations, the cross-bridge interactions are maintained longer and the elasticity of other tissues in muscle is overcome to a maximum extent, leading to maximum tension development during sustained tetanic contraction.

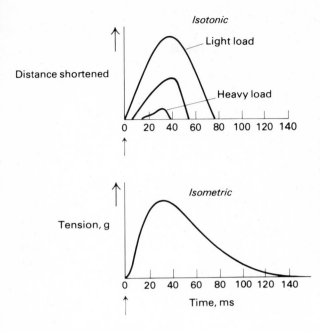

Figure 18.18
Characteristics of single-twitch isometric and isotonic contractions. The arrows indicate single depolarization stimuli at time zero. Note the delay in shortening for isotonic contractions that is dependent on the magnitude of the load moved (after Vander *et al.*, 1975).

Control of Muscle Tension

The pattern shown in Fig. 18.19 is for a set of muscle fibers of a vertebrate innervated by one nerve cell. As action-potential frequency in the motor neuron increases, tension in the muscle fibers innervated by *that* neuron increases. Within a whole muscle of a vertebrate there are hundreds of neuron-muscle-fiber units called *motor units*, and the total tension developed by an entire muscle depends on the number of individual motor units that are active and the extent of action-potential activity in each.

Variation in the number of active motor units produces some *graded control* over total muscle tension. Contraction requiring little tension is accomplished by the activation of relatively few motor units; contraction requiring high tension involves recruitment of more motor units to add to the total tension developed by a muscle. The stretch receptors within muscles function to detect the degree of stretch, which is related to the load on a muscle; different numbers of motor units are recruited for different demands for the development of tension.

Among invertebrates, control of muscle tension depends more on the extent of postsynaptic muscle depolarization. Instead of large numbers of motor units per muscle, invertebrate muscles are extensively innervated by single motor neurons. However, each action potential in the motor neuron does not produce an all-or-none twitch in invertebrate muscle. The postsynaptic muscle membrane potential is instead graded depending on the number of action

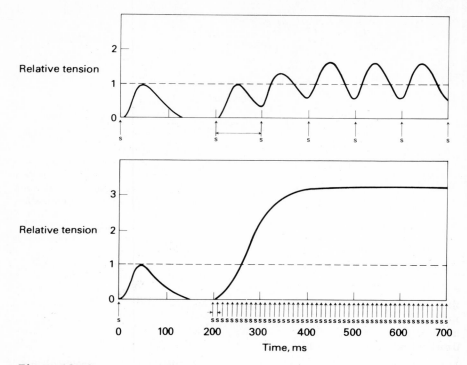

Figure 18.19
Summation of tension from successive stimuli—the development of tetany. Tension developed by a single twitch is indicated as 1, and stimuli are shown as arrows (from Vander *et al.*, 1975).

potentials, and the amount of calcium released into sarcomeres depends on the graded potential.

The major difference between invertebrate and vertebrate muscles is their size. In invertebrate muscles, which have comparatively few muscle fibers per muscle, graded control of muscle tension is achieved with an economy of neuron-effector interactions. In vertebrates, which have larger numbers of muscle fibers per muscle, adequate control of muscle tension involves larger numbers of neuron-effector units.

Energy for Contraction

ATP is required for two important events during contraction: myosin-actin detachment, and active transport of calcium into the sarcoplasmic reticulum. There are three sources of ATP supply within most muscle cells that differ in the *rate* at which ATP is formed to meet demands for contraction. Variation in the occurrence of these three mechanisms within muscles accounts for some differences in types of muscle function both within and among animals.

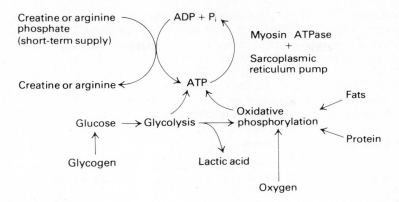

Figure 18.20
Schematic summary of the three sources for ATP supply in muscle cells.

The three ATP supplies are shown in Fig. 18.20. The most rapid supply for forming ATP from ADP is from creatine phosphate (in vertebrates) or arginine phosphate (in invertebrates). A single enzyme mediates the reaction for phosphate transfer to ADP, and this is the most immediate source for providing energy during the early phase of a tetanic contraction. However, the supplies of arginine or creatine phosphate are usually only sufficient to provide a source of ATP for 2-3 seconds of contraction.

Oxidative phosphorylation of ADP in mitochondria results in ATP production by means of oxidation of energy substrates (Fig. 18.20). This mechanism is metabolically efficient in the use of substrate energy, but the rate of production of ATP through the several enzymatic steps of the Krebs cycle is relatively slow (see Chapter 4). This mechanism also depends on the rate of supply of oxygen to muscle cells by circulatory transport. Rate of ATP use during contraction may exceed the rate of oxygen supply via blood transport, but this limitation can be overcome somewhat from internal cellular supplies of oxygen associated with myoglobin, which dissociates oxygen at low intracellular P_{O_2} (see Chapter 3).

The third source of ATP supply is a result of anaerobic glycolysis in the cytoplasm (Fig. 18.20). Recall that this mechanism is 1.8 times faster than oxidative phosphorylation but is 1/18 as efficient for immediate utilization of glucose substrates (Chapter 4). Thus for this mechanism the supply of glucose substrates within muscle cells from glycogen stores could become limiting to the demands for rapid rates of ATP production in some muscle fibers.

If the rate of supply of ATP from these three sources is less than the demands for ATP use in contraction, a muscle would fatigue. Tension would decrease despite repeated neural stimulation, and the extent to which a muscle would maintain contraction (the rate at which it fatigued) could depend on its rate of use of ATP as compared with supplies.

Although continued supplies of ATP are required for contraction, several studies indicate that short-term fatigue produced from tetanic contraction is not due to exhaustion of all ATP supplies. Chemical analysis of muscle fibers following tetanic fatigue indicates that only about 15% of the ATP has been used. Some current hypotheses to explain fatigue suggest that certain chemical products related to energy production (such as H^+ accumulation) may interfere with the action of Ca^{++}. These effects should in turn be related to the rate at which ATP is produced and used for the demands of contraction. Not all ATP is used during contraction nor should it be, since movement of filaments is not the only energy-demanding process within muscles.

TYPES OF MUSCLE FIBERS

There is great diversity in the functions of contractile tissues. In some cases differences are apparent in the extent to which a particular muscle obtains ATP from various sources. In other cases there are differences reflected at the level of morphology of cellular mechanisms for contraction (see below for smooth muscle and cardiac muscle). Finally, there are differences in the coupling of muscles to limbs to produce different types of movements.

Slow, Fast, Red, and White Muscle-fiber Types

There are at least three types of muscle fibers, characterized partly by their supply-demand relationships involving the rate of ATP production and use and factors related to this; they are illustrated in Fig. 18.21. In red muscle fibers the rate of cross-bridge formation is relatively *slow*, and tetanic contraction is maintained for relatively long periods at low tension without fatigue. The red color comes from high concentrations of intracellular supplies of myoglobin. Because the rate of contraction is relatively slow, the rate of production of ATP by oxidative phosphorylation is sufficient to meet demands for long periods. Many postural muscles have a high proportion of red muscle fibers that provide slow but sustained levels of contraction.

In white muscle fibers the rate of cross-bridge formation is *rapid*, tension developed during contraction is high, and fatigue occurs soon during tetanic contraction (Fig. 18.21). White muscle fibers lack oxygen-supply pigments but have relatively high concentrations of glycogen (white muscles taste sweeter). Fatigue occurs soon when chemical products producing fatigue are formed at high rates. Many muscles involved in rapid movements with high power production have a high proportion of white muscle fibers, such as muscles involved in limb movements for predator avoidance or prey capture.

Different major groups of animals have different proportions of white versus red muscle-fiber types. For example, those reptiles, amphibians, and fishes that depend primarily on anaerobic glycolysis for supplying increased energy expenditures for activity (see Chapters 5 and 9) have relatively high proportions of white muscle-fiber types. Recall that these animals are capable of increasing power production to a large extent (compare the tensions developed in Fig. 18.21), but the increase can be sustained only for relatively short periods as compared with animals that have evolved to utilize high-oxidative mechanisms for power production (Fig. 18.21).

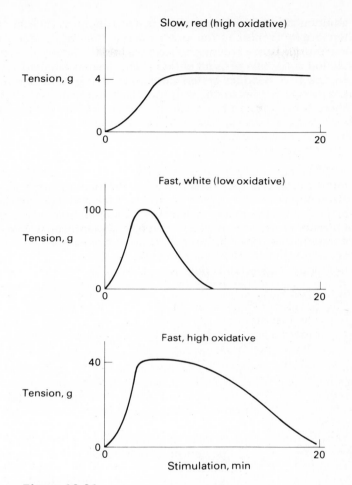

Figure 18.21
Three types of skeletal muscle fibers, based on rate of supply of energy for contraction and rate of use of ATP. Maximum tension developed by each type of fiber is different and is indicated on the y axis (based on Burke *et al.*, 1971).

A third category of muscle fibers exhibit *rapid* contraction with high oxidative activity (Fig. 18.21); it is intermediate between red and white muscle-fiber types. The difference with respect to slow, red muscle fibers is in the rate of use of ATP. Fast, high-oxidative muscle fibers have rich blood supplies as well as intracellular myoglobin oxygen supplies. The high rate of oxidation results in events leading to fatigue relatively soon as compared with high-oxidative, slow muscle-fiber types (Fig. 18.21).

In addition to differences in sources for supply of ATP within muscle fibers, there are differences in the motor neurons that innervate fast and slow fibers. Slow muscle fibers are innervated by relatively small neurons that conduct few action potentials per unit of time; fast muscle fibers are innervated by larger neurons that conduct action potentials at higher frequencies. In experiments in which nerves to different muscle-fiber types were switched, slow muscle fibers became more rapid and fast muscle fibers became more slow. These results suggest some synaptic effects between neurons and muscle fibers that determine the type of muscle function.

Smooth Muscles

Smooth-muscle cells lack the regular order of striation seen in striated muscle. Myosin and actin filaments are present, but they are distributed throughout the cytoplasm so that there usually is some overlap of filaments regardless of the degree of stretch of smooth muscle. Therefore, smooth muscle contracts at a variety of lengths, and the tension developed depends less on stretch than it does for striated muscle. However, because the ordering of filaments is less regular, the total tension that is developed in smooth muscle is less for a given cross-sectional area because fewer total cross bridges can be formed as compared with striated muscle.

Smooth-muscle cells occur in tubular internal organs (arteries, veins, the digestive and reproductive tracts), in association with hairs or feathers, or dispersed within some organs such as the spleen. Individual smooth-muscle cells are smaller than striated muscle fibers, and an organized sarcoplasmic reticulum is usually absent. Depolarization of the individual surface membranes of smooth-muscle cells produces calcium influx sufficient for the smaller cells to produce cross-bridge linkage, and there are two general types of smooth-muscle cells, depending on the mechanism for producing surface-membrane depolarization.

Multiunit smooth-muscle cells are innervated by neurons of the autonomic nervous system. A single neuron may innervate several smooth-muscle cells (in a unit) such that the activity in neurons results in synchronized contraction or relaxation (inhibition) via chemical-transmitter effects on postsynaptic membranes. These smooth-muscle cells are also sensitive to hormonal secretions of the adrenal gland. In general, multiunit smooth-muscle effectors are involved in localized controls following neural integration.

Single-unit smooth muscle cells exhibit contraction without any neural or hormonal action. These muscle cells exhibit a "spontaneous" contraction in response to a local stimulus. The cells are joined together with electrical junctions (see Chapter 17) such that depolarization in one cell rapidly spreads through adjacent cells. Single-unit smooth muscles are common where local stimuli provide sufficient information for effective responses. For example, many single-unit smooth-muscle types are found in parts of the digestive tract where local stimuli from food (pH, distension, osmotic pressure) result in contraction to move food for digestive processing (Chapter 10).

Cardiac Muscle

Cardiac muscle differs from striated skeletal muscle primarily in the characteristics of the membranes surrounding the myofibrils. The mechanism of contraction is the same, with sliding of thin filaments in highly organized and striated sarcomeres. However, the muscle-membrane depolarization that leads to
calcium release from the sarcoplasmic reticulum is extremely prolonged. In
skeletal and smooth muscle, a membrane depolarization for twitch contraction
lasts 2-50 milliseconds. In cardiac muscle membranes, the depolarization lasts
200-300 milliseconds.

In addition, the membranes between cardiac myofibrils have numerous
electrical junctions. These electrical interconnections provide a rapid and synchronous spread of the long membrane depolarization throughout the heart
such that contraction of sarcomeres produces a synchronized maximum tension.

The cardiac muscle membranes also have the different characteristic of
very long refractory periods following depolarization as compared with other
muscle types. This characteristic prevents summation of separate contractions
in time (tetany) so that cardiac muscle relaxes following a heartbeat while the
ventricles are filling.

Asynchronous Flight Muscles

In some insects the wings beat up to 1000 times per second, faster than action
potentials could arrive at the muscles to produce depolarizations. These muscle
types occur in flies, mosquitoes, wasps, bees, and some beetles and bugs. They
are called "asynchronous" because contractions and neural action potentials
are not one-for-one.

Asynchronous muscles differ from other types of muscles by their arrangement in animals. They are arranged to distort the exoskeleton and stretch
the antagonistic set of muscles when they contract. The principal set of muscle
pairs is shown diagramatically in Fig. 18.22. The central muscles, shown in

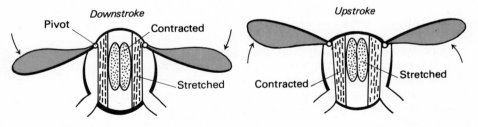

Figure 18.22
Relationship between antagonistic pairs of asynchronous flight muscles that distort
the exoskeleton and stretch the antagonistic muscle pairs. The stretch serves as a
stimulus for muscle contraction such that the rate of contraction exceeds the rate of
arrival of action potentials in efferent neurons (based on Schmidt-Nielsen, 1979).

cross section, are attached longitudinally to the exoskeleton such that their contraction produces leverage and forces the exoskeleton upward (upper part of Fig. 18.22). This attachment arrangement also stretches the more lateral, vertical set of muscles, which contract when they are stretched, swinging the wings upward from their lever attachment to the exoskeleton, which in turn stretches the longitudinal muscles, leading to their contraction, etc. By utilizing the stretch of antagonistic pairs of muscles by means of their morphological arrangement, local stimuli will result in contraction out of synchrony with the arrival of information in efferent neurons once the alternating stretching has been initiated. In addition, some neural input is required because the system is not friction-free.

MOVEMENT WITH ACCESSORY STRUCTURES

The contraction of muscle fibers is only the site of origin of forces for movement or tension. As in the case of asynchronous flight muscles, the muscles are associated with other structures that are not themselves contractile but which serve to translate the forces developed during contraction into appropriate responses. These accessory structures account for the vast majority of differences in patterns of movement by animals, and the diversity of movement patterns among animals is as great as the differences in their structures. The importance of accessory structures can be illustrated by considering some examples for two general categories of structures: *levers* and *elastic springs*.

Skeletal Levers

An example of a muscle-skeletal lever system is shown in Fig. 18.23 for your biceps muscle, connected between a shoulder blade and forearm. The biceps inserts onto (is connected to) the forearm about 5 cm from the elbow pivot. This arrangement results in a requirement for producing a force greater than a force exerted some farther distance from the fulcrum of the lever for the arm to remain stationary. Force is mass times distance, meaning that force must vary for masses over the different distances of the lever for the arm to remain stationary.

The upper part of Fig. 18.23 illustrates this point for a 10-kg weight held stationary in the hand. The downward force at the elbow pivot due to the weight acting over a distance of 30 cm must be balanced by biceps contraction forces in the opposite direction at the 5-cm insertion of the muscle such that the forces (mass times distances) in opposite directions are equal. In this case a force of 60 kg must be exerted by the biceps muscle, one considerably greater than that exerted by the mass of the object in the hand. This requirement is met among muscles associated with levers by their thickness or the number of sarcomere units in parallel.

The mechanical disadvantage in lever systems is partially offset by most forces from large masses acting along the lengths of bones rather than at angles to them through joints. For example, when you are standing the weight of your body is transmitted through the interlocking bones of legs to your skeletal system such that muscular contraction for the maintenance of posture is

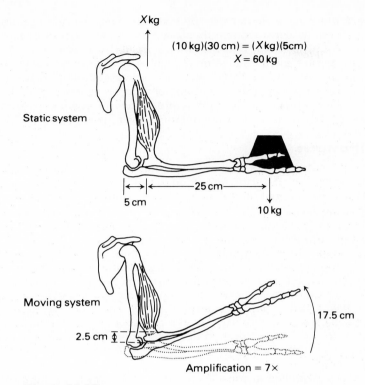

Figure 18.23
The muscle-skeletal lever system for the biceps muscle in a static
load situation (upper) in which contraction force at the muscle
exceeds load at the hand, and for isotonic contraction (lower) during which the lever system amplifies velocity of movement at the
hand. For the static-load case, force times distance must be equal
in opposite directions at the elbow pivot for the arm to remain
stationary (based on Vander *et al.*, 1975).

minimized. When you are running, your legs hit the ground in a relatively
unflexed position such that most of the force of the impact is transmitted to
your skeleton.

In predators that leap from heights, such as leopards and other cats, the
force of impact could be sufficient to shatter bone. Their forelimbs are isolated
from the rest of the skeleton; the shoulder bone (clavicle) is unattached and surrounded by muscle and connective tissues that function as a shock absorber.
This arrangement produces a somewhat higher requirement for postural support by means of their front limbs. In animals that swing from trees, much of
the mass of the animal is distributed through the forelimbs to the general
skeleton from the attached shoulder bone.

An advantage to muscular-skeletal levers is *amplification* of movements for speed in isotonic contractions, illustrated in the lower part of Fig. 18.23. When the biceps muscle shortens 2.5 cm, the tip of the hand moves 17.5 cm in the same time period, an amplification of 7 times in velocity of movement along the length of the lever. From these types of mechanical relationships it is possible to examine the comparative anatomy of animals in terms of their muscle functions and skeletal structures for maintenance of posture (work against gravity) and speed and force of movements of lever systems.

Elastic Structures

The elastic elements in series with muscles must be stretched before maximum tension can be developed during tetanic contraction. The recoil from stretching these "springs" can contribute to movement of a limb in an opposite direction. In addition, external forces that stretch elastic tendons and connective tissue can produce energy storage in these components that can be recovered as kinetic energy during movement.

This principle is the one involved in the hopping of kangaroos (see Chapter 9). When the oxygen consumption of hopping kangaroos was measured, their rate of oxygen use remained essentially constant despite increased speed of hopping. Thus some form of energy not requiring oxygen use contributed to the increased efficiency of locomotion as speed increased. Elastic stretching of the Achilles tendon system apparently provides this source and accounts for saving 40 percent of the cost of locomotion at moderate speeds. As speed increases, more energy is saved by elastic storage because the forces involved with ground impact are greater.

The forces produced at muscles, the translation of those forces through levers, and the contribution of elastic storage and recoil all combine to determine the patterns of movement of animals. Differences in movement patterns reflect modifications of any of these components for specific types of effector responses. Moreover, the "design" of movement patterns by means of evolutionary changes reflect a variety of mechanisms for efficient resource use in different envrionments.

SUMMARY

Effectors for producing responses for feedback involve either chemical or mechanical actions on an environment. The basic mechanisms of secretion and movement are similar at the cellular level; differences in patterns of responses depend primarily on the type of chemicals produced in different glands or on the type of muscle that is activated and its arrangement in animals that translates into movements.

Endocrine glands secrete internally, and exocrine glands secrete either externally or into the digestive tract. Differences in response to different environmental stimuli depend on which types of glands are activated via neural or hormonal mechanisms. Chemicals are manufactured in cells on a rough endoplasmic reticulum and in a Golgi apparatus and packaged in secretory vesicles at one end of a cell; the contents are released by exocytosis in response to a calcium-mediated membrane depolarization. The response (secretion) is related to the stimulus within a cell type via a relationship between the amount of Ca^{++} influx (depolarization) and the amount of secretory exocytosis.

The specificity of chemical activation of different tissues is due to the presence of membrane receptors. Following a chemical-receptor interaction, three types of mechanisms can lead to changes in cell function. For β-type receptors, enzyme function is influenced following the release of adenylate cyclase from the inner membrane surface, and this leads to cAMP production and effects on cytoplasmic protein kinases. For α-type receptors, enzyme function is influenced following changes in internal Ca^{++} concentration. For steroid-type receptors, a receptor complex influences nuclear gene expression.

Most mechanisms for producing movement involve interactions between filaments within cells. Three steps are usually involved in generating forces for movement: (1) linkage between two filament types, (2) sliding of one filament with respect to the other, and (3) detachment of the filaments. One or more of these steps usually involves divalent cations and/or hydrolysis of ATP as a source of energy.

Protoplasmic streaming (cyclosis) and amoeboid movements may involve interactions between actin and myosin filaments in the transformation of fluid (sol) endoplasm to less fluid (gel) ectoplasm, but the transformations have not yet been visualized at a microstructure level, and little is known of the control of movements by protoplasmic transformations.

Ciliary and flagellar movements are associated with more obvious structural changes. The forces for axoneme bending arise locally from interactions between sets of microtubules, and bending is propagated along the organelles. Magnesium and/or calcium are required for linkage of the dynein arms of the A microtubules on the 9 peripheral sets of microtubular pairs to the B microtubule on an adjacent pair. Dynein arms exhibit ATPase activity, and sliding occurs after linkage if ATP is present and if a mechanical force occurs by means of the bending of an adjacent region of an axoneme. The control of axoneme-bending initiation is poorly understood, but the direction of ciliary beating appears to depend on internal calcium concentrations. Cilia of protozoans reverse the direction of beating following calcium influx depolarizations. Synchronization of cilia beating in waves (metachronism) results in swimming transport with minimization of energy expenditures for bending.

The mechanism of movement for muscle is most completely understood for striated muscle, in which the changes in myofibril sarcomere structure during contraction indicate the sliding of thin filaments with respect to thick filaments. Thin filaments extend in each direction from a Z-line at each end of a sarcomere and overlap parts of thick filaments, forming the A-band. The I-band is composed of the Z-line and thin filaments on each side, and this band shortens during contraction such that the Z-lines move closer together.

Thin filaments are composed of helical polymers of G-actin together with the proteins tropomyosin and troponin. Thick filaments are composed of polymers of a light meromyosin "tail" and a heavy meromyosin "head" arranged such that the "heads" project at intervals from the thick filaments toward six adjacent thin filaments. Each thin filament is associated with three thick filaments, and the heavy meromyosin heads form cross-bridge attachments to the thin filaments at the tropomyosin-troponin complex region during contraction.

The sliding-filament theory of contraction was confirmed by experiments that related the tension developed during contraction to the number of cross bridges that could be formed when sarcomeres were experimentally pushed or stretched to influence thin-thick filament overlap. Maximum tension occurred during the maximum degree of filament overlap (maximum cross-bridge linkage); minimum tension occurred when sarcomeres were stretched such that no overlap could occur or when they were pushed together such that few attachments could occur.

Cross-bridge attachment depends on an increase in calcium concentration in the region of the filaments. The calcium combines with troponin and changes the configuration of the troponin-tropomyosin complex, permitting the attachment of HMM to actin. This action leads to the rotation of the cross bridge toward the center of the thick filaments and the sliding of actin filaments with respect to the thick filaments. ATP hydrolysis is then required for detachment of a cross bridge.

Filament sliding is coupled to effector neuron action by means of the influence of muscle-membrane depolarization on the release of calcium from the sarcoplasmic reticulum membranes surrounding individual sarcomeres. Membrane depolarization at the surface is propagated internally via transverse-tubule invaginations of the surface membrane. This depolarization leads to depolarization of the sarcoplasmic membrane and the release of sequestered calcium. Relaxation following contraction occurs as a consequence of membrane active transport of calcium into the sarcoplasmic reticulum from the region of the filaments.

Differences among muscles are apparent at the level of muscle-fiber contraction. A muscle will shorten during isotonic contraction if an external load is moved, or tension will develop during isometric contraction if an external load is not moved. Velocity of isotonic twitch contraction depends on the magnitude of an external load and on the length of a muscle (the number of sarcomeres in series within a muscle). The latency between a depolarizing stimulus and the beginning of an isotonic twitch contraction also depends on the external load due to the stretching of series elastic components of muscle (connective tissues, tendons).

Several isometric contractions in succession result in the summation of tension because of the stretching of series elastic components to a greater degree (tetanic contraction). Maximum tetanic-contraction tension within a muscle depends on its cross-sectional area (the number of sarcomeres in parallel). The control of muscle tension in vertebrate muscles depends on the number of successive action-potential-membrane depolarizations within motor units and the number of recruited motor units. In smaller invertebrate muscles, control of tension depends more on the magnitude of a graded muscle-membrane depolarization rather than on the number of motor units activated.

There are three sources of ATP supply for muscle contraction. Creatine or arginine phosphate supplies phosphate for ATP production at a rapid rate for

2-3 seconds. Oxidative phosphorylation utilizes energy substrates efficiently but is limited in speed and by the rate of supply of oxygen. Anaerobic glycolysis yields ATP 1.8 times more rapidly than does oxidative phosphorylation, but it may be limited by available supplies of less efficiently utilized glucose substrates.

Slow, high-oxidative red muscle fibers maintain tetanic contraction at low tensions for prolonged periods because intracellular oxygen supplies are sufficient to meet relatively low demands for ATP production. Fast, low-oxidative white muscle fibers fatigue sooner. Fast, high-oxidative muscle fibers also fatigue rapidly if the rate of use of ATP exceeds the oxygen-related supplies.

In addition to these differences in the chemistry of the supply of energy for contraction, muscles differ either in cellular mechanisms for filament interactions or at a more gross level of structural arrangement within animals. Smooth muscle fibers lack the ordered arrangement of thick and thin filaments. Actin and myosin filaments are present, but they are more dispersed so that tension can be developed at a variety of muscle lengths at the expense of production of high tensions during contraction. Multiunit smooth muscles are depolarized by effector neurons of the autonomic nervous system. Single-unit smooth muscles are depolarized by local stimuli, and depolarization spreads between muscle cells via electrical junctions. Cardiac muscle differs from voluntary striated muscle in the electrical chracteristics of membrane depolarization, with prolonged depolarization spread through electrical junctions followed by a relatively long refractory period between contractions. Some flight muscles in insects contract in response to local stretch stimuli as a consequence of contraction of antagonistic sets of muscles; contraction occurs more rapidly than does the arrival of neural action potentials at the muscles.

Attachment of muscles to accessory structures produces great diversity in patterns of movement caused by translation of contraction forces by means of skeletal levers and/or the storage of energy in elastic tissues. Skeletal levers are usually arranged to amplify velocity of movement at the lever tip and to minimize contractile energy required for the sustained maintenance of posture. Elastic recoil of tissues associated with muscles and levers, such as the Achilles tendon complex of kangaroos, can also minimize contractile energy expenditures required for locomotion.

ANNOTATED REFERENCES

Chapter 16: Information Transmission and Resource Detection

Aidley, D.J. (1979). *The Physiology of Excitable Cells*. Cambridge University Press, New York. (Textbook on neurophysiology.)

Brown, P.K., and **G. Wald** (1964). Visual pigments in single rods and cones of the human retina. *Science* 144:45-52. (Sensitivity of photopigments to light wavelengths matches sensitivity of photoreceptors.)

Bullock, T.H., and **R.B. Cowles** (1952). Physiology of an infrared receptor: the facial pit of pit vipers. *Science* 115:541-543.

Bullock, T.H., and **G. Horridge** (1965). *Structure and Function in the Ner-*

vous System of Invertebrates (2 vols.). W.H. Freeman, San Francisco. (Elaborate and detailed account of invertebrate neurophysiology.)

Case, J. (1966). *Sensory Mechanisms*. Macmillan, New York. (Introductory account of receptors and sensory physiology.)

Fuortes, M.C.F. (1959). Initiation of impulses in visual cells of *Limulus. Jour. Physiol.* (London) 148:14-28. (Properties of the generator potential in ommatidia.)

Griffin, D.R. (1958). *Listening in the Dark: The Acoustic Orientation of Bats and Man*. Yale University Press, New Haven. (Delightful account of research on bat echolocation mechanisms.)

Gould, J.L. (1980). The case for magnetic sensitivity in birds and bees (such as it is). *Amer. Sci.* 68:256-267. (Summary of evidence on sensitivity of pigeons and honeybees to magnetic fields.)

Hille, B. (1975). An essential ionized acid group in sodium channels. *Fed. Proc.* 34:1318-1321. (Description of a negatively charged acid group within membrane channels that influences cation permeability.)

Hodgkin, A.L. (1964). *The Conduction of the Nervous Impulse*. CC Thomas, Springfield, Ill. (Details of action-potential propagation in excitable tissues.)

Hodgkin, A.L., and **A.F. Huxley** (1952). A quantitative description of membrane current and its application to conduction and excitation in nerve. *Jour. Physiol.* (London) 117:500-544. (Detailed mathematical model of nerve action potentials based on studies of squid giant axons.)

Katz, B. (1966). *Nerve, Muscle, and Synapses*. McGraw-Hill, New York. (Excellent summary of bioelectric phenomena applied to nerve and muscle and their interaction.)

Lissmann, H.W. (1963). Electric location by fishes. *Sci. Amer.* 208:50-59. (Introductory summary of mechanisms for electronavigation in weakly electric fishes.)

Loewenstein, W.R. (1960). Biological transducers. *Sci. Amer.* 203:98-108. (Introductory summary of filtration and amplification functions of receptors.)

Loewenstein, W.R. (1971). *Handbook of Sensory Physiology,* Vol. 1, *Principles of Receptor Physiology*. Springer-Verlag, New York. (Details of receptor function for all major groups of receptors.)

Thomas, R.C. (1972). Electrogenic sodium pump in nerve and muscle cells. *Physiol. Rev.* 52:563-594. (Details of the characteristics of Na^+-K^+ active transport in excitable tissues.)

von Békésy, G. (1960). *Experiments in Hearing*. McGraw-Hill, New York. (Classic description of analysis of function of the human ear.)

Wald, G. (1964). The receptors of human color vision. *Science* 145:1007-1016. (Visual basis for trichromatic color detection.)

Waterman, T.H. (1950). A light polarization analyzer in the compound eye of *Limulus. Science* 111:252-254. (Structural basis for detection of plane of polarization in ommatidia.)

Young, R.W. (1970). Visual cells. *Sci. Amer.* 223:89-91. (Introductory summary of structure and function of photoreceptors.)

Chapter 17: Information Integration and Behavior ✕ ✕

Birks, R.I., H.E. Huxley, and **B. Katz** (1960). The fine structure of the neuromuscular junction of the frog. *Jour. Physiol.* (London) 150:134-144. (Ultrastructure of the neuromuscular synapse.)

Burn, J.H. (1963). *The Autonomic Nervous System.* Blackwell, Oxford. (Emphasis on the pharmacology of function in the autonomic nervous system.)

Creed, R.S., D. Denny-Brown, J.C. Eccles, E.G.T. Lidell, and **C.S. Sherrington** (1932). *Reflex Activity of the Spinal Cord.* Oxford University Press, London.

Dowling, J.E. (1970). Organization of vertebrate retinas. *Invest. Ophthal.* 9:655-680. (Structural and functional properties related to information processing within the retina.)

Eccles, J.C. (1964). *The Physiology of Synapses.* Springer-Verlag, New York. (Details of neurophysiology of synapses, particularly for motor neurons.)

Fatt, P., and **B. Katz** (1952). Spontaneous subthreshold activity at motor nerve endings. *Jour. Physiol.* (London) 117:109-128. (Description of mEPSP properties in neuromuscular synapses.)

Hall, Z., J. Hildebrand, and **E. Kravitz** (1974). *The Chemistry of Synaptic Transmission.* Chiron Press, Newton, Mass. (Details of chemical interactions for information transmission.)

Hartline, H.K., and **F. Ratliff** (1957). Inhibitory interaction of receptor units in the eye of *Limulus. Jour. Gen. Physiol.* 40:357-376. (Lateral-inhibition characteristics for edge detection and amplification.)

Hubel, D.H., and **T.N. Wiesel** (1962). Receptive fields, binocular interaction and functional architecture in the cat's visual cortex. *Jour. Physiol.* (London) 160:106-154. (Characteristics of receptive fields at the level of the cortex.)

Lettvin, J.Y., H.R. Maturana, W.H. Pitts, and **W.S. McCulloch** (1961). Two remarks on the visual system of the frog. In W.A. Rosenblith, (ed.), *Sensory Communication.* MIT Press, Cambridge. (Summary of the functional basis for species-specific "bug-detector" receptive fields in the retina of frogs.)

Lloyd, D.P.C. (1946). Facilitation and inhibition of spinal motor neurons. *Jour. Neurophysiol.* 9:421-444. (Early description of processes of synaptic facilitation and inhibition.)

Michael, C.R. (1969). Retinal processing of visual images. *Sci. Amer.* 220:104-114. (Introductory summary of function of receptive fields in the retina.)

Miledi, R. (1973). Transmitter release induced by injection of calcium ions into nerve terminals. *Proc. Roy. Soc.* (B) 183:421-425. (Dependence of transmitter exocytosis on calcium potential at presynaptic nerve terminals.)

Roeder, K.D. (1967). *Nerve Cells and Insect Behavior,* revised edition. Harvard University Press, Cambridge. (Delightful account of moth-bat interactions in the context of neural mechanisms for information processing in relatively simple neural systems.)

616 ANNOTATED REFERENCES—PART VI

Sherrington, C.S. (1906). *Integrative Action of the Nervous System.* Yale University Press, New Haven. (Dated but classic deductions of mechanisms involved in spinal reflexes.)

Tinbergen, N. (1951). *The Study of Instinct.* Clarendon Press, Oxford. (Excellent summary of species-specific fixed-action patterns and their releasing stimuli.)

von Euler, U.S. (1961). Neurotransmission in the adrenergic nervous system. *Harvey Lectures* Ser. 55:43-65. (Summary of synaptic effects in the sympathetic nervous system.)

Chapter 18: Effectors and Responses

Alexander, R. McN. (1968). *Animal Mechanics.* Sidgwick & Jackson, London. (Physical principles applied to the comparative anatomy of animal levers and elastic components.)

Alexander, R. McN., and **A. Vernon** (1975). The mechanics of hopping by kangaroos (Macropodidae). *Jour. Zool.* (London) 177:265-303. (Detailed mechanical analysis of lever-and-spring systems for hopping and their relationship with the energetics of locomotion.)

Eckert, R. (1972). Bioelectric control of ciliary activity. *Science* 176:473-481. (Dependence of direction of ciliary beating on calcium membrane potentials.)

Exton, J.H. (1980). Mechanisms involved in α-adrenergic phenomena: role of calcium ions in actions of catecholamines in liver and other tissues. *Amer. Jour. Physiol.* 238:E3-E12. (Summary of evidence for mode of action of β-type and α-type membrane receptors.)

Gordon, A.M., **A.F. Huxley**, and **F.J. Julian** (1966). The variation in isometric tension with sarcomere length in vertebrate muscle fibers. *Jour. Physiol.* (London) 184:170-192. (Length-tension relationship and extent of possible cross-bridge attachments.)

Gray, J. (1968). *Animal Locomotion.* Weidenfield & Nicolson, London. (Detailed analysis of movement patterns of animals.)

Huxley, A.F., and **R.M. Simmons** (1971). Proposed mechanism of force generation in striated muscle. *Nature* 233:533-538. (Discussion of crossbridge linkage and rotation during contraction.)

Huxley, H.E. (1969). The mechanism of muscular contraction. *Science* 164: 1356-1366. (Summary of evidence for the sliding-filament theory of contraction.)

Katzenellenbogen, B.S. (1980). Dynamics of steroid hormone receptor action. *Ann. Rev. Physiol.* 42:17-35.

Lüttgau, H.C., and **G.D. Moisescu** (1978). Ion movements in skeletal muscle in relation to the activation of contraction. In T.E. Andreoli, J.F. Hoffman, and D.D. Fanestil (eds.), *Physiology of Membrane Disorders.* Plenum, New York. (Summary of ionic events in excitation-contraction coupling.)

Mommaerts, W.F.H.M. (1969). Energetics of muscle contraction. *Physiol. Rev.* 49:427-508. (Relationship between sources of energy supply and generation of forces in muscle fibers.)

Nassar-Gentina, V., J.V. Passonneau, J.L. Vergara, and **S.I. Rapoport** (1978). Metabolic correlates of fatigue and recovery from fatigue in single frog muscle fibers. *Jour. Gen. Physiol.* 72:593-606.

Peachy, L.D. (1965). Transverse tubules in excitation-contraction coupling. *Fed. Proc.* 24:1124-1134. (Ultrastructural basis for conduction of the action potential to individual sarcomeres.)

Pringle, J.W.S. (1949). The excitation and contraction of the flight muscles of insects. *Jour. Physiol.* (London) 108:226-232. (Mechanism of asynchronous contraction in insect flight muscles.)

Satir, P. (1974). How cilia move. *Sci. Amer.* 231:44-52. (Introductory summary of theories of ciliary bending.)

Warner, F.D., and **D.R. Mitchell** (1978). Structural conformation of ciliary dynein arms and the generation of sliding forces in *Tetrahymena* cilia. *Jour. Cell Biol.* 76:261-277. (Changes in dynein arms with ATP hydrolysis.)

Weber, A., and **J.M. Murray** (1973). Molecular control mechanisms in muscle contraction. *Physiol. Rev.* 53:612-673. (Review of information about proteins on actin filaments that influence cross-bridge attachment.)

Zanetti, N.C., D.R. Mitchell, and **F.D. Warner** (1979). Effects of divalent cations on dynein cross-bridging and ciliary microtubule sliding. *Jour. Cell Biol.* 80:573-588.

Appendix I

Dimensions and Units for Measurements

Throughout this book we are concerned with measuring variables and interpreting the measurements. Physiology has a strong basis in physical measurements. This appendix is designed to refamiliarize you with some fundamental rules that, if followed, will increase your ability to deal with the more "mathematical" nature of the subject.

Two pieces of information are required to precisely describe any variable: (1) a *qualitative* description of what kind of variable it is, and (2) a *quantitative* description of how much of that variable there is. The first involves the *dimensions* of a variable, whereas the second involves the *units* or system of units used to measure the variable.

DIMENSIONS

Except for electromagnetic variables, all physical quantities can be described qualitatively with one or any combination of four dimensions: length (l), time (t), mass (m), and temperature (T). Table AI.1 lists a number of common physical variables and their dimensions in terms of this "fundamental" four-component system. The list in Table AI.1 is not inclusive; a complete list can be found in the CRC *Handbook of Chemistry and Physics*.

Table AI.1

PHYSICAL QUANTITY	DIMENSIONS IN TERMS OF LENGTH (l), TIME (t), MASS (m), AND TEMPERATURE (T)
Distance	l
Area	l^2
Volume	l^3
Velocity	l/t
Acceleration (rate of change of velocity)	l/t^2
Flow (volume/time)	l^3/t
Force (acceleration of mass)	$(m)(l)/t^2$
Pressure (force/area)	$m/(l)(t^2)$
Work or energy (force × distance)	$(m)(l^2)/t^2$
Power (energy/time)	$(m)(l^2)/t^3$
Density (mass/volume)	m/l^3
Thermal capacity	$l^2/(t^2)(T)$

The importance of knowing the dimensions for any variable is related to a set of rules that specify how dimensions can be related to each other. Any equation that states a functional relationship between variables must follow the rules for dimensional correctness, and these rules are independent of those for the mathematical correctness of the numbers associated with dimensions. For example, respiratory ventilation (a measure of flow in respiratory structures; see Chapter 1) can be dimensionally written as

$$l^3/t = \text{breaths}/t \times l^3/\text{breath},$$

or

$$\text{Flow} = \text{Rate} \times \text{Volume},$$

regardless of the numbers associated with each component of the equation.

In order to be correct, an equation like this must be written so that dimensions on each side of the equality are in fact equal. If this equality does not occur the equation will be fundamentally wrong.

The following two rules are useful for checking the correctness of dimensions for an equation.

1. Quantities can be added and subtracted *only* when dimensions are the same.

2. Quantities can be multiplied and divided without regard to dimensional similarity, but the multiplication or division must be algebraically correct.

The best procedure to follow is to rewrite an equation with only dimensional symbols, check rule 1 and replace all symbols so that only division and multiplication need to be done, perform the manipulations, and examine the result for equality.

Checking the dimensional correctness of functional relationships can be very important for understanding the physical basis of a relationship that may not be immediately obvious. For example, respiratory "ventilation" is the product of a rate and a volume, and thus ventilation is fundamentally *flow* (see Table AI.1). You will find many different terms for similar basic relationships. For example, "ventilation" (Chapter 1 and Chapter 2), "cardiac output" (Chapter 4), and renal "clearance" (Chapter 13) are all terms which have the dimensions of flow. By paying careful attention to the dimensions defining terms such as "output," "ventilation," and "clearance" in equations, you can strip them of the mystery surrounding the different words. You will then be dealing with the simpler "fundamentals" of common physical dimensions.

UNITS

Dimensions are most important in the *theoretical* derivation of equations, in which units are unimportant to show the physical basis for a function. However, to test a theory it is necessary to characterize the amount of different physical variables as they change. To do this, some decision has to be made about which arbitrary system of units is to be used to measure magnitudes.

It is important to note that all systems of units are arbitrary; there is nothing inherently more or less correct about using a meter stick or a yardstick to measure distance. However, it is important to follow two guidelines: (1) use the *same* system of units for all measurements in any operation to avoid confusion, and (2) use a system with which other people are familiar to minimize problems of information communication. If you devised a system to measure things with which only you were familiar (or only you knew the rules), it would be useless to others, and you could not communicate information to others without translating your rules into rules with which others were familiar. Systems of units are simply sets of rules that a large number of people have agreed to use.

The most commonly used system of units for communicating scientific information is the Meter-Kilogram-Second (MKS) system. Over time and in usage it has proved to be somewhat more convenient than other systems, and this is the system with which most scientists are familiar and thereby use for communication with each other. However, a modified system of units has recently been introduced into common usage, called the Système International (SI), the International System of Units. This system is a modification of the traditional MKS system and has been recommended by a committee of scientists to reduce some confusion in terminology and symbolism associated with the MKS system.

All systems of units can be converted into different systems of units simply by multiplying or dividing by appropriate conversion factors. I have decided to use primarily the unrevised MKS system of units in this book because it is likely that more people will be familiar with it. However, the two systems can easily be interconverted simply by using the factors shown in Table AI.2. Electrical dimensions and units are listed in Table AI.3 and the units are common to both the MKS and the SI systems of units.

In this text, the term "weight" is often used in the traditional, *informal* sense and means "mass." However, please note that weight and mass have different meanings in physics. Weight is *strictly* defined as the force of gravity—the mass of an object times the local acceleration of gravity—and is expressed in newtons (N) in the SI system of units (Table AI.2).

Table AI.2   The meter-kilogram-second and SI systems of units with conversion factors

PHYSICAL QUANTITY	MKS	SI	CONVERSIONS
Force $(m)(l)/t^2$ or $kg(m)/s^2$*	kilogram (kg)	newton (N)	1 kg = 9.807 N 1 N = 0.102 kg
Pressure $m/(l)(t^2)$ or $kg/(m)s^2$	atmosphere (atm) mmHg (Torr)	pascal (Pa)	1 atm = 1.013×10^5 Pa 1 Torr = 1.333×10^2 Pa 1 Pa = 0.0075 mmHg 1 Pa = 9.869×10^{-3} atm
Energy $(m)(l^2)/t^2$ or $kg(m^2)/s^2$	calorie (cal)	joule (J)	1 cal = 4.184 J 1 J = 0.239 cal
Power $(m)(l^2)/t^3$ or $kg(m^2)/s^3$	cal/time	watt (W)	1 cal/s = 4.184 W 1 W = 0.239 cal/s

Dimensions to the left are converted to *units* on the right such that mass (m to the left) = Kg, length (l) = meters (m to the right), and time (t) = seconds (s).

Table AI.3   Electrical dimensions and units

PHYSICAL QUANTITY	DIMENSIONS	UNITS
Electric current	A	ampere (A)
Electric potential difference	$(m)(l^2)/(t^3)A$	volt (V)
Electric resistance	$(m)(l^2)/(t^3)A^2$	ohm (Ω)
Electric charge	$(t)(A)$	coulomb (C)

Dimensions: l = length, m = mass, t = time,   and A = ampere.

Again, to avoid much confusion it is imperative that you use the *same system* of units throughout any series of calculations. For example, you may use either pascals or mmHg for pressure, but not both because the result would be a confused mixture of different quantitative amounts for the same variable.

Finally, Table AI.4 presents a series of basic and derived physical constants that occur extensively in fundamental physical equations. The values for the constants are useful for solving equations for specific magnitudes or amounts.

Table AI.4 Values of physical constants

A. *Basic constants*

Velocity of light (c)	2.997902×10^{10} cm/s
Planck constant (h)	6.62377×10^{-27} erg s
Avogadro number (N_0)	6.0254×10^{23}/mol
Faraday constant (F)	96,493 coulombs/equivalent
	23,062.4 cal/volt equivalent
Absolute temperature of the "ice" point, or 0°C ($T_{0°}$)	273.16 °K (Kelvin degrees)

B. *Derived constants*

Gas constant: $R = \dfrac{(PV)}{T_{0°}}$	1.9872 cal/degree mol
	82.057 cm^3 atm/degree mol
	8.3144 joules/degree mol
Boltzmann constant: $k = \dfrac{R}{N_0}$	1.38026×10^{-16} erg/degree
Electronic charge: $e = \dfrac{F}{N_0}$	1.60186×10^{-19} coulomb

---

Appendix II

Types of Equations and their Solutions

The interpretation of physiological principles depends on quantitative relationships between variables. These relationships fall into specific categories, and attention to some general types of functions will help to understand them in the specific cases encountered throughout the book.

The simplest relationship between two variables is a *linear* function, or

$$y = a + bx.$$

This function has the graphical representation of a straight line (Fig. A.1) with an *intercept* (value of y when $x = 0$) of a and a *slope* (change in y with a change in x) of b. From this type of relationship we would conclude that the value of y depends on the value of x in a linear manner that is characterized by the slope and intercept.

There are several physiological functions that are linear. An example is the linear dependence of heat production in endotherms on the product of thermal conductance (C) and the difference between body and environmental tem-

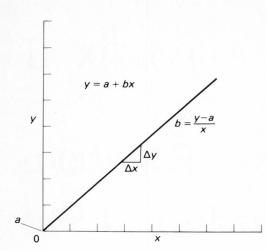

Figure A.1
The general graphical representation of a linear equation.

perature, where C is the slope of the function in heat production $= C(T_B - T_A)$ (see Chapter 7). In general, most proportional control functions in physiology are linear (see Introduction), including proportional control of ventilation (Chapters 1 and 2), cardiac output (Chapter 4), and body temperature (Chapters 6 and 7).

EXPONENTIAL FUNCTIONS

Many biological functions are exponential. They pervade all aspects of biology, and it is very important to know how to interpret them and how to solve them with ease.

If a is greater than zero, the exponential function f with "base" a is defined as

$$f(x) = a^x.$$

The graphical representation of the general function $y = a^x$ ($a > 1$) is shown in Fig. A.2. Note that y depends disproportionately (nonlinearly) on x such that a slight change in x results in a large change in y. In many cases these types of functions set limits in the physiology of animals because of the disproportionate relationships between variables.

Some examples of exponential functions important in physiology include the relationships between rates of reactions and temperature (Q_{10}; see Chapter 6), relationships between energy expense and concentration differences for osmoregulation by active transport (see Chapter 12), and the relationship between water vapor pressure and air temperature (see Chapter 13).

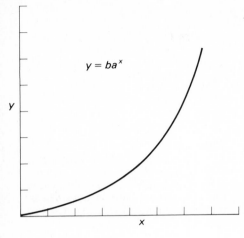

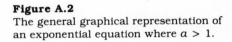

Figure A.2
The general graphical representation of
an exponential equation where $a > 1$.

LOGARITHMIC FUNCTIONS

According to the laws of exponents (see below), a unique number x exists such
that for a given "base" a ($a > 0$ and $\neq 1$) in an exponential function,

$$a^x = y.$$

In this exponential function x is called the *logarithm* to the base a of y, or
$x = \log_a y$.

There are two common "bases" for logarithmic functions, i.e., the numbers that are raised to an exponent to equal another number. The "natural"
logarithmic functions (denoted ln) have the base e ($e = 2.71828$), whereas
"common" logarithmic functions (denoted Log) have the base 10.

Figure A.3 shows the graphical representation for a logarithmic function
$y = \log_a x$ or $a^y = x$. Note in the equation that x and y are reversed as compared
with exponential functions, and logarithmic functions have inverse properties
as compared with exponential functions (compare Fig. A.2 and Fig. A.3). For a
logarithmic function, a change in y depends to a great extent on a change in x at
low values of x. A common example of a logarithmic function in physiology is
the dependence of receptor generator potentials on the logarithm of stimulus-
energy strength (see Chapter 16).

THE LAWS OF EXPONENTS AND LOGARITHMS

There are sets of complementary rules used in manipulating either exponential
or logarithmic equations. They are reviewed here and are followed by some examples of the utility of logarithms for ease of solutions of these equations.

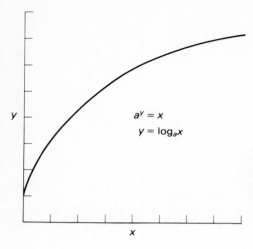

$$a^y = x$$
$$y = \log_a x$$

Figure A.3
The general graphical representation of a logarithmic equation. Note that this is the reverse of the general shape shown in Fig. A.2 because here we have $a^y = x$ instead of $a^x = y$.

The Laws of Exponents

If a and b are positive numbers, then for any numbers x and y the following equations hold.

1. $a^x a^y = a^{x+y}$

2. $\dfrac{a^x}{a^y} = a^{x-y}$

3. $a^x b^x = (ab)^x$

4. $\dfrac{a^x}{b^x} = \left(\dfrac{a}{b}\right)^x$

5. $(a^x)^y = a^{xy}$

The Laws of Logarithms

If x and y are positive numbers, then for any base, the following are true.

1. $\log(xy) = \log x + \log y$

2. $\log\left(\dfrac{x}{y}\right) = \log x - \log y$

3. $\log(x)^c = c \log x$

4. $\log_a 1.0 = 0; \log_a a = 1.0$

The Computational Uses of Logarithms

These "laws" permit an exponential function, a logarithmic function, or *any* function with some *power* or exponent to be rewritten in a linear form by taking logarithms of both sides of the equation. For example, the general exponential equation

$$y = ba^x$$

can be transformed to

$$\text{Log } y = \text{Log } b + (x) \text{ Log } a.$$

In this case, Log y is a linear function of x with slope Log a. For the general logarithmic function

$$a^y = x,$$

taking logarithms of both sides yields

$$y \text{ Log } a = \text{Log } x,$$
$$y = \text{Log } x - \text{Log } a.$$

In this case y is a linear function of Log x.

Transformation of equations with power terms or exponents into linear equations permits solutions by addition, subtraction, multiplication, and division by following the laws of exponents and logarithms. After these manipulations, taking *antilogarithms*, that is, finding the values of exponents of a base (from tables for that purpose), permits determination of the values of y.

Calculated Examples

Some examples will illustrate the procedure. In Chapter 6 the following equation is given for the dependence of reaction rates on temperature:

$$R_2 = R_1 Q_{10}^{\frac{(T_2 - T_1)}{10}}$$

where R_2 = rate at temperature T_2 R_1 = rate at temperature T_1, and Q_{10} is a "constant." If $Q_{10} = 2.0$, and $R_1 = 7.0$ at a temperature $T_1 = 24\,°C$, what is the rate (R_2) when $T_2 = 30\,°C$?

$$R_2 = (7.0)(2.0)^{\frac{(30-24)}{10}}$$
$$= (7.0)(2.0)^{0.6}$$

or

$$\text{Log } R_2 = \text{Log } 7.0 + 0.6(\text{Log } 2.0)$$
$$= 0.8451 + 0.6(0.3010)$$
$$= 0.8451 + 0.1806$$
$$= 1.0251.$$

Because Log $R_2 = 1.0251$, we wish to determine the value of x for $10^{1.0251} = x$. This is called finding the antilog of a logarithm. By inspection of the exponent the number is between 10^1 and 10^2 and closer to 10^1. The number 0.0251, the fractional amount between 1.0 and 2.0 for the exponent, is the *mantissa*. Tables of logarithms list the values of 10 raised to the fractional mantissa amounts given in the body of the tables. For this example, $10^{1.0251} = 10.595 = R_2$.

In Chapter 5 the following equation is given for the relationship between rate of energy use and body size in mammals:

$$\text{kcal/day} = 70(\text{kg})^{0.75}.$$

Strictly speaking, this equation is neither exponential nor logarithmic because it is of the form $y = x^a$. This type of equation is called a *power function*. However, it still contains an exponent; thus we can use logarithms for its solution.

Consider a mammal that weighs 12 kg. What is its energy requirement per day?

$$
\begin{aligned}
\text{Log kcal/day} &= \text{Log } 70 + 0.75(\text{Log kg}) \\
&= 1.8451 + 0.75\,(1.0792) \\
&= 1.8451 + 0.8094 \\
&= 2.6545 \\
\text{kcal/day} &= 10^{2.6545} = 451.336.
\end{aligned}
$$

Credit References

Adolph, E.F. (1939). Measurement of water drinking in dogs. *Amer. Jour. Physiol.* 125:75-86.

Adrian, E.D. (1928). *The Basis of Sensation: The Action of the Sense Organs.* Christophers, London.

Alberts, J.R. (1978). Huddling by rat pups: Group behavioral mechanisms of temperature regulation and energy conservation. *Jour. Comp. Physiol. Psychol.* 92:231-245.

Andrewartha, H.G., and L.C. Birch (1954). *The Distribution and Abundance of Animals.* University of Chicago Press, Chicago.

Aschoff, J., and H. Pohl (1970). Rhythmic variations in energy metabolsm. *Fed. Proc.* 29:1541-1552.

Balmer, R.T., and A.D. Strobusch (1977). Critical size of newborn homeotherms. *Jour. Appl. Physiol.* 42:571-577.

Bartholomew, G.A. (1964). The roles of physiology and behavior in the maintenance of homeostasis in the desert environment. *Symp. Soc. Exp. Biol.* 18:7-29.

Bartholomew, G.A. and T.J. Cade (1963). The water economy of land birds. *Auk* 80:504-539.

Beadle, L.C. (1943). Osmotic regulation and the faunas of inland waters. *Biol. Rev.* 18:172-183.

Beament, J.W.L. (1958). The effect of temperature on the waterproofing mechanism of an insect. *Jour. Exp. Biol.* 35:494-519.

Bentley, P.J. (1971). *Endocrines and Osmoregulation: A Comparative Account of the Regulation of Water and Salt in Vertebrates.* Springer-Verlag, New York.

Bill, R.G., and W.F. Herrnkind (1976). Drag reduction by formation movement in spiny lobsters. *Science* 193:1146-1148.

Bloom, W., and D.W. Fawcett (1975). *A Textbook of Histology.* W.B. Saunders, Philadelphia.

Bohr, C., K. Hasselbach, and A. Krogh (1904). Über einen biologischer beziehung wichtigen Einfluss, den die Kohnlen-Saurespannung des Blutes auf dessen Sauerstoff bingung ubt. *Skand. Arch. Physiol.* 16:402.

Bonner, J.T. (1965). *Size and Cycle: An Essay on the Structure of Biology.* Princeton University Press, Princeton.

Brett, J.R. (1971). Energetic response of salmon to temperature. A study of some thermal relations in the physiology and freshwater ecology of sockeye salmon (*Oncorhynchus nerka*). *Amer. Zool.* 11:99-114.

Bretz, W.L., and K. Schmidt-Nielsen (1971). Bird respiration: Flow patterns in the duck lung. *Jour. Exp. Biol.* 54:103-118.

Brower, J.Z. (1960). Experimental studies of mimicry. IV. The reactions of starlings to different proportions of models and mimics. *Amer. Nat.* 94:271-282.

Brower, L.P., W.N. Ryerson, L.L. Coppinger, and S.C. Glazier (1968). Ecological chemistry and the palatability spectrum. *Science* 161:1349-1351.

Brown, J.H., and R.C. Lasiewski (1972). Metabolism of weasels: The cost of being long and thin. *Ecology* 53:939-943.

Buchsbaum, R. (1948). *Animals Without Backbones: An Introduction to the Invertebrates,* Rev. ed. University of Chicago Press, Chicago.

Bullock, T.H., and F.P.J. Diecke (1956). Properties of an infrared receptor. *Jour. Physiol.* (London) 134:47-87.

Bunge, M.B., R.P. Bunge, and H. Ris (1961). Ultrastructural study of remyelination in an experimental lesion in adult cat spinal cord. *Jour. Biophys. Biochem. Cytol.* 10:67-94.

Burke, R.E., D.N. Levine, F.E. Zajack III, P. Tasiris, and W.K. Engel (1971). Mammalian motor units: Physiological-histochemical correlation of three types in cat gastrocnemius. *Science* 174:709-712.

Calder, W.A., and J.R. King (1974). Thermal and caloric relations of birds. In D.S. Farner and J.R. King (eds.), *Avian Biology,* Vol. IV. Academic Press, New York.

Calloway, N.O. (1976). Body temperature: Thermodynamics of homeothermism. *Jour. Theor. Biol.* 57:331-344.

Carey, F.G., and J.M. Teal (1969). Regulation of temperature by the bluefin tuna. *Comp. Biochem. Physiol.* 28:205-213.

Cavagna, G.A., N.C. Heglund, and C.R. Taylor (1977). Mechanical work in terrestrial locomotion: two basic mechanisms for minimizing energy expenditure. *Amer. Jour. Physiol.* 233:243-261.

Chiodi, H. (1971). Comparative study of the blood gas transport in high altitude and sea level Camelidae and goats. *Resp. Physiol.* 11:84-93.

Cohen, C. (1975). The protein switch of muscle contraction. *Sci. Amer.* 233: 36-45.

Crawford, E.C., Jr. (1962). Mechanical aspects of panting in dogs. *Jour. Appl. Physiol.* 17:249-251.

Croghan, P.C. (1958). The osmotic and ionic regulation of *Artemia salina* (L.). *Jour. Exp. Biol.* 35:219-233.

Davies, N.B. (1977). Prey selection and social behavior in wagtails (Aves: Motacillidae). *Jour. Anim. Ecol.* 46:37-57.

Dawson, W.R. (1975). On the physiological significance of the preferred body temperature of reptiles. In D.M. Gates and R.B. Schmerl (eds.), *Perspectives of Biophysical Ecology, Ecological Studies*, Vol. 12. Springer-Verlag, New York.

Denton, E.J. (1961). The buoyancy of fish and cephalopods. *Prog. Biophys. and Biophys. Chem.* 11:177-234.

Dethier, V.G. (1976). *The Hungry Fly.* Harvard University Press, Cambridge.

DeWitt, C.B., and R.M. Friedman (1979). Significance of skewness in ectotherm thermoregulation. *Amer. Zool.* 19:195-209.

DiBona, D.R., and J.W. Mills (1979). Distribution of Na^+- pump sites in transporting epithelia. *Fed. Proc.* 38:134-143.

Doi, Y. (1920). Studies on muscular contraction. *Jour. Physiol.* (London) 54: 218-226.

Donnelly, P., and A.J. Woolcock (1977). Ventilation and gas exchange in the carpet python, *Morelia spilotes variegata. Jour. Comp. Physiol.* 122: 403-418.

Eckert, R., and D. Randall (1978). *Animal Physiology.* W.H. Freeman, San Francisco.

Epstein, A.N., and E. Stellar (1955). The control of salt preference in the adrenalectomized rat. *Jour. Comp. Physiol. Psychol.* 48:167-172.

Exton, J.H. (1980). Mechanisms involved in α-adrenergic phenomena: role of calcium ions in actions of catecholamines in liver and other tissues. *Amer. Jour. Physiol.* 238:E3-E12.

Farlow, J.O., C.V. Thompson, and D.E. Rosner (1976). Plates of the dinosaur *Stegosaurus*: Forced convective heat loss fins? *Science* 192:1123-1125.

Farner, D.S. (1955). The annual stimulus for migration: Experimental and physiological aspects. In A. Wolfson (ed.), *Recent Studies in Avian Biology.* University of Illinois Press, Urbana.

Fisher, K.C., and J.F. Manery (1967). Water and electrolyte metabolism in heterotherms. In K.C. Fisher, A.R. Dawe, C.P. Lyman, E. Schönbaum, and

F.E. South (eds.), *Mammalian Hibernation III*. American Elsevier, New York.

Flock, A. (1971). Sensory transduction in hair cells. In W.R. Lowenstein (ed.), *Handbook of Sensory Physiology*, Vol. 1, *Principles of Receptor Physiology*. Springer-Verlag, Berlin.

Florey, E. (1966). *General and Comparative Physiology*. W.B. Saunders, Philadelphia.

Friedman, M.I., and E.M. Stricker (1976). The physiological psychology of hunger: A physiological perspective. *Psychol. Rev.* 83:409-431.

Fuortes, M.C.F. (1959). Initiation of impulse in visual cells of *Limulus*. *Jour. Physiol.* (London) 148:14-28.

Gans, C., H.J. de Jongh, and J. Farber (1969). Bullfrog (*Rana catesbeiana*) ventilation: How does the frog breathe? *Science* 163:1223-1225.

Gates, D.M. (1965). Spectral properties of plants. *Appl. Optics* 4:11-20.

Gates, D.M., and W.P. Porter (1970). The energy budget of animals. In J.D. Hardy, A.P. Gagge, and J.A.J. Stolwijk (eds.) *Physiological and Behavioral Temperature Regulation*. CC Thomas, Springfield, Ill.

Gelperin, A. (1966). Control of crop emptying in the blowfly. *Jour. Insect Physiol.* 12:331-345.

Gessaman, J.A. (1973). The effect of solar radiation and wind on energy metabolism. In J.A. Gessaman (ed.), *Ecological Energetics of Homeotherms*. Utah State University Press, Logan.

Gilbert, L.I. (1964). Physiology of growth and development. In M. Rockstein (ed.), *The Physiology of Insects*, Vol. 1. Academic Press, New York.

Gordon, A.M., A.F. Huxley, and F.J. Julian (1966). The variation in isometric tension with sarcomere length in vertebrate muscle fibers. *Jour. Physiol.* (London) 184:170-192.

Gordon, M.S., G.A. Bartholomew, A.D. Grinnell, C.B. Jorgensen, and F.N. White. (1972). *Animal Physiology: Principles and Adaptations*. 2d ed. Macmillan, New York.

Gordon, M.S., K. Schmidt-Nielsen, and H.M. Kelley (1961). Osmotic regulation in the crab-eating frog (*Rana cancrivora*). *Jour. Exp. Biol.* 38:659-678.

Grey, I.E. (1954). Comparative study of the gill area of marine fishes. *Biol. Bull.* 107:219-225.

Grigg, G.C. (1969). Temperature-induced changes in the oxygen equilibrium curve of the blood of the brown bullhead, *Ictalurus nebulosus*. *Comp. Biochem. Physiol.* 28:1202-1223.

Gunderson, H.L. (1976). *Mammalogy*. McGraw-Hill, New York.

Hainsworth, F.R. (1978). Feeding: Models of costs and benefits in energy regulation. *Amer. Zool.* 18:701-714.

Hainsworth, F.R., B.G. Collins, and L.L. Wolf (1977). The function of torpor in hummingbirds. *Physiol. Zool.* 50:215-222.

Hainsworth, F.R., and E.M. Stricker (1970). Salivary cooling by rats in the heat. In J.D. Hardy, A.P. Gagge, and J.A.J. Stolwijk (eds.), *Physiological and Behavioral Temperature Regulation*. CC Thomas, Springfield, Ill.

Hainsworth, F.R., and L.L. Wolf (1979). Feeding: An ecological approach. *Adv. Study Behav.* 9:53-96.

Hall, F.G., and F.H. McCutcheon (1938). The affinity of hemoglobin for oxygen in marine fishes. *Jour. Cell. Comp. Physiol.* 11:205-212.

Hammel, H.T. (1962). Thermal and metabolic measurements on a reindeer at rest and in exercise. *Arctic Aeromed. Lab. Report* AAL-TDR-61-54.

Hammel, H.T., F.T. Caldwell, Jr., and R.M. Abrams (1967). Regulation of body temperature in the blue-tongued lizard. *Science* 156:1260-1262.

Hardy, J.D., and G.F. Soderstrom (1938). Heat loss from the nude body and peripheral blood flow at temperatures of 22°C to 35°C. *Jour. Nutr.* 16: 493-510.

Hardy, J.D., and J.T. Stitt (1976). Interaction of hypothalamic and skin temperature in cold thermogenesis in the rabbit. *Israel Jour. Med. Sci.* 12:1052-1055.

Hayward, J.S., and C.P. Lyman (1967). Nonshivering heat production during arousal from hibernation and evidence for the contribution of brown fat. In K.C. Fisher, A.R. Dawe, C.P. Lyman, E. Schönbaum, and F.E. South (eds.), *Mammalian Hibernation III.* American Elsevier, New York.

Heath, J.E. (1965). Temperature regulation and diurnal activity in horned lizards. *University of California* (Berkeley) *Publs. Zool.* 64:97-136.

Heinrich, B. (1972). Temperature regulation in the bumblebee, *Bombus vagans. Science* 175:185-187.

Heinrich, B. (1974). Thermoregulation in endothermic insects. *Science* 185: 747-756.

Heller, H.C., and G.W. Colliver (1974). CNS regulation of body temperature during hibernation. *Amer. Jour. Physiol.* 277:583-589.

Hemmingsen, A.M. (1960). Energy metabolism as related to body size and respiratory surfaces, and its evolution. *Rep. Steno. Mem. Hosp.* 9:1-110.

Hendricks, S.B. (1956). Control of growth and reproduction by light and darkness. *Amer. Scientist* 44:229-247.

Henshaw, R.E., L.S. Underwood, and T.C. Casey (1972). Peripheral thermoregulation: Foot temperature in two Arctic canines. *Science* 175:988-990.

Herreid, C.F., and B. Kessel (1967). Thermal conductance of birds and mammals. *Comp. Biochem. Physiol.* 21:405-414.

Hill, R.W. (1976). *Comparative Physiology of Animals: An Environmental Approach.* Harper & Row, New York.

Hodgman, C.D. (ed.) (1954). *Handbook of Chemistry and Physics.* 36th ed. Chemical Rubber Publ. Co., Cleveland, Ohio.

Horn, H. (1978). Optimal tactics of reproduction and life-history. In J.R. Krebs and N.B. Davies (eds.), *Behavioural Ecology.* Sinauer Associates, Sunderland, Mass.

Horwitz, B.A. (1978). Neurohumoral regulation of nonshivering thermogenesis in mammals. In L.C.H. Wang and J.W. Hudson (eds.), *Strategies in Cold: Natural Torpidity and Thermogensis.* Academic Press, New York.

Houpt, R.T. (1977). Digestion and nutrition. In L. Goldstein (ed.), *Introduction to Comparative Physiology.* Holt, Rinehart & Winston, New York.

Hughes, G.M. (1964). *Comparative Physiology of Vertebrate Respiration.* Harvard University Press, Cambridge.

Hughes, G.M., and G. Shelton (1958). The mechanism of gill ventilation in three freshwater teleosts. *Jour. Exp. Biol.* 35:807-823.

Jackson, D.C., and K. Schmidt-Nielsen (1964). Countercurrent heat exchange in the respiratory passages. *Proc. Nat. Acad. Sci.* 51:1192-1197.

Jahn, T.L., and E.C. Bovee (1967). Motile behavior of protozoa. In T.T. Chen, (ed.), *Research in Protozoology*, Vol. I. Pergamon Press, New York.

Jahn, T.L., M.D. Landman, and J.R. Fanseca (1964). The mechanism of locomotion of flagellates, II. Function of mastigonemes of *Ochromonas*. *Jour. Protozool.* 11:291-296.

Javaid, M.Y., and J.M. Anderson (1967). Influence of starvation on selected temperature of some salmonids. *Jour. Fish. Res. Bd. Canada* 24:1515-1519.

Johansen, K. (1964). Regional distribution of circulatory blood during submersion asphyxia in the duck. *Acta Physiol. Scand.* 62:1-9.

Johansen, K., and A.W. Martin (1962). Circulation in the cephalopod, *Octopus dofleini*. *Comp. Biochem. Physiol.* 5:161-176.

Katz, B. (1950). Depolarization of sensory terminals and the initiation of impulses in the muscle spindle. *Jour. Physiol.* (London) 111:261-282.

Katz, B., and R. Miledi (1967). A study of synaptic transmission in the absence of nerve impulses. *Jour. Physiol.* (London) 192:407-436.

Kendeigh, S.C., J.E. Kontogiannus, A. Malzac, and R.R. Roth (1969). Environmental regulation of food intake by birds. *Comp. Biochem. Physiol.* 31:941-957.

Keys, A., and E.N. Willmer (1932). "Chloride secreting cells" in the gills of fishes, with special reference to the common eel. *Jour. Physiol.* (London) 76:368-378.

King, J.R. (1957). Comments on the theory of indirect calorimetry as applied to birds. *Northwest Science* 31:155-169.

Kleiber, M. (1961). *The Fire of Life*. Wiley, New York.

Kluger, M.J. (1978). The evolution and adaptive value of fever. *Amer. Scientist* 66:38-43.

Krebs, J.R. (1978). Optimal Foraging: decision rules for predators. In J.R. Krebs and N.B. Davies (eds.) *Behavioral Ecology*. Sinauer Associates, Sunderland, Mass.

Krebs, J.R., J.C. Ryan, and E.L. Charnov (1974). Hunting by expectation or optimal foraging? A study of patch use by chickadees. *Anim. Behav.* 22:953-964.

Kuffler, S.W., and J.G. Nicholls (1976). *From Neuron to Brain: A Cellular Approach to the Function of the Nervous System*. Sinauer Associates, Sunderland, Mass.

Lang, F., and H. Atwood (1973). Crustacean neuromuscular mechanisms, functional morphology of nerve terminals and the mechanism of facilitation. *Amer. Zool.* 13:337-338.

Laver, M.B., E. Jackson, M. Scherperel, C. Tung, W. Tung, and E.P. Radford (1977). Hemoglobin-O_2 affinity regulation: DPG, monovalent ions, andhemoglobin concentration. *Jour. Appl. Physiol.* 43:632-642.

Lehrman, D.S. (1964). The reproductive behavior of ring doves. *Sci. Amer.* 211:48-54.

Lehrman, D.S., P.N. Brody, and R.P. Wortis (1961). The presence of the mate and of nesting material as stimuli for the development of incubation behavior and for gonadotrophin secretion in the ring dove (*Streptopelia risoria*). *Endocrinology* 68:507-516.

LeMagnen, J., M. Devos, J.P. Gaudilliere, J. Louis-Sylvestre, and S. Tallon (1973). Role of a lipostatic mechanism in regulation by feeding of energy balance in rats. *Jour. Comp. Physiol. Psychol.* 84:1-23.

Lenfant, C., K. Johansen, and D. Hanson (1970). Bimodal gas exchange and ventilation-perfusion relationship in lower vertebrates. *Fed. Proc.* 29:1124-1129.

Lettvin, J.Y., H.R. Maturana, W.H. Pitts, and W.S. McCulloch (1961). Two remarks on the visual system of the frog. In W.A. Rosenblith (ed.), *Sensory Communication*. MIT Press, Cambridge.

Levy, R.I., and H.A. Schneiderman (1966). Discontinuous respiration in insects IV. Changes in intratracheal pressure during the respiratory cycle of silkworm pupae. *Jour. Insect Physiol.* 12:465-492.

Lillywhite, H.B., P. Licht, and P. Chelgren (1973). The role of behavioral thermoregulation in the growth energetics of the toad, *Bufo boreas. Ecology* 54:375-383.

Lissaman, P.B.S., and C.A. Shollenberger (1970). Formation flight of birds. *Science* 168:1003-1005.

Lissmann, H.W. (1963). Electric location by fishes. *Sci. Amer.* 208:50-59.

Lloyd, M., and H.S. Dybas (1966). The periodical cicada problem I. Population ecology. *Evolution* 20:133-149.

Lockwood, A.P.M. (1966). *Animal Body Fluids and Their Regulation.* Harvard University Press, Cambridge.

Loewi, O. (1921). Über humorale Ubertragbarkeit der Herznervenwirkung. *Pflug. Arch. Ges. Physiol.* 189:239-242.

Lowenstein, W.R. (1960). Biological transducers. *Sci. Amer.* 203:98-108.

Magilton, J.H., and C.S. Swift (1969). Response of veins draining the nose to alar-fold temperature changes in the dog. *Jour. Appl. Physiol.* 27:18-20.

Malan, A., H. Arens, and A. Waechter (1973). Pulmonary respiration and acid-base state in hibernating marmots and hamsters. *Resp. Physiol.* 17:45-61.

Mayhew, W.W. (1968). Biology of desert amphibians and reptiles. In G.W. Brown, Jr. (ed.), *Desert Biology.* Academic Press, New York.

McCauley, W.J. (1971). *Vertebrate Physiology.* Saunders, Philadelphia.

McMahon, B.R. (1970). The relative efficiency of gaseous exchange across the lungs and gills of an African lungfish, *Protopterus aethiopicus. Jour. Exp. Biol.* 52:1-15.

McMahon, T. (1973). Size and shape in biology. *Science* 179:1201-1204.

Mellanby, K. (1934). The site of water loss from insects. *Proc. Roy. Soc.* (B) 116:139-149.

Menaker, M., and N. Zimmerman (1976). Role of the pineal in the circadian system of birds. *Amer. Zool.* 16:45-55.

Miller, W., F. Ratliff, and H.K. Hartline (1961). How cells receive stimuli. *Sci. Amer.* 205:223-238.

Murrish, D.E., and K. Schmidt-Nielsen (1970). Exhaled air temperature and water conservation in lizards. *Resp. Physiol.* 10:151-158.

Nagy, K.A., D.K. Odell, and R.S. Seymour (1972). Temperature regulation by the inflorescence of *Philodendron. Science* 178:1195-1197.

Nalbandov, A.J. (1958). *Reproductive Physiology: Comparative Reproductive Physiology of Domestic Animals, Laboratory Animals, and Man.* W.H. Freeman, San Francisco.

Ostrom, J.H. (1978). A new look at dinosaurs. *National Geog.* 154:152-185.

Otis, A.B. (1964). Quantitative relationships in steady-state gas exchange. In W.O. Fenn and H. Rahn (eds.), *Handbook of Physiology,* Section 3, *Respiration,* Vol. I. American Physiological Society, Washington, D.C.

Peachy, L.D. (1965). The sarcoplasmic reticulum and transverse tubules of the frog's sartorius. *Jour. Cell Biol.* 25:209-231.

Pengelley, E.T., and K.C. Fisher (1963). The effect of temperature and photoperiod on the yearly hibernating behavior of captive golden-mantled ground squirrels (*Citellus lateralis tescorum*). *Can. Jour. Zool.* 41:1103-1120.

Pennycuick, C.J. (1969). The mechanics of bird migration. *Ibis* 111:525-556.

Peper, K., F. Dreyer, C. Sandri, K. Ackert, and H. Moor (1974). Structure and ultrastructure of the frog endplate. *Cell Tissue Res.* 149:437-455.

Piiper, J., and P. Scheid (1973). Gas exchange in avian lungs: Models and experimental evidence. In L. Bolis, K. Schmidt-Nielsen, and S.H.P. Maddrell (eds.), *Comparative Physiology: Locomotion, Respiration, Transport, and Blood.* American Elsevier, New York.

Pitts, R.F. (1968). *Physiology of the Kidney and Body Fluids.* Year Book Medical Publs., Chicago.

Potts, W.T.W., and G. Parry (1963). *Osmotic and Ionic Regulation in Animals.* Pergamon Press, New York.

Precht, H., J. Christophersen, H. Hensel, and W. Larcher (1973). *Temperature and Life.* Springer-Verlag, New York.

Rahn, H. (1966). Aquatic gas exchange: Theory. *Resp. Physiol.* 1:1-12.

Randall, D.J. (1968). Functional morphology of the heart in fishes. *Amer. Zool.* 8: 179-190.

Randall, D.J., and J.N. Cameron (1973). Respiratory control of arterial pH as temperature changes in rainbow trout, *Salmo gairdneri. Amer. Jour. Physiol.* 225:997-1002.

Reeves, R.B. (1977). The interaction of body temperature and acid-base balance in ectothermic vertebrates. *Ann. Rev. Physiol.* 39:559-586.

Regal, P.J. (1967). Voluntary hypothermia in reptiles. *Science* 155:1551-1553.

Reynolds, W.W., and M.E. Casterlin (1979). Behavioral thermoregulation and the "final preferendum" paradigm. *Amer. Zool.* 19:211-224.

Richter, C.P., and K.K. Rice (1945). Self-selection studies on coprophagy as a source of vitamin B complex. *Amer. Jour. Physiol.* 143:344-354.

Rieck, A.F., J.A. Belli, and M.E. Blaskovics (1960). Oxygen consumption of whole animal tissues in temperature acclimated amphibians. *Proc. Soc. Exper. Biol. Med.* 103:436-439.

Riggs, A. (1951). The metamorphosis of hemoglobin in the bullfrog. *Jour. Gen. Physiol.* 35:23-44.

Roeder, K.D. (1967). *Nerve Cells and Insect Behavior.* Harvard University Press, Cambridge.

Rosenthal, G.A., D.L. Dahlman, and D.H. Janzen (1978). L-canaline detoxification: A seed predator's biochemical mechanism. *Science* 202:528-529.

Rozin, P. (1965). Specific hunger for thiamine: Recovery from deficiency and thiamine preference. *Jour. Comp. Physiol. Psychol.* 59:98-101.

Rozin, P. (1969). Adaptive food sampling patterns in vitamin deficient rats. *Jour. Comp. Physiol. Psychol.* 69:126-132.

Rushmer, R.F. (1960). Control of cardiac output. In T.C. Ruch and H.D. Patton (eds.), *Physiology and Biophysics*. 18th ed. Saunders, Philadelphia.

Rushton, W.A.H. (1951). A theory of the effects of fibre size in the medulated nerve. *Jour. Physiol.* (London) 115:101-122.

Satir, P. (1968). Studies on cilia III. Further studies on the cilium tip and a "sliding filament" model of ciliary motility. *Jour. Cell Biol.* 39:77-94.

Schmidt-Nielsen, B. (1964). Organ systems in adaptation: The excretory system. In D.B. Dill (ed.), *Handbook of Physiology*, Section 4, *Adaptation to the Environment*. American Physiological Society, Washington, D.C.

Schmidt-Nielsen, B. (1972). Mechanisms of urea excretion by the vertebrate kidney. In J.M. Campbell and L. Goldstein (eds.), *Nitrogen Metabolism and the Environment*. Academic Press, New York.

Schmidt-Nielsen, B., and K. Schmidt-Nielsen (1951). A complete account of the water metabolism in kangaroo rats and an experimental verification. *Jour. Cell. Comp. Physiol.* 38:165-181.

Schmidt-Nielsen, K. (1964a). *Animal Physiology*. Prentice-Hall, New York.

Schmidt-Nielsen, K. (1964b). *Desert Animals*. Oxford University Press, New York.

Schmidt-Nielsen, K. (1970). Energy metabolism, body size, and problems of scaling, *Fed. Proc.* 29:1524-1532.

Schmidt-Nielsen, K. (1972). Locomotion: Energy cost of swimming, flying, and running. *Science* 177:222-228.

Schmidt-Nielsen, K. (1975). *Animal Physiology: Adaptation and Environment*. Cambridge University Press, New York.

Schmidt-Nielsen, K. (1979). *Animal Physiology: Adapation and Environment*. 2d ed. Cambridge University Press, New York.

Schmidt-Nielsen, K., W.L. Bretz, and C.R. Taylor (1970). Panting in dogs: unidirectional air flow over evaporative surfaces. *Science* 169:1102-1104.

Schmidt-Nielsen, K., F.R. Hainsworth, and D.E. Murrish (1970). Countercurrent heat exchange in the respiratory passages: Effect on water and heat balance. *Resp. Physiol.* 9:263-276.

Schmidt-Nielsen, K., S.A. Jarnum, and T.R. Houpt (1957). Body temperature of the camel and its relation to water economy. *Amer. Jour. Physiol.* 188:103-112.

Scholander, P.F. (1963). The master switch of life. *Sci. Amer.* 208:92-105.

Scholander, P.F., V. Walters, R. Hock, and L. Irving (1950). Body insulation of some arctic and tropical mammals and birds. *Bio. Bull.* 99:225-236.

Shapiro, B. (1977). Neurophysiology. In L. Goldstein (ed.), *Introduction to Comparative Physiology*. Holt, Rinehart & Winston, New York.

Short, R.V. (1972). Species differences. In C.R. Austin and R.V. Short (eds.), *Reproduction in Mammals*, Vol. 4, *Reproductive Patterns*. Cambridge University Press, New York.

Smith, R.E., and B.A. Horwitz (1969). Brown fat and thermogenesis. *Physiol. Rev.* 49:330-425.

Snyder, G.K. (1977). Blood corpuscles and blood hemoglobins: A possible example of coevolution. *Science* 195:412-413.

Somero, G.N., and P.W. Hochachka (1971). Biochemical adaptation to the environment. *Amer. Zool.* 11:159-167.

Stahl, W.R. (1967). Scaling of respiratory variables in mammals. *Jour. Appl. Physiol.* 22:453-460.

Stricker, E.M. (1973). Thirst, sodium appetite, and complementary physiological contributions to the regulation of intravascular fluid volume. In A. Epstein, H. Kissileff, and E. Stellar (eds.), *The Neuropsychology of Thirst.* V.H. Winston, Washington, D.C.

Taylor, C.R. (1970). Dehydration and heat: effects on temperature regulation of East African ungulates. *Amer. Jour. Physiol.* 219:1136-1139.

Taylor, C.R., and V.J. Rowntree (1973). Temperature regulation and heat balance in running cheetahs: a strategy for sprinters? *Amer. Jour. Physiol.* 224:848-851.

Taylor, C.R., K. Schmidt-Nielsen, R. Dmi'el, and M. Fedak (1971). Effect of hyperthermia on heat balance during running in the African dog. *Amer. Jour. Physiol.* 220:823-827.

Taylor, C.R., K. Schmidt-Nielsen, and J.L. Raab (1970). Scaling of energetic cost of running to body size in mammals. *Amer. Jour. Physiol.* 219: 1104-1107.

Tepperman, J. (1962). *Metabolic and Endocrine Physiology.* Year Book Medical Pubs., Chicago.

Thorson, T.B., C.M. Cowan, and D.E. Watson (1967). *Potamotrygon* spp: elasmobranchs with low urea content. *Science* 158:375-377.

Tinbergen, N. (1948). Social releasers and the experimental method required for their study. *Wilson Bull.* 60:6-52.

Torke, K.G., and J.W. Twente (1977). Behavior of *Spermophilus lateralis* between periods of hibernation. *Jour. Mammal.* 58:385-390.

Trivers, R.L. (1974). Parent-offspring conflict. *Amer. Zool.* 14:249-264.

Tucker, V.A. (1965). The relation between the torpor cycle and heat exchange in the California pocket mouse *Perognathus californicus. Jour. Cell. Comp. Physiol.* 65:405-414.

Tucker, V.A. (1968). Respiratory exchange and evaporative water loss in the flying budgerigar. *Jour. Exp. Biol.* 48:67-87.

Tyuma, I., and K. Shimizu (1970). Effect of organic phosphates on the difference in oxygen affinity between fetal and adult hemoglobin. *Fed. Proc.* 29:1112-1114.

Vander, A.J., J.H. Sherman, and D.S. Luciano (1975). *Human Physiology: The Mechanisms of Body Function.* 2d ed. McGraw-Hill, New York.

Wald, G., and P.K. Brown (1958). Human rhodopsin. *Science* 127:222-226.

Walsberg, G.E., G.S. Cambell, and J.R. King (1978). Animal coat color and radiative heat gain: a re-evaluation. *Jour. Comp. Physiol.* 126:211-222.

Ware, D.M. (1975). Growth, metabolism, and optimal swimming speed of a pelagic fish. *Jour. Fish. Res. Bd. Canada* 32:33-41.

Warner, F. (1972). Macromolecular organization of eucaryotic cilia and flagella. In E.J. DuPraw (ed.), *Advances in Cell & Molecular Biology*, Vol. 2. Academic Press, New York.

Waterman, A.J. (1971). *Chordate Structure and Function*. Macmillan, New York.

Weathers, W.W., and D.C. Schoenbaechler (1976). Contribution of gular flutter to evaporative cooling in Japanese quail. *Jour. Appl. Physiol.* 40:521-524.

Webb, P.W. (1971). The swimming energetics of trout II. Oxygen consumption and swimming efficiency. *Jour. Exp. Biol.* 55:521-540.

Weihs, D. (1973). Hydromechanics of fish schooling. *Nature* 241:290-291.

Wells, L.A. (1971). Circulatory patterns of hibernators. *Amer. Jour. Physiol.* 221:1517-1520.

Wheeler, P.E. (1978). Elaborate CNS cooling structures in large dinosaurs. *Nature* 275:441-443.

Whitford, W.G., and V.H. Hutchinson (1963). Cutaneous and pulmonary gas exchange in the spotted salamander, *Ambystoma maculatum*. *Biol. Bull.* 124:344-354.

Wigglesworth, V.B. (1972). *The Principles of Insect Physiology*, 7th ed. Chapman & Hall, London.

Wilkie, D.R. (1968). *Muscle*. Edward Arnold, London.

Wohlbarsht, M.L., and F.E. Hanson (1965). Electrical activity in the chemoreceptors of the blowfly. III. Dendritic action potentials. *Jour. Gen. Physiol.* 48:673-683.

Wolf, L.L., and F.R. Hainsworth (1972). Environmental influence on regulated body temperature in torpid hummingbirds. *Comp. Biochem. Physiol.* 41:167-173.

Wolf, L.L., and F.R. Hainsworth (1977). Temporal patterning of feeding by hummingbirds. *Anim. Behav.* 25:976-989.

Wood, S.C., and C.J.M. Lenfant (1977). Circulation and respiration. In L. Goldstein (ed.). *Introduction to Comparative Physiology*. Holt, Rinehart & Winston, New York.

Young, R.W. (1970). Visual cells. *Sci. Amer.* 223:89-91.

Zotterman, Y. (1953). Special senses: thermal receptors. *Ann. Rev. Physiol.* 15:357-372.

Figure and Photo Acknowledgments

(Figs. 1.9, 2.8, 4.24, 5.2, 10.14, and 15.19 acknowledged in text as required by certain publishers)

Chapter 1

Fig. 1.5   Reprinted from *Animals Without Backbones* (rev. ed.) by Ralph Buchsbaum by permission of The University of Chicago Press. Copyright 1938 and 1948 by The University of Chicago. All rights reserved.

Fig. 1.6   Reprinted from *Animals Without Backbones* (rev. ed.) by Ralph Buchsbaum by permission of The University of Chicago Press. Copyright 1938 and 1948 by The University of Chicago. All rights reserved.

Fig. 1.12   From K. Schmidt-Nielsen, *Animal Physiology: Adaptation and Environment* (2nd ed.), Cambridge University Press, 1979. Reproduced by permission of the publisher.

Chapter 2

Fig. 2.15   From *Chordate Structure and Function* by Allyn J. Waterman. Copyright © 1971 Allyn J. Waterman. Reprinted by permission of Macmillan Publishing Co., Inc.

Fig. 2.20   Reprinted from *Animals Without Backbones* (rev. ed.) by Ralph Buchsbaum by permission of The University of Chicago Press. Copyright 1938 and 1948 by The University of Chicago. All rights reserved.

Fig. 2.21   Reprinted with permission from *J. Insect Physiol.* 12, R.I. Levy and H.A. Schneiderman, "Discontinuous respiration in insects—IV. Changes in intratracheal pressure during the respiratory cycle of silkworm pupae," Copyright 1966, Pergamon Press, Ltd.

Chapter 3

Fig. 3.2   From G.K. Snyder, *Science* 195:412-413, 28 January 1977. Copyright 1977 by the American Association for the Advancement of Science.

Fig. 3.3   From G.K. Snyder, *Science* 195:412-413, 28 January 1977. Copyright 1977 by the American Association for the Advancement of Science.

Fig. 3.10   Reprinted from K. Schmidt-Nielsen, *Federation Proceedings* 29:1524-1532, 1970. Reproduced by permission of the publisher.

Fig. 3.13   Reproduced from A. Riggs, *J. Gen. Physiol.* 35:23-44, 1951 by copyright permission of The Rockefeller University Press.

Chapter 4

Fig. 4.1   Reprinted from *Animals Without Backbones* (rev.ed.) by Ralph Buchsbaum by permission of The University of Chicago Press. Copyright 1938 and 1948 by The University of Chicago. All rights reserved.

Fig. 4.3   Reprinted from *Animals Without Backbones* (rev. ed.) by Ralph Buchsbaum by permission of The University of Chicago Press. Copyright 1938 and 1948 by The University of Chicago. All rights reserved.

Fig. 4.10   From A.J. Vander, J.H. Sherman, and D.S. Luciano, *Human Physiology: The Mechanisms of Body Function* (2nd ed.), McGraw-Hill, 1975. Reproduced by permission of the publisher.

Fig. 4.11   From A.J. Vander, J.H. Sherman, and D.S. Luciano, *Human Physiology: The Mechanisms of Body Function* (2nd ed.), McGraw-Hill, 1975. Reproduced by permission of the publisher.

Fig. 4.17   From D.J. Randall, *Amer. Zool.* 8:179-190, 1968. Reproduced by permission of the publisher.

Fig. 4.20   From A.J. Vander, J.H. Sherman, and D.S. Luciano, *Human Physiology: The Mechanisms of Body Function* (2nd ed.), McGraw-Hill, 1975. Reproduced by permission of the publisher.

Chapter 5

Fig. 5.3   Reprinted from J. Aschoff and H. Pohl, *Federation Proceedings* 29:1541-1552, 1970. Reproduced by permission of the publisher.

Chapter 6

Fig. 6.7   From K. Schmidt-Nielsen, *Animal Physiology: Adaptation and Environment*, Cambridge University Press, 1975. Reproduced by permission of the publisher.

Fig. 6.9   From A.F. Rieck, J.A. Belli, and M.E. Blaskovics, *Proc. Soc. Exp. Biol. Med.* 103:436-439, 1960. Reproduced by permission of the publisher.

Chapter 7

Fig. 7.11   From R.T. Balmer and A.D. Strobusch, *J. Appl. Physiol.* 42:571-577, 1977. Reproduced by permission of the publisher.

Fig. 7.14   From K. Schmidt-Nielsen, F.R. Hainsworth, and D.E. Murrish, *Resp. Physiol.* 9:263-276, 1970. Reproduced by permission of the publisher.

Fig. 7.15   From K. Schmidt-Nielsen, F.R. Hainsworth, and D.E. Murrish, *Resp. Physiol.* 9:263-276, 1970. Reproduced by permission of the publisher.

Fig. 7.19   From B. Heinrich, *Science* 185:747-756, 30 August 1974. Copyright 1974 by the American Association for the Advancement of Science.

Chapter 8

Fig. 8.7   Reprinted from "The function of torpor in hummingbirds" in *Physiological*

Zoology 50 by F.R. Hainsworth, B.G. Collins, and L.L. Wolf by permission of The University of Chicago Press. © 1977 by the University of Chicago. All rights reserved.

Fig. 8.10   Reproduced by permission of the National Research Council of Canada from E.T. Pengelley and K.C. Fisher, *Can. J. Zool.* 41:1103-1120, 1963.

Fig. 8.11   From K.G. Torke and J.W. Twente, *J. Mammal.* 58:385-390, 1977. Reproduced by permission of the publisher.

Fig. 8.12   From V.A. Tucker, *J. Cell. Comp. Physiol.* 65:405-414, 1965. Reproduced by permission of the publisher.

Fig. 8.14   From H.C. Heller and G.W. Colliver, *Amer. J. Physiol.* 227:583-589, 1974. Reproduced by permission of the publisher.

Fig. 8.15   From J.S. Hayward and C.P. Lyman in K.C. Fisher, A.R. Dawe, C.P. Lyman, E. Schönbaum, and F.E. South (eds.), *Mammalian Hibernation* III, American Elsevier, 1967. Reproduced by permission of the author (Lyman).

Fig. 8.16   From J.S. Hayward and C.P. Lyman in K.C. Fisher, A.R. Dawe, C.P. Lyman, E. Schönbaum, and F.E. South (eds.), *Mammalian Hibernation* III, American Elsevier, 1967. Reproduced by permission of the author (Lyman).

Chapter 9

Fig. 9.9   From V.A. Tucker, *J. Exp. Biol.* 48, 1968. Reproduced by permission of the publisher.

Fig. 9.16   From P.B.S. Lissaman and C.A. Schollenberger, *Science* 168:1003-1005, 22 May 1970. Copyright 1970 by the American Association for the Advancement of Science.

Fig. 9.18   From P.B.S. Lissaman and C.A. Schollenberger, *Science* 168:1003-1005, 22 May 1970. Copyright 1970 by the American Association for the Advancement of Science.

Fig. 9.22   From D. Weihs, *Nature* 241:290-291, 1973. Reproduced by permission of the publisher.

Fig. 9.24   From K. Schmidt-Nielsen, *Science* 177:222-228, 21 July 1972. Copyright 1972 by the American Association for the Advancement of Science.

Chapter 10

Fig. 10.6   From L.L. Wolf and F.R. Hainsworth, *Anim. Behav.* 25:976-989, 1977. Reproduced by permission of the publisher.

Fig. 10.7   Reprinted with permission from *Comp. Biochem. Physiol.* 31, S.C. Kendeigh, J.E. Kontogiannis, A. Malzac, and R.R. Roth, "Environmental regulation of food intake by birds," Copyright 1969, Pergamon Press, Ltd.

Fig. 10.10   From J.R. Krebs and N.B. Davies (eds.), *Behavioural Ecology: An Evolutionary Approach*, Blackwell Scientific Publications, 1978. Reproduced by permission of the publisher.

Fig. 10.12   From E. Florey, *An Introduction to General and Comparative Animal Physiology*, W.B. Saunders, 1966. Reproduced by permission of the publisher.

Fig. 10.20   From M.I. Friedman and E.M. Stricker, "The physiological psychology of hunger: A physiological perspective," *Psychol. Rev.* 83:409-431, 1976. Copyright 1976 by the American Psychological Association. Reprinted by permission.

Fig. 10.21   From M.I. Friedman and E.M. Stricker, "The physiological psychology of hunger: A physiological perspective," *Psychol. Rev.* 83:409-431, 1976. Copyright 1976 by the American Psychological Association. Reprinted by permission.

Fig. 10.22   Reprinted by permission of the author and publishers from *The Hungry Fly* by V.G. Dethier. Cambridge, Mass.: Harvard University Press. Copyright © 1976 by the President and Fellows of Harvard College.

Chapter 11

Fig. 11.6   From L.P. Brower, W.N. Ryerson, L.L. Coppinger, and S.C. Glazier, *Science* 161:1349-1351, 27 September 1968. Copyright 1968 by the American Association for the Advancement of Science.

Chapter 13

Fig. 13.3   From K. Schmidt-Nielsen, F.R. Hainsworth, and D.E. Murrish, *Resp. Physiol.* 9: 263-276, 1970. Reproduced by permission of the publisher.

Fig. 13.4   From K. Schmidt-Nielsen, F.R. Hainsworth, and D.E. Murrish, *Resp. Physiol.* 9: 263-276, 1970. Reproduced by permission of the publisher.

Fig. 13.8   From B. Schmidt-Nielsen and K. Schmidt-Nielsen, *J. Cell. Comp. Physiol.* 38:165-181, 1951. Reproduced by permission of the publisher.

Fig. 13.9   From G.A. Bartholomew and T.J. Cade, *Auk* 80:504-539, 1963. Reproduced by permission of the publisher.

Fig. 13.10 From K. Mellanby, *Proc. Roy. Soc.* (London), Ser. B., 116:139-149, 1934. Reproduced by permission of the publisher.

Fig. 13.11 From J.W.L. Beament, *J. Exp. Biol.* 35, 1958. Reproduced by permission of the publisher.

Fig. 13.20 From K. Schmidt-Nielsen, *Animal Physiology: Adaptation and Environment* (2nd ed.), Cambridge University Press, 1979. Reproduced by permission of the publisher.

Fig. 13.22 From E.F. Adolph, *Amer. J. Physiol.* 125:75-86, 1939. Reproduced by permission of the publisher.

Chapter 14

Fig. 14.2   From E.C. Crawford, Jr., *J. Appl. Physiol.* 17:249-251, 1962. Reproduced by permission of the publisher.

Fig. 14.4   From K. Schmidt-Nielsen, W.L. Bretz, and C.R. Taylor, *Science* 169:1102-1104, 11 September 1970. Copyright 1970 by the American Association for the Advancement of Science.

Fig. 14.10 From G.A. Bartholomew, *Symp. Soc. Exp. Biol.* 18:7-29, 1964. Reproduced by permission of the publisher.

Chapter 15

Fig. 15.4   From M. Lloyd and H.S. Dybas, *Evolution* 20:133-149, 1966. Reproduced by permission of the publisher.

Fig. 15.8   From K. Schmidt-Nielsen, *Animal Physiology: Adaptation and Environment* (2nd ed.), Cambridge University Press, 1979. Reproduced by permission of the publisher.

Fig. 15.20 From *Mammalogy* by Harvey L. Gunderson. Copyright © 1976 McGraw-Hill. Used with the permission of McGraw-Hill Book Company.

Chapter 16

Fig. 16.9   From W.A.H. Rushton, *J. Physiol.* (London) 115:101-122, 1951. Reproduced by permission of the publisher.

Fig. 16.18 Reproduced from M.L. Wohlbarsht and F.E. Hanson, *J. Gen. Physiol.* 48:673-683, 1965 by copyright permission of The Rockefeller University Press.

Chapter 17

Fig. 17.9   From A.J. Vander, J.H. Sherman, and D.S. Luciano, *Human Physiology: The Mechanisms of Body Function* (2nd ed.), McGraw-Hill, 1975. Reproduced by permission of the publisher.

Fig. 17.10 From A.J. Vander, J.H. Sherman, and D.S. Luciano, *Human Physiology: The Mechanisms of Body Function* (2nd ed.), McGraw-Hill, 1975. Reproduced by permission of the publisher.

Chapter 18

Fig. 18.19 From A.J. Vander, J.H. Sherman, and D.S. Luciano, *Human Physiology: The Mechanisms of Body Function* (2nd ed.), McGraw-Hill, 1975. Reproduced by permission of the publisher.

Index